A TRIP THROUGH TIME:
Principles of
Historical Geology

A TRIP THROUGH TIME: Principles of Historical Geology

JOHN D. COOPER
California State University, Fullerton

RICHARD H. MILLER
San Diego State University

JACQUELINE PATTERSON
California State University, Fullerton

MERRILL PUBLISHING COMPANY
A Bell & Howell Company
Columbus   Toronto   London   Sydney

Published by Merrill Publishing Co.
A Bell & Howell Company
Columbus, Ohio 43216

This book was set in Garamond Light
Cover Design Coordination: Cathy Watterson
Production Coordination: Gnomi Schrift Gouldin
Cover Illustration: Michael Linley. This is the artist's impression of Figure 13–18 and is not
intended to reflect technical detail.

Library of Congress Catalog Card Number: 85-062492
International Standard Book Number: 0-675-20140-3
Printed in the United States of America
1 2 3 4 5 6 7 8 9—91 90 89 88 87 86

CONTENTS

PREFACE

Historical geology—the study of the earth's evolution, including changes in its crust, surface, atmosphere, and life—has experienced its own evolutionary burst during the past two decades as new models for interpreting earth history have been developed. The *plate tectonics* doctrine has become the new central theme of historical geology and has caused a monumental shift in how earth scientists view the earth's evolution.

Traditionally Historical Geology, as a formal course, has been taught at the first- or second-year college level, either in tandem with Physical Geology or as a separate course appealing to a general education audience. Teachers of historical geology always have faced the twofold problem of how to boil down an enormous body of factual information about *what* happened during the earth's past history and at the same time present a meaningful dose of guiding principles and concepts to facilitate appreciation and understanding of *how* we know about geologic events and their chronology. With the advent of plate tectonic theory and its many ramifications for earth history, this dilemma of information level and balance has been compounded. Historical geology has become a more complex subject because of greater synthesis of data and integration of ideas from many diverse disciplines. Consequently, in recent years, both teaching and development of text material for historical geology have faced new challenges as well as frustrations—the key elements of plate tectonic concepts and interpretations must be woven into the fabric of geologic time, evolution, ancient seas, mountain building, changing climates, dinosaurs, and extinctions, without losing the student.

Some geology programs have abandoned the concept of a beginning course in historical geology, instead deferring it to an upper division part of the major and requiring a solid background in structural geology, geophysics, stratigraphy and sedimentation, paleontology, and petrology. An advanced Historical Geology course set in a regional or global format is a desirable capstone course for the undergraduate major. Many geology programs, however, no longer include a lower division historical geology course. We regret this and are committed to its revival.

Such a course for the geology major, early in the program, provides an opportunity to interpret, to relate cause and effect, and to learn some important principles and methods as a prelude to more advanced courses. It also provides an opportunity to begin to *think geologically,* to develop the resourcefulness and detective approach required to read and decode the rock record, and to develop appreciation of past geologic events set in their chronologic sequence. Such a course for the nonscience, general education student (with or without the benefit of background in physical geology) provides an overview of prehistory and of mankind's evolutionary heritage, promotes an appreciation for the earth's natural resources (how they form and how long they take to form), and helps in development of a perspective on the earth's cosmic connection and how the earth was born. A well-designed course also has the potential for acquainting students with the unique aspects of the geological sciences as well as for showing how historical geology relates to the breadth of human knowledge and problems.

Our mission in writing this text has been in part to meet the challenges that historical geology presents for a beginning course, but, more importantly, to share the excitement, beauty, intrigue, fascination, and achievement of historical geology. In our attempt to accomplish this mission we have set the following principal goals:

1. To show the thought processes and methodologies of historical geology—what the guiding principles are and how they are applied to the interpretation of earth history. The text is structured so that working principles and concepts are developed in the first half. Plate tectonics, as a unifying theme, is introduced in the first chapter within the context of the grand *geologic cycle* (how the earth works).

2. To make clear the very important concepts of geologic time. Principles and concepts of relative and radiometric time, presented in Chapter 2, are initially set apart from discussions of the birth and development of the geologic time scale and fundamental stratigraphic principles, which are presented in Chapter 3. This separation is intended to focus on the working principles and concepts first in order to facilitate clear understanding of how interpretations of earth history are made.

3. To strike a healthy balance between *what happened when* (geologic events and their chronology) and *how we know* (the guiding principles employed in the interpretation of earth history). Historical events are outlined in the second half of the text, with emphasis on the geological evolution of North America. In order to avoid isolating principles and events phases in their own separate vacuums, we have endeavored to tie them together through extensive cross-referencing and by providing case histories from the geologic record.

4. To stimulate the reader by punctuating the text with scientific controversies, personal feuds, amusing anecdotes, and other human interest items. Historical geology has had its own fascinating evolution and many colorful, imaginative, and inventive people have made important contributions to its development. Human interest aspects and the evolution of ideas provide extra dimension to the science and furnish themes for the various chapter openings.

5. To develop as a unifying theme the interaction between physical and organic changes in the history of the earth and the cyclic nature of many geologic phenomena. The geologic events part of the text has been developed in such a fashion as to illustrate and exemplify important principles and concepts, to downplay detailed facts and figures, and to concentrate on cause and effect. We have tried to emphasize major highlights and how they have been described and interpreted within the limits of state-of-the-art scientific research.

6. Finally, to present information in a logical sequence and to make it appealing and interesting, but also to challenge the mind with a blend of depth and rigor that is in keeping with a college-level learning experience.

Notable departures from most other texts on historical geology are

1. separate chapters on the evolutionary history of plants and mammals;

2. a somewhat modified geologic time scale that combines the newly proposed Ediacarian Period with the Paleozoic (and thus the Phanerozoic). This combination inspires a separate chapter on Ediacarian and Cambrian history, suggesting the legitimacy of a separate era designation—the Eopaleozoic (Ediacarian and Cambrian Periods)—that more faithfully reflects the importance of this time interval in the history of life;

3. chapter openings that focus on people, particularly on pioneers in the development of geologic concepts;

4. downplaying of formal stratigraphic nomenclature, with emphasis on natural stratigraphic patterns (formations versus facies); and

5. progressively more comprehensive treatment of regional tectonics and plate tectonic models from pre-Paleozoic through Cenozoic history (Chapters 9–13).

We feel our text is appropriate for a general education course at the junior college and four-year university level as well as for the geology major early in the program. We hope our text will provide a thread of continuity in learning about the fascinating history of the earth and about the guiding principles and methods that allow interpretation of that history. But we also realize no text can provide the last word on the subject. It is up to the individual instructor to fill the gaps and catalyze the learning experience, and it is up to the individual student to accept responsibility for the learning.

ACKNOWLEDGMENTS

Throughout this project, we have had the critiques of our colleagues to help us prepare a text that would meet our goals. Karl J. Koenig of Texas A & M University, Raymond Sullivan of San Francisco State University, C. John Mann of the University of Illinois, and Charles Rockwell of Nassau Community College reviewed an early version of the manuscript. A later version of the manuscript was further reviewed by Charles J. Mott of St. Petersburg Junior College, Clearwater, William I. Ausich of Ohio State University, William N. Orr of the University of Oregon, Thomas H. Dunham of Old Dominion University, Lawrence H. Balthaser of California Polytechnic State University, San Luis Obispo, and E. Joan Baldwin of El Camino College. We appreciate their thoughtful comments and suggestions. The text benefitted from their careful attention.

1

HOW THE EARTH WORKS

CONTENTS

KEY TERMS

Plate tectonics
Lithosphere
Mesosphere
Asthenosphere
Geologic cycle
Hydrologic cycle
Rock cycle
Tectonic cycle
Isostasy
Divergent plate boundary
Magnetic anomaly
Convergent plate boundary
Subduction zone
Andesite
Volcanic island arc
Magmatic arc
Transform fault
Triple junction
Orogenic belt
Terrane

HISTORICAL GEOLOGY

Geology is the science of the solid earth—its composition, processes, and history. *Historical geology* is the science of the earth's evolution and involves interpretation of rocks and geologic structures for an understanding of changes and events in the earth's dynamic past. This historical synthesis and perspective

Table 1-1. Major subdivisions of the geologic time scale

Eon	Era	Period		Age in m.y. B.P.*
PHANEROZOIC	Cenozoic	Quaternary		2
		Tertiary	Neogene	24
			Paleogene	65
	Mesozoic	Cretaceous		144
		Jurassic		208
		Triassic		245
	Paleozoic	Permian		286
		Carboniferous	Pennsylvania	320
			Mississippian	360
		Devonian		408
		Silurian		438
		Ordovician		505
		Cambrian		570
		Ediacarian		700
CRYPTOZOIC	PROTEROZOIC	Late Proterozoic		900
		Medial Proterozoic		1600
		Early Proterozoic		2500
	ARCHEAN	Late Archean		3000
		Medial Archean		3400
		Early Archean		~3800
-	Pregeologic history of the earth; origin of earth			4600

*Numerical ages in millions of years before present. Ages for bases of Ediacarian and Cambrian Periods are currently in a state of flux.
Data from A.R. Palmer, comp., 1983, The Decade of North American Geology 1983 Time Scale, p. 504: *Geology,* vol. 11, no. 9; G.V. Cohee, M.F. Glaessner, and H.D. Hedberg, 1978, Contributions to the Geologic Time Scale: *Am. Assoc. Petroleum Geologists, Studies in Geology,* no. 6; P.E. Cloud and M.F. Glaessner, 1982, The Ediacarian Period and System: Metazoa Inherit the Earth: *Science,* vol. 217, no. 4562, pp. 193–206; W.B. Harland et al., 1982, *A Geologic Time Scale,* Charts 1.1, 1.2, pp. 4, 5: Cambridge University Press, Cambridge, England.

involves consideration of the solid earth, of the earth's *atmosphere, hydrosphere* (water of the continents and oceans), and *biosphere* (life), and also the solar system to which the earth belongs. The underpinnings of the science of historical geology are furnished by the *geological time scale* (Table 1-1), a calendar that subdivides the 4600-million (4.6 billion)- year history of the earth and provides a meaningful time frame within which events of earth history are arranged.

This textbook of historical geology treats the principles and concepts used in the interpretation of earth history and illustrates their application with models of the North American continent. The earth has experienced many changes throughout its long, eventful history. What has been the driving force behind these changes? The answer to this question is complex, but it touches on a comparatively new revolution in the earth sciences—a revolution that has provided a unifying principle for comprehending how the earth works and how physical, chemical, and biological processes have influenced the evolution of our planet.

THE NEW REVOLUTION

During the 1960s, earth scientists reached a new plateau in their understanding of the earth. An exciting revolution, called **plate tectonics,** has given substance to the once-outrageous hypothesis that the continents have moved—at times separating, at other times colliding—during much of the earth's history. The nub of plate tectonics theory is that the earth's outer rind, the **lithosphere,** is fragmented, consisting of a mosaic of rigid pieces, called plates (Fig. 1-1), that move and interact. The implications of this theory for evolution of the earth's lithosphere and surface are great. Plate tectonics provides at last a viable unifying concept that can explain such diverse phenomena as origin of the Rocky Mountains, formation of the Atlantic Ocean, why and where earthquakes and volcanoes occur, and some of the major patterns in the evolution of life. It is perhaps the most important breakthrough in the geological sciences since development of the time scale. The knowledge that has come with the plate tectonics theory is continually interwoven with the fabric of this text. The purpose of this introductory chapter is to explain what plate tectonics is and its relationship to how the earth

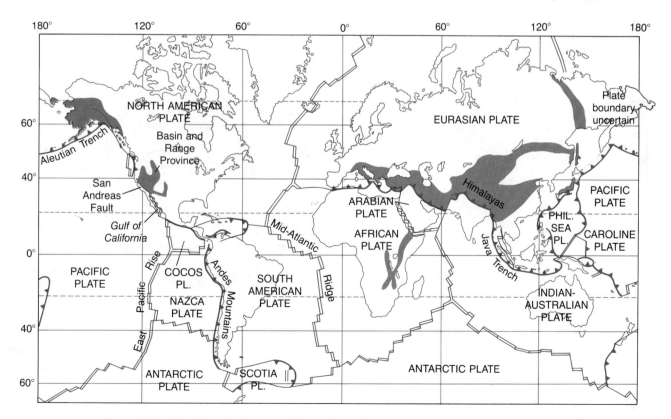

Figure 1-1. Major lithospheric plates of the world.
(From Warren B. Hamilton, 1979, Tectonics of the Indonesian Region, Fig. 2, p. 8: *U.S. Geol. Survey Prof. Paper* 1978)

works. Before examining the main tenets of the theory, it is appropriate to become acquainted with the earth's composition and architecture and the *cycles* that mold its character.

ARCHITECTURE OF THE EARTH

The Earth's Interior

The earth has an average density of about 5.5 grams per cubic centimeter (g/cm^3) and is believed to have a solid inner core and a liquid outer core (both of which are composed predominantly of iron and nickel). Outside the core, the earth has a solid **mesosphere,** a mushy, almost plastic **asthenosphere,** where partial melting occurs, and a rigid **lithosphere** (Fig. 1-2). Most of this knowledge has come from the study of seismic (earthquake) waves. Two principal wave types—shear (S waves) and compressional (P waves)—have characteristic velocities in solid media. P waves travel faster than S waves (Fig. 1-3) and the velocities of both increase with increasing density of the earth's interior. S waves, however, are not transmitted in a medium (such as a liquid) that does not have resistance to shearing deformation.

Thus, due to its failure to transmit S waves, together with the dramatic decrease in P-wave activity, the outer core is believed to be in a molten state.

Study of seismic waves and of volcanoes in the oceanic regions (e.g., the Hawaiian chain in the Pacific) and direct sampling by means of deep-sea drilling show that the lithosphere that forms the floor of the ocean basins is composed of *basalt.* Basalt is an igneous rock rich in magnesium-bearing *silicate minerals* such as *olivine* and *pyroxene* (Fig. 1-9A) and has a density of about 3.0 g/cm^3.

Continents have been found, through seismic-wave studies and other geological investigations, to have an average composition of *granite,* an igneous rock that crystallizes from *magma* below the surface. However, in their composition and structure, continents are complex, generally more so than oceanic lithosphere. Granites and their close relatives are composed predominantly of silicate minerals that consist mainly of oxides of silicon and aluminum. Continental lithosphere has a density that averages about 2.7 g/cm^3, significantly less than that of basalt. The density of the bottom part of the continental masses appears to approach the density of basalt (Fig. 1-4). Below the lithosphere lies a zone 100–200 km thick (Fig. 1-5), in which the velocities of earthquake waves are lower than in the more rigid lithosphere. This low-velocity zone constitutes the upper part of the asthenosphere.

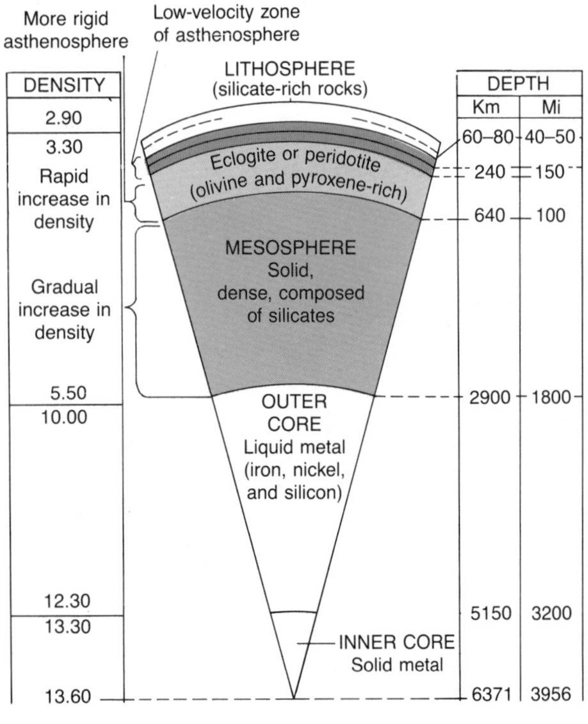

Figure 1-2. Internal zones, composition, and density of the earth. (Modified from William Glen, 1975, *Continental Drift and Plate Tectonics,* Fig. 1.7, p. 10: Charles E. Merrill Publishing Co., Columbus OH)

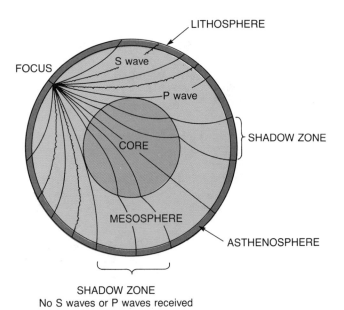

Figure 1-3. Effect of the earth's interior zones on earthquake waves. The S waves (shear waves) have a transverse motion and thus are not able to penetrate the molten outer core. The shadow zone is an expression of the inward bending of P waves (compressional waves) where they enter a zone in which they are slowed. (Modified from William Glen, 1975, *Continental Drift and Plate Tectonics,* Fig. 1-6, p. 9: Charles E. Merrill Publishing Co., Columbus OH)

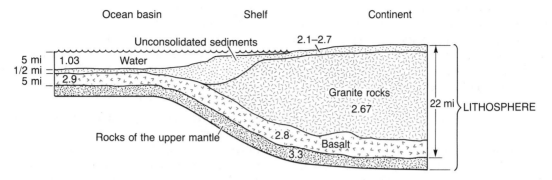

Figure 1-4. The earth's crust and upper mantle showing relative densities.
(From A. Leigh Mintz, 1981, *Historical Geology: The Science of a Dynamic Earth,* 3rd ed. Fig. 8-11, p. 199:
Charles E. Merrill Publishing Co., Columbus OH)

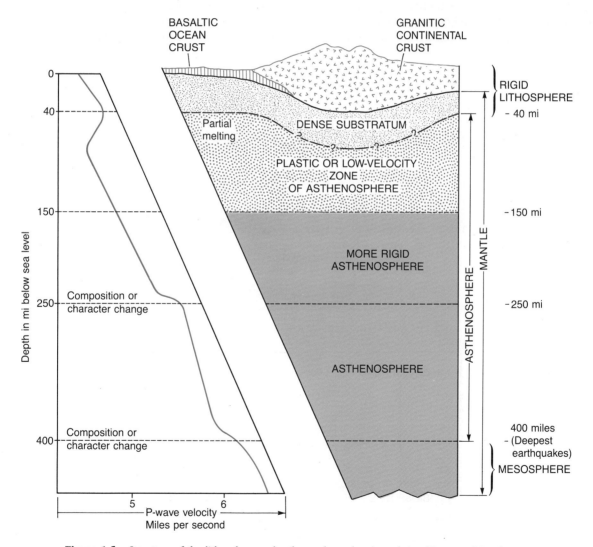

Figure 1-5. Structure of the lithosphere and asthenosphere showing relationship to traditional crust
and mantle. Note the decrease in seismic-wave velocity in the upper part of the asthenosphere between
40 and about 90 miles depth. Seismic-wave velocity depends greatly on the density, rigidity, and elas-
ticity of rock. The low-velocity zone of the upper asthenosphere behaves plastically.
(Modified from William Glen, 1975, *Continental Drift and Plate Tectonics,* Fig. 1.8, p. 14: Charles E.
Merrill Publishing Co., Columbus OH)

It is the lithosphere, essentially "floating" on the more plastic asthenosphere below, that figures most prominently in our later discussion of how the earth works. Before examining some of the key elements of the plate tectonics theory, let us consider the lithosphere in a broader geologic context.

The Geologic Cycle

The lithosphere is composed of a complex assortment of rock types with a wide range of compositions. These different rocks are the products of physical, chemical, and biological processes occurring within the **geologic cycle** (Fig. 1-6), which schematically portrays the way the earth works. The components of the geologic cycle are the **hydrologic cycle** (Fig. 1-7), the **rock cycle** (Fig. 1-8), and the **tectonic cycle,** all involving interactions of the lithosphere with the atmosphere, hydrosphere, and biosphere. The interrelation-

ship of these cycles and spheres is a very important concept in earth history.

The rock cycle is the hub of the geologic cycle. Figure 1-8 graphically portrays the interrelationships among the three families of rock that make up the lithosphere: igneous (Fig. 1-9), sedimentary (Fig. 1-10), and metamorphic (Fig. 1-11).

Igneous rocks are those that have solidified (crystallized) from a hot silicate melt (Fig. 1-8). This molten material, generated within the asthenosphere and lithosphere, is magma below the earth's surface and lava on the surface (Fig. 1-9A). As the melt cools through a succession of critical temperatures, various silicate minerals crystallize and aggregate to form rocks. Common types of igneous rocks are granite, basalt, rhyolite, and andesite (Fig. 9B).

Sedimentary rocks have resulted from the accumulation of particles on the earth's surface through the combined action of the hydrosphere, atmosphere, and

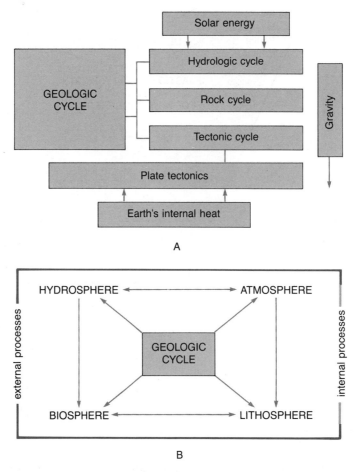

Figure 1-6. A. Relationship of the geologic cycle to plate tectonics and energy sources. B. Relationship of the geologic cycle to the four earth spheres within the framework of internal (igneous and tectonic) and external (weathering and erosion) processes.
(Modified from Peter J. Wyllie, *The Way The Earth Works,* Fig. 6-1, p. 74. Copyright © 1976 by John Wiley & Sons, Inc., New York. Reprinted by permission of John Wiley & Sons, Inc.)

commonly the biosphere (Figs. 1-8; 1-10). Most sedimentary particles are *deposited* by the action of water, but some are deposited directly by the wind or by ice. Sedimentary rocks are developed characteristically in stratified (layered) sequences, the result of spreading and working by the depositional agent. Some common sedimentary rock types are sandstone, shale, conglomerate, and limestone (Table 1-2). It is the layered arrangement, together with the occurrence of fossils (re-

mains of ancient life) in many sedimentary sequences that make this family of rocks most critical for interpreting earth history. More detailed discussions of sedimentary rocks are presented in Chapters 2 and 6.

Metamorphic rocks are those that have been changed from their original state by the individual or combined action of high pressure, high temperature, and chemically active fluids and as a consequence of deep burial, igneous activity, and deformation within the

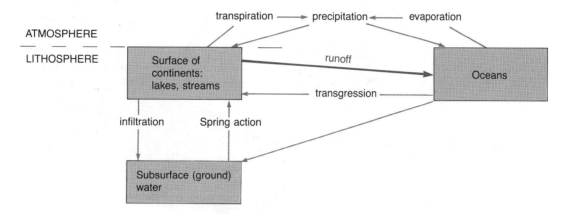

Figure 1-7. The hydrologic cycle, showing the myriad pathways of water through the atmosphere, hydrosphere, and lithosphere.

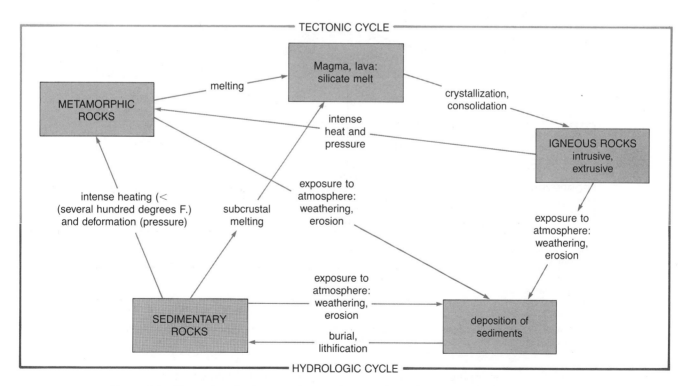

Figure 1-8. The rock cycle, depicting the complex interrelationships among the three main families of rocks. The rock cycle is intimately related to both the tectonic cycle, which involves the generation of magmas, formation of igneous rocks, and dynamic metamorphism, and the hydrologic cycle, whose main influence is in the weathering and eroding of rock and in providing the aqueous medium of sedimentation.

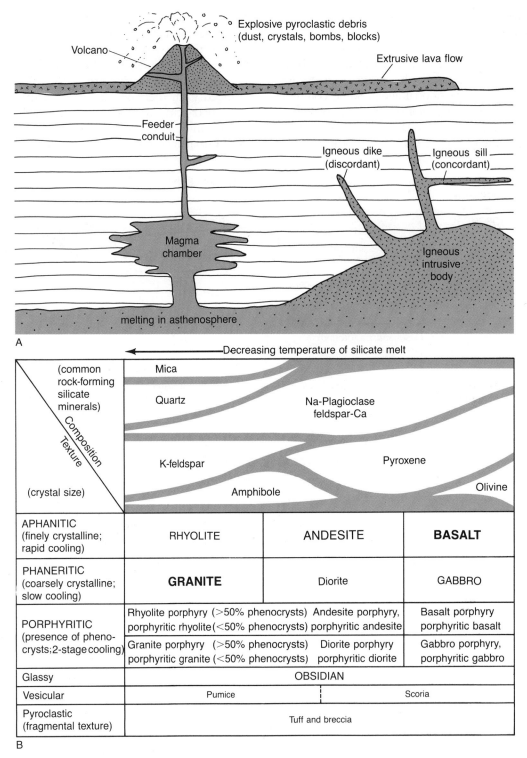

Figure 1-9. A. Kinds of igneous activity and some of the forms taken by igneous rocks. B. Simplified classification of igneous rocks showing relationship between mineral composition and texture.

lithosphere (Fig. 1-8). The metamorphic part of the rock cycle occurs only when and where the tectonic cycle is operating. Some of the more common metamorphic rocks are slate, schist, gneiss, and marble (Fig. 1-11C).

The great bulk of the lithosphere is composed of *intrusive* igneous rocks (Fig. 1-9) and, secondarily, meta-

morphic rocks. Sedimentary rocks are confined to the surface and to the moderately shallow subsurface. Through geologic time all three rock types have been continually recycled: melting of rock produces magma for the formation of new igneous rocks; weathering and erosion of igneous, metamorphic, and sedimentary

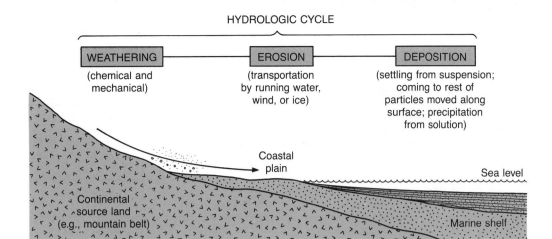

Figure 1-10. Processes active in formation of sedimentary deposits. Composition of sedimentary particles is related to extrabasinal (transported solids derived from preexisting rocks) and intrabasinal (formed within basin of deposition) origin.

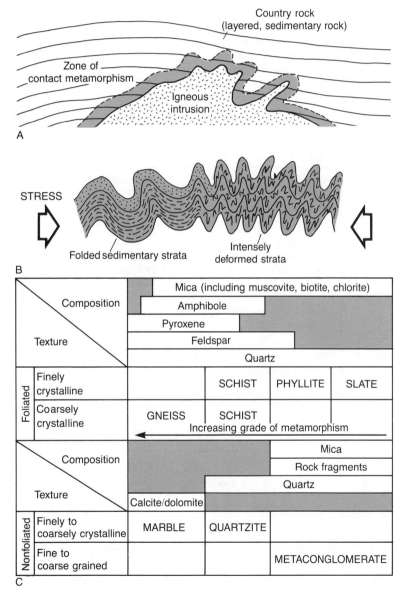

Figure 1-11. A. Contact metamorphism (thermal) of preexisting (country) rock around margins of igneous intrusion. B. Regional, dynamic metamorphism (regional high pressure and temperature) resulting from deep burial and intense deformation produced by compression. C. Classification of metamorphic rocks showing relationship between texture and composition for foliated (parallel to subparallel layering produced by mineral segregation and orientation of platy minerals such as mica) and nonfoliated (generally involving recrystallization of monomineralic rocks such as quartz sandstone, limestone, or dolomite) rocks. Metaconglomerates have elongate pebbles (produced by stretching).

Table 1-2. Simplified classification of sedimentary rocks*

Composition \ Texture	Gravel	2 mm Sand	1/16 mm Mud (or silt, clay)
TERRIGENOUS CLASTIC (extrabasinal)	CONGLOMERATE; BRECCIA if angular	SANDSTONE	MUDSTONE, SILTSTONE, CLAYSTONE; SHALE if fissile
Quartz	Quartz-pebble conglomerate/breccia	If > 95% quartz: QUARTZ ARENIFE	Quartz siltstone
Feldspar	Granite rock fragments + coarse feldspar Arkosic conglomerate	If 25% feldspar-ARKOSE	Arkosic siltstone
Rock fragments	Polymictic conglomerate if mixed composition; single source + conglomerate if one clast type; e.g., limestone-pebble conglomerate, volcanic-pebble conglomerate	If 25% rock fragments: LITHIC SANDSTONE	
Mica and clay minerals		Micaceous ankose or micaceous lithic sandstone	Micaceous siltstone
NONTERRIGENOUS (intrabasinal)	Express in terms of grain size if rock is composed largely of transported particles; crystal size if texture is crystalline		
Calcium carbonate: LIMESTONE	CALCIRUDITE; modify with dominant grain type (e.g., skeletal calcirudite)	CALCARENITE; modify with dominant grain type (e.g., oolitic calcerenite)	CALCILUTITE or MICRITE
Calcium-magnesium carbonate: DOLOMITE	Coarsely crystalline DOLOMITE	Medium crystalline DOLOMITE	Finely crystalline DOLOMITE
Silica-crypocrystalline quartz: CHERT	Includes green and red jasper; white novaculite, and gray/black flint varieties; can form as primary precipitate or secondary replacement		
Calcium sulfate: GYPSUM and ANHYDRITE Sodium chloride: HALITE	Coarsely crystalline gypsum, anhydrite, halite	Medium crystalline gypsum, anhydrite, halite	Finely crystalline gypsum, anhydrite, halite
Carbon: COAL	Includes anthracite, bituminous, subbituminous, and lignite varieties		

Simplified classification of sedimentary rocks showing relationship between texture and composition for terrigenous clastic and nonterrigenous (chemical) sediments. Terrigenous clastic sediments are composed of particles derived from preexisting rocks. The particles are mainly the more stable silicate minerals such as quartz and K-feldspar, but include rock fragments (grains that maintain the full identity of their parent rocks), particularly in the conglomerates (and breccias). Nonterrigenous sediments are composed of transported particles and minerals precipitated from solution within the basin of deposition. Composition of individual nonterrigenous rocks is generally less complex than that of terrigenous clastic rocks. Most of the other chemical sediments (with the exception of coal) have a crystalline texture.

rocks produce new sedimentary rocks; and intense heating and deformation of igneous and sedimentary rocks result in new metamorphic rocks (Fig. 1-8).

The Earth's Surface

The surface of our planet can be directly observed and sampled. Two of the most apparent features of the earth's surface are the continents and ocean basins. As mentioned previously, the continental and oceanic lithospheres are made of rock of very different composition. The continents represent blocks of lighter rock that essentially "float" on denser subjacent materials, much as an iceberg floats in water. This so-called "flotational bobbing" is in accordance with the principle of **isostasy,** which, like Archimedes' principle of buoyancy, says that rock masses sink in more dense substratum until the weight of substratum displaced becomes equal to the weight of the rock mass (Fig. 1-12). This creates a state of balance, or equilibrium, illustrated in Figure 1-12 by the blocks of copper floating in mercury.

Approximately 70 percent of the earth's surface is covered by the oceans. Figure 1-13 shows the relative percentages of the surface areas that lie at various elevation intervals above or below sea level. Of particular note among the submerged regions of the earth are submarine mountains (the ocean ridges), the abyssal plains of the ocean basin floor, and the continental shelves, slopes, and rises, which are major areas of sedimentation in the modern world's oceans. Sediments and sedimentary rocks that veneer the ocean bottoms together with the sedimentary rocks on the continents account for about 75 percent of the rocks exposed at the earth's surface. This wide surface distribution is yet another feature that makes the sedimentary rock record so important for understanding earth history.

Within the context of continent and ocean-basin structure, there are various kinds and scales of relief features. The tectonic cycle causes large areas of the earth's surface to subside slowly during geologic time and subsequently to rise slowly through uplift and mountain building. Without intervals of continental uplift, all exposed rocks would have been worn down by weathering and erosion and redeposited in the oceans long ago. Therefore the surface morphology of the earth, at any given time in history, is the net result of internal and external processes operating within the *geologic cycle*.

The hydrologic cycle and the sedimentary part of the rock cycle are driven by external (surface) processes. The igneous and metamorphic parts of the rock cycle and the tectonic cycle are driven by internal (subsurface) processes. The explanation of the tectonic cycle as well as of parts of the rock cycle lies in the theory of plate tectonics.

The earth is a dynamic planet with an atmosphere that weathers and erodes, a hydrosphere that transports and collects particles and chemical compounds, a bio-

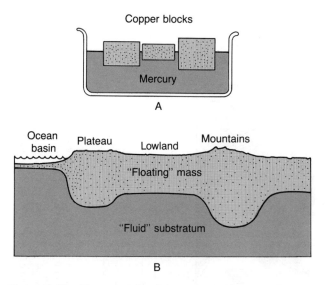

Figure 1-12. The principle of isostasy. A. Flotation of copper blocks in more dense mercury. B. Flotation of lighter continental crustal blocks in "more fluid" denser substratum.
(From H. Takeuchi, S. Uyeda, and H. Kanamori, 1970, *Debate about the Earth,* rev. ed., Fig. 1-8, p. 33. Reproduced by permission of Freeman, Cooper & Company, San Francisco)

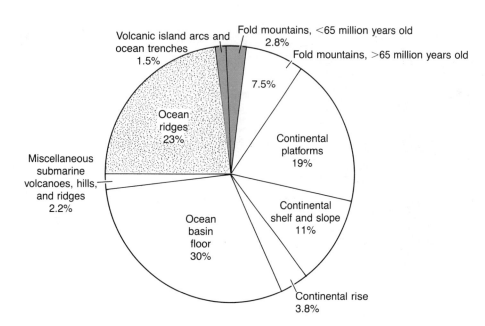

Figure 1-13. Major features of the earth's solid surface shown as percentages of the total world surface.
(From Peter J. Wyllie, *The Way the Earth Works,* Fig. 3-11, p. 38. Copyright © 1976 by John Wiley & Sons, Inc., New York. Reprinted by permission of John Wiley & Sons, Inc.

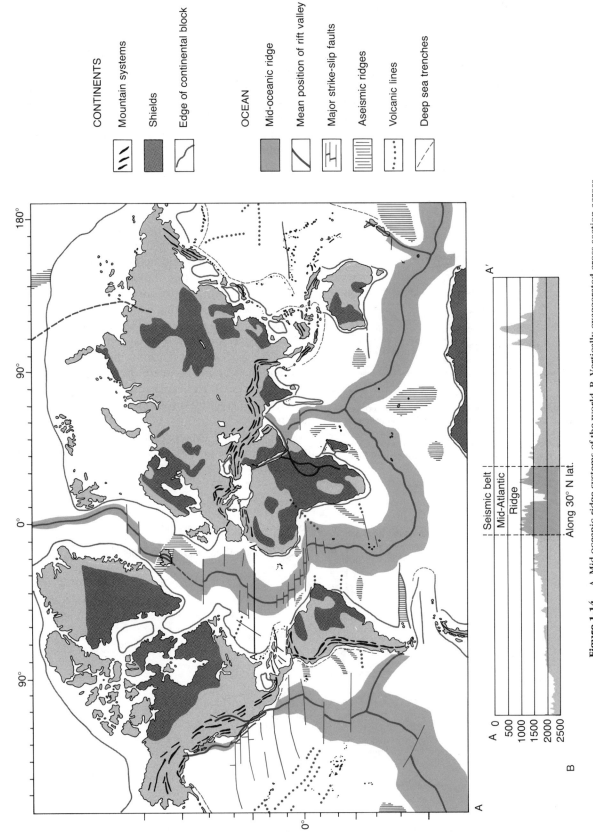

CONTINENTS

Mountain systems

Shields

Edge of continental block

OCEAN

Mid-oceanic ridge

Mean position of rift valley

Major strike-slip faults

Aseismic ridges

Volcanic lines

Deep sea trenches

Figure 1-14. A. Mid-oceanic ridge systems of the world. B. Vertically exaggerated cross section across the Atlantic ocean basin showing Mid-Atlantic Ridge and rift as well as topography of adjacent seafloor. See reference line A–A′ in A for location and orientation. (From B.C. Heezen, 1962, The Deep Sea Floor, Figs. 19, 20, pp. 260, 262, *in Continental Drift,* International Geophysics Series, vol. 3, S.K. Runcorn, , ed.: Academic Press, New York. Reproduced by permission of Academic Press, Inc.)

sphere that includes all life processes, and an internal heat engine that is responsible for driving the lithospheric plates. Because of the geologic cycle and particularly because of plate tectonics, ocean basins and continents have changed dramatically during geologic history. We will examine the outline of the unifying plate tectonics concept by taking a more detailed look at composition, structure, and surface features of the ocean basins and continents as they relate to the geologic cycle.

PLATE TECTONICS

Ocean Basins

The most prominent feature of the ocean basins is a 70,000-km-long ridge system that constitutes the most extensive mountain range on earth. This ridge system is fundamentally and significantly different in composition, origin, and structure from the mountain ranges that exist on continents. In the Mid-Atlantic Ridge, there are on both sides of the ridge abyssal plains (Fig 1-14), which

are immense flat regions blanketed by fine-grained sediments at depths of 5000 to 6000 m or more. The sides of the ridge rise gradually from the abyssal plains to the crest of the ridge, which is generally 1500 m or less below the surface of the ocean, but in places projects to the surface as islands such as Iceland and the Azores. Along the crest is a rift or structural valley (Fig. 1-14) which has abnormally high heat flow, abundant earthquakes, and submarine volcanism.

The volcanism results from upwelling basaltic material that solidifies and becomes welded onto the inner sides of the rift. This activity creates new lithosphere along the rift. The rift itself continues to exist because newly formed lithosphere separates and moves away on both sides of the rift.

In the language of plate tectonics, the Mid-Atlantic Ridge and the rest of the oceanic ridge system are *ocean-floor spreading centers.* They represent **divergent plate boundaries,** where two major lithospheric plates move apart at the rate of a few centimeters per year (Fig. 1-15). This spreading of oceanic lithosphere along the Mid-Atlantic Ridge has been responsible for

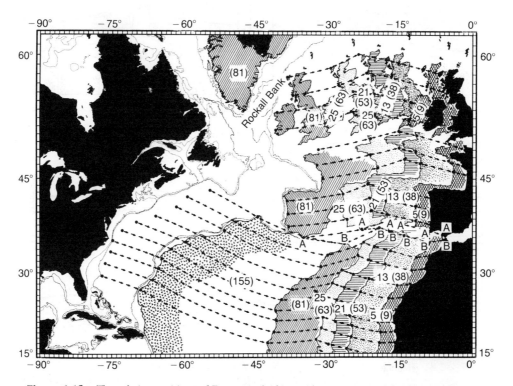

Figure 1-15. The relative positions of Europe and Africa with respect to North America for specific times. Blacked-in continents represent present positions. Dates for earlier positions are indicated on the map. Greenland, for example, is shown at 81 million years B.P. and in its present position. At 81 million years B.P. Greenland is assumed to have been attached to Europe. The 200-m and 1000-m isobaths (lines of equal depth) are shown for the positions at 81 million years. The arrows show the path of drift of Africa and Europe away from North America. A and B, which represent a point in Spain and a point in Africa, respectively, are shown as guides to the relative motion of Eurasia and Africa. (From W.C. Pitman III and M. Talwani, 1972, Sea-floor Spreading in the North Atlantic, Fig. 6, p. 629: *Geol. Soc. America Bull.,* vol. 83, no. 3. Reproduced by permission of authors and Geological Society of America).

separation of the opposite shores of the Atlantic Ocean. As a divergent plate boundary, the ridge system has played an instrumental role in global tectonics history and patterns.

The signature of spreading is the pattern of paired **magnetic anomalies** (irregularities) arranged like stripes on both sides of the oceanic ridge system (Fig. 1-16). These anomalies are detected by sensitive magnetometers: instruments that record the *remnant magnetism* of rocks on the ocean floor. Remnant, or "fossil-

ized," magnetism is preserved in many rocks of the earth's lithosphere, particularly in oceanic basalts, and is an expression of the earth's magnetic field at the time of formation of the rocks (Fig. 1-17A). Iron-bearing minerals in the rocks were oriented parallel to the magnetic field dipole at the time of final consolidation (Fig. 1-17B). The orientation of the iron-bearing particles is an index to the paleolatitude in which the rocks formed. Magnetized particles in volcanic rocks and sediments (Figs 1-17C–E) forming at the present time show a pro-

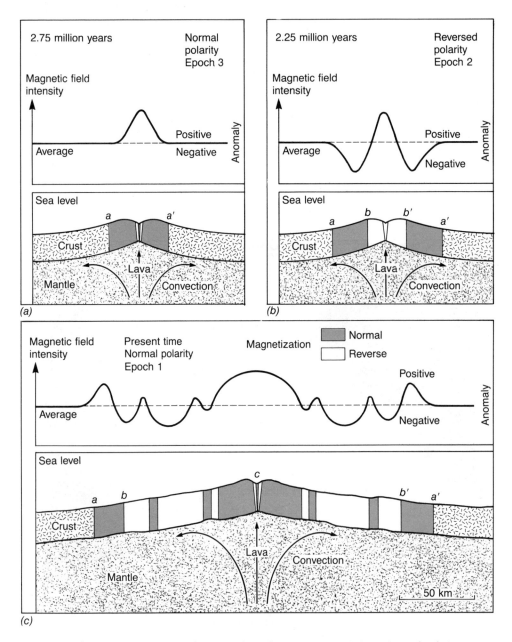

Figure 1-16. Process whereby sea-floor spreading, changes in magnetic intensity, and polarity reversals produce a series of magnetized lava strips parallel to an ocean-ridge crest. The magnetized strips are symmetrically paired magnetic anomalies parallel to the ridge crest. Symbols, *a, a′, b, b′,* and *c* represent magnetic anomaly boundaries.
(From Peter J. Wyllie, *The Way the Earth Works,* Fig. 10-5, p. 140. Copyright © 1976 by John Wiley & Sons, Inc. Reproduced by permission of John Wiley & Sons, Inc.)

gressive range of inclinations from horizontal at the equator to vertical at the poles. The magnetic anomaly stripes record not only differences in magnetic intensity (the initial indication of their existence), but, perhaps even more importantly, reversals in polarity of the earth's magnetic field.

These positive (present magnetic pole orientation) and negative (reversed magnetic pole orientation) anomalies (Fig. 1-16) are clearly the records of succes-sive polarity reversals—magnetic north became mag-netic south and vice versa—at intervals of several hundred thousand or millions of years. The anomalies provide a kind of "tape recording" of the succession of polarity reversals: oceanic lithosphere continually formed along the spreading centers and moved away from the ridge as new material took its place. The actual dating of sea-floor rocks by various methods (Chapter 2) has demonstrated conclusively that the rocks (and

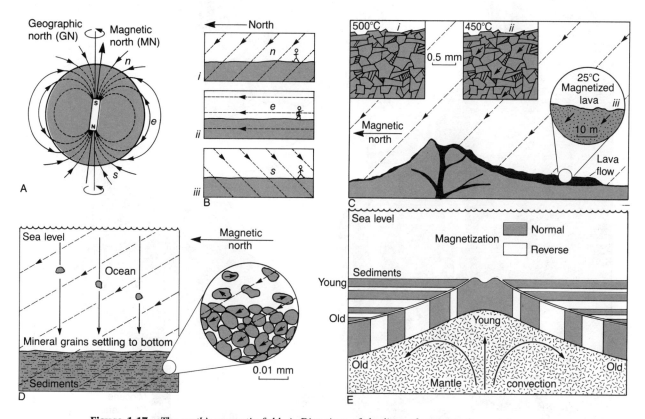

Figure 1-17. The earth's magnetic field. A. Directions of the lines of magnetic force at the earth's surface and in space around the earth (as measured from satellites) are consistent with the presence of a "magnet" within the earth. The earth's magnetic field is probably produced by electromagnetic forces generated in the earth's outer, liquid core. The lines of magnetic force, as they would be measured by a person standing on the earth's surface at points *n, e,* and *s,* respectively, are illustrated in larger scale in B, *i, ii,* and *iii.* C. Schematic vertical cross section through a volcano showing directions of the earth's lines of magnetic force. In this sequence, lava erupts, flows, cools, and crystallizes. Inset *(i)* represents the interlocking crystals in the lava after it has solidified but while it is still hot; the minerals are not magnetized. Inset *(ii)* shows the same rock when it has cooled to 450° C., below the melting point for some minerals. These minerals have become magnetized in the direction of the earth's mag-netic field. Inset *(iii)* represents a larger part of the lava flow after it has cooled further; it contains many minerals, all cooled below their melting points and magnetized as shown in *(ii).* Consequently, the entire rock is magnetized in the direction shown in *(iii).* D. Schematic vertical cross section through ocean and submarine sediments, showing mineral grains settling slowly to the ocean floor and the direction of the earth's lines of magnetic force. The inset shows that some of these falling particles have been magnetized at some earlier stage in their history (as in C); as these particles settle on the sediment surface, they become oriented in the direction of the earth's magnetic field. This magnetic direction is recorded by the sediment layer as a whole when it becomes compacted and consolidated into sedimentary rock. The best results in reading the magnetic signature of sediments are obtained from very fine-grained deposits that accumulated under quiet (free from disturbance by currents) water conditions of settling from suspension. E. Schematic cross section depicting the effect of polarity rever-sals on magnetization of deep-sea sediments with sea-floor spreading. Each magnetized sedimentary layer overlies the older sediments and rests on a portion of the lava crust that was generated at the ridge crest during the same time interval and subsequently migrated away from the ridge. (From Peter J. Wyllie, *The Way the Earth Works,* Figs. 8-5, 8-10, 10-7, pp. 105, 113, 144. Copyright © 1976 by John Wiley & Sons, Inc. Reproduced by permission of John Wiley & Sons, Inc.)

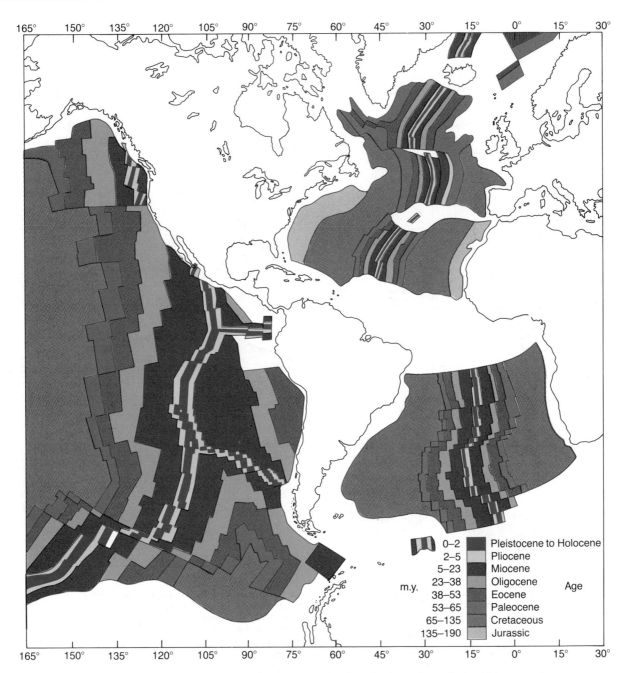

Figure 1-18. Age of the ocean floors in the East Pacific and Atlantic Ocean basins. This pattern is an expression of sea-floor spreading since the Mesozoic.
(From W.C. Pitman III, R.L. Larson, and Ellen M. Herron, compilers, 1974, Age of the Ocean Floors: Geological Society of America Map. Reproduced by permission of authors and Geological Society of America)

their remnant magnetic anomaly patterns) increase in age away from the ridge crests and that the oldest rocks in the Atlantic Ocean basin are less than 180 million years of age. This age is only one-twentieth that of the oldest known rocks on the continents. All of the oceanic ridge spreading centers are flanked by magnetic anomalies that can be matched from one side to the other.

The most symmetrical pattern is in the Atlantic Ocean, where it appears that the ocean basin has been steadily widening (and the bordering continents separating) since Jurassic time (Fig. 1-18). The continental blocks on both sides of the Atlantic are linked to the spreading Atlantic Ocean lithosphere and thus are separating as a consequence of spreading.

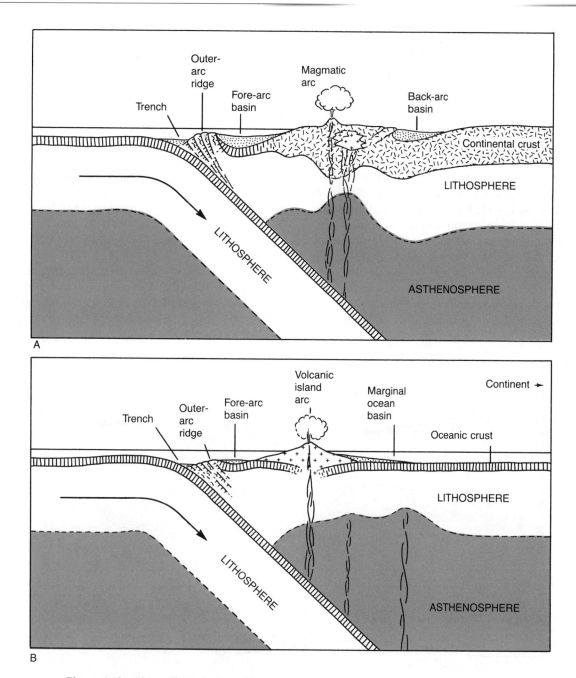

Figure 1-19. Plate collision in the Pacific Ocean. A. Continental margin magmatic arc–trench system. B. Intraoceanic volcanic island arc–trench system. Note the igneous activity generated by subduction into the asthenosphere.

(From Ben M. Page, 1977, Subduction Tectonics, in *Late Mesozoic and Cenozoic Sedimentation and Tectonics in California,* Fig. 1-1B, p. 21, Short Course Syllabus published by San Joaquin Geological Society. Reproduced by permission of San Joaquin Geological Society)

In the Pacific region, the situation is more complex. There the continents are either colliding with or sideswiping oceanic lithosphere instead of drifting apart, as they are in the Atlantic. In this region, we see the results of actual interaction between plates of different density: those composed of the relatively light, granitic continental lithosphere ($d \approx 2.7$ g/cm^3) and those of the heavier, basaltic oceanic lithosphere ($d \approx 3.0$ g/cm^3). Because of its greater density, the spreading oceanic lithosphere gradually moves beneath the edge of the bordering continental block in the fashion depicted in Figure 1-19A.

Such collision margins constitute a second type of plate boundary called a **convergent boundary.** In the situation described (Fig. 1-19A), the older, leading edge of the oceanic lithospheric plate plunges to destruction at depth as it impinges against the leading edge of the continent. This interface of underthrusting of oceanic lithosphere beneath continental lithosphere is called a **subduction zone;** the underthrusting itself is most commonly referred to as subduction and occurs along a zone of high-pressure metamorphism, deep subcrustal melting, and deep-focus seismic activity. Deep-sea trenches (Figs. 1-13, 1-14, 1-19), such as the Peru–Chile trench in the East Pacific, represent the deepest ocean-bottom environments and are the sea-floor topographic expressions of subduction zones. Subduction, causing subcrustal melting and mixing of continental and oceanic lithosphere, is responsible for considerable volcanism and is the most prevalent mechanism for generating a hybrid type of igneous rock called **andesite** (Fig.

1-9). Andesite forms in (1) **volcanic island arc** chains (Fig. 1-19B), such as the Japanese and Aleutian archipelagos, where two oceanic plates collide, and (2) **magmatic arcs,** which are volcanic and intrusive igneous mountain chains along continental margins; the Cascades of the northwestern United States and the Andes of western South America are examples of such mountain chains. Both volcanic island arcs and magmatic arcs bear testimony to the igneous activity resulting from subduction. The arc configuration results from the intersection of the planar subduction zone with the curvature of the earth's surface. In subduction old oceanic lithosphere is consumed essentially as fast as new lithosphere is created (rates for both generally less than 10 cm/year).

The mechanisms of sea-floor spreading and subduction demonstrate that the continents are not independent blocks moving across more dense lithosphere, but rather are parts of much larger plates. In a sense, the

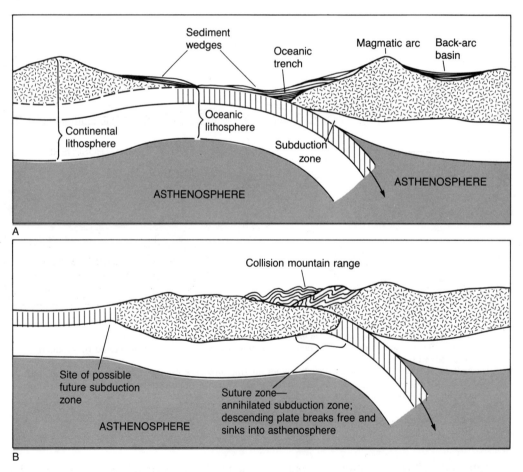

Figure 1-20. Continent-to-continent collision. A. Subduction, at the leading edge of a continent, of an oceanic plate carrying another continent. B. Collision of two continents. Because of isostasy, the continental block attached to the subducting plate is not subducted; the block is too buoyant (owing to its being less dense) to be carried down into the asthenosphere. Such a collision produces a mountain range along the suture zone. The mountains will be composed mainly of highly deformed sedimentary rocks that accumulated along the collision margins of both continents. The lofty Himalayan mountain chain formed in just such a setting during the last 40 million years: an oceanic plate carrying India collided with the Asian continental plate.

continents are passengers that are carried along on giant slabs of spreading oceanic lithosphere. North America, in such a context, is one part of a colossal lithospheric plate that extends to the Mid-Atlantic Ridge. As new lithosphere is continuously added along the spreading ridge, the North American Plate slowly migrates relatively westward. Along its leading western edge, it impinges upon, and in places overrides, the several plates that comprise the Pacific oceanic lithosphere (Fig. 1-1).

Another manifestation of plate convergence involves the collision of continental masses. Continent-to-continent collisions have occurred when continental lithospheric blocks have interacted with other continental slabs along a subduction margin. Once the continent-to-continent collision occurs, the oceanic lithosphere as well as the subduction zone are annihilated. One continent is not subducted beneath the margin of the other because of the density similarities between the two masses, although some partial subduction and thickening of continental lithosphere along the suture zone (zone of contact) does occur. Continental plate collisions (Fig. 1-20), such as that between India and Asia during Cenozoic time, have resulted in the uplift of mountain belts along the sutured margins and the growth of supercontinents.

A third type of plate boundary involves a lateral or sideswiping motion between rigid plates. Such a boundary, caused primarily by *shearing stress* (Fig. 1-21A), is developed most characteristically in oceanic areas where gigantic fractures cut across and offset oceanic ridges. The formation of new lithosphere does not occur at a constant rate: Different segments of an oceanic ridge are volcanically active at varying rates at different times. The variations in spreading motion and differential movement have produced numerous fracture systems (Fig. 1-21B) that are generally quite different from those that appear on geologic maps of continental regions. These fractures were first recognized and named **transform faults** in 1965 by J. Tuzo Wilson, one of the pioneers of plate tectonics theory. The term "transform" refers to the transforming of one type of plate boundary into another. The famous San Andreas Fault in California represents a most unusual and significant type of transform fault that will be discussed in Chapter 13. The San Andreas fault zone separates, or transforms, two segments of the East Pacific spreading center (Fig. 1-22B) and therefore forms the boundary between the North American and Pacific Plates.

Plate boundaries, in addition to being described as divergent, convergent, or transform, show more complex interactions at some places. Those points that mark the intersection of three lithospheric plates are called **triple junctions.** Examples include those off the coast of Central America and northern California (Fig. 1-22B), where sea-floor spreading ridges intersect both a subduction zone and a transform fault. It is important to

realize that the three types of plate boundaries can occur in three principal styles of physiographic and lithospheric settings: ocean basin-to-ocean basin, continent-to-ocean basin, and continent-to-continent (Fig. 1-23). Plate margins are named for the geographic settings where the different kinds of plate interactions occur.

Continents: The North American Example

In contrast to the oceanic lithosphere, which is geologically young (rocks less than 180 million years old), the continental lithosphere includes rocks ranging in age from roughly 3800 million years to yesterday's sand and gravel. The continents, as a result, are a mosaic of many different kinds of rocks of different ages and different origins.

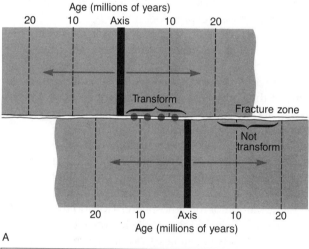

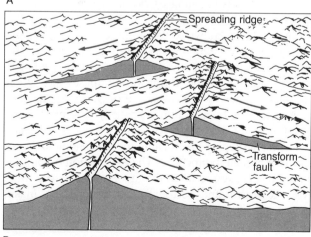

Figure 1-21. Transform fault plate boundary. A. Note difference between direction of displacement of oceanic ridge segment (transform) and direction of sea-floor spreading (not transform). B. Perspective view of situation illustrated in A.
(From Leigh W. Mintz, 1981, *Historical Geology: Science of a Dynamic Earth,* 3rd ed., Fig. 8.27, p. 113: Charles E. Merrill Publishing Co., Columbus OH)

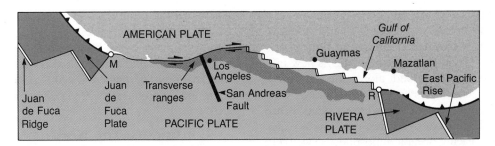

Figure 1-22. The San Andreas fault zone as an oblique transform plate boundary between the North American and Pacific Plates. M is the Mendocino triple junction between the North American, Juan de Fuca, and Pacific Plates and R is the Rivera triple junction between the North American, Rivera, and Pacific Plates.
(From Tanya Atwater, 1970, Implications of Plate Tectonics for the Cenozoic Tectonic Evolution of Western North America, Fig. 1, p. 3514: *Geol. Soc. America Bull.,* vol. 81, no. 12. Reproduced by permission of the author and the Geological Society of America)

Figure 1-24 shows the major geologic subdivisions of the North American continent: shield, interior lowlands, mountains, and coastal plains, each with its own definable set of rock types, rock ages, structural styles, and physiographic expression.

The regions from which structural mountains are born are called **orogenic belts.** These are regions of lithospheric mobility and deformation during intervals of geologic time. Ancient orogenic belts are significant parts of the continents. We shall have a good deal more to say about them when we investigate the geological evolution of North America in Chapters 9–13.

What is the cause of structural mountains? What produces the forces within the earth that bend and break rocks? How do mountains form and grow? Why are they located where they are? How important have they been in the evolution of the lithosphere? How do they tie into the tectonic cycle and the geologic cycle? The theory of plate tectonics is bringing us closer to some of the answers.

It has been known for some time that the geologically young, active mountain chains of the world are narrow, elongate belts. Reconstruction of the geologic history leading to the formation of these and older mountain chains indicates that the operation of the tectonic cycle has been confined spatially to similar elongate belts. Nowadays, armed with the concept of plate tectonics, geologists are relating these orogenic belts to convergent plate boundaries. The explanation of the tectonic cycle is seen to lie in the theory of plate tectonics: The volcanism and tectonic deformation that make mountains are generated through lithospheric plate interactions.

The main structural framework of the continent of North America is somewhat symmetrical (Fig. 1-24). In the north-central part, mainly in Canada, the areally extensive Canadian Shield, composed of a patchwork of ancient lavas, igneous plutons (large bodies of intrusive rock), metamorphic complexes, and sedimentary successions of Cryptozoic (Table 1-1) age, contains the roots of several ancient orogenic belts. During the last 1000 million years of earth history, the shield has been worn down into a low, rolling surface that plunges southward beneath a cover of younger sedimentary rocks of the interior lowlands (Fig. 1-24). Nearly encircling the shield and interior lowlands are various mountain systems formed at various times during the Phanerozoic (Table 1-1). These include the Appalachian–Ouachita systems to the east and south, the wide Cordilleran system extending the entire western length of the continent, and a prominent system along the oceanward side of the Arctic islands and Greenland.

Along the continental margin, especially along the east and southeast, coastal plains have been constructed from erosional debris of the mountain systems as those systems have been worn down. The coastal plains are composed of geologically young sediments deposited across the eroded surfaces of deformed mountain structures (Fig. 1-25). The continental shelves, submerged parts of the same prism of sediments that make up the coastal plain (Fig. 1-25), are inclined seaward at less than 50 m/km (less than 1°) and include the area generally between sea level and about 200 m below sea level. Although now submerged beneath marine waters, the shelves are parts of the continental structure. Seaward of the shelves (Fig. 1-21), and composed of major volumes of sediment, are the continental slopes (inclined at an angle of 1° to 3°) and continental rises (inclined at an angle of generally less than 1°); the transitional boundary between true continental granitic lithosphere and oceanic basaltic lithosphere is beneath the continental slope–rise wedge (Fig. 1-25).

The generalized plan of continental structure is: shield–interior lowlands–mountains–coastal plains–continental shelf–continental slope (Fig. 1-24). From the perspective of the tectonic cycle viewed within the model of plate tectonics, the shield represents the forged-together **terranes** of ancient orogenic belts, inte-

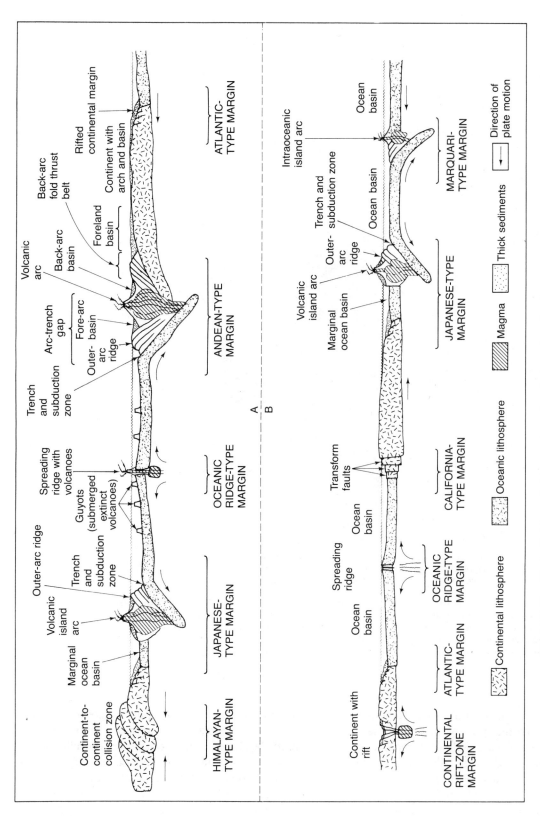

Figure 1-23. Kinds of plates and plate margins and geologic settings for plate interactions. (From Leigh W. Mintz, 1981, *Historical Geology: The Science of a Dynamic Earth*, 3rd ed., Fig. 8.26, p. 211: Charles E. Merrill Publishing Co, Columbus OH)

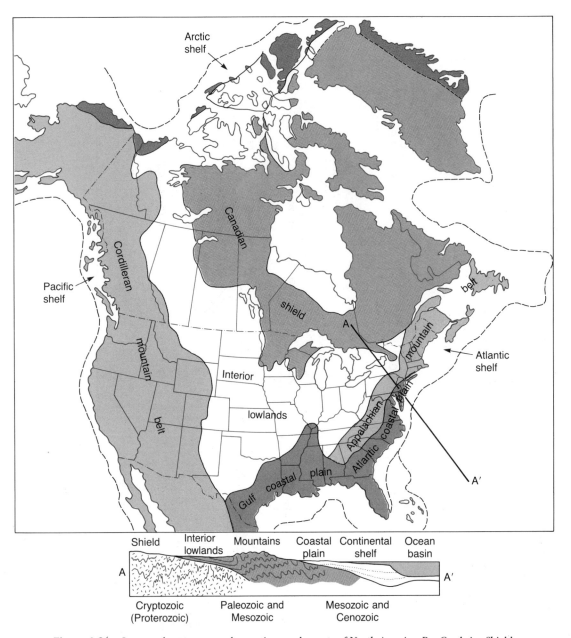

Figure 1-24. Large-scale structure and accretionary elements of North America: Pre-Cambrian Shield, Paleozoic–Mesozoic interior lowlands, Paleozoic–Mesozoic mountain chains, Mesozoic and Cenozoic coastal plain–continental shelf–continental rise and slope sedimentary wedge. Mountains represent a deformed earlier trailing (Atlantic-type) margin sedimentary wedge.
(Data from Tectonic Map of North America, 1969, U.S. Geological Survey)

rior lowlands, and coastal plains and shelves, together with pieces of ancient oceanic crust. As such, the shield is a patchwork of pieces of ancient lithospheric plates. The interior lowlands include generally undeformed sedimentary rocks that accumulated in shallow seas that periodically invaded the tectonically more stable continental interior. The mountain systems record the geologically more recent and, consequently, more apparent collisions between lithospheric plates. The shelf–slope–

rise sedimentary prisms making up the continental margin represent the most recent additions to the continent and have developed mainly on the tectonically passive, trailing edge of the moving continental block during the last 180 million years. In the geologic future the trailing eastern margin of North America may become an active collision margin similar to the present-day Pacific borderland. If this happens, the continental shelf–slope–rise prism of sediments will be deformed into a moun-

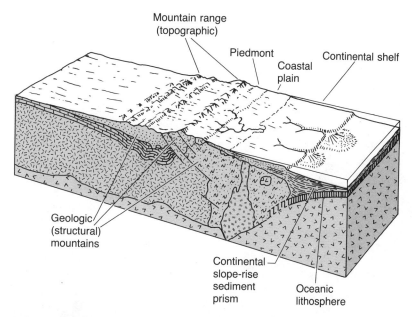

Figure 1-25. Relationship between oceanic and continental lithosphere and the coastal plain–continental shelf–continental slope and rise prism of sediments. The continental shelf sediments have accumulated along the passive, trailing margin of the continental block and on older rocks that were deformed during earlier plate collisions.
(From R.S. Dietz and J.C. Holden, 1974, Collapsing Continental Rises: Actualistic Concept of Geosynclines—a Review, Fig. 6a, p. 22, in *Ancient and Modern Geosynclines, Soc. of Econ. Paleontologists and Mineralogists Spec. Public.* 19. Reproduced by permission of Society of Economic Paleontologists and Mineralogists)

tain belt and *accreted* to the continental structure. Thus, through lateral accretion of orogenic belts during much of its long history, the continent of North America has grown.

Central Theme of Geology

Plate tectonics has become a unifying concept that is providing a powerful model with which to interpret and reinterpret a vast storehouse of descriptive geology that has accumulated during the last two centuries. Let us mention briefly some of the more noteworthy examples that illustrate how plate tectonics provides a central theme in the geological sciences.

Interaction of lithospheric plates has produced orogenic belts, which, in turn, have been the sites of active mountain building. In addition, the distribution of major earthquake epicenters and active volcanoes on the earth (Fig. 1-26) is directly associated with the locations of lithospheric spreading centers and especially with deep-sea trenches (ocean-floor manifestations of subduction zones). Plate boundaries have been loci for the formation of economic mineral deposits, particularly where igneous activity is involved. The occurrences of certain rock types, such as andesite, are related directly to plate tectonic activity.

Plate tectonics also has been the determining influence in the volumetric importance, size, and location of

sedimentary basins (Fig. 1-23) and those depositional and postdepositional environments where fossil fuels (oil, gas, and coal) have formed. Still another exciting revelation and an area of active multidisciplinary research concerns the biological consequences of plate tectonics. The lithosphere and hydrosphere have been the stage of life, and changes in that stage—separation and coming together of continents, opening and closing of ocean basins, and changes in environments and climates—have had far-reaching effects on the diversification and extinction of species through time. Finally, plate tectonics also has had a profound influence on world climatic patterns, determined as they are largely by the cycle and periodicity of mountain building, the amount of volcanism, the position of continents with respect to the poles, and the degree of clustering or fragmentation of continental blocks.

Plate tectonics is a theory that has emerged from a controversy about moving continents. Once the theory became a working principle, the phenomenon of "continental drift" was regarded to be merely a byproduct of sea-floor spreading. In this light the continents are viewed as parts of much larger, rigid lithospheric slabs that move across the plastic asthenosphere. For nearly 100 years after its first serious proposal, the notion of moving continents was a subject of heated controversy. To appreciate how far this theory has come to take its place among the great breakthroughs in science, let us

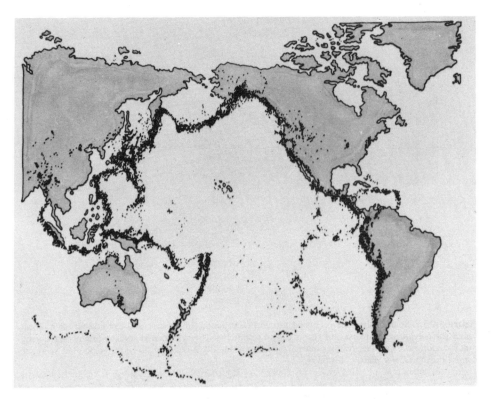

Figure 1-26. Distribution of earthquake epicenters and volcanic activity in the Pacific region. Most earthquake and volcanic activity is concentrated along collision plate margins around the Pacific Ocean Basin—the so-called Pacific "Ring of Fire."
(Photo courtesy of Critter Creations, Inc., San Diego CA)

look at a brief historical summary. The evolution of great ideas is a necessary and important part of the great ideas themselves.

THE EVOLUTION OF A CONCEPT

Historical Perspective

In 1858, a publication by Antonio Snider-Pelligrini hypothesized that Europe and North America as well as Africa and South America had been joined during the geologic past. Some continental drift historians credit Snider-Pelligrini as being the originator of modern continental drift theory. This is probably an overstatement of his contributions because Snider-Pelligrini envisioned a catastrophic rupturing of his described continental masses coincident with the great deluge (Noachian flood). In this light, he applied geologic reasoning that is no longer considered a viable explanation of the way the earth works (see Chapter 3).

Perhaps the distinction of being the "first drifter" belongs to a little-known but nonetheless remarkable French naturalist, Elisée Reclus, who set forth a clearly enunciated and brilliantly conceived notion of continental drift in his book, *The Earth* (1872).* Continents in

motion was a dominant theme in this treatise, and Reclus invoked it as a major part of some unifying scheme for the earth that he confidently felt would be discovered by future generations of geologists. Perhaps better known as a political anarchist, Reclus was clearly decades ahead of his time in his capacity to relate continental drift to mountain building, volcanic island arcs, and earthquakes and in his strong convictions about slow natural change of a very old earth.

During the later part of the nineteenth century, it was observed that many rock sequences of southern hemisphere continents matched, but, in turn, were quite different from those of northern hemisphere continents. Coupled with recognition of this similarity in geologic sequences was the recognition of evidence for late Paleozoic (Table 1-1) glaciation on the southern hemisphere continents. During the last two decades of the nineteenth century these geologic similarities inspired a brilliant Austrian geologist and synthesist, Edward Suess, to postulate the former existence of a vast southern supercontinent that he called *Gondwanaland*. To account for the fragmentation of Gondwanaland, he suggested that the former connections between its present-day fragments might have foundered to great depths to become ocean-basin crust. Of course now it is known that the very different composition and density of granite and basalt preclude the foundering of continental crust

---

*Harper and Bros., New York. American ed. trans. by B. Woodward and H. Woodward.

into ocean basins. It is inconceivable through any known process to convert granite to basalt.

An American geologist, F.B. Taylor, inspired by the writings of Suess on global tectonics, published a detailed account of his own comprehensive continental drift theory in 1910. Taylor explained the formation of geologically young mountain ranges throughout the world as the products of continental drift. His proposed mechanism for drift involved tidal action and rotational forces generated when the earth, according to his calculations, captured the moon during Cretaceous (Table 1-1) time.

The recognized father of the moving continent idea is Alfred Wegener, a German geographer and meteorologist. In a remarkable book, *The Origin of the Continents and Oceans,** first published in 1912, Wegener made a cogent case for the nonpermanency of the positions of continents, and "continental drift" was proposed as an actual theory. Wegener based his theory on similarities recognized from a comprehensive comparison of rocks, structures, stratigraphy, paleontology, paleoclimatology, and the geometrical fit of continental margins (particularly the fit between the outer edges of the continental shelves of eastern South America and western Africa). He proposed the name *Pangaea* for a supercontinent composed of today's separated continental masses—a supercontinent that was assumed to have existed prior to the beginning of fragmentation about 180 million years ago. This was a truly brilliant synthesis.

In spite of all the evidence marshaled by Wegener in support of continental drift, the theory suffered badly from not being explained by a viable mechanism. The geophysicists of the day vehemently attacked the notion as absurd and contrary to the physical properties and laws of the solid earth. They considered it a theory that represented more fiction than truth and one that appealed to extremists in geology. How, the geophysicists and most geologists maintained, could the lighter continental masses move headlong across the heavier ocean basin crust? Preposterous! It seems that each line of evidence proposed was squelched by a more definitive counterargument. Wegener and his followers were convinced they were right; they felt the onus of finding a mechanism was to be borne by the geophysicists. Their counter-counterargument was that the geophysicist doubters simply were not smart enough to discover the mechanism for drifting. And thus raged the debate— from the early 1900s on through the teens and twenties. A contingent of southern hemisphere geologists, proponents of drift, continued to assemble evidence in support of their radical views. The most vociferous skeptics continued to be the geophysicists, who summarily dismissed what the drifters considered to be geologic facts as inferences based on inconclusive data.

---

*1966 translation by John Biran from 4th German ed. (1929): Dover Publications, New York.

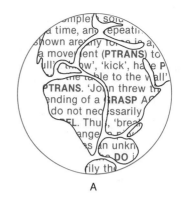

A

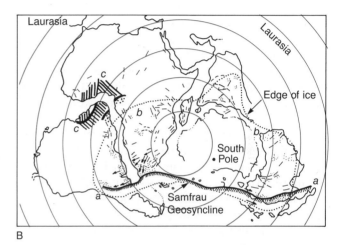

B

Figure 1-27. A. Printed lines across the continents, depicted as pieces of a torn page. B. Reconstruction of Gondwanaland on the basis of alignment of mountain belts *(a)*, glacial deposits *(b)*, and matching of geologic terranes *(c)*.
(A from H. Takeuchi, S. Uyeda, and H. Kanamori, 1970, *Debate about the Earth,* rev. ed., Fig. 1-16, p. 46. Reproduced by permission of Freeman, Cooper & Company, San Francisco. B from A.L. du Toit, 1937, *Our Wandering Continents,* Fig. 11, p. 93. Oliver and Boyd, Edinburgh. Reproduced by permission of Longman Group Ltd., London)

For example, the drifters maintained that the continents had moved into different climatic belts at different times. The skeptics retaliated by saying that different climatic regimes had fluctuated and migrated during the past. The fitting together of the continents on opposite sides of the Atlantic Ocean, as exemplified by the snug fit between South America and Africa, was an impressive piece of evidence that the proponents of drift felt could not be ignored. The antidrifters dismissed this as fortuitous happenstance. But what if the goodness-of-fit hypothesis could be supported by matching up geologic features across the ocean abyss—much like reading the newsprint across two pieces of torn newspaper (Fig.1-27)?

Early proponents of continental drift did cite similarities between rocks of various ages from Brazil and Africa. Alex du Toit, an imaginative southern hemisphere geologist, recognized an ancient distinct orogenic belt (which he called the Samfrau) present in South America, Africa, and Australia.

In 1928 the American Association of Petroleum Geologists organized a historic international meeting in New York City, where prominent earth scientists from around the world gathered to review the theory of continental drift. Both sides assembled their entire arsenal of ammunition and fired volley after volley. But when the smoke had cleared, the opponents of drift held the upper hand and carried the victory banner. It seemed that their serious objections were articulated more strongly and more convincingly than the evidence in favor of drift.

The next year, Alfred Wegener disappeared on the Greenland ice cap while engaged in historic research on echo sounding to determine the thickness of polar ice. During the 1930s and 1940s, for lack of new or otherwise stimulating evidence, the "continental drift" controversy began to wind down. It would not be until the decade of the 1950s that the controversy would be renewed.

During the twenty year period of dormancy following the historic 1928 symposium, serious debate on the subject was continued in only a few enclaves as some of the more revolutionary geologists continued to support the idea of footloose continents. One of the most brilliant adherents of the outrageous hypothesis was Arthur Holmes, a British geologist who first proposed the existence of convection currents in the solid material of the earth. Could the large-scale convective motion of heat move continents?

What was the prevailing attitude among North American geologists regarding the continental drift hypothesis? According to P.B. King, in his book, *The Evolution of North America,** most North American geologists were not so much hostile to the idea as indifferent. Continental drift had seemingly little consequence for the kinds of local and regional problems to which they were devoting attention. Moreover, the busy North American geologists were somewhat repelled by both the enthusiasts for drift and the more vocal opponents. Such thinking may seem rather provincial in hindsight; at the time, however, continental drift just did not fit into the scheme of their geologic research.

After World War II, a great surge of oceanographic research commenced and a flood of new data became available. One of the most significant revelations came from the charting of great mountain ranges beneath the surface of the oceans. Investigators identified a 70,000-km-long mountain system girdling the world—the oceanic ridge system. Mapping of the ocean basins demonstrated that the floors of the oceans are not flat, monotonous plains covered by water, but in fact contain

more extensive mountains and deeper valleys than can be found on the more familiar continents.

As new research, very different from that involved in the earlier debate, progressed during the 1950s, there was a revival of interest in the continental drift theory. One of the principal arenas for the new research involved paleomagnetism. Studies on the paleomagnetism of rocks on the continents revealed some rather astonishing results. For example, in 1953 it was suggested that England was at a low latitude during the Triassic Period (Table 1-1; Fig. 1-28) and, since the Triassic, not only had moved northward but had rotated 30° clockwise! Sets of curves were generated that showed the relationship of the continents to the earth's magnetic poles during successive intervals of time in geologic history (Fig. 1-29). These *polar wandering curves* suggested either that the poles had changed positions significantly during geologic history or that the continents had assumed various positions with respect to essentially fixed poles. Slowly, more and more geologists and geophysicists began to look upon the theory of continents in motion with favor.

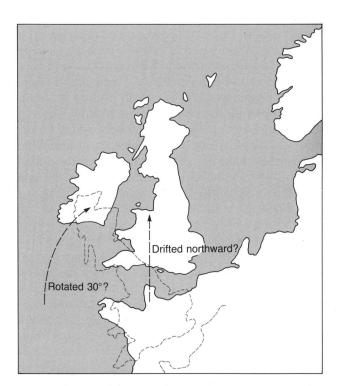

Figure 1-28. Presumed northward migration and clockwise rotation of England since the Triassic, determined on the basis of considerable deviation in Triassic geomagnetic field directions from present field direction. In 1954, this phenomenon was interpreted by a group of University of London geophysicists to represent not a change in geomagnetic field direction, but rather a change in England's position with respect to essentially fixed poles. This study reopened the question of the possibility of continental drift.
(From H. Takeuchi, S. Uyeda, and H. Kanamori, 1979, *Debate about the Earth,* rev. ed., Fig. 5-8, p. 167. Reproduced by permission of Freeman, Cooper & Company, San Francisco)

---

*1977, rev. ed.: Princeton University Press, Princeton NJ.

We have already recounted how paleomagnetic data from the ocean floors provide a kind of tape recording of the production of new oceanic lithosphere at ridge-spreading centers and destruction of that lithosphere in subduction zones. In 1960 a great American geologist, the late Harry Hess of Princeton University, postulated the hypothesis of sea-floor spreading, but the actual documentation of the phenomenon was not presented until 1963. F. J. Vine, a graduate student at Cambridge University, and D.H. Matthews, a senior colleague, collaborated on a paper that gave an explanation of anomalies in the earth's magnetic field. They suggested that the anomalies (Fig. 1-30), discovered in the Pacific Ocean some five years previously, were caused by changes in magnetic intensity. The ocean-floor magnetic stripes were successively generated by sea-floor spreading and carried the indelible imprint of periodic reversals of the earth's magnetic field. It was the confirmation of periodic magnetic field polarity reversals and the associaton of these with the system of submarine ridges (spreading centers) that transformed what Harry Hess had called geopoetry into geologic fact.

Vine and Matthews are credited with the model combining sea-floor spreading with magnetic anomalies and polarity reversals, but a Canadian, L. W. Morley, independently worked out the same model at about the same time. In fact, Morley's paper on the subject was rejected for publication by two journals in early 1963! This illustrates the prevailing attitude against acceptance of radical new ideas; even the Vine–Matthews scheme was not immediately accepted. But by 1966, nearly forty years after the memorable New York meeting that almost laid the "continental drift" theory to rest, a group of experts reviewed all available evidence on magnetic anomalies, magnetized rocks, earthquakes, and matching continental age provinces across the oceans. The evidence was persuasive, and the new revolution was proclaimed. Plate tectonics became the geological buzzword of the late 1960s and early 1970s.

Let us recapture for a moment the revival of the continental drift theory during the late 1950s. It is ironic how the dissenting opinions of one period may become the majority opinions of another. Therein lies the beauty of the highly interpretive science of geology. It is ironic also that leading the way in the revival were the geophysicists, originally the most vocal opponents. Yet another irony relates to Alfred Wegener, the "father of continental drift" and the central figure in the debate.

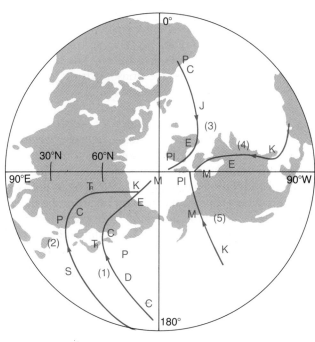

Figure 1-29. Postulated paleomagnetic "polar-wandering" curves for Europe (1), North America (2), Australia (3), India (4), and Japan (5). G = Cambrian, S = Silurian, D = Devonian, C = Carboniferous, P = Permian, R = Triassic, J = Jurassic, K = Cretaceous, E = Eocene, M = Miocene, Pl = Pliocene. These curves represent the paths of movement of the five continental areas during the Phanerozoic with respect to essentially fixed poles.
(From Allan Cox and Richard Doehl, 1960, Review of Paleomagnetism. *Geol. Soc. America Bull.*, vol. 71, no. 6, Fig. 34, p. 759. Reproduced by permission of Geological Society of America)

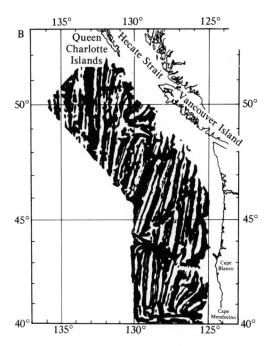

Figure 1-30. Index anomaly map of the total magnetic field for the Northeast Pacific. Positive magnetic anomalies are shown in black. Data from this study led the Canadian geologist L.W. Morley to his "radical" idea that mid-oceanic ridges are loci for creation of new ocean bottom crust. Morley surmised, quite rightly and brilliantly, that the mid-ocean ridge, which should be off California, was, in fact, under California at the location of the San Andreas Fault, having been overridden in California by continental drift.
(From Arthur D. Raff and Ronald G. Mason, 1961, Magnetic Survey off the West Coast of North America, 32° N. latitude to 42° N latitude. *Geol. Soc. America Bull.*, vol. 72, no. 8)

When he died in 1930, Wegener was engaged in pioneering echo-sounding techniques that later would give the concept of continent in motion strong support. Although Wegener did not live to see the general acceptance of his theory, justice has been served through recognition of his accomplishments.

A Unifying Principle

In 1960 a comprehensive program of deep sea drilling was inaugurated by utilizing the research vessel *Glomar Challenger*. This has proved to be one of the most successful long-term scientific ventures of all time. Hundreds of drill cores of deep-sea sediments and oceanic crust, together with magnetic and other geophysical surveys, have provided important data on spreading rates and directions and have continuously substantiated the theory that the present ocean basins are indeed comparatively young geologic features. During the past decade the theory of plate tectonics has catapulted into the role of a master scheme for the whole earth—a unifying principle that gives meaning to and helps explain many diverse phenomena. It is a model that seems to accommodate and link together many aspects of earth science that previously had been treated as subjects for independent study. One of the exciting outgrowths of this new theory is the need for information from a number of disciplines in geology. We are seeing a most significant synthesis of data from traditional areas such as paleontology, stratigraphy, petrography, structural geology, geophysics, geochemistry, and oceanography. Application of the theory to many geological problems has fostered healthy communication among the subdisciplines of geology. How truly exciting for our science is this time when there is such healthy interfacing and cooperation among its practitioners.

Certainly one of the most challenging questions in the entire plate tectonics theory is "What drives the plates?" Does in fact a convective stirring within the asthenosphere drag the lithosphere? Hess in 1960 had considered drifting continents to be a direct consequence of convection currents, whereby the continents were carried along as if on a conveyer belt (Fig. 1-31). Can we really invoke the convention cells that first

Holmes and later Hess envisioned? If so, what drives the mantle convection? Could motion be generated by a kind of gravity sliding, whereby lateral movement on the gently inclined boundary between lithosphere and asthenosphere produces downward motion? Could the rising of material from beneath the crest of the oceanic ridge, with partial melting and eruption of lavas, force the plates apart? Are the lithosphere and asthenosphere linked together in such a fashion that one cannot move without the other? Could rising hot plumes from the mantle, as proposed by J. Tuzo Wilson, be a realistic mechanism for driving the lithospheric plates?

Despite the rapid acceptance of plate tectonics, some skeptics of the theory, in part or in entirety, play the devil's advocate. This skepticism brings to mind the debate about the earth between J. Tuzo Wilson and a famous Russian tectonophysicist, V. V. Beloussov. Wilson was an early champion of the revolution in the earth sciences. He wrote that the revolution was similar in its impact on the earth sciences to great breakthroughs— the theory of relativity in physics; microbiology and evolution in the biological sciences—made in some of the other sciences. Wilson firmly believed that the revolution would unite formerly fragmented branches of the earth sciences into a new unified science of the dynamic earth. Beloussov is a respected skeptic who has steadfastly maintained that the concepts of moving continents and sea-floor spreading should serve only as working hypotheses along with other hypotheses. This and other challenges are not impeding progress, but rather are asking legitimate, serious questions. Such skepticism is a necessary ingredient of our science.

As you progress through this text, particularly the section on the geological evolution of North America (Chapters 9–13), the central theme of plate tectonics will become apparent. It is our hope that you will think about this central theme often as you become acquainted with how our continent came to be as it is. We do not begin to have all the answers; many of the major implications and applications of the plate tectonics theory still await study. But it is the ability to begin asking the right questions and to search for the elusive answers that provides the real excitement in probing the mysteries of historical geology—the science of a dynamic earth.

SUMMARY

The planet earth has a density of 5.5 g/cm^3 and an internal structure that consists of (1) a central iron-nickel core whose inner part is solid and outer zone is molten,

(2) a thick, dense silicate mesosphere, (3) a mushy, almost plastic silicate asthenosphere, and (4) a thin, rigid lithosphere. Historical geology is concerned primarily with evolution of the lithosphere and interaction between lithosphere, atmosphere, hydrosphere, and bio-

sphere during the 4600-million-year history of the earth.

The lithosphere has evolved within the context of the geologic cycle, which consists of three subcycles and their interactions: (1) The rock cycle involves the interrelationships between igneous, sedimentary, and metamorphic rocks, which give the lithosphere its composition. (2) The hydrologic cycle traces the myriad pathways of water and is instrumental in weathering and eroding of the lithosphere. (3) The tectonic cycle relates to the mobility of the lithosphere and involves the deformation of rock.

A new revolution in the earth sciences has brought the tectonic cycle into clearer focus and has given earth scientists a better understanding of how the earth works. Only during the last twenty years has it been appreciated that the earth's lithosphere is fragmented into a mosaic of plates, which interact by spreading apart at divergent boundaries, colliding at convergent boundaries, and sideswiping at transform boundaries. This dynamic aspect of the earth's lithosphere is called plate tectonics.

New oceanic lithosphere is formed along mid-ocean ridges at sea-floor spreading centers where fresh basalt, derived from the asthenosphere, wells up into a central rift zone. This new lithosphere is continuously accommodated as older sea-floor material glides away on both sides of the rifts at rates of several cm/year. The continents, which "bob" higher than the ocean basins because they are lighter, move apart or approach one another as passive passengers attached to spreading oceanic lithosphere. The mid-ocean ridges rise up from the abyssal plains and form an interconnected system some 70,000 km in length—the mountain chains of the ocean basins.

Old oceanic lithosphere is destroyed in subduction zones, where it dives (subducts) beneath the edges of continental lithosphere at rates of several cm/year. Oceanic–continental lithosphere collisions result in magmatic arcs, like the Cascade and Andes mountain chains, which form along the edges of continents. Oceanic–oceanic lithosphere collisions result in volcanic island-arc chains, like the East Indies and Japanese archipelagos, which form seaward of continental margins. Both island arcs and magmatic arcs produce hybrids of basalt and granite called andesite. Deep-sea trenches are the sea-floor topographic expressions of subduction zones.

New sea-floor crust is created at spreading centers

and old crust is destroyed in subduction zones. Through geologic history, oceanic crust probably has been totally replaced by new material every 300 to 400 million years. The oldest rocks yet recovered from any of the present-day ocean basins are approximately 180 million years old. The locations of most geological observations, however, are on the continents, where the oldest rocks discovered are about 3800 million years old—more than twenty times the age of the oldest ocean-basin rocks.

The continents consist of complex patchworks of igneous, sedimentary, and metamorphic rocks. The nuclei of continents are shield areas that consist of Cryptozoic basement forged from earlier plate collisions. Mountain ranges like the Appalachians and Rockies represent younger plate collision boundaries where thick piles of sedimentary rock were folded and faulted, intruded by igneous plutons, and regionally metamorphosed during the tectonic cycle. Emergent coastal plains and submergent continental shelves, slopes, and rises are the surfaces of giant prisms of sediment deposited along the passive, trailing continental margins. In the context of the tectonic cycle, these trailing margins with their thick sediment wedges are likely to become the collision margins and resulting mountain belts of the future. Continents have grown by accretion of orogenic belts around continental nuclei; ocean basins have opened and closed through cycles of sea-floor spreading and subduction.

The theory of plate tectonics provides a unifying concept and model that explains such diverse phenomena as origin and evolution of mountain belts, distribution and cause of earthquakes and volcanoes, occurrence of economic mineral deposits, and major patterns in the history of life. Plate tectonics came as a scientific revolution in the decade of the 1960s and represented an about-face in geologic thinking—from belief in a static crust with fixed continents to advocacy of a dynamic lithosphere with footloose continents. The turning point came with the discovery of sea-floor magnetic anomalies: paired patterns of magnetic "stripes" parallel to mid-ocean ridges. These anomalies represent differences in magnetic intensity, reflecting a history of changes in the earth's magnetic field. The symmetrically paired magnetic anomaly patterns documented a calendar of sea-floor spreading, thus providing a viable mechanism to explain continental drift—an old theory clearly enunciated near the turn of the century by Alfred Wegener.

SUGGESTIONS FOR FURTHER READING

Condie, K.C., 1976, *Plate Tectonics and Crustal Evolution:* Pergamon, London.

Hallam, A., 1973, *A Revolution in the Earth Sciences:* Clarendon, Oxford, England.

Marsh, B.D., 1979, Island-Arc Volcanism: *Am. Scientist,* vol. 67, pp. 161-172.

Molnar, Peter, and Tapponier, Paul, 1977, The Collision between India and Eurasia: *Sci. American* Offprint 923, W.H. Freeman, San Francisco.

Uyeda, Seiya, 1978, *The New View of the Earth: Moving Continents and Moving Oceans:* W.H. Freeman, San Francisco.

Wilson, J.T., ed., 1976, Continents Adrift and Continents Aground, *in* Readings from *Scientific American:* W.H. Freeman, San Francisco.

Wyllie, P.J., 1976, *The Way the Earth Works: An Introduction to the New Global Geology and its Revolutionary Development:* John Wiley & Sons, New York.

2

THE ABYSS OF TIME:
Concepts and Principles
of Geologic Time

CONTENTS

KEY TERMS

Stratigraphy
Superposition
Original horizontality
Lateral continuity
Correlation
Cross-cutting relationships
Unconformity
Hiatus
Nonconformity
Angular unconformity
Diastem
Disconformity
Law of inclusions
Fossil succession
Radiometric
Radioactive decay
Radioactive emission
Nuclide
Atomic number
Isotope
Radiogenic

THE BOOK OF TIME

> High up in the North in the land called Svithjod, there stands a rock. It is a hundred miles high and a hundred miles wide. Once every thousand years a little bird comes to this rock to sharpen its beak. When the rock has thus been worn away, then a single day of eternity will have gone by.*

This quote from Hendrick Van Loon may be overstating the case somewhat, but the point to be made is that geologic time is enormous. Time is on the side of all geologic processes and has accommodated many changes and cycles in the earth's lithosphere, hydrosphere, atmosphere, and biosphere. We are but some of the latest inhabitants of the third planet from our sun, which occupies the center of a solar system that was born some 5 billion years ago. The earth and its companion planets belong to a great galactic spiral that formed from a big-bang explosion some 15 to 20 billion years ago and now moves silently through the universe.

Arthur Holmes once said, "Perhaps it is indelicate of us to ask Mother Earth her age." Nonetheless, our latest scientific findings tell us she is some 4600 million years old and we have learned much about her

---

*Hendrick Van Loon, 1951, *The Story of Mankind,* p. 2 of 1962 Black and Gold ed.: Liveright Publishers, New York.

secretive past. We have learned that a giant supercontinent began to split apart on the earth's lithosphere about 200 million years ago and that upright beasts who might reasonably lay claim to the name *Homo* lived along the shores of a lake in what is now Ethiopia about 3 million years ago. In 1492, Christopher Columbus planted his footprint on the shores of Hispaniola in the New World, and in 1969 Neil Armstrong left his (Fig. 2-1) on the surface of the moon. The last two dates we can comprehend; the others are incomprehensible. It stretches the imagination to ponder intervals of time measured in millions, tens of millions, and even hundreds of millions or billions of years. But this vast abyss of time is reality, not fantasy. What makes intervals of geologic time so staggering is that our personal, everyday calendars deal with increments of time measured in hours, days, weeks, months, and years. Our history books relate events set in a time frame of centuries, but even a century is difficult to appreciate fully.

Our individual lifetimes, or for that matter the longevity of our species on this planet, are indeed short by comparison with a time interval such as the Cambrian Period (Table 2-1), which lasted for 70 million years. It is an exciting, yet humbling experience to split open a piece of shale and expose to the light of the world for the first time in 550 million years the fossilized remains of a trilobite (Fig. 2-2), a

Figure 2-1. Footprint of U.S. astronaut Neil Armstrong on surface of moon.
(**Photo courtesy of NASA**)

Figure 2-2. Cluster of Cambrian trilobite casts and molds.
(Photo courtesy of Ward's Natural Science Establishment, Rochester, NY)

Table 2-1. Geologic time scale and bar scale showing relative proportions of Cryptozoic and Phanerozoic time

Eon	Era	Period	Age	Geologic event
PHANEROZOIC (Age of manifest animal life)	CENOZOIC (modern life)	Neogene	24	Glaciation in northern hemisphere; Dawn of man
		Paleogene	65	Evolutionary expansion of mammals
	MESOZOIC (middle life)	Cretaceous	144	Extinction of dinosaurs; faunal crisis; first angiosperm plants
		Jurassic	208	Origin of birds
		Triassic	245	Origin of mammals; origin of dinosaurs
	PALEOZOIC (ancient life)	Permian	286	Complete formation of Pangaea, great crisis in history of life; uplift of Appalachians
		Carboniferous — Pennsylvanian	320	Major coal-forming swamps; oldest reptiles
		Carboniferous — Mississippian		
		Devonian	380	Age of fish; oldest amphibians
		Silurian	408	Oldest land plants
		Ordovician	438	Glaciation in Gondwana
		Cambrian	505	Origin of vertebrate animals; Origin of exoskeletons
		Ediacarian	570	Origin of multicellular life
			700	
CRYPTOZOIC (age of microscopic life) — PROTEROZOIC	Late Proterozoic		900	Sexual reproduction; oldest nucleated cells
	Medial Proterozoic		1600	Attainment of 1% present atmospheric level of O_2
	Early Proterozoic		2500	
ARCHEAN	Late Archean		3000	Oldest definitive fossils of single nonnucleated cells
	Medial Archean		3400	
	Early Archean		~3800	—— oldest known rocks / Meteoric bombardment; volcanic outgassing; formation of early lithosphere; origin of earth atmosphere, biosphere
	Pregeologic history of earth		4600	

Bar scale (to scale), m.y. B.P.: 65, 245, 700, 2500, 3800, 4600

PHANEROZOIC (CENOZOIC, MESOZOIC, PALEOZOIC) — CRYPTOZOIC (PROTEROZOIC, ARCHEAN — Oldest known rocks, Pregeologic history of the earth) — Origin of the earth

crustaceanlike arthropod that once scuttled across a Cambrian mudflat in a world we can never truly fathom. But therein lies the intrigue, the challenge, and the beauty of historical geology—to probe the record of the rocks in search of clues of what the world was like long ago.

If all of tangible earth history, from 4600 million years ago to the present, were condensed and portrayed in the form of a 460-page book, we would see something like the following: Each original page would correspond to 10 million years, but far fewer than one-half the pages would even be present, much less easy to read. Certainly the first seventy pages of the book would be missing. On a tattered and smudged page 71 there would be a sketchy description of the formation of the oldest rocks yet discovered on earth, and on page 110 we would see the first mention of evidence of primitive life on our planet. The origin of life itself, unless possibly decoded from a few badly smudged lines of type several pages back, is probably lost forever on one of the missing pages. The attainment of a significant level of free oxygen in our atmosphere, perhaps 1 percent of the present-day level, would be mentioned on page 260. The first true animals would be discussed on page 390, and the first animals with hard external coverings capable of being preserved would make their appearance on page 400.

Up to this point, seven-eighths of the way through, the book would be dilapidated. Many pages, even whole chapters, would be missing. What remained would be crinkled fragments of pages with smudged type; the decipherable words would be in cryptic code. But the last sixty pages, although difficult and at times impossible to read, would be easier. A primitive fish, the earliest animal with a backbone, would be described on page 410. The oldest land plants would be mentioned on page 415, and extensive coal-forming swamps inhabited by the earliest reptiles would be described on page 430. The breakup of a giant supercontinent would be discussed on pages 442 to 450; dinosaurs would roam the continents on pages 440 to 455; and the formation of the Rocky Mountains would be mentioned on pages 450 to 456. Man would not appear until three-fourths of the way down page 460, and Columbus' voyage to the New World and the first walk on the moon by the *Apollo 11* crew would be mentioned on the last line of page 460. But the book of earth history continues to be written: 10 million years hence another page will be added. What will the earth be like 10 million or 100 million years from now? Will members of our species be here to chronicle the events? But enough of analogies; let us investigate the fundamental principles that provide the basis for our appreciation of geologic time.

RELATIVE TIME

History and Prehistory

One of the most unique and fascinating aspects of any of the sciences is how geologists measure time. The methods of reckoning geologic time are very different indeed from time determinations made during the history of human experience. Human events are chronicled through written testimony, recorded and passed down from generation to generation. We are all familiar with various kinds of historical time scales. At one time or another we have had to learn important dates in human history: the Battle of Hastings, the Boston Tea Party, the treaty signing at Appomattox, and many others. Such events could be arranged on a time scale that shows chronological succession as well as absolute dates.

Geologic time is illustrated by a *prehistoric* time scale that attempts to accomplish the same thing. Events from the distant past—the origin of the earth, the first appearance of multicellular life, the formation of the Rocky Mountains, the dawn of humans (Table 2-1)—are arranged in their relative order of occurrence with dates in years before present (B.P.) assigned to the events. This record of prehistoric time is written in the rocks of the earth's lithosphere. The rock record, like the pages and chapters in the book-of-time analogy, contains the secrets of the earth-shaping events of the past. How do earth scientists read and decode the history recorded in the rocks and how do they place geologic events in a meaningful time frame? In this chapter we shall examine the principal methods by which earth scientists tell time. In the following chapter, we shall become acquainted with the historical development of geologic time concepts and the birth and subsequent development of the geologic time scale.

The geologic time scale actually incorporates two scales (Table 2-1): The first is a relative scale that expresses the order of geologic events as determined from their positions in the rock record. Intervals of geologic time—major chapters in earth history—have been given names such as Cambrian, Cretaceous, and Miocene. The second is an absolute scale, which designates ages in years before present. These ages are based on the natural radioactive decay of various chemical elements that are present in trace amounts in certain minerals in some rocks. Knowledge of the principles of relative and absolute geologic time is essential for understanding and appreciating earth history. First we shall consider aspects of relative time in geology.

A   B

Figure 2-3. Stratification of sedimentary rocks. A. Jurassic Summerville Formation, Colorado National Monument, Colorado. B. View looking west at Permian section exposed near Valley of the Gods, southern Utah.
(Photos by J.D. Cooper)

A Science Called Stratigraphy

Of the three main kinds of rocks (igneous, sedimentary, metamorphic) that make up the earth's lithosphere, sedimentary rocks provide the most complete record of earth history. Although igneous and metamorphic rocks make up more than 90 percent of the total volume of the earth's lithosphere, sedimentary rocks make up more than 75 percent of the rock record exposed at the surface or present in the uppermost few kilometers. Of singular importance for reconstructing geologic history is the layering or *stratification* of sedimentary rocks. Stratification allows for the ordering, arrangement, and determination of sequence (Fig. 2-3).

The science of layered rocks or strata is **stratigraphy:** the study of spatial and time relationships of bod-

ies of rock to one another and the dynamic depositional patterns that can be observed and interpreted. Stratification of sedimentary rocks results from deposition and net accumulation of solid particles that settle through water or air in response to gravity. Particles of individual minerals or fragments of rocks settle according to size, shape, density, and the velocity of the transporting medium (water or air). The resulting layers, also referred to as beds or strata, range in thickness from a few millimeters to several meters or more (Fig. 2-4) and generally are separated by well-defined surfaces called *bedding planes.* Sedimentation tends to occur in pulses, with depositional events being separated by quiescent intervals. Such episodic activity is largely responsible for the textural differences observed between successive strata and for the bedding planes that separate them.

Deposition occurs in sedimentary basins of various sizes. Compaction and lithification of the sediments result from postdepositional burial beneath subsequent layers and tend to enhance the bedding and the *contacts* between beds.

Superposition

Perhaps the most fundamental principle of stratigraphy is **superposition.** In a sequence of layered rocks, the oldest layer is at the bottom and the youngest layer is at the top (Fig. 2-3). Consideration of superposition is the first step in developing relative time sequences in layered rocks.

Superposition is rather straightforward if the rock sequence has not been inverted. In this regard it is appropriate to mention another fundamental law or principle of stratigraphy, that of **original horizontality.** This principle states that not only does sedimentation (and thus the net accumulation of sediment in superposed layers) proceed from the bottom upward, but also that the depositional surfaces, throughout their extent, are essentially flat, generally not departing more than a few degrees from horizontal. Thus sedimentary strata are originally horizontal because the surfaces (interfaces between sediments and water or air) on which they are deposited are essentially flat, in response to gravity. When we see sedimentary successions, such as those shown in Figure 2-5, that depart appreciably from the horizontal it is reasonable to assume that postdepositional events have repositioned the strata. Sedimentary strata tilted from the horizontal at some measurable angle are said to have *dip,* which is expressed as both a direction and an angle. The direction of dip is the compass direction toward which the bedding plane faces (direction of inclination); the angle of dip is the acute angle of inclination measured downward from the horizontal (Fig. 2-6A). Dip is perpendicular to the *strike* of

the bed, which is the compass direction of a horizontal line on the bedding plane.

The angle of dip can range from 0° to 90°. If a succession of strata is bent or tilted beyond vertical (Fig. 2-7), the direction of dip is reversed (as before, the angle is measured downward from the horizontal), and the succession's superposition is inverted. Such sequences are *overturned.* Tectonic forces within the earth's lithosphere bend and break rock strata and have been responsible for the tilting and folding of sedimentary rock sequences, particularly in the mountain belts of the world.

How can the superposition of a sequence of strata dipping at a steep or even vertical angle be determined? How can an overturned sequence be recognized? Why is it important to make this determination? The working out of superposition in layered rocks is the essential first step in unraveling the geologic history of sedimentary

A

B

Figure 2-5. Dipping successions of sedimentary strata. A. Lower Paleozoic succession in Appalachians of southwestern Virginia. B. Pliocene section in southern California. In both photos the superposition is from left to right; the succession is oldest to youngest in the direction of dip.
(A, photo by B.N. Cooper; B, photo by J.D. Cooper)

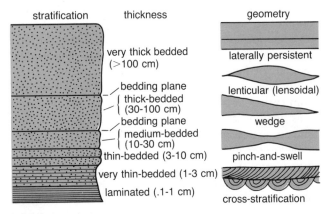

stratification   thickness   geometry

very thick bedded (>100 cm)

laterally persistent

bedding plane
thick-bedded (30-100 cm)

lenticular (lensoidal)

bedding plane
medium-bedded (10-30 cm)

wedge

thin-bedded (3-10 cm)

pinch-and-swell

very thin-bedded (1-3 cm)

laminated (.1-1 cm)

cross-stratification

Figure 2-4. Stratification, thickness, and style (geometry) of layered rocks.

sequences. But recognizing correct superposition of layered rocks is not always automatic. For example, how might one determine superposition in a stratigraphic succession such as that shown in Figure 2-8? What are the clues that tell sequence and how are they expressed?

The determination of sequence in layered rocks involves the recognition and correct interpretation of bottom versus top indicators. Such indicators are *primary*

sedimentary structures: megascopic features incorporated in the sediment in the original environment of deposition. Primary sedimentary structures are external if they are present on the bedding surfaces and internal if they are present within the bed.

An example of an external sedimentary structure is shown in Figure 2-9A. These peculiar structures, called *flute casts,* represent the sand fillings or castings of

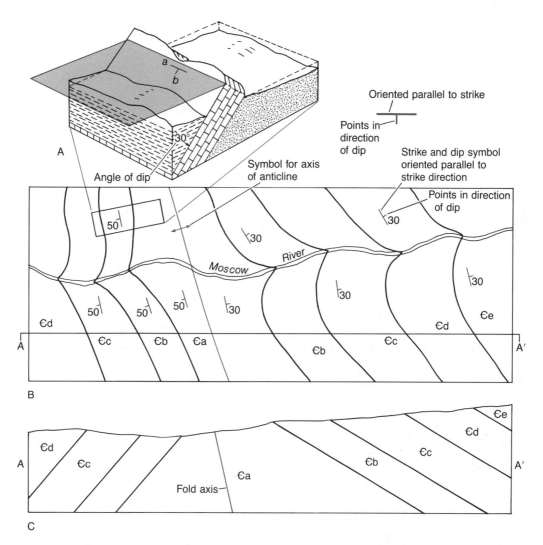

Figure 2-6. Strike and dip *(attitude)* of sedimentary strata. Є a, Є b, Є c, Є d, and Є e represent a succession of Cambrian strata. A. Strike *(a)* is the compass direction (generally expressed in terms of the northerly direction; e.g., N 50° W) of a horizontal line on the bedding plane of a tilted (dipping) bed; direction of dip *(b)* is the compass direction (measured in a horizontal plane that intersects the bedding surface) toward which the inclined bedding plane faces. True dip direction is *always* perpendicular to the strike. Angle of dip is the angle (generally acute; maximum of 90°) of inclination of the bedding plane downward from the horizontal. The maximum dip angle is measured in the plane of the true dip direction. B. Geologic map view showing configuration of geologic contacts (lines of intersection of bedding planes with surface of map) and strike and dip symbols. Note that the strike part of the symbol is parallel to the geologic contacts. This is because the contacts are lines representing the intersection of the map surface (a horizontal plane) with the tilted stratification planes. Note also that the contacts form a V where they intersect the stream. The apex of the V points in the direction of dip of the beds. C. Geologic cross section along reference line A–A′ (shown in B) showing a fold in which the oldest rocks are in the middle and successively younger rocks are repeated systematically on the flanks (limbs). This structure is an *anticline.* A *syncline* is a fold in which the youngest rocks are in the center and successively older rocks are systematically repeated outward on the flanks of the structure.

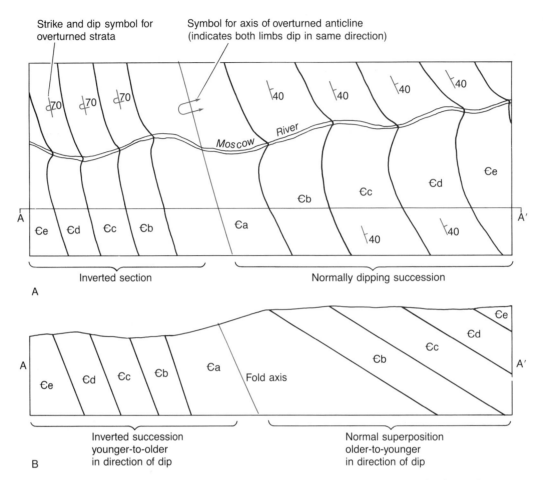

Figure 2-7. Overturned (inverted) strata in west limb of anticline shown in Figure 2-6. Єa, Єb, Єc, Єd, and Єe represent a succession of Cambrian strata. A. Overturning is indicated by special attitude symbol. B. Geologic cross section along reference line A–A′ (shown in A) showing vertical profile view of both limbs of anticline dipping in same direction. This is an overturned anticline because the superposition in one limb (left or west) has been inverted.

Figure 2-8. View looking north at road cut along U.S. 90 near Marathon, Texas. A vertical sequence of upper Paleozoic sedimentary rocks is exposed.
(Photo courtesy of Critter Creations, Inc., San Diego CA)

scours (flutes) formed on the original depositional surface (Fig. 2-9B). Such structures are indicative of the soles or bottoms of beds. Their shape and orientation provide information on the direction of movement of the original scouring current. Flute casts and other related sole marks are most common in successions of interbedded sandstone and mudstone. Depressions formed by currents that scoured the surface of a muddy bottom served as molds for the filling by sand during deposition of the immediately overlying layer. These structures, present on the soles of many of the beds in the sequence of Figure 2-8, indicate the top of the sequence is to the left. A number of additional sedimentary structures can be used to determine stratigraphic top versus stratigraphic bottom (Figs. 2-10, 2-11). The paleoenvironmental implications of some of the more common sedimentary structures are discussed in Chapter 6.

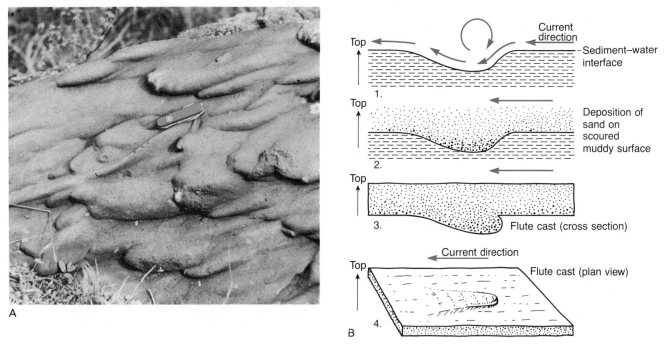

Figure 2-9. A. Flute casts on bottom (sole) of sandstone bed, upper Paleozoic, Ouachita Mountains, southern Oklahoma. B. Sequence (1–4) showing formation of flute cast on sole of sandstone bed. 1. A flute or scour mark is eroded into the surface of a muddy bottom at the sediment–water interface, producing an asymmetrical depression. 2. The scour is filled by deposition of sand of the overlying bed. 3. The sandstone layer is separated from the underlying mudstone, displaying an asymmetrical flute cast on the sole of the sandstone bed. 4. A plan view of the flute cast on the sole of the sandstone bed. (Photo by J.D. Cooper)

Lateral Continuity

Another stratigraphic principle is that of **lateral continuity:** Sedimentary strata, when originally formed, are three-dimensional and extend laterally in all directions until they thin to a zero-thickness edge, terminate against the edge of the *depositional basin* in which they accumulate, or change character by merging into laterally adjacent deposits. Strata can be identified laterally by tracing and **correlation** (matching). At the surface, isolated exposures of the same stratigraphic unit, when correctly recognized and correlated, indicate that the exposures are parts of what was once a laterally continuous unit (Fig. 2-12). Position in a superpositional sequence that can be recognized at each outcrop section in Figure 2-12B helps substantiate the physical correlation of the units.

Individual, distinctive, thin stratigraphic intervals or beds can commonly be demonstrated to have time-significant value and can be used to define *time lines.* Such physical units are considered to have formed geologically instantaneously over the area of lateral continuity. Thin volcanic ash beds, for example, are fallout blankets deposited by volcanic eruptions (Fig. 2-13). Such key beds (marker beds) not only provide local to subre-

gional time lines, but their distinctive character allows them to be recognized in separate sections, thus facilitating correlation.

Cross-Cutting Relationships

A most important principle used in the determination of relative geologic time involves **cross-cutting relationships.** Intrusive igneous rock bodies (such as dikes) and faults (fractures in the earth's crust allowing movement of adjoining blocks of rock) cut across preexisting rocks and structures and thus are younger than the geologic features they cut. Figure 2-14 illustrates cross cutting and superposition. Using these two fundamental principles can you arrange the rocks and other geologic features in their relative chronological older?

A very important stratigraphic feature known as an **unconformity** is critical to our discussion of principles of relative time and geologic history. Unconformities are a type of cross-cutting feature (Fig. 2-15), but unlike igneous intrusions and faults, which are structural (tectonic) features, unconformities are stratigraphic surfaces. Simply stated, an unconformity is *a buried surface of erosion or nonaccumulation.* As such, the buried

Structure	Description	Graphic
External Stratigraphic top		
Ripple marks	Sharp crests indicating top of bed	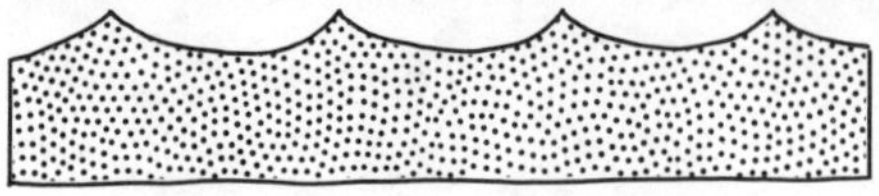
Desiccation cracks	Developed on cohesive surface; area between cracks commonly concave upward	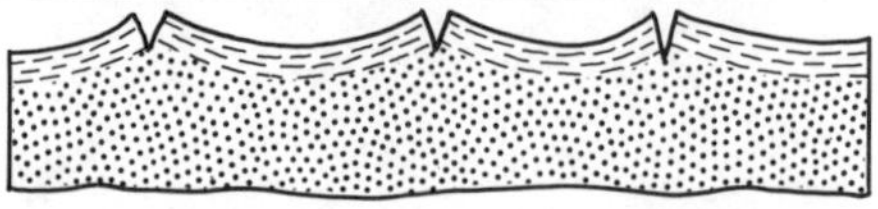
Vertical animal burrows	Burrow truncated at top of bed	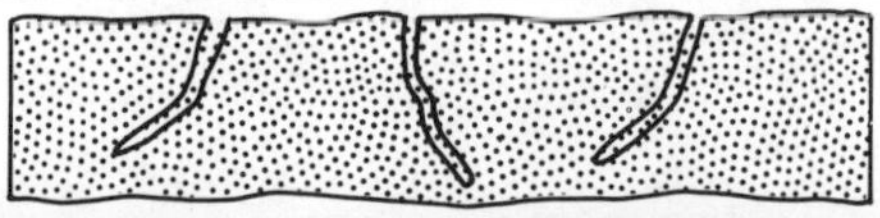
Sole (bottom) structures		
Flute casts	Positive relief on bottom surface of bed	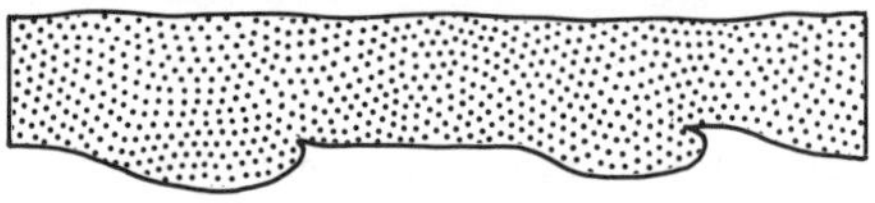
Groove casts		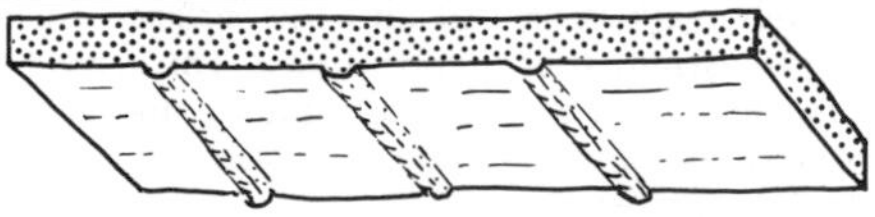
Load casts		
Ripple mark casts	Sharp crests developed as negative impression on bottom of bed	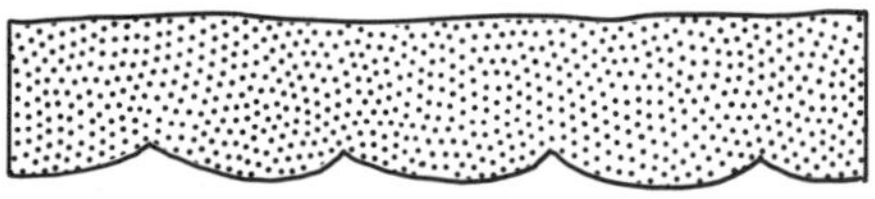
Trace fossils	Casts of grazing trails preserved on bottom of bed	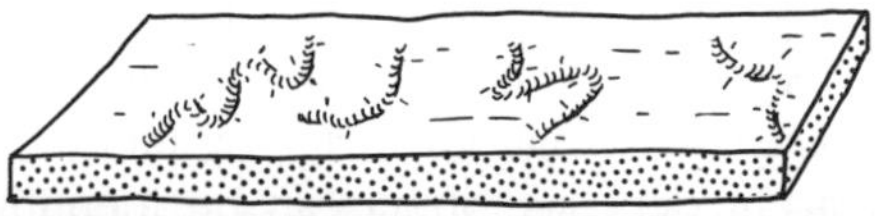
Internal		
Cross-stratification	Cross-stratification sets generally sharply truncated toward top of bed	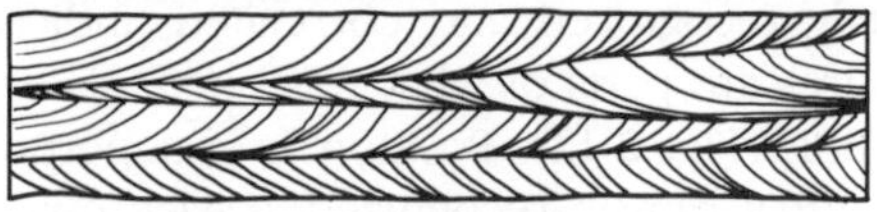
Normal graded bedding	Coarser at bottom, finer at top; gradational change	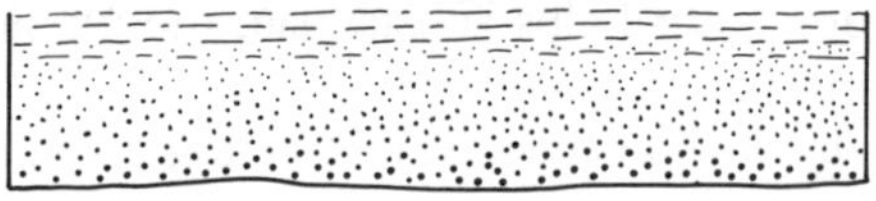
Flame structures	Flames of mud projecting toward top of bed	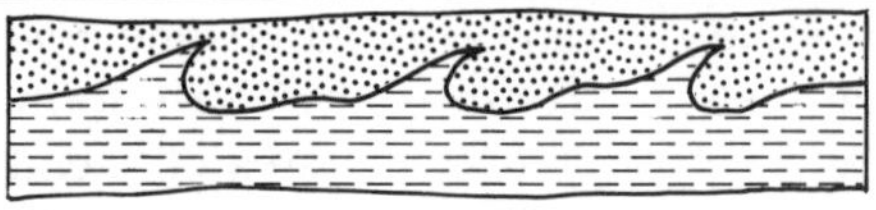

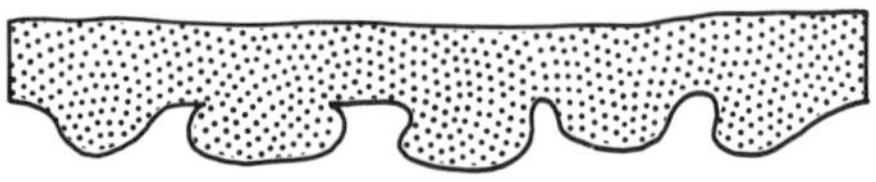

Figure 2-10. Primary sedimentary structures useful for determination of stratigraphic bottom versus top.

40

A

B

C

Figure 2-11. A. Camel hoofprint casts on bottom of sandstone bed, Pliocene of southern California. B. Recumbent anticline (axial plane horizontal). Strata in the lower limb are overturned, as indicated by graded bedding (see Fig. 2-10) and cross-stratification (see Figs. 2-10, 2-11C). C. Upside-down truncated cross-stratification from lower limb of anticline in Figure 2-11B.
(A, B, photos by J.D. Cooper; C, photo by C.P. Buckley)

A

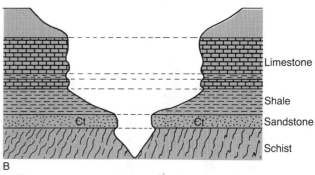

Limestone

Shale

Sandstone

Schist

B

Figure 2-12. A. View looking north in Grand Canyon. Pre-Cambrian schist (at bottom of gorge) is overlain by a layered sequence of Paleozoic sedimentary rocks. B. Correlation of strata from one wall of erosional valley to opposite wall. €t = Cambrian Tapeats Sandstone.
(Photo by J.D. Cooper)

surface represents a significant gap in the geologic record. The time value of an unconformity is called a hiatus, which is the difference in age between the rocks directly above and directly below the unconformity surface.

Three main kinds of unconformities occur in the stratigraphic record and can be recognized on the basis of physical relationships. One kind, called **nonconformity** (Fig. 2-15A), is a stratigraphic surface that sep-

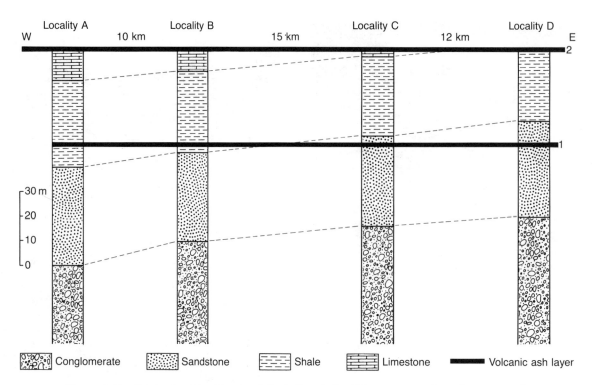

Figure 2-13. Distinctive thin volcanic ash beds (1 and 2) within sequence of conglomerate–sandstone–shale–limestone. The volcanic ash key or marker beds were deposited as fallout of volcanic ejecta that blanketed the area in a geologic "instant" of time. The volcanic ash beds cross the correlated boundaries (dashed lines) separating the sandstone and shale and shale and limestone units, suggesting that these contacts are *not* time equivalent from section to section.

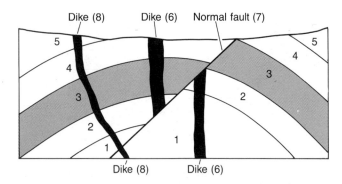

Figure 2-14. Cross-cutting relationships of two dikes, a *normal* fault, and sedimentary strata. Faults divide the broken rock masses into a hanging wall and footwall side. Faults always dip toward the hanging wall. If the hanging wall appears to have moved relatively *down* with respect to the footwall, the fault is a *normal* fault; if the hanging wall appears to have moved up, the fault is a *reverse* fault. The sedimentary sequence formed first (1 = oldest, 5 = youngest in order of superposition), followed by the intrusion of a dike (6), then the occurrence of the normal fault (7) displacing the strata and dike, and finally the intrusion of the dike (8) that crosses the fault. Folding of the strata into an anticline probably occurred after deposition of 5 and before intrusion of 6.

arates older crystalline rock (igenous or metamorphic rocks) from younger (overlying) sedimentary strata. A second type of unconformity, the **angular unconformity** (Fig. 2-15B), has an angular relationship between older, deformed sedimentary strata and younger, less deformed strata. In the example shown in Figure 2-15B, the underlying succession appears internally conformable but is tilted at a sharp angle and is abruptly truncated by the unconformity surface. The younger, overlying sequence also appears internally conformable, and the stratification is essentially parallel to the unconformity.

The term "unconformity" should be reserved for buried stratigraphic surfaces representing major hiatuses. A large amount of geologic time is required for deposition of a sequence, lithification into coherent rock, uplift and tilting, erosion to develop a surface across the upturned strata, subsidence, and resumption of sedimentation. The surfaces separating successive individual units within conformable sequences represent comparatively very short geologic time durations and are called **diastems.** Diastems are part of the normal, continuous sedimentation process whereby deposition alternates with nondeposition and erosional removal to result in a net accumulation of sediments. Unconformities generally represent missing record amounting to millions or hundreds of millions of years; diastems represent time periods of weeks, months, years, perhaps even centuries—comparatively short intervals of geologic time.

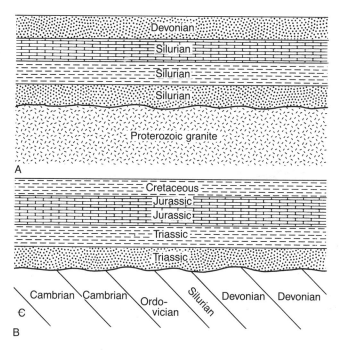

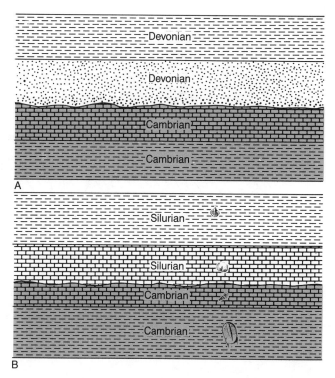

Figure 2-15. A. Nonconformity (wavy line). Note the age difference between the crystalline basement rocks and the overlying sedimentary sequence. If this were an intrusive igneous contact, the igneous rock would have to be younger than the sedimentary sequence. B. Angular unconformity. Note that the unconformity separates two superpositional packages of rock, each with its own internal superpositional arrangement.

Figure 2-16. Erosional disconformities (wavy lines). The hiatus (time value of the unconformity) is the difference in age between the oldest rock above the erosional surface and the youngest rock below that surface. A. In this cross section note the absence of at least Ordovician and Silurian rocks and the inclusions of Cambrian limestone as pebbles in the basal part of the Devonian section. B. The most subtle type of disconformity, showing no change in basic lithology and no inclusions of older rock. The absence of Ordovician strata is determined by comparison of fossils above and below the erosional surface.

The third kind of unconformity, called a **disconformity,** is more subtle than the other two because it occurs between essentially parallel strata (Fig. 2-16). Disconformities, being not so obvious as angular unconformities or nonconformities, generally require more careful examination for recognition. Evidence is generally in the form of physical criteria that suggest erosion such as fragments of underlying rock in the directly overlying strata (Fig. 2-16A). This situation involves yet another stratigraphic principle, the **law of inclusions,** which states that the fragments or particles in a terrigenous clastic sedimentary deposit are older than the deposit itself. This is especially significant for the detection of disconformities in the stratigraphic record because inclusions of underlying strata incorporated in overlying strata represent erosional products. Cross-cutting relationships (Fig. 2-17) also can aid in the recognition of disconformities.

A good illustration of the three kinds of unconformities can be seen in the Grand Canyon of the Colorado River (Fig. 2-18). This spectacular canyon exposes a stratigraphic succession spanning more than 1000 million years. An impressive *nonconformity* occurs between the Tapeats Sandstone and the Vishnu Schist; an *angular unconformity* is present between the Tapeats Sandstone and the tilted units of upper Proterozoic

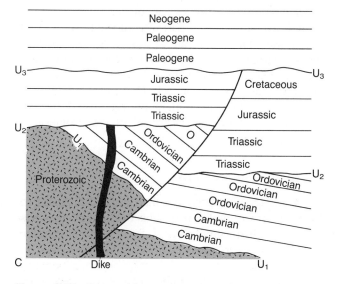

Figure 2-17. Superposition, cross-cutting relationships, and the three main kinds of unconformities. U_1 indicates a nonconformity, U_2 an angular unconformity, and U_3 a disconformity.

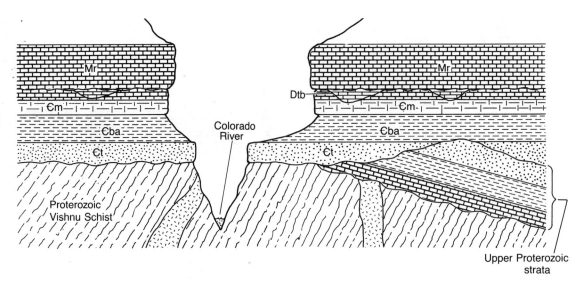

Figure 2-18.   Rock succession in Grand Canyon illustrating three main kinds of unconformities. Locate and identify them. Єt = Cambrian Tapeats Formation, Єba = Cambrian Bright Angel Formation; Єm = Cambrian Muav Formation; Dtb = Devonian Temple Butte Formation; Mr = Mississippian Redwall Formation.

strata. A *disconformity* occurs between the Redwall Limestone and the Bright Angel Shale; the Redwall is Mississippian in age and the Bright Angel is Cambrian. Reference to Table 2-1 indicates that strata of Ordovician, Silurian, and Devonian ages are missing at this boundary.

Unconformities have their advantageous and disadvantageous aspects. The fact that they represent missing geologic record and loss of information for certain localities and regions would seem to be a negative factor. This is tantamount to the absence of some pages, sections, or entire chapters from our book-of-time analogy. In reality, unconformities, as evidence of erosional events, are an important part of geologic history. In addition, they commonly can be used to bracket parts of the stratigraphic record and thus provide convenient, recognizable stratigraphic boundaries for large-scale superpositional packages of rock. The use of unconformities in organizing the stratigraphic record will become more apparent in our discussion of the evolution of North America.

Fossil Succession

A stratigraphic principle that has had a most profound effect on the science of historical geology is **fossil succession.** Simply stated, fossil succession refers to the changes in fossil content (kinds of organic remains or evidence) upward through superpositional sequences of rock. As documents of life (Chapters 4 and 5) and the products of organic evolution (Chapter 7), fossils provide the key to recognizing the relative ages

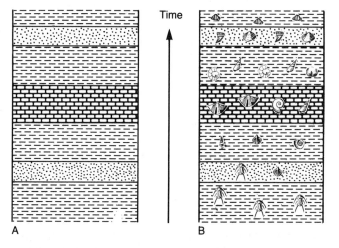

Figure 2-19. Lithologic and fossil succession. A. The vertical change in rock type and repetition of rock type is an expression of the patterns of changing yet repeating environments through time. A bed of a particular lithology might be thought to correlate with a later bed of the same lithology in another location. B. Note the change in fossil assemblages through the vertically repeating succession of lithologies shown in A. Even though some forms persist through more than one lithology, the combination of forms in any particular lithology is definitive.

of sedimentary strata. As mentioned previously, sedimentary rocks provide the most complete record of earth history, especially because of the layered, superpositional arrangement of strata. The presence of fossils in sedimentary rocks contributes greatly to recognition of relative time sequence.

The principle of fossil succession transcends the principle of superposition. Fossils, unlike inorganic particles such as rock fragments or quartz grains, are objects that do not occur in random or haphazard fashion, but rather in a regular, irreversible order. Rock types can be repeated many times in vertical succession (Fig. 2-19) because of recurrence of the sedimentary environments that produced the rocks; fossil assemblages, however, change progressively in vertical succession, and assemblage compositions are nonrepetitive because of evolution (see Chapter 7).

With few exceptions, fossils represent the preserved remains of organisms that lived and died and became preserved in the same or nearby contemporaneous environments in which the containing sediments accumulated. Some exceptions involve fossils that represent inclusions derived from the erosion of older fossil-bearing rock (Fig. 2-20).

Correlation of sedimentary strata can be established using both physical (rocks) and biologic (fossils) criteria, but fossils provide the most reliable tools for making time-significant correlations in sedimentary rocks. Fossils of the same kind of organism commonly can be recognized well beyond the limits of individual sedimentary basins, which determine the lateral continuity of rock units (Fig. 2-21).

Fossil succession and correlation have paved the way to development of a geologic time scale and to interpretation of earth history. Superposition and cross-cutting relationships allow for the determination of relative sequence of geologic events on a local to subregional scale, but fossil succession permits the recognition of relative age and the correlation of isolated sections of rock on a larger scale. In Chapter 3, we will

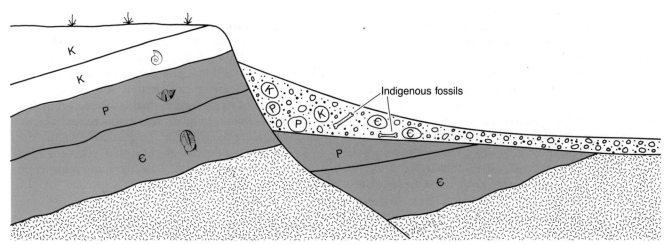

Figure 2-20. The law of inclusions as exemplified by reworked fossils. A Pliocene nonmarine alluvial deposit contains marine Cambrian, Permian, and Cretaceous, and nonmarine Pliocene fossils. A deposit can be no older than its youngest fossil; thus the deposit in question is Pliocene, as determined from the indigenous Pliocene fossils. The Cambrian (Є), Permian (P), and Cretaceous (K) fossils have been reworked into the Pliocene alluvial deposit as erosional debris from older source rocks.

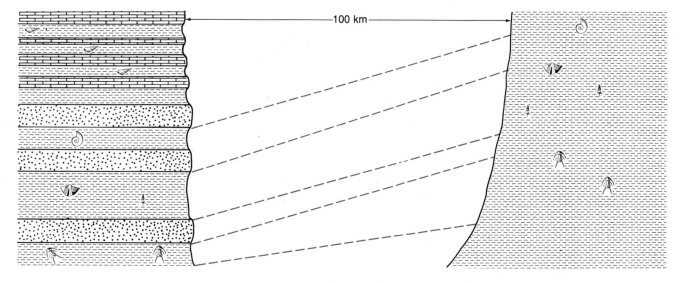

Figure 2-21. Biostratigraphic correlation between widely separated sections of different lithologies.

trace the development of these fundamental principles and their historic influence on the birth and development of the geologic time scale.

GEOCHRONOLOGY

How Long Ago?

Up to this point, the emphasis has been on concepts of relative time: the determination of sequence in layered rocks and the relative order of geologic events as they can be read from a rock record that involves superposition, cross-cutting relationships, and fossil succession. As shown in Table 2-1, the geologic time scale includes not only a relative scale, but also an absolute scale, expressed by dates in years before present and superimposed on the relative scale. Although expressed in years (generally millions of years) before present, the dates are not absolute in the strict sense because there is a small percent of error in their calculation. Absolute dates such as 4600 million years for the age of the earth and 245 million years for the boundary between Paleozoic and Mesozoic Eras give us an appreciation of the antiquity and duration of the relative time-scale subdivisions. As we shall explore further in Chapter 3, the relative time scale was gradually pieced together and was completed, in essentially its modern form, by the end of the nineteenth century; the absolute time scale, developed through the science of **geochronology,** became a reality in the early decades of the twentieth century, spurred by the discovery of radioactivity and its application to mineralogy. Refinements of both scales have continued to the present day. Relative and absolute geologic time scales represent two of the great achievements in the history of science.

Principles of Radiometric Dating

Radiometric age determination is based on the phenomenon that many kinds of atoms are unstable and, consequently, change spontaneously to a more stable, lower energy state. The process of change involves **radioactive decay,** which results in **radioactive emissions.** Particular atoms called **nuclides** are different from others according to the number of *protons* (positively charged particles) and *neutrons* (neutral particles) in the atomic *nucleus* (Fig. 2-22). Each chemical element of the standard periodic table (see Appendix A) is defined by the number of protons in its nucleus, which gives the **atomic number.** For example, helium (He), the number 2 element in the period chart, has two protons in its nucleus; uranium, element number 92, has ninety-two protons. The *mass number* of an atom is the number of protons plus the number of neutrons (Fig. 2-22). Orbitals around the nucleus are filled with electrons (negatively charged particles), whose number

equals the number of protons. Therefore each individual nuclide has a unique atomic number. A chemical element, which is a nuclide with fixed atomic number, can have alternate forms, called **isotopes,** which are distinguished on the basis of the number of neutrons; different isotopes of the same element have different mass numbers (Fig. 2-22).

Uranium-235 and uranium-238 are examples of nuclides that contain the same number of protons but different numbers of neutrons (and therefore different masses). These two isotopes of the element uranium figure importantly in the concept of radioactive decay and the radiometric dating of some kinds of igneous rocks.

Radioactive isotopes decay to more stable nuclides: A nuclide, called the *parent,* changes through a series of steps to a more stable end product, the *daughter nuclide.* Radioactive decay involves *alpha decay* (the emission of two protons and two neutrons from the nucleus); *beta decay* (the emission of a high-speed electron from the nucleus); and, in some cases, *electron capture* (Fig. 2-23). In alpha decay, the nucleus of the parent atom, by losing two protons and two neutrons, assumes a new mass number that is decreased by four and a new atomic number that is decreased by two, making it a different element. In beta decay, the nucleus, by emitting a high-speed electron, has one of its neutrons turned into a proton. Thus the mass remains un-

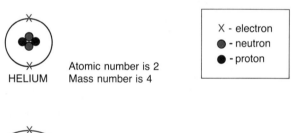

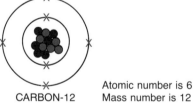

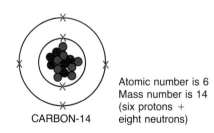

Figure 2-22. Atomic structure, atomic number, mass number, and isotope.

changed, but the atomic number increases by one, and a new element results. In electron capture, a proton in the nucleus picks up an orbital electron and changes into a neutron, thereby decreasing the atomic number by one (forming a new element), while the mass number remains unchanged.

Individual radioactive isotopes, such as uranium-238, have particular decay products (e.g., uranium-238 decays to lead-206 through eight alpha decay steps and six beta decay steps) and unique, constant, unchanging rates of decay (decay constant). Ideally then, as applied to mineralogy, if a radioactive nuclide becomes incorporated into a mineral upon crystallization, the amount of radioactive parent that decays to **radiogenic** daughter (e.g., uranium-238 to lead-206) is a function only of passage of time. However, for accurate data, it is absolutely essential that both parent and daughter nuclides be fully retained within the lattice structure of the mineral; if accurate daughter:parent ratios are to be observed, the mineral must be a closed system. But what is so significant about daughter:parent ratios, and how is the amount of elapsed time since the radiometric clock started ticking determined?

A time interval called *half-life* is unique to each radioactive nuclide. Half-life is the time required for one-half of the atoms of a particular radioactive nuclide to decay (Fig. 2-24). Radioactive decay occurs at a geometric rate: For a given quantity of radioactive nuclide (N_0) half the original number of atoms ($N_0/2$) remain after one half-life, half of those or one-fourth of the original ($N_0/4$) are present after the next half-life, half of those or one-eighth of the original ($N_0/8$) after still another half-life, and so on, theoretically to infinity. Simply stated, the radiometric age of a geologic sample is the amount of elapsed time since crystallization of the mineral lattice that contains the radioactive atoms. The starting point is age zero and the radiogenic daughter:radioactive parent nuclide ratio is zero at the setting of the radiometric clock. The amount of elapsed time since crystallization is determined by measuring the ratio of radiogenic daughter and radioactive parent nuclides in the mineral. Of course, the half-life of the radioactive parent must be known; this is multiplied by the ratio. For example, in a particular sample in which the uranium-238–lead-206 decay series is used, the lead-206:uranium-238 ratio is found to be 1:1. That means that half the original parent has decayed to daughter product (i.e., one half-life has elapsed); because ura-

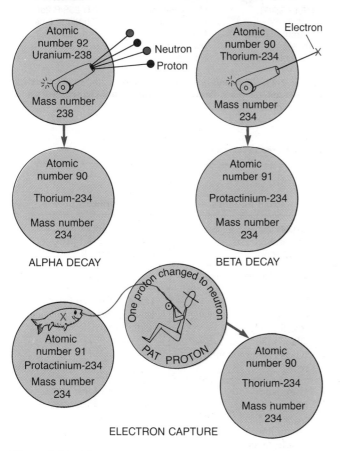

Figure 2-23. Three kinds of radioactive decay: alpha decay, beta decay, and electron capture.

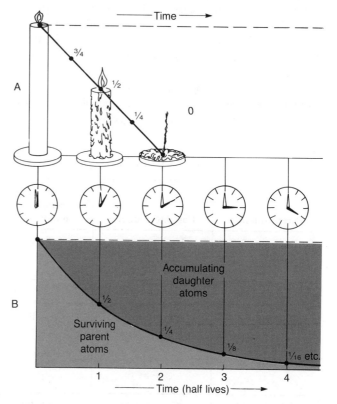

Figure 2-24. A. Uniform, straight-line depletion characteristic of most everyday processes. B. The radioactive decay curve, which approaches the zero line asymptotically: The end of one half-life is the beginning of another.
(From Don L. Eicher, GEOLOGIC TIME, 2nd ed., ©1976, Fig. 6-1, p. 120. Reprinted by permission of Prentice-Hall, Inc., Englewood Cliffs NJ.)

nium-238 has a half-life of 4510 million years, that is the age of the sample (Fig. 2-25).

The half-lives of radioactive nuclides have been determined with sensitive analytical instruments in the laboratory; some half-lives are only fractions of a second long and some are tens, hundreds, and even thousands of millions of years long. The dating of most geologic events by radiometric methods necessarily involves those radioactive isotopes with long half-lives; the choice of specific dating method should be tailored to the particular problem that must be solved.

Table 2-2 lists some radioactive isotopes, their decay processes, their half-lives, and their daughter nuclides. These isotopes constitute the chief decay series used in determining radiometric ages of ancient rocks. Also included are the minerals and rocks that most commonly provide the raw materials for dating and some of the most important applications. Just four radioactive nuclides of a great multitude that have existed since the origin of the earth have provided nearly all the radiometric ages for ancient rocks. Igneous rocks generally yield the best results because rocks of this family are the products of crystallization of a silicate melt. As such they are primary rocks. Metamorphic rocks can yield radiometric dates, but the ages obtained are ages from the time of metamorphism: Such dates generally do not give the ages of the original, unmetamorphosed rock. Metamorphic changes generally include recrystallization of existing minerals as well as formation of new minerals and result in a resetting of the radiometric clocks. Sedimentary rocks, in general, are not amenable to radiometric dating because the detrital grains have been de-

rived ultimately from older igneous or metamorphic rocks. Dating of detrital zircon or microcline, for example, would give the age of the original rock rather than the age of the sedimentary deposit itself. Glauconite, a green potassium-iron silicate mineral that forms as a primary mineral in certain marine sedimentary environments, can yield reliable potassium-argon dates for some sedimentary rocks.

Figure 2-26 is a schematic geologic cross section that illustrates the fundamental stratigraphic principles of superposition and cross-cutting relationships. By employing these principles, one can work out the sequence of geologic events, thus providing a relative time frame. If radiometric dates can be determined for the metamorphic basement complex and the three igneous rock bodies, the sedimentary depositional sequences, whose relative ages are determined by fossil succession, can be bracketed with dates in years before present, telling the duration and antiquity of the relative time sequence. By

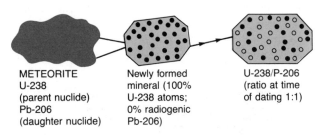

METEORITE
U-238
(parent nuclide)
Pb-206
(daughter nuclide)

Newly formed
mineral (100%
U-238 atoms;
0% radiogenic
Pb-206)

U-238/P-206
(ratio at time
of dating 1:1)

Figure 2-25. Radiometric dating of meteorite. U-238 has a half-life of 4510 million years. Because one half-life has elapsed, the age of the meteorite is 4510 million years.

Table 2-2. The most important radiometric decay series used in age dating of ancient rocks

Parent nuclide	Daughter nuclide	Half-life of parent nuclide (million years)	Chief minerals	Applications
Uranium-238	Lead-206	4510	Zircon Uraninite Pitchblende	Dating of lunar samples, meteorites, old pre-Cambrian rocks
Uranium-235	Lead-207	713	Same as U-238	Same as U-238
Potassium-40	Argon-40	1300	Muscovite Biotite Hornblende Glauconite Whole metamorphic rock	Dating of ocean-floor basalts, lava flows, some sedimentary deposits
Rubidium-87	Strontium-87	47,000	Muscovite Biotite Lepidolite Microcline Whole metamorphic rock	Dating of oldest rocks on earth: meteorites, ancient pre-Cambrian rocks

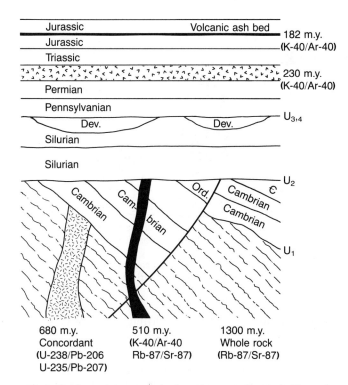

Figure 2-26. Relative and absolute dating applications. The sedimentary sequences are assigned relative ages such as Cambrian, Silurian, Permian, or Jurassic on the basis of their fossil content. Radiometric dates on the bracketing igneous rocks provide information on the antiquity and duration of various parts of the sedimentary sequences. For example, the Paleozoic sections can be no younger than 250 million years and no older than 680 million years. $U_{1,2,3,4}$ represent unconformities.

this method, radiometric dates have been determined for the boundaries of the relative geologic time scale.

Sources of Error

The most reliable radiometric dates are provided by the concordant results of two decay series. For example, if a uranium-bearing crystal has remained a closed system, the uranium-238:lead-206 and uranium-235:lead-207 ages should agree (Fig. 2-27). Perhaps the greatest source of inaccuracy in *geochronology* is the failure of rocks and minerals to remain closed systems. In particular, loss of radiogenic daughter product such as argon-40 can be detected only by checking results from more than one dating method. In addition, original nonradiogenic daughter nuclide is commonly present as a contaminant in the mineral to be dated. This daughter nuclide was incorporated when the mineral first crystallized; its presence must be detected and its exact amount precisely determined.

Yet another source of error is in the laboratory analyses. The determination of ratio of daughter product to parent nuclide is generally accomplished by a *mass spectrometer*, a highly sensitive analytical instrument that has the capacity to separate and measure the proportions of minute particles according to their mass differences. Intermediate steps that attempt to calculate presence of original daughter product can, for example, introduce small errors that may become compounded during successive steps. The amount of error depends

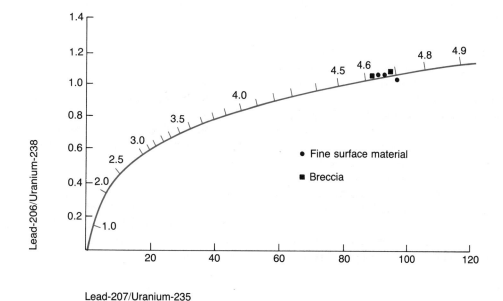

Figure 2-27. Uraniam—lead concordia curve showing the essentially concordant ages of lunar soil and breccia collected during the *Apollo 11* landing. The age of between 4600 and 4700 million years agrees closely with the age of the solar system as determined from dating of meteorites.
(From G.W. Wetherill, Of Time and the Moon, Fig. 4, p. 386: *Science,* vol. 173. pp. 383–392. Copyright 1971 by the American Association for the Advancement of Science.)

on the amounts of radioactive parent, radiogenic daughter product, nonradiogenic original daughter, half-life of the parent, and true age of the sample.

Thus radiometric ages commonly are expressed as a number with amended plus and minus values. The date of 475 ± 15 million years incorporates the stipulated total error (insofar as it can be determined); the range of 30 million years is a statistically determined expression of the precision of the measurement. In addition to the systematic and analytical errors, the spread also reflects the scatter of dates obtained in conducting several analyses. The range 460 to 490 million years for our example expresses the statistical precision of the age determination. Should another sample of the same rock be analyzed, there is a very high statistical probability that it will give a date that falls within the 30-million-year time range. This statistical repeatability, however, may not be an accurate measure of the true age of the rock. Accuracy is a measure of the departure of the determined age from the true age.

Radiocarbon Dating

Carbon is an important element, both in nature and for radiometric dating of geologically very young organic material. The most common carbon atom has six protons and six neutrons in its nucleus, thus an atomic number of 6 and a mass number of 12. Carbon has two isotopes, carbon-13 and carbon-14, both of which behave chemically like carbon-12. The significance of this is that plants and animals do not distinguish among the various forms of carbon and consequently use them all in the manufacture of such diverse organic substances as cellulose, the calcium phosphate of bones and teeth, and the calcium carbonate of shells (Fig. 2-28). Carbon-14, however, is unstable and decays by losing a beta particle from its nucleus. The result is the formation of a daughter nuclide, nitrogen-14.

The basic justification for all radiometric techniques is that decay proceeds at a uniform rate, unique to each radioactive nuclide, and is not changed by variations of heat, pressure, or chemical reactions. Therefore, knowing the half-life of carbon-14 (5730 years) and the decay constant, it should be a relatively simple matter to calculate when an animal or plant died by measuring the amount of carbon-14 in the fossil remains. Carbon-14 decays to nitrogen-14, but the age of the carbon-bearing material is not determined from the daughter:parent ra-

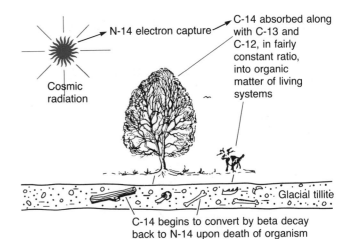

Figure 2-28. Part of the carbon cycle and the formation, incorporation into living systems, and decay of carbon-14.

tio as in uranium–lead age determinations. Instead, age is computed from the ratio of carbon-14 to other carbon in the sample. Therefore if the original proportion of carbon-14 to carbon-12 and carbon-13 is known, the fossil can be dated according to the amount of carbon-14 that remains. The critical measurements are not accomplished with a mass spectrometer, but rather with a highly sensitive geiger counter.

With carbon-14's half-life of 5730 years, radiocarbon dating is limited to a maximum age of about 50,000 years. Remember that a decay curve is an exponential curve; accordingly, the amount of carbon-14 diminishes rapidly with time. After about 17,000 years only one-eighth of the original carbon-14 remains; after about 50,000 years, the amount becomes too small to measure accurately. Despite this age constraint, radiocarbon dating has become indispensible for archeology and a valuable technique for Pleistocene geology. One of the earliest applications, after the technique was devised in 1947, was to date accurately the time of the last advance of continental glacial ice in North America. The results showed that the ice advanced 11,400 years ago, a figure less than half that estimated previously by use of stratigraphic criteria.

Other techniques used in dating events in more recent time (the last several million to several tens of thousands of years) include thorium-230 and protactinium-231 methods, amino acid racemization, and nuclear fission tracks.

SUMMARY

Relative geologic time relates to the order of occurrence and sequence of geologic events and stratigraphic units. The fundamental principles used to place rocks and structures in a relative time frame are

1. superposition, which means the oldest strata are at the bottom and the youngest at the top of a layered sequence;

2. original horizontality and lateral continuity, which provide for a better understanding and visualization of the superpositional arrangement of strata;

3. cross-cutting relationships, which give the relative age of faults, intrusive igneous rock bodies, unconformities, and the sedimentary sequences that are cut by them;

4. unconformities, which are cross-cutting features, but, as buried surfaces of erosion or nonaccumulation, not only are the products of missing record, but also provide convenient stratigraphic boundaries that aid in organization of the record;

5. law of inclusions, which gives the relative age relationships between inclusions in sedimentary rocks and the parent rocks that represent the source of the inclusions;

6. fossil succession, which is the change in fossil assemblages from bottom to top (the time dimension) in superpositional sedimentary sequences; superposition and cross-cutting relationships allow determination of relative sequence of geologic events on a local scale, but fossil succession permits the recognition of relative age and the correlation of isolated sections of rock on a larger scale; fossil succession also provides the basis for the relative geologic time scale.

Radiometric time involves the age in years before present of rocks and geologic events. In radiometric determinations the age of a datable mineral or rock is determined by analyzing the original number of parent atoms of a particular radioactive nuclide, the number of such atoms remaining at present, and the half-life of the parent radioactive isotope. The age is the elapsed time since the decay process began at the time of crystallization.

Four principal methods of radiometric dating are used to determine the age of ancient igneous and metamorphic rocks: uranium-238 : lead-206; uranium-235 : lead-207; potassium-40 : argon-40; rubidium-87 : strontium-87. Thorium-232 : lead-208 is an additional method that occasionally can be used in conjunction with the uranium–lead techniques. The most reliable dates come from concordant results using several different methods. Applications of radiometric dating have included determining the age of the earth, dating the earth's oldest rocks, dating lunar rocks, and measuring sea-floor spreading rates and lithospheric plate movements.

Radioactive isotopes with short half-lives include carbon-14, thorium-230, and protactinium-231. Ages are determined from the amounts of these nuclides that remain in the sample relative to the amounts present originally. Techniques involving these isotopes have been successful in dating rocks less than several million years old, sea-floor sediments, glacial deposits, and archeologic sites. Nuclear fission tracks and amino acid racemization are other absolute dating techniques.

Techniques of relative and radiometric dating and the geologic time scale that has evolved represent major scientific achievements. Knowledge of these techniques and of the time scale that is based on them are fundamental to interpretation of earth history.

SUGGESTIONS FOR FURTHER READING

Cloud, Preston, ed., 1970, *Adventures in Earth History:* W.H. Freeman, San Francisco.

Dunbar, C.O., and Rodger, John, 1597, *Principles of Stratigraphy:* John Wiley & Sons, New York.

Eicher, D.L., 1976, *Geologic Time,* 2nd ed.: Foundations of Earth Science Series, Prentice-Hall, Englewood Cliffs N.J.

Fleisher, R.L., 1979, Where do nuclear tracks lead?: *Am. Scientist,* vol. 67, p. 194–203.

Harbaugh, J.W., 1968, *Stratigraphy and Geologic Time:* Foundations of Earth Science Series, Wm. C. Brown, Dubuque IA.

Harper, C.T. (ed.), 1973, *Geochronology: Radiometric Dating of Rocks and Minerals:* Dowden, Hutchinson and Ross, Stroudsburg PA.

Krumbein, W.C., and Sloss, L.L., 1963, *Stratigraphy and Sedimentation,* 2nd ed.: W.H. Freeman, San Francisco.

Laporte, Leo, 1979 *Ancient Environments* (2nd ed.): Foundations of Earth Science Series, Prentice-Hall, Englewood Cliffs NJ.

Pettijohn, F.J., and Potter, P.E., 1964, *Atlas and Glossary of Primary Sedimentary Structures:* Springer-Verlag, New York.

Ralph, E.K., and Michael, Henry N., 1974, Twenty-five years of radiocarbon dating: *Am. Scientist,* vol. 62, p. 553-560.

Schaeffer, O.A., and Zahringer, J., eds., 1966, *Potassium-Argon Dating:* Springer-Verlag, New York.

Schrock, R.R., 1948, *Sequence in Layered Rocks:* McGraw-Hill, New York.

3

A MOST AMAZING CALENDAR:
Birth and Development of
the Geologic Time Scale

CONTENTS

KEY TERMS

Catastrophism
Neptunism
Uniformitarianism
System
Stratotype
Erathem
Era
Series
Stage
Zone
Facies
Time-stratigraphic unit
Time unit
Period
Eonothem
Lithostratigraphic unit
Formation
Member
Group
Supergroup
Biostratigraphic unit
Biozone
Chronozone
Actualism

CLASH OF OPINIONS: WERNER AND HUTTON

"In the year of our Lord 4004 B.C., he created the earth." With these words, Archbishop Ussher of Ireland, from research on biblical scriptures and literal acceptance of ancient Hebrew writings, proclaimed in 1654, that the earth was about 6000 years old. Blind faith in this pronouncement together with the belief that the earth's surface had been shaped by various major catastrophes conditioned the thinking of most seventeenth and eighteenth century naturalists. Natural processes were assumed to be controlled by providence, and the history of the earth and processes of nature were concluded to have neither great antiquity nor any appreciable invariability. All things—animal, vegetable, and mineral—had been placed by the creator upon the earth following the last great catastrophe, the Noachian flood.

One of the key figures during this age of **catastrophism** was Abraham Gottlob Werner (1749–1817) (Fig. 3-1A), professor of mineralogy at the Freiberg Mining Academy in the German province of Saxony. Through the influence of his father, Werner grew up with a love for minerals. To nurture his interest in mineralogy he attended the small academy at Freiberg and later went to the University of Leipzig. In 1775 he accepted an invitation to become a member of the teaching faculty at the Freiberg Academy and thus began a career of unparalleled influence in the field of geology. He spent the remainder of his life at Freiberg and changed the academy from a small training school for local miners to a bastion of learning known to geologists throughout Europe.

Werner became famous for his theories about the earth. He was a master at projecting his kind and energetic personality and boundless enthusiasm for geology; a large number of impressionable students were captivated. He unselfishly shared his mineral collection and library with students and graciously repeated his stirring lectures for the benefit of those who were not able to squeeze into his crowded lecture halls. He was a powerful and convincing lecturer who had an uncanny ability to impress his opinions indelibly on the minds of his listeners. Consequently he built up a loyal cadre of devoted followers who clung to his every word and considered his opinions as final dictates.

Figure 3-1. Abraham Gottlob Werner.

For his time, Werner was a great geologist. Unfortunately he was disposed to teaching narrow dogmatic theory: He championed a view of the earth that all rocks, regardless of composition—basalt, granite, schist, limestone, or sandstone—were the products of an original, primeval ocean. Because water was invoked to explain the formation of all rocks, this school of thought became known as **neptunism.** Werner viewed the earth's crust as like an onionskin, with the different layers representing successive development in the earth-encompassing ocean. These layers became the fundamental subdivisions of a time scale of earth history. Even though there was no mention of Noah or the great flood, Werner's philosophy could be reconciled with biblical scripture and thus had considerable appeal, even among the clergy.

Werner himself published little; most of the modifications and amplifications of neptunism were disseminated by publications of his students. Werner discussed as gospel the geology of the entire globe, yet he rarely ventured from the security of his position of influence at Frieberg. He disdained the field testing of his conclusions, preferring instead to proceed on blind faith in his beliefs. This style commonly brought Werner and his followers into heated debate with those who challenged his assumptions through study of rock relationships in outcrop. In his later years, he even refused to debate his ideas and remained until his death a dogmatic theorist blindly intolerant of any view but his own. Even so, the influence and authority of this one man dominated geologic thinking for almost a half century; Werner's doctrinaire teachings were spread throughout Europe by several generations of his students.

James Hutton (1726–1797) (Fig. 3-2), a Scottish physician–farmer turned geologist, was not bound by the arithmetic of Bishop Ussher's biblical chronology. From his studies of rocks along the Scottish coast, he observed that every rock formation, no matter how old, appeared to be derived from other rocks, still older. Hutton's departure from biblical chronology was prompted by his approach to the history of the earth: he was convinced that geologic processes in ages past were no different from the processes now active. Based on phenomena open to observation, geology could be interpreted without recourse to postulated catastrophes such as the Noachian flood. Hutton openly challenged these ideas with the view that the earth was constantly changing although its basic nature remained the same. Observable processes such as weathering, erosion, deposition, and volcanic activity were recognized by Hutton as recorded in the rocks of his native Scotland. He believed that the present provided the key to understanding the past.

Hutton was a member of the Royal Society of Edinburgh, one of a number of scientific societies that had sprung up in Britain as interest in scientific investigation burgeoned during the latter part of the eighteenth century. Much of this scientific interest was geologic, but most geologists paid obeisance to the Book of Genesis, and the accepted conclusions about the history of the earth were tied to church dogma. Despite the biblical constraints, a considerable body of observations and conclusions pertaining to uniformity of natural processes through time had been recorded. Aware of this work, Hutton synthesized it with his own observations and conclusions and prepared himself for

Figure 3-2. James Hutton, trusty rock hammer in hand, contemplating a rock outcrop whose weathered profile simulates the faces of some of his antagonists.
(From John Kay, 1842, Edinburgh Portraits)

a presentation to the Royal Society of Edinburgh on the results of years of geologic investigation. His lecture, delivered in 1785, and its printed version, published in the Transactions of the Society in 1788, documented the principle that natural laws may be derived from studying present-day processes and that nature's past operations can be appreciated from observations of present natural relationships. Because of its then-radical nature, this theory came under sharp attack; in many circles, including some cliques within the Royal Society itself, Hutton's views were labeled untheological and Hutton himself was branded an athiest.

James Hutton died in 1797, two years after a two-volume expansion of his philosophy entitled *Theory of the Earth, with Proofs and Illustrations** was published. Unfortunately his writing was cumbersome, and much of the thrust of his arguments was lost in excess verbiage and detail. John Playfair, a devoted friend and colleague of the Royal Society, assumed the task of clarifying and defending Hutton's views. In a clear, concise style, he set forth the observations and conclusions in a brilliant summary published in 1802: *Illustrations of the Huttonian Theory of the Earth.*† It was through this effort that Hutton's views became widely known, but acceptance was slow. Ironically, less

---

*W. Creech, Edinburgh.

†W. Creech, Edinburgh.

than a decade after Hutton's death, the University of Edinburgh became a wellspring of Wernerian ideology.

Eventually, however, the death knell began to sound for the Wernerian neptunism philosophy as devotees of Hutton amassed opposing views based on field observation. One of the telling blows involved a controversy over the origin of basalt. The neptunists regarded basalt as a chemical precipitate of the primeval ocean. The Auvergne Province in France served as an instructive region for the open-minded observer to view and understand the relationship between basalt and volcanic activity (Fig. 3-3). Those who maintained that basalt had an igneous origin became known as the plutonists. They slowly convinced some of the disciples of Werner that basalt did not precipitate chemically from water and that volcanic activity had played an important role in the formation of the earth's crust. Charles Lyell (1797–1875) was the principal developer and enunciator of Hutton's uniformitarian approach and through his writings championed the doctrine of **uniformitarianism,** which gradually displaced catastrophism.

James Hutton was the embodiment of the "age of reason" philosophy that began to condition much of

Figure 3-3. Geologically recent volcanic activity in the Auvergne region of south-central France: an early graphic testimonial to the igneous origin of basalt.
(From Faujas de Saint-Fond, 1778)

the scientific thinking of the late eighteenth and early nineteenth centuries. His theory of the earth, based on inductive principles, laid the foundation for the science of geology and paved the way for development of the modern geologic time scale.

BEGINNINGS OF HISTORICAL GEOLOGY

Early Ideas

In the preceding chapter we reviewed the fundamental principles and methods used in telling geologic time: superposition, original horizontality and lateral continuity, inclusions, cross-cutting relationships, fossil succession, correlation, and radiometric dating. In this chapter we shall consider these principles in historical perspective in reference to the birth and development of the geologic time scale and the evolution of historical geology as a science.

The beginning of scientific thought and recording of scientific observations in the western world probably dates back to the sixth century B.C. when Greek scholars began to record and analyze data within the context of assumed principles that governed nature. Some of these early classical Greek thinkers optimistically assumed that natural phenomena within their world could be understood. This attitude kindled the search for guiding principles and natural laws and led to some interesting conclusions regarding causes and effects.

Some of the earliest recorded observations were geological and even stratigraphical. The observation that shells, not unlike those of clams and snails then living along coasts, were found on mountain tops and other places far inland inspired the conclusion that the ocean had once extended to the places where the fossils were found. Herodotus, a Greek historian who lived just prior to 400 B.C., concluded from scientific observation that the Mediterranean Sea was at one time more widespread and that the Nile delta was constructed from great volumes of sediment that had been transported and deposited by the Nile River. Such students of natural history were impressed by the repeatability of certain processes that shape the face of the earth as well as by the impermanence of details of the earth's surface. This style of thinking about natural phenomena presaged Hutton's philosophy.

Grecian theories on scientific inquiry transcended time and cultures and were inherited by Roman scholars. Some of the more notable Greek-influenced Roman scholars, such as Pliny the Elder (A.D. 23–79), came to conclusions similar to those of the Greeks about past positions of land and sea, comparisons of modern sea shells with fossils, volcanic eruptions, earthquakes, and floods. But after the decline of the Roman Empire, the principle of uniformity of nature's processes, along with scientific inquiry in general, suffered a critical blow. The Dark and Middle Ages witnessed a major recession in scholarly activity, and interpretation of natural history took a giant step backward. A few Byzantine and Arab scholars borrowed some ideas from the Greeks and Romans, but did not advance them. Other scientific writings during this nearly thousand-year span of western history were limited to monasteries and various cloisters. Religious dogma held sway, and departures from the strictures of the time were considered serious offenses.

The Renaissance witnessed advances in many different fields of science, spawned mainly by development of scientific principles. The stimulus provided by such scientific thinkers as Copernicus, Kepler, and Galileo put the earth in a new cosmic context. Unfortunately, the religious climate of the times created a barrier to rapid advances in geology, the science of the earth itself. The Book of Genesis portrayed an earth that was only a few thousand years old and that was formed during the creation. The creationist viewpoint considered the earth to have been in a stable state ever since a single great catastrophe, the Noachian flood.

Principles of Stratigraphy

Geology and its subdiscipline, stratigraphy, like other sciences, are based on principles that have been derived by induction from the masses of data that have accumulated from observations of natural phenomena. For example, the modern concept of plate tectonics represents a major principle developed from observations of many aspects of the oceanic lithosphere. Principles are the very essence of science and provide the basis for interpretation. The first stratigraphic principles were formulated and clearly enunciated in the last half of the seventeenth century by Nils Stensen, better known as Nicholas Steno, a Danish physician who settled in Florence, Italy. Steno observed layered strata in Tuscany in northern Italy and also witnessed the process of deposition of layers in an aqueous medium. He noted that when any layer was forming, only the fluid from which its particles came was above it, and therefore no overlying layers could have been present when the lower layers were formed. From this observation he concluded that the lower layers must be older than the upper layers in any sequence of strata. Steno arrived at the principle of *superposition* through induction (see Chapter 2). In time, a set of criteria involving recognition and interpretation of sedimentary structures (Chapter 2) was gradually established, allowing stratigraphers to recognize the tops and bottoms of beds in deformed sequences. Many naturalists before Steno had observed that certain rocks were layered and perhaps some of these naturalists tacitly assumed the correct superposition. But Steno is credited with being the first to put the idea of sequence in perspective. He also stated the principles of *lateral continuity* and *original horizontality* of

strata (Chapter 2) and recognized that tilted and deformed strata were the results of postdepositional earth movements such as might be caused by volcanic eruption and cave-ins. Steno's principles laid the foundation for placing rocks in relative order, a first step toward unraveling the history of the earth. His observations also led him to identify objects known as "tongue-stones" as the teeth of ancient sharks (Fig. 3-4).

First Attempts at Subdivision: An Economic Incentive

Geology, as a science based on inductive principles, received its earliest stimulus during the Renaissance because of interest in mining and mineralogy. Mining had been practiced for centuries throughout much of the western world, and many techniques for extracting metals had been devised. As the economic aspects and uses of metals and gems broadened during the beginning of the industrial revolution in the eighteenth century, more and more emphasis was placed on natural occurrences of minable rocks and minerals. The science of mineralogy itself had its greatest early stimulus from the writings of George Bauer, better known as Agricola, a sixteenth century student of the earth. He is commonly referred to as the father of mineralogy because of his classification scheme and vivid descriptions of minerals and mining activity (Fig. 3-5).

During the seventeenth and eighteenth centuries, as more attention was focused on the study of minerals and rocks, mining academies were established with teaching positions in mineralogy. Knowledge of the origin, occurrence, and relationships of ore minerals expanded. In time there developed a critical need for a scheme by which the time of formation of economically important rocks in one mining area might be related to the time of formation in other areas. And so it was that early attempts to subdivide and organize the rock record had an economic incentive.

The earliest attempts to unravel earth history used only superposition as a guide. The targets for study were mainly sedimentary rock successions in areas of mining activity (Fig. 3-6). Early eighteenth century geologists began to use the local sequences in attempting to develop a history of the entire earth. This was a noble ambition and led not only to some good descriptive generalizations, but also to some faulty hypotheses. Superposition was a valid and necessary principle, but it alone would not be the tool for working out earth history. One of the generalizations that resulted from this attempt to order the rocks of the entire earth was that stratified younger looking sedimentary rocks that formed low mountains and foothills overlie older looking more complex igneous and metamorphic rocks that make up the cores of topographically higher mountain

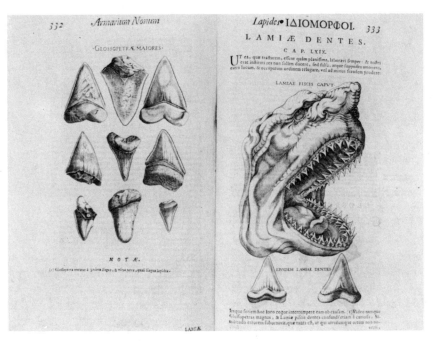

Figure 3-4. A. Head of the lamia fish (shark). B. Glossopetrae ("tongue-stones"). Steno used these plates to demonstrate that the "tongue-stones," which were found in great abundance on the island of Malta, were, in fact, the teeth of ancient sharks that lived prior to Malta's being uplifted as an island (a former sea bottom now land). The plates first appeared in Steno's work, but had been prepared in bronze in the previous century by Michele Mercati (1541–1593), who saw no scientific relationship between the "tongue-stones" and the teeth of modern sharks.
(From Nicolas Steno, 1667, The Dissection of the Head of a Carcharias Shark)

ranges. This led to a stratigraphic subdivision that carried the use of superposition beyond the local scale to a larger scope. Use of large-scale superposition was valid for individual areas, but it led to the fallacious assumption that *all* the crystalline rocks formed at one time, and *all* the layered, sedimentary rocks formed later. What started out as a good descriptive generalization, applicable to individual mountain belts, was stretched to a faulty, unfounded principle and perpetuated.

In the mid-1750s, Johann Lehmann, who taught mining and mineralogy at a mining academy in Berlin, proposed a broad threefold subdivision of the rocks of the earth's crust (Fig. 3-7). He referred to each subdivision as a mountain, a designation that included the topographic expression, kinds of rocks, and sequence of events that had acted collectively to form it. According to Lehmann's scheme, the cores of the highest mountains were made up of the oldest rocks—crystalline rocks that he called Ore Mountains because they were the source of so many valuable mineral deposits. These rocks were judged to have formed at the time of the earth's origin. Outward from the Ore Mountains and superpositionally above them were a second class of mountains termed Stratified Mountains. The rocks of the Stratified Mountains consisted of layered sedimentary

rocks such as limestone, sandstone, and shale (many with fossils), some beds of coal and marble, and mineral veins. The stratified rocks were assumed to have formed at the time of the Noachian flood. Lehmann interpreted the fossils as the remains of animals and plants that had inhabited the slopes of the older Ore Mountains and subsequently were swept up by the floodwaters and deposited along with the sediment as the water receded from the mountain flanks. The third class of mountains, called the Alluvial Mountains, consisted of loosely consolidated rocks that Lehmann presumed to represent accidents of nature—volcanic eruptions, earthquakes, storms, landslides, cave-ins, and so forth—that occurred after the flood.

Lehmann's scheme, although simple, was based on arduous field work and research and was truly a monumental work for its time. The mountains, with their individualistic topographic expression and rock makeup, were interpreted as representing a set of geologic events unique in time. To apply the scheme, all one had to do to determine relative age of a rock was to identify the rock and relate it to one of the three mountains. As a means of subdividing the history of the earth there are inherent problems with such a scheme because similar rock types can form at vastly different times. It is a point of interest, however, that Lehmann correctly observed

Figure 3-5. Agricola's observations on the occurrence and mining of mineral veins. A. A and C, vena profunda (fissure veins); B, intervenium (nonvein material or country rock); B. A, B, D, E, vena dilatata (stratified deposits); C, intevenium. C. A–D, the mountain or intervenium; E–K, vena cumulata (impregnation or replacement veins). D. Building fires in tunnel along vena dilatata to loosen ore material. (Redrawn from Georgius Agricola, 1556, De Re Metallica, Translated 1950 from first Latin ed. by Herbert C. Hoover and Lou H. Hoover: 1950, Dover, New York)

the general large-scale stratigraphic relationships that exist in many of the mountain ranges and adjacent foothills and plains of the world: namely, a central core of metamorphic and intrusive igneous rocks, commonly referred to as *basement complex;* these overlain nonconformably by folded and faulted, well-indurated sedimentary strata; and these in turn separated by an angular unconformity from overlying, more loosely consolidated sediments that flank the mountains and commonly rep-

resent the erosional debris of the mountains themselves. Lehmann's scheme, however, perpetuated the erroneous principle that geologic time could be told by what rocks look like.

Giovanni Arduino, a contemporary of Lehmann, was professor of mineralogy at the University of Padua, Italy. He was an ardent student of the rocks in Tuscany, where Steno, a century before, had formulated the principle of superposition of strata. Arduino, an active researcher,

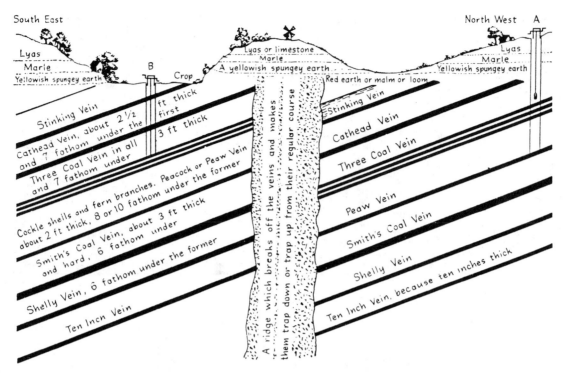

Fig. 3-6. One of the earliest published geologic cross sections, drawn by John Strachey across part of
Somerset, south of Bristol, England, in 1719.
(From John Strachey, 1719, A Curious Description of the Strata Observed in the Coal-mines of Mendip in
Sommersetshire; Being a Letter of John Strachey Esq. to Dr. Robert Welsted, M.D. and Roy. Soc. and by
Him Communicated to the Society: *Royal Soc. London Philos. Trans.,* vol. 30, no. 360)

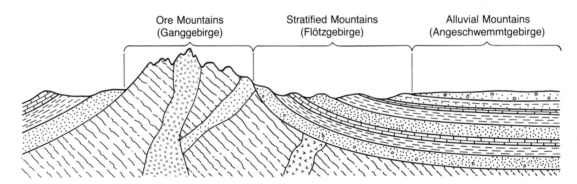

Figure 3-7. Johann Lehmann's subdivisions of rocks of the earth's crust.

was influential among other geologists and during the
1760s and 1770s, as had Lehmann before him, described
three kinds of mountains formed from different suites
of rock that he called *Primary, Secondary,* and *Tertiary.*
The primary rocks, analogous to the ore mountains of
Lehmann, included igneous and metamorphic rocks
making up the cores of high mountains (Fig. 3-8). The
Secondary rocks were described as fossiliferous limestone, claystone, and other kinds of sedimentary rocks
exposed on the flanks of the high mountains. The *Tertiary* rocks consisted of a younger, generally less consolidated succession of fossiliferous limestone, sandstone, clay, and marl (clayey limestone) forming lower
mountains and hills. Arduino also included a fourth

class of rocks, *Volcanic,* as a subdivision of the Tertiary
mountains; these rocks included lavas and volcaniclastic
sediments.

Arduino did not consider this scheme to be applicable universally, but only to the area he studied. He
maintained that rock type alone was not a sufficient criterion to make a relative age assignment and that rocks
had to be studied and sequences worked out on a local
basis. His open-mindedness is to be admired considering the prevailing tendency to organize the rocks of the
world solely on the basis of their appearance.

The point of view that all crystalline rocks formed
at one time and all sedimentary rocks at successive later
times was developed to an extreme during the last

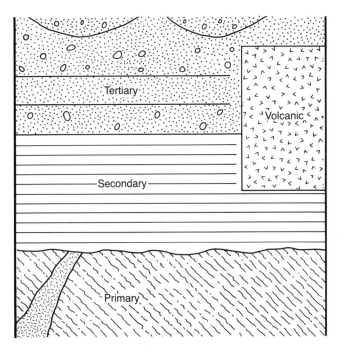

Figure 3-8. Arduino's subdivision of rocks of the earth's crust.

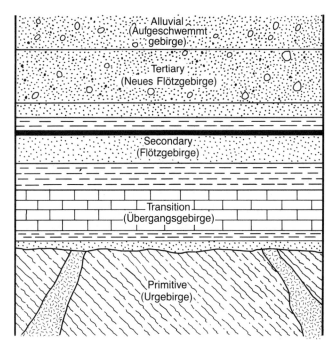

Fig. 3-9. Werner's subdivision of rocks of the earth's crust (corresponding names of Arduino's subdivision are shown in parentheses). Although Werner was misguided in many of his geological interpretations, he was, nonetheless, a great pioneer in the geological sciences and made an important contribution in his definition of the Transition series.

quarter of the eighteenth century. The foremost individual associated with attempts to determine the earth's history from superposition alone and to tell age of rocks by their composition was Werner. He zealously nurtured the idea that all rocks of the earth's crust, regardless of composition, were precipitates of a world-encompassing primeval ocean. As this original ocean gradually shrank, it left behind precipitated layers arranged in superpositional order. Each major layer was considered to have its own unique composition and to be the same age everywhere.

Werner and Neptunism

Werner recognized a succession of four main groupings of rocks (Fig. 3-9) that he believed revealed the four primary stages in formation of the earth's crust. He visualized the earth's crust as being granite at the bottom of the stack followed in superpositional order by a thick sequence of gneisses, schists, and other crystalline rocks, together constituting the *Primitive* series. Above these primary precipitates of the earth-encompassing ocean, Werner recognized a sequence of slates, graywackes, quartzites, and limestones that he called the *Transition* series. As we shall see later, this recognition of the Transition series was an important contribution. Fossils in these rocks were considered the earliest forms of life. In terms of original lateral continuity, Werner considered the Primitive and Transition rocks to have covered the entire global surface at one time. Next in order were the stratified rocks of Lehmann (Fig. 3-7). Werner believed that this fossiliferous succession of sandstone, limestone, slate, and coal, called the *Second-*

ary series, formed by chemical precipitation and settling after the sea had begun to shrink in size, with some of the rock being reworked by surface running water.

Werner's universal sequence was topped by the *Alluvial* rocks, consisting of sand, gravel, clay, peat, and ash and cinder beds, interpreted as the deposits formed as the primeval ocean waters retreated well below the mountaintops, leaving major areas of bare land. The deposits themselves were formed in the remaining parts of the original ocean, in large part from materials delivered to the sea by running water that coursed over the land. Werner's concept of the sequence of formation is shown in Figure 3-10.

The appeal of the neptunist philosophy was that, even without mentioning Noah or the Bible, a picture of the history of the earth was presented that anyone could reconcile with prevailing religious opinion.

Hutton and the Roots of Uniformitarianism

Werner's fourfold division of the rocks of the earth's crust became widely known, but with time its application to separate rock sequences ran into difficulty. Neptunist ideas were extremely vulnerable to field scrutiny and commonly came into sharp conflict with the observations of less provincial, more far-sighted followers of James Hutton. These observers were able to demonstrate on field relationships that rock units were not worldwide and that each region had its own sequence. To exemplify the conflict, basalt, considered by the neptunists to have formed during a particular time in the Primitive series as a precipitate from the primeval ocean, was proved to have an igneous origin. Arduino had appreciated the igneous origin of basalts when he named the Volcanic subdivision of the Tertiary series (Fig. 3-8). And Hutton himself demonstrated that basalt could form as an intrusive body. He correctly interpreted the picturesque *Salisbury Craig* (Fig. 3-11) in his native Scotland as a basalt sill. In addition, Hutton's observations of granite dikes cutting through stratified rocks led him to the conclusion that they had an igneous origin and were molten when they forced their way into fissures and cracks in other rocks. Most certainly plutonism represented a contradiction of the tenets of neptunism.

James Hutton was a remarkably perceptive observer who clearly recognized the stratigraphic and historical significance of unconformities (Fig. 3-12) and other cross-cutting features. He also appreciated the relationship between the grains in sedimentary rocks and the parent rocks from which they were derived; he stands as the first geologist to clearly enunciate the principles of cross-cutting relationships and inclusions. He was able to offer many proofs that past changes in the earth were real and had been brought about by everyday, observable processes. Also inherent in the Huttonian theory of the earth was the concept of long geologic time. His pronouncement (1788), "The result, therefore, of our present inquiry is that we find no vestige of a beginning and no prospect of an end,"* was unsettling for those who subscribed to the notion of a 6000-year-old earth.

---

*Theory of the Earth: *Royal Soc. Edinburgh Trans.*

As geologic evidence accumulated, the biblical concept of time faded and the Wernerian viewpoint lost ground because assumed principles had not been verified by experience. The Lehmann–Werner time scales were no more valid than the assumptions on which they were based. There still remained the problem of relating the rocks and structures of one region to those of another on any kind of meaningful time basis. How could geologic events be correlated from place to place and put into a viable, universal scheme of chronology? Development of a time scale beyond one based on the layer-cake approach of appearance would have to wait the development of a new principle—the recognition in rocks of ingredients that had time significance.

Smith, Cuvier, and Fossil Succession

The answer lay in the fossil record. Fossils had been observed and were objects of curiosity for centuries. They were also the center of controversy between believers in uniformity in nature and creationists (see Chapter 4). Fossils were mentioned in the rock subdivision schemes of Lehmann and Werner, but only as adjuncts to the compositions of the series. Interestingly, Robert Hooke, the brilliant seventeenth century British physicist and inventor, was a keen student of nature. His studies of fossils led him to the reasoning that fossils might be used to determine a record of past ages; he believed that extinctions of old life forms and appearance of new life forms had taken place. At the time such an idea received little serious attention, but in retrospect it amounts to an initial step toward fossil succession as a valid principle. The discovery by William Smith that fossil assemblages change upward through sequences of strata and that isolated sequences of strata, no matter how different they might appear, are of the same age

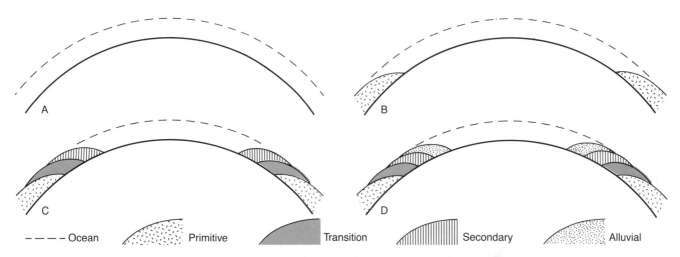

Figure 3-10. Werner's neptunist view concerning history of formation of the earth's crust. A. Original, earth-encompassing primeval ocean. B. Formation of Primitive series. C. Formation of Transition and Secondary series and gradually shrinking ocean. D. Formation of Alluvial series.

when they contain the same fossils provided an efficient means of establishing sequence. Smith not only independently formulated the hypothesis of fossil succession (see Chapter 5); he also demonstrated its broad applicability through his 1815 geologic map of Britain (Fig. 3-13) and thus established it as a working principle.

Figure 3-11. Salisbury Crag, a classic Hutton locality, near Edinburgh, Scotland, where Hutton demonstrated the intrusive nature of a sill.
(Photos by Lynn Coppel)

The time significant aspects of fossil succession were put into even clearer perspective for early nineteenth century geologists by George Cuvier in France. Cuvier, from his stratigraphic work in the Paris Basin, also appreciated the reality of fossil succession, but he related the individual faunas to a succession of catastrophic annihilations (extinctions) and subsequent new creations. This view, no doubt, was greatly influenced by the presence of numerous unconformities in the stratigraphic succession. Cuvier's viewpoint was compatible with the still-prevailing Wernerian philosophy of the times, which also could be more comfortably accommodated by religious opinion. But regardless of how fossil succession was interpreted, it demonstrated that vertical changes in fossil content in a stratigraphic sequence could be recognized from place to place and that rocks could be correlated on the basis of contained fossils. During the early decades of the nineteenth century in western Europe, fossil succession stood the test of repeated observation and experience. Fossil succession and correlation using fossils became the keys that unlocked the door to developing a chronology of earth history.

The widespread application of the principle of fossil succession was immediate. It provided a real point of departure from the confusion resulting from diversity of geologic opinion that existed at the end of the eighteenth century and on into the early part of the nineteenth century. Geologists all over Europe were engaged in tracing strata, collecting and describing fossils from them, and establishing fossil succession and lithologic detail of local stratigraphic sequences (Fig. 3-14). These studies generated new interest, geology became popular, and geologic knowledge expanded rapidly. It was through such endeavors that the relative geologic time scale was gradually pieced together.

GROWTH OF THE RELATIVE TIME SCALE

Stratigraphic Systems

As William Smith had shown, fossils could be used to establish the time equivalency of rocks even where lithologies differed. The principle of fossil succession led gradually to the recognition of large-scale aggregates of fossils upon which the major stratigraphic divisions of the time scale were based. With this recognition we see the transition from purely descriptive stratigraphic units to interpretive units. Today we refer to these main divisions as **systems,** but as the geologic column was being assembled, various other designations such as group, formation, order, and series were used to denote these major packages of rock. These stratigraphic systems were established one by one (Fig. 3-15, Table 3-1) by different people working in western Europe. The systems were based on particular stratigraphic sections, called *type sections,* which today are referred to as **stratotypes.** The geographic areas or districts that contain the stratotypes are called type areas. Some of the systems originally were established on lithologic grounds, reflecting the influence of Wernerian philosophy; some were based on unconformities in the stratigraphic record, an artifact of catastrophist thinking; and some were based on distinctive fossil content. But regardless of how they were defined originally, they were subsequently refined and recognized on every continent

Figure 3-12. Siccar Point, along the North Sea coast, where Hutton correctly identified and interpreted the unconformity relationship between underlying highly deformed Silurian slates and graywackes and overlying, less deformed Old Red Sandstone. Upon visiting this site in 1805, John Playfair stated, "The mind seemed to grow giddy by looking so far into the abyss of time." (Photos by J. Patterson)

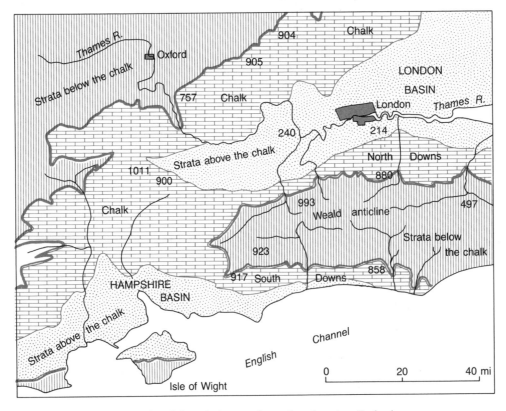

Figure 3-13. Simplified version of Wiliam Smith's 1815 geologic map of part of southwestern England. Numbers are spot elevations in feet.
(From A. D. Woodford, 1965, *Historical Geology,* Fig. 3-4, p. 48: W. H. Freeman, San Francisco. Reproduced by permission of author)

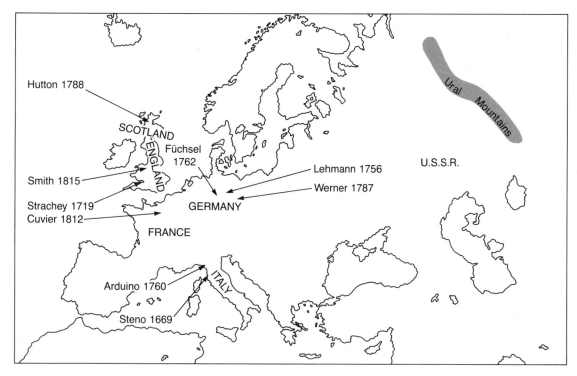

Figure 3-14. Map of Europe showing localities of major pioneer geologic work and dates of publication.
(From Leigh W. Mintz, 1981, *Historical Geology: The Science of a Dynamic Earth,* 3rd ed., Fig. 2-5, p. 15: Charles E. Merrill, Columbus OH)

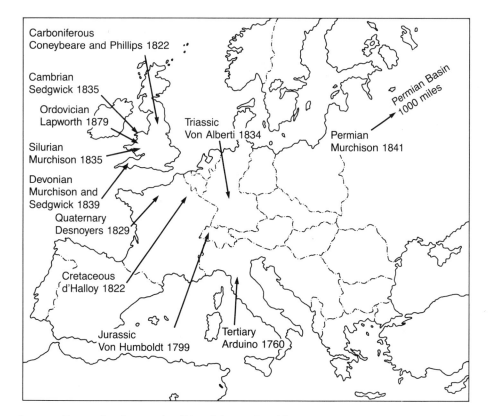

Figure 3-15. Map of western Europe showing type localities of the stratigraphic systems.
(From Leigh W. Mintz, 1981, *Historical Geology: The Science of a Dynamic Earth,* Fig. 2-18, p. 27: Charles E. Merrill Publishing Company, Columbus OH)

Table 3-1. Different vintages of geologic time scales

Early Subdivisions				Modern Usage*			
Arduino 1760	Lehmann 1756 Füchsel 1760–1773	Werner ca. 1800	English Equivalents	Eras	Periods	Epochs	Alternate periods
Volcanic / Alluvium	Angeschwemmtgebirge	Aufgeschwemmtgebirge / Neues Flötzgebirge	Alluvium	CENOZOIC Phillips 1841	NEOGENE Hoernes 1853	Holocene / Pleistocene	QUATERNARY Desnoyers 1829
Tertiary			Tertiary			Pliocene / Miocene	TERTIARY Arduino 1760
					PALEOGENE Naumann 1866	Oligocene / Eocene / Paleocene	
Secondary	Flotzgebirge	Flötzgebirge	Secondary	MESOZOIC Phillips 1841	CRETACEOUS d'Halloy 1822		
					JURASSIC von Humboldt 1799		
					TRIASSIC von Alberti 1834		
				PALEOZOIC Sedgwick 1838	PERMIAN Murchison 1841		
					PENNSYLVANIAN Williams 1891	CARBONIFEROUS Conybeare and Philips 1822	
					MISSISSIPPIAN Winchell 1870		
	Ganggebirge	Übergangsgebirge	Transition		DEVONIAN Murchison, Sedgwick 1839		
Primary					SILURIAN Murchison 1835		
					ORDOVICIAN Lapworth 1879		
					CAMBRIAN Sedgwick 1835		
		Urgebirge	Primitive		PRECAMBRIAN		

*The differences between this relative time scale and the one used in the text (Table 2-1) are that Precambrian is replaced by Cryptozoic and a new, earliest Paleozoic period, the Ediacaran (named by Cloud and Glaesner, 1982).

through application of fossil succession. They became interpretive units based on unique fossil content.

The modern relative geologic time scale (Table 3-1) was firmly established by the beginning of the twentieth century as a scale based on a succession of fossiliferous units, each with its own type section and type area and unique aggregate of faunas. The system names in common use today are essentially the same as those employed at the turn of the century and, with few exceptions, have received worldwide usage. The large-scale stratigraphic names have remained stable, testimony to universal confidence in the validity of the units and their worldwide applicability.

In the early 1840s it was proposed that the systems be lumped together into three larger scale units that represented the major subdivisions of the fossil record and thus fundamental chapters in the history of life. Today these subdivisions are called **erathems,** which correspond to the time term **era.** The names of the eras—Paleozoic, Mesozoic, and Cenozoic—are derived from the Greek verb meaning "to live" combined with the Greek words for ancient; middle, and new, respectively. The eras are convenient large-scale generalized units useful for distinguishing large segments of prehistoric time. It is interesting to note how the major subdivisions of the Arduino–Lehmann–Werner time scale correspond, in a broad sense, with the eras of the modern time scale (Table 3-1). Note also the correspondence of the Transition series with the lower Paleozoic, and the Primitive series (= Primary of Arduino) with the pre-Cambrian. However, it is important to bear in mind that application of the principle of fossil succession has dem-

onstrated that many rock units have turned out to be either older or younger than would have been assumed solely on the basis of their appearance.

Other Subdivisions of the Time Scale: Series, Stage, Zone

Once the large-scale systems were established, they were, in turn, subdivided into smaller units. As the systems became more easily recognized from place to place and correlation of fossil-bearing sections established a reliable web of continuity, relative age assignments were made away from type sections, and smaller scale subdivisions were recognized.

In 1833 the English geologist Charles Lyell defined four subdivisions of the Tertiary system and called them the Eocene, Miocene, older Pliocene, and new Pliocene periods, in ascending order (Table 3-2). This represents one of the earliest attempts at recognizing time-significant stratigraphic units defined on fossil content. The definition was based on the relative proportion of living and extinct species of fossils that each unit contained. This ingenious technique served to separate the units more conveniently. Lyell, although having confidence in the technique, stated that more detailed work on the type-Tertiary beds of France very likely would refine the subdivisions and additional ones would be defined. In 1854 the Oligocene and in 1874 the Paleocene were added to the succession (Table 3-2). Subsequent workers demoted the Tertiary subdivisions to the rank of **series** because they felt the magnitudes were too small compared to magnitudes of other systems of the geo-

Table 3-2. Cenozoic subdivisions—Lyell and modern*

Era	Period	Lyell's scheme	Modern usage	
CENOZOIC (1841)	QUATERNARY (1829)	Recent (time since first appearance of humans)	NEOGENE (1853)	Holocene (1885)
		New Pliocene (1833) 90% living species		Pleistocene (1839)
	TERTIARY (1760)	Older Pliocene (1833) 33–50% living species		Pliocene
		Miocene (1833) 18% living species		Miocene
		Eocene (1833) 3.5% living species	PALEOGENE (1866)	Oligocene (1854)
				Eocene
				Paleocene (1874)

*Lyell's units defined on percentage of living species found as fossils. Paleogene and Neogene are preferred to Tertiary and Quaternary, although in certain contexts the latter are acceptable. Dates show when names were established.
Charles Lyell, 1830, 1832, 1833, *Principles of Geology*, vols. 1–3, J. Murray, London.

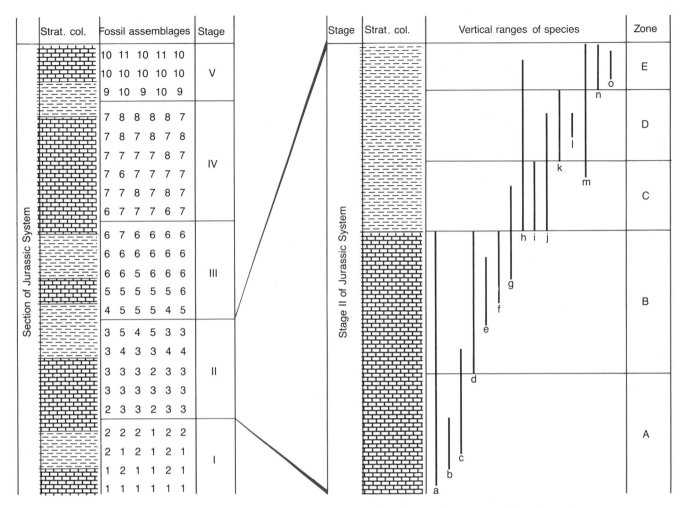

Figure 3-16. A. Subdivision of Jurassic System into stages based on smaller aggregates (assemblages) of fossils. Concept developed by A. d'Orbigny in 1842. Numbers 1–11 represent assemblages of fossils composed of species (a, b, c, etc.) shown in B. B. Subdivision of stage into zones, based on still smaller aggregates of fossils defined by overlapping ranges of individual fossil species. Concept developed by A. Oppel in 1856–1858. Species a, d, f, g, h, and m make up assemblage 3; species b, c, and e make up assemblage 2; species i, j, k, and l make up assemblage 4; species n and o make up assemblage 5.

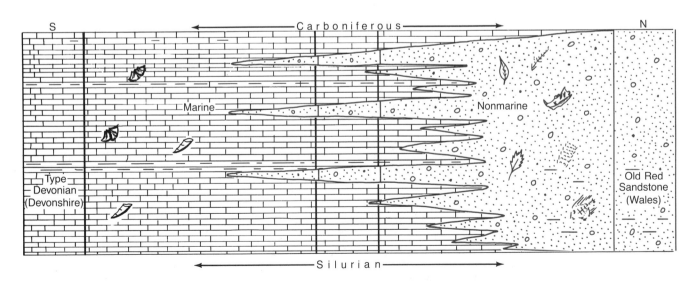

Figure 3-17. General facies relationship between the Old Red Sandstone and the type marine Devonian in Devonshire.

Table 3-3. Principal categories of formal stratigraphic units*

Time-stratigraphic	Time	Lithostratigraphic	Biostratigraphic
Eonothem	Eon	Supergroup	
Erathem	Era	Group	
System	*Period*	*Formation*	*Biozone*†
Series	Epoch	Member	Subbiozone
Stages	Age	Bed	
Chronozone	Phase		

*Fundamental units italicized.
†Biozone replaces zone.

logic time scale. These series subdivisions of the Tertiary System have had essentially worldwide application. Other stratigraphic systems have been subdivided into formal series, most commonly designated by stratigraphic position terms such as Lower, Middle, and Upper (e.g., Lower Cambrian Series; Upper Cretaceous Series). Some series have names based on *type* sections and type areas (e.g., Guadalupian Series of the Permian System, with type locality being the Guadalupe Mountains of southeastern New Mexico). Although there is general universal agreement in concept, the individual series names themselves, with the exception of those of the Tertiary, have only regional to continentwide application.

As work continued on a wider scale, it was realized that understanding the history of the earth demanded even smaller scale time-stratigraphic units. Many local stratigraphic sections that had been worked out represented only isolated pieces of the composite puzzle. How could more refined time correlations be effected, and how could the vast number of local formations and fossiliferous sequences that had been studied be related to one another? Careful work in France, Germany, Switzerland, and England demonstrated the reliability of plotting the vertical ranges of fossil species for the recognition of smaller scale paleontologically defined units that could be correlated. From such painstaking work emerged the concepts of faunal **stage** and **zone** (Fig. 3-16). The recognition that rock types change laterally within the same time-stratigraphic interval spawned the concept of sedimentary **facies** (which will be discussed in Chapter 6), which are deposits produced by different, but laterally adjacent environments of deposition (Fig. 3-17).

By the beginning of the twentieth century the relative geologic time scale had been laboriously pieced together, and the stratigraphic principles for working out earth history had been formulated. Notions about a 6000-year-old earth appealed only to those who had (and still do have) a fundamentalist interpretation of biblical scriptures. However, it would not be until after the discovery of radioactivity and its application to age dating that realistic determinations of the age of the earth would be made and the antiquity and duration of the relative subdivisions of the geologic time scale would be appreciated.

HISTORICAL GEOLOGY MATURES
Stratigraphic Classification

Through the late nineteenth and into the twentieth century, establishment of a worldwide stratigraphic correlation network and refinements in the geologic time scale continued, but various stratigraphic concepts and terms such as series, stage, group, formation, and zone were used interchangeably and became confusing. Through the years, a number of symposia have been held and various national and international stratigraphic commissions have been organized in an attempt to achieve consistency in formal stratigraphic nomenclature and stratigraphic practice. In the spirit of such concern, the North American Commission on Stratigraphic Nomenclature, in collaboration with the American Association of Petroleum Geologists, has published several editions of a Code of Stratigraphic Nomenclature. This code has evolved primarily to clarify the unequivocal distinctions among four principal categories of formally named stratigraphic units (time-stratigraphic, time, lithostratigraphic, and biostratigraphic) and has resulted in a classification (Table 3-3) whose utility involves the organization of the stratigraphic record so that geologists can communicate and nomenclature is consistent. The 1983 edition of the North American Stratigraphic Code has twelve categories of formally named stratigraphic units, including magnetostratigraphic, intrusive igneous and metamorphic, unconformity-bound, and geochronometric as additions to the traditional units of Table 3-3.

Time-stratigraphic units are intervals of strata formed during specific intervals of time and demarcated by *isochronous* (synchronous) boundaries as defined by synchronous events in the fossil record.

Geologic **time units** are abstract units that correspond to time-stratigraphic units (Table 3-3). The time concept of the Cretaceous **Period,** for example, is based on the tangible time-stratigraphic record of the Cretaceous System. Time terms are necessary to refer accurately to historical events and circumstances.

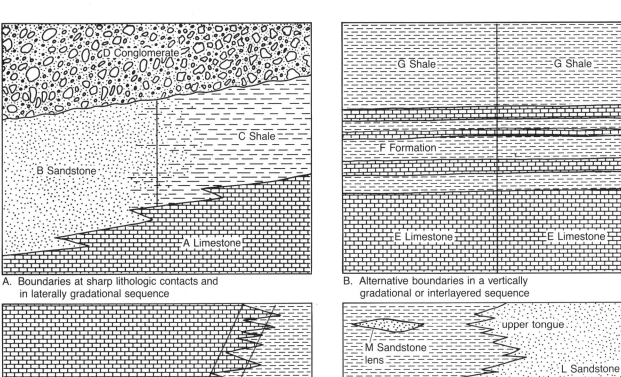

A. Boundaries at sharp lithologic contacts and in laterally gradational sequence

B. Alternative boundaries in a vertically gradational or interlayered sequence

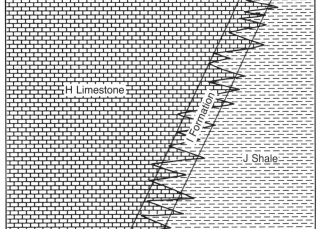

C. Possible boundaries for a laterally intertonguing sequence

D. Possible classification of parts of an intertonguing sequence

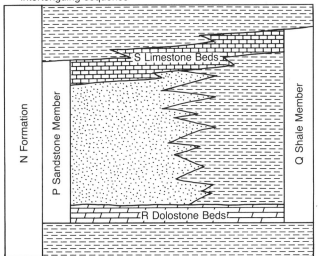

E. Key beds, here designated the R Dolostone Beds and the S Limestone Beds, are used as boundaries to distinguish the Q Shale Member from the other parts of the N Formation. A lateral change in composition between the key beds requires that another name, P Sandstone Member, be applied. The key beds are part of each member.

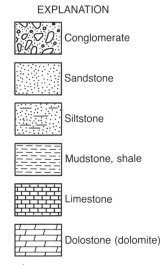

EXPLANATION

Conglomerate

Sandstone

Siltstone

Mudstone, shale

Limestone

Dolostone (dolomite)

Figure 3-18. Lithostratigraphic boundaries and classification.
(From North American Commission on Stratigraphic Nomenclature, 1983, North American Stratigraphic Code, Fig. 2, p. 857: *Am. Assoc. Petroleum Geologists Bull.,* vol. 67, no. 5. Reproduced by permission of American Association of Petroleum Geologists)

A third class of formal stratigraphic unit, called **lithostratigraphic,** is based on the physical (nonbiological) aspects of rocks. The fundamental unit in this category is the **formation,** defined as a body of strata that has characteristic properties by which it can be identified. This definable character is based on mappable physical criteria such as color, mineral composition, texture, thickness and geometry of contained beds, sedimentary structures, and topographic expression. Fossils are included in the set of identifying criteria only if they are used as lithologic adjuncts (e.g., a fossiliferous limestone formation) with no regard paid to the ranges of individual species. Formations are formally named for geographic locales where the stratotypes are defined. In terms of lithologic homogeneity, formations are descriptive units, recognized for their individuality because they are different from adjacent bodies of rock. They represent parts of or complete sedimentary *facies,* and their boundaries, also defined on physical criteria, may be sharp or gradational (Fig. 3-18). Formations are the fundamental stratigraphic units on geologic maps of conventional ($7^{1}/_{2}$- and 15-minute) quadrangle scale. They are units of convenience, designated to aid in organization and communication of the stratigraphic record in a physical sense. Formations commonly are subdivided into **members** (Table 3-3) or, conversely, may be lumped together into **groups.** Some closely related groups are combined into very thick **supergroups.** Such procedure and practice enhance recognition and correlation of lithostratigraphic units.

A lithostratigraphic unit of any rank can be recognized away from the type section. More precisely stated, isolated outcrop or subsurface sections can be assigned to the same lithostratigraphic unit if it can be demonstrated that they were once part of the same contiguous body of rock and bear sufficient similarity to the stratotype (Fig. 3-18).

The boundaries of lithostratigraphic units, selected on lithologic criteria, are not intended to be isochronous. Such units are fundamentallly different from the interpretive time-stratigraphic units and are defined by different criteria. This fundamental distinction may seem obvious to you, but the tendency of many stratigraphers to recognize formations on fossil content or to consider formations and groups as time-stratigraphic units has resulted in much poor stratigraphic practice and has prompted the necessity of making the distinction clearer.

A fourth category of formal stratigraphic classification is the **biostratigraphic unit** (Table 3-3). The fundamental unit is the **biozone** (see Chapter 5), which is simply a body of rock whose boundaries are defined as fossil content. Most biozones are formally named after a particular characterizing species. Biozones are descriptive units; however, a biozone that has time-stratigraphic utility is an interpretive unit and the designation **chronozone** is commonly used (Table 3-3). As originally defined, such biozones are the smallest formal times stratigraphic units that can be recognized. Again, biozones are descriptive units first. However, certain biozones can be interpreted to have chronostratigraphic value; that is why the distinction is necessary between time-stratigraphic and biostratigraphic units.

The Meaning of Correlation

The classification of stratigraphic units has evolved as a result of the progressive awareness by stratigraphers that there are fundamentally different kinds of units: different rock types (facies) can be found within a time-stratigraphic unit; formations can be of different ages in different places; and only those biostratigraphic units that have become interpretive in a time sense can be designated time-stratigraphic units. This philosophy of unequivocal distinction among the kinds of stratigraphic units has important overtones for correlation. Correlations can be based on material units and on temporal and related chronostratigraphic units (Table 3-3). Most practical time-stratigraphic correlations are biostratigraphic in nature, but they are interpretive biostratigraphic, whereby fossils are used in an interpretive sense to establish time equivalency (Fig. 3-19). Lithostratigraphic correlation refers to the matching of physical lithologic units which may be, but most generally are not, time equivalent (Fig. 3-19).

The fundamental principles of stratigraphy (Chapter 2) and stratigraphic classification apply to subsurface sequences as well as to surface outcrop exposures. Subsurface geology presents some special problems because the rocks are seldom observed directly. Countless

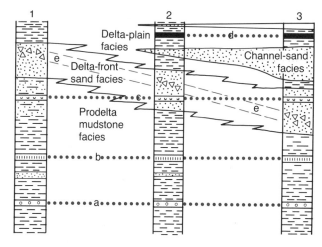

Figure 3-19. Time-stratigraphic correlation with interpreted time-significant fossil horizons (a, b) and with blanketlike geologically "instantaneous" volcanic ash beds (c, d). Note that the correlated lithofacies boundaries cross the time lines defined by the time-stratigraphic markers. Note also that the biostratigraphic correlation (e) is not time stratigraphic—probably a result of the lithofacies control on the distribution of the fossils.

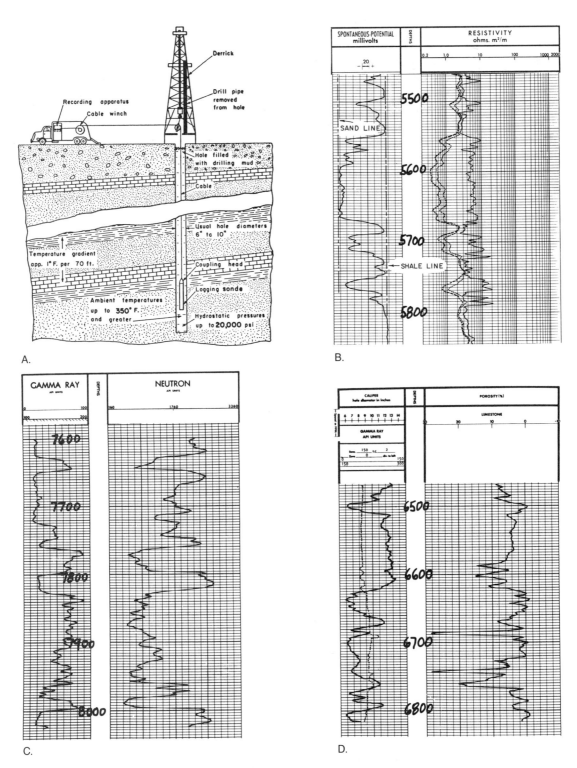

Figure 3-20. A. Instrument truck logging a well. Specially equipped truck lowers electrical recording device (sondes) by cable down the borehole. As sondes are slowly raised back to the surface, electrical impulses are recorded on a sensitive drum and fed to a computer, which gives a printout as in B. B. Portion of an electric well log showing lithology in a sandstone–shale sequence. The spontaneous potential curve is a recording versus depth of the difference between the potential of a movable electrode in the borehole and the fixed potential of a surface electrode. The resistivity curve is a measure of the resistivity of contained fluids in strata in the borehole. C. Portion of a well log showing natural radioactivity (gamma ray curve) and amount of hydrogen (neutron curve) in strata in borehole. D. Portion of well log showing borehole diameter (caliper) and porosity of strata in walls of borehole. Such properties simulate lithology and provide important information about the fluid content of subsurface strata.

(A from F. Segesman, S. Soloway, and M. Watson, 1962, Well logging—the Exploration of Subsurface Geology, p. 2228: *Proc. IRE,* vol. 50, no. 11; B–D modified from Schlumberger Ltd., 1972, *Log Interpretation,* vol. 1, *Principles,* pp. 7, 50)

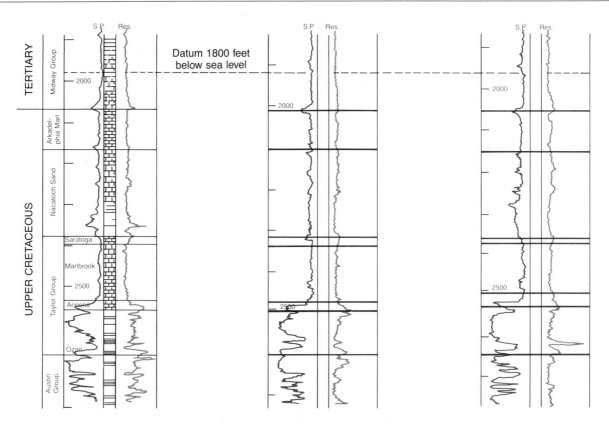

Figure 3-21. Subsurface correlation of upper Cretaceous and Paleogene rocks in southern Arkansas using electric logs from wells.
(From Peterson, Morris S., and J. Keith Rigby, INTERPRETING EARTH HISTORY, 3d ed. (c) 1982, Wm. C. Brown Publishers, Dubuque, Iowa. All Rights Reserved. Reprinted by permission)

thousands of wells have probed the subsurface realm, mainly to depths of less than 10,000 meters below the surface, in the search for and production of oil and gas. Well cuttings, cores, and various kinds of electrical and radioactivity logs (Fig. 3-20) have provided a twentieth-century dimension of stratigraphic information not available to the pioneer architects of the geologic time scale. Logs can be matched from locality to locality just as surface sections are matched (Fig. 3-21). Subsurface basin analysis and correlation have been greatly enhanced also by the application of seismology. Seismic profiles such as the one shown in Figure 3-22 have greatly facilitated the interpretation of stratigraphy and structure of subsurface sedimentary basins within and marginal to the present-day continents.

Although formal classification has facilitated necessary and improved organization of the stratigraphic record and consistency in terminology, the most important advances in sedimentary geology, from the standpoint of understanding sedimentary processes and natural stratigraphic patterns, have been in the area of informal stratigraphy. This has involved the development of sophisticated models for analysis of sedimentary facies and for interpretation of ancient environments. Here the emphasis is on genetic units rather than on formal ones,

although formal stratigraphic classification provides the framework for correlation and analysis. Much of this work has centered around developing models for exploration for oil and gas and other mineral deposits—the rationale being that formation of many economic mineral deposits is critically influenced by depositional and postdepositional environments. Sedimentary facies and ancient environments are discussed in Chapter 6.

Uniformitarianism and Actualism

In the year 1830, while the modern geologic time scale was beginning to take shape, a milestone volume in geology was published: *Principles of Geology.** Its author, Charles Lyell, clearly and forcefully presented all the documentation he could marshall in support of the doctrine of uniformitarianism. James Hutton's view of the earth and Playfair's brilliant clarification of it were slow in being totally acceptable. Lyell launched a crusade to lay to rest once and for all the idea that the earth and all things on it were the product of divine creation. His attack advanced on two fronts: (1) establishment of uniformitarianism at the expense of catastrophism as the

---

*J. Murray, London.

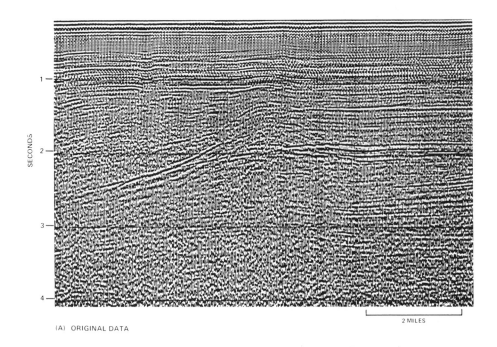

(A) ORIGINAL DATA

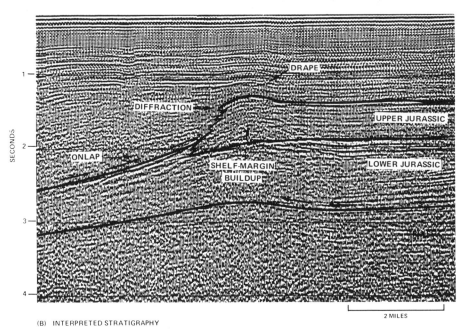

(B) INTERPRETED STRATIGRAPHY

Figure 3-22. A. Reflection seismogram panel. Energy pulses, whose travel times are computed, are reflected from various stratigraphic horizons. The computations are computerized and printed out. Note that the depth dimension is in seconds, which correspond to various depths based on rock type. Remember from Chapter 1 that seismic wave velocity is a function of lithology, which determines density—the more dense the rock, the faster the travel time. B. Interpreted stratigraphy.
(From N. J. Bubb, and W. G. Hatfield, 1977, Seismic Stratigraphy and Global Changes of Sea Level, part 10: Seismic recognition of carbonate build-ups, Fig. 6, p. 191: *Am. Assoc. Petroleum Geologists Mem.* 26. Reprinted by permission of American Association of Petroleum Geologists)

acceptable philosophy for interpreting the history of the earth and (2) establishment of geology among the sciences as a discipline based on inductive principles. Lyell gave the uniformitarian theory its biggest boost, and by effectively reintroducing the concept of unlimited time,

he founded modern historical geology. Uniformitarianism then successfully nourished progress in historical geology.

However, as so often happens when one philosophy displaces another, the pendulum swung a bit too far

Figure 3-23. The Natural Bridge of Virginia. Thomas Jefferson described this "natural wonder" as having formed during a sudden convulsion of nature, a view compatible with the doctrine of catastrophism that reigned during the late eighteenth century. In reality, the natural bridge was formed over a period of time by the action of running water, perhaps punctuated by natural catastrophes in the form of intense storms and floods.
(Photo by J. Cooper)

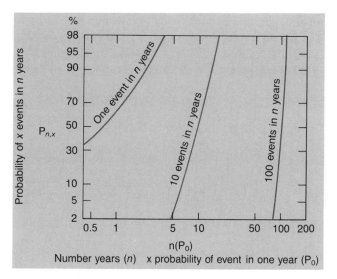

Figure 3-24. Plot showing probability ($P_{n,x}$) of a rare event to occur at least x times in n years. P_O is the probability that the event will take place in a single year. Such a plot lends credence to the certainty of occurrence of natural catastrophes (volcanic eruptions, earthquakes, intense storms, and floods) during certain intervals of geologic time. Such episodic events have been instrumental in the evolution of the earth's lithosphere.
(From P. E. Gretener, 1967, Significance of the Rare Event in Geology, Fig. 1, p. 2198: *Am. Assoc. Petroleum Geologists Bull.*, vol. 51, no. 11. Reproduced by permission of American Association of Petroleum Geologists)

from catastrophism to uniformitarianism. Lyell, in refuting the catastrophist view of earth history, argued, quite rightly, for the invariability of natural laws. Unfortunately, he also argued for uniformity in rate of change. He held steadfast to the notion that rates of change or relative importance of geologic events had never been different from what they were perceived to be within the context of human experience. This strict interpretation of uniformitarianism stressed uniformity in rate and magnitude and somewhat overstated the theme of "the present is the key to the past."

Today the basic tenets of uniformitarianism are viable: The present does give us insight into past history, and the physical, chemical, and biological laws that govern activity and change are viewed as permanent. However, we no longer demand an earth model where rates, intensities, and relative importance of processes have prevailed unwaveringly through geologic history. Many

earth scientists have dropped the word uniformitarianism and have adopted in its place the term **actualism,** in accordance with viewing the earth in terms of actuality (Fig. 3-23). Actualism removes the "uniformity" from uniformitarianism and views the earth in the more realistic perspective of the *geologic cycle* (see Chapter 1).

Lyellian uniformitarianism lulled geologists into accepting an earth undergoing gradual, almost imperceptible change. It was assumed that major changes resulted only from accumulation of small changes. Today most earth scientists believe in punctuational change and appreciate that uniformity of natural law does not preclude *natural* catastrophes or *episodes* such as floods, storms, volcanic eruptions, asteroid impacts, and others. Actualism accommodates such natural punctuations in geologic history (Fig. 3-24). As Derek Ager, a British geologist, writes: "The history of any one part of the earth, like the life of a soldier, consists of long periods of boredom and short periods of terror."*

Rewarding applications of actualism have involved, among others, the study of present-day depositional environments and their deposits in order to better understand ancient facies patterns (see Chapter 6). Actualistic models provide first-hand knowledge of environmental processes and products. What better way, for example, to appreciate how an ancient limestone deposit formed

---

*1981, pp. 106–107.

than to go to a tropical paradise such as the Bahamas (Fig. 3-25) and monitor a calcium carbonate-producing environment in action. The Bahamas provide an instructive model for beginning to realize what the midconti-

nent of the United States might have been like during the Late Ordovician. Actualism also underscores the much greater magnitude of calcium carbonate production during the Late Ordovician than now (Fig. 3-26). Yet

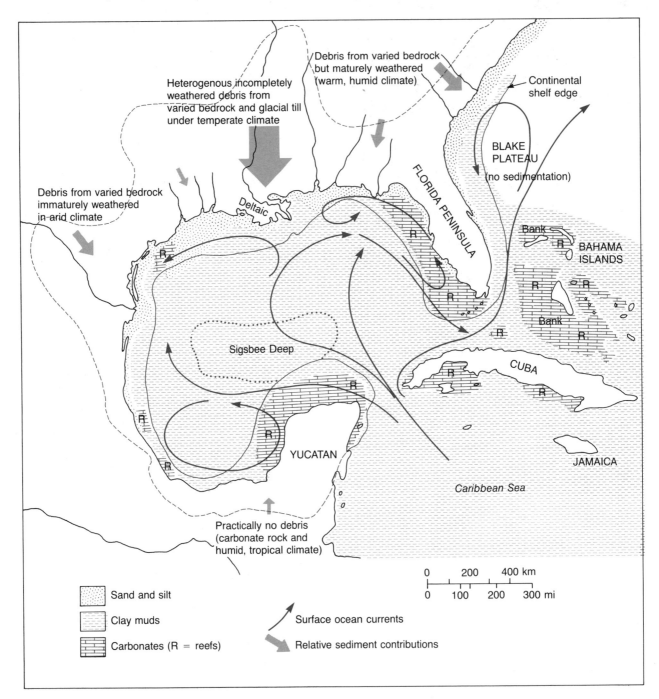

Figure 3-25. Map of present-day southeastern United States and Gulf of Mexico–Caribbean region showing major sediment dispersal patterns. Most of the sediment being deposited in the marine environments is terrigenous clastic, derived from erosion of continental land masses and delivered by major trunk streams. Carbonate sediments are restricted to places such as the Bahama platform and the west Florida shelf that are relatively free of terrigenous influx. A strict uniformitarian view would hold that such a situation has existed throughout earth history. Reference to Figure 3-26 shows this is not the case.

(From R. H. Dott, Jr., and R. L. Batten, *Evolution of the Earth,* 3rd ed., Fig. 11.22, p. 247: Copyright 1981 by McGraw Hill Book Company. Reproduced by permission of publisher)

another application of actualism is the appreciation of varying accumulation rates of sedimentary strata in different environments of deposition (Fig. 3-27).

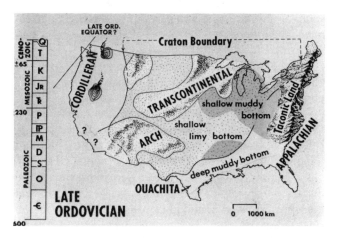

Figure 3-26. Paleogeographic map of U.S. during Late Ordovician time. Compare this situation of widespread shallow-marine carbonate deposition within continental platform areas covered by shallow seas with the present-day model shown in Figure 3-25. Although the geochemical, physicochemical, and biochemical laws governing deposition of carbonate sediments have changed little, if any, the rates, magnitudes, and intensities of carbonate sedimentation have changed the actualistic view.
(Courtesy of Critter Creations, Inc., San Diego CA)

Age of the Earth: A Uniformitarium Approach

The eminent Lord Kelvin, perhaps the most esteemed physicist of the nineteenth century, proposed estimates of geologic time expressed with elaborate and impressive mathematical detail. His estimates, based on calculations of heat loss from the earth, were always less than 100 million years. Kelvin's calculations, employing physical measurements, used known temperature gradients from the earth's surface into the progressively warmer subsurface (as observed in deep mines in many areas) to arrive at the rate of heat loss from the earth's interior to the surface. This rate of heat flow and subsequent loss was then extrapolated back in time to when Kelvin believed the earth to have been in a completely molten condition—less than 100 million years ago.

Lord Kelvin and a number of other scientists believed that the sun's energy was burning out at a uniform rate. Lord Kelvin himself reasoned that the amount of solar energy reaching the earth had been much greater only a few tens of millions of years ago and that several million years hence it would be much less than now. Accordingly, the amount of time available for the earth to have been sufficiently cooled to support life was judged to be between 20 and 40 million years, much

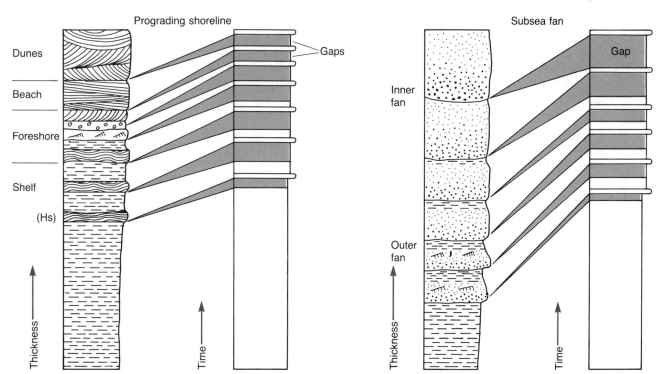

Figure 3-27. Two depositional models emphasizing the importance of gaps in depositional sequences and illustrating the disparity of time represented by different types of strata. Note that gaps represent more than preserved rock in the upper halves of both sequences. Such gaps are hiatuses resulting from diastems.
(From R. H. Dott, Jr., 1983, Episodic sedimentation—how normal is average? Fig. 3, p. 8: *Jour. Sedimentary Petrology*, vol. 53. Reproduced by permission of Society of Economic Paleontologists and Mineralogists)

less than Darwin needed to defend his theory of evolution. Darwin's biological reasoning about geologic time was mostly intuitive and not amenable to quantitative verification. Paleontologists and evolutionary biologists could offer only qualitative estimates in reply to Kelvin's calculations on heat dissipation from the earth and on the sun's history. By the turn of the century, most geologists reluctantly adjusted their ideas on the age of the earth to conform more with Kelvin's estimates.

Age of the Earth: Radioactivity and Radiometric Dating

No principle or theory is any more valid than the basic assumptions on which it is founded. As it turned out, the seemingly vague intuitive hunches of the evolutionists would be vindicated; the tide would turn against the quantitative proofs of Lord Kelvin. One of the first serious challenges to Lord Kelvin's elegantly quantified experiments on the age of the earth was presented in 1899 by T. C. Chamberlin, professor of geology at the University of Chicago. Chamberlin had been working on the hypothesis that the earth and other planets of our solar system had never been completely molten but rather had accumulated from small cold pieces that he called *planetesimals*. In addition he threw open the question about whether the internal constitution and energy aspects of atoms were completely known. The thrust of this argument concerned the thermal possibilities for keeping the sun alive for much longer than Lord Kelvin had calculated. This turned out to be a rather prophetic good guess about the nuclear energy source for the sun.

Prior to Chamberlin's postulation, Henri Becquerel discovered the radioactive property of uranium. In 1903, Pierre and Marie Curie discovered that radioactive radium was a source of heat. Three years later, R. J. Strutt, in England, showed that the amount of helium in uranium minerals was larger than what would have accumulated from alpha decay within the time frame of Lord Kelvin's calculations. Strutt estimated the quantity of heat that is continuously generated by radioactive minerals in the earth's crust and showed that this could easily account for the flow of heat from the surface. This was indeed a monumental scientific breakthrough. With it came the realization that the heat escaping from the earth's surface no longer needed to be considered residual; instead, it is always being produced within the earth by radioactive decay. The fact that heat loss from the earth's surface probably had been about the same for a very long time totally destroyed the accuracy of Kelvin's measurements. In 1906, Lord Rutherford, in England, made the first attempt to measure the age of minerals from their helium–uranium ratio. This was the real beginning of the science of *geochronology*. Quan-

titative measurements suggested that the earth was considerably older and that the length of geologic time during which the earth was capable of supporting life was considerably greater than Lord Kelvin's estimates. In spite of the overwhelming evidence, Lord Kelvin remained unimpressed; he died in 1907 convinced it was all wrong.

In 1907, Bertram Boltwood, an American chemist at Yale University, discovered that lead was also a stable end product of uranium decay. He demonstrated that the ratio of daughter lead to parent uranium increased with increased age of the mineral. He collected a large amount of data on published analyses of pure uranium minerals and calculated age from lead content. His uranium half-life value of 10^{10} years was more than twice that known at present, the uranium decay series was still incompletely known, and chemical techniques were severely limited, but Boltwood's age determinations clearly established that the geologic time scale was very long.

In 1911, Arthur Holmes, a brilliant young English geologist and student of Strutt, summarized all available data on radiometric age determinations, presented evidence that radioactive decay rates are constant for particular nuclides, and pointed out the tremendous potential of radiometric dating for unraveling pre-Cambrian history. He also clearly predicted that periods of the relative geologic time scale might one day be defined by accurate radiometric dates. Years later, he published such a scale, with subsequent revisions, and was one of the most instrumental scientists in combining the relative and radiometric time scales.

A major breakthrough came in the late 1930s with the invention of the *mass spectrometer,* which permitted the mass spectra of uranium and lead to be clearly resolved. This allowed for calculation of the uranium-238:uranium-235 ratio, the decay rate of the rarer uranium isotope uranium-235, and the ratio of lead-206:lead-207. One of the main byproducts of this discovery was measurement of the age of the earth. Our present knowledge about the age of the earth is derived directly from interpretations made during the 1940s of the primordial ratios of lead-207:lead-206:lead-204.

All common lead contains a blend of four lead isotopes: lead-204, lead-206, lead-207, and lead-208. Fortunately, in most minerals used for radiometric dating, the proportions of these lead isotopes are nearly constant. Although the quantities of radiogenic lead-206 and lead-207 have increased during geologic time (through decay of uranium-238 and uranium-235) in relation to unchanging nonradiogenic lead-204, these three lead isotopes have changed progressively in ratio with respect to one another because of the different decay rates of their parent isotopes. Regarding the age of the earth, we must know what the lead-206:lead-207 isotopic ratio was at the time the earth formed. Extrapolation of ratios back in time to the point where, theoretically, there was no lead-206 or lead-207 shows a date of 5600 million years ago for zero lead-207 and 6800 million years ago for zero lead-206 (Fig. 3-28). The discordance of these dates indicates that some portion of the earth's accumulated lead-206 and lead-207 was furnished along with other nonradiogenic nuclides when the earth formed. Thus the earth must be younger than 5600 million years (the theoretical time of accumulation of lead-207) yet older than the oldest lithospheric rocks (gneisses from Greenland), which have yielded radiometric dates slightly younger than 4000 million years. Data from iron and stony *meteorites,* which are believed to represent the primordial materials from which our solar system is made (and thus are the same age as the earth), have filled the gap. Meteorite samples containing trace quantities of lead but no uranium have been compared to those containing both lead and uranium (mixtures of primordial lead and radiogenic lead). Thus primordial lead is shown to have isotopic ratios of 1.0 for lead-204, 9.3 for lead-206, and 10.3 for lead-207, indicating an age of 4600 million years on lead-evolution curves (Fig. 3-29).

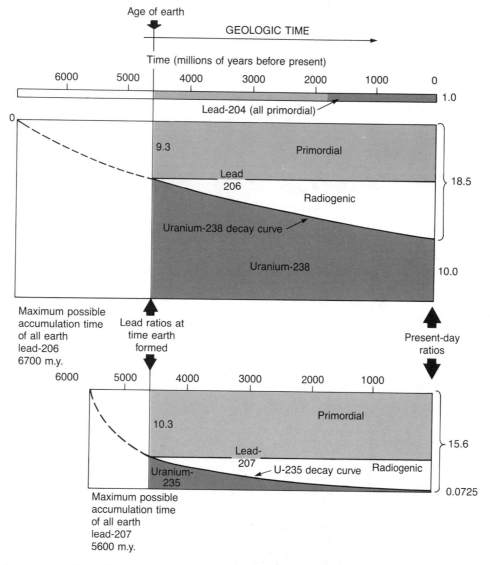

Figure 3-28. The radioactive decay of the earth's uranium has added significant lead-206 and lead-207 and has changed their proportions throughout geologic time. Left margin of lead-207 curve shows that it cannot have been accumulating for more than 5600 million years and, similarly, lead-206 cannot have been accumulating for more than 6700 million years. Indicated ratios are based on lead-204 = 1.0. Extrapolating back to the appropriate lead ratios at the time the earth formed gives an age of approximately 4600 million years.
(From Don L. Eicher, GEOLOGIC TIME, 2nd ed. © 1976, Fig. 6-11, p. 137. Reprinted by permission of Prentice-Hall, Inc., Englewood Cliffs, NJ)

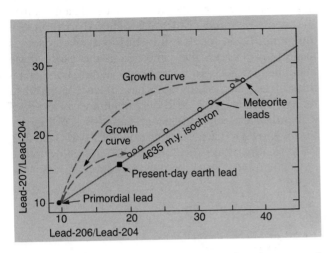

Figure 3-29. Lead isochron diagram for meteorites gives an age of 4635 million years. Present-day earth lead falls on the isochron, indicating it came from the same primordial meteorite source and at the same time.
(From G. R. Tilton, 1973, Isotopic Lead Ages of Chondritic Meteorites, p. 325: *Earth and Planetary Science Letters,* vol. 19. Reprinted by permission of Elsevier Science Publishers)

Supercharged research on radioactivity during World War II boosted nuclear analytical technology to a point where the very minute quantities of radiogenic daughter products in radioactive minerals could be handled. Accurate mass spectrometers and other sophisticated tools were developed and refined. After the war, when information pertaining to radioactivity was not so restricted, it became possible to measure accumulation of argon from the decay of potassium and accumulation of strontium from the decay of rubidium for determination of the age of ancient rocks. By the early 1950s laboratories for isotopic age determination were established in research centers in the United States and abroad. By the late 1950s, isotopic age determinations on rocks securely tied to the fossil record were becoming more and more available, and refinements have continued to the present, highlighted by radiometric dating of lunar samples and of the oldest known rocks on earth (Fig. 3-30A).

In 1947, Nobel laureate W. R. Libby and associates devised radiocarbon dating, which provided the first

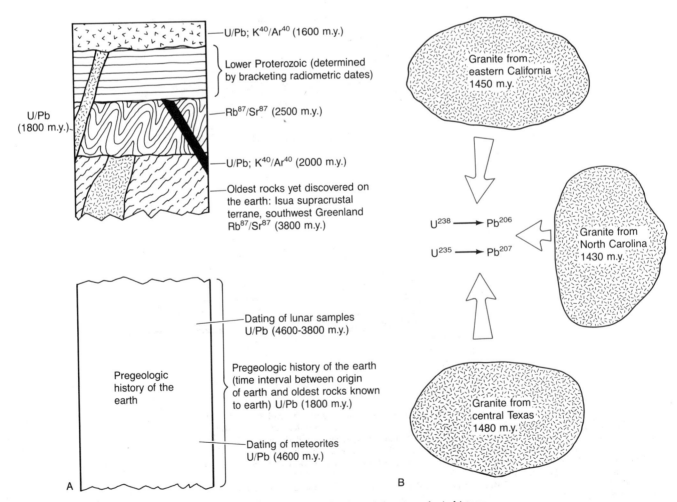

Figure 3-30. A. Applications of radiometric dating of pre-Cambrian rocks and the pregeologic history of the earth. B. Radiometric correlations between separated pre-Cambrian terranes.

technique for obtaining absolute dates from the upper-most part of the geologic column. Subsequently, techniques involving thorium-230, and protactinium-231, as well as amino acid racemization (not a radiometric method) and fission-track dating have allowed filling of significant age-determination gaps between 50,000 years B.P. (the lower limit for radiocarbon dating) and about 3 million years B.P. (the upper limit of potassium-argon dating). Thus quantitative dating techniques have spanned the entire geologic column.

Not only has the network of radiometric dates improved the radiometric time scale for the Phanerozoic part of the geologic column; it has facilitated age correlations within the very complex pre-Cambrian rocks (Fig. 3-30B), which are widely distributed in many parts of the world and comprise more than 15 percent of the surface exposures. As radiometric dating became a reality, we soon learned that the pre-Cambrian was a much longer geologic time interval than previously had been estimated.

James Hutton would probably be delighted to know that we have found some "vestige of a beginning": We have reliable dates on the age of the earth (4600 million years), on the oldest exposed rocks of the earth's lithosphere (3800 million years), and on lunar samples (dates ranging between 3800 and 4600 million years). We also have 4200 million year dates on detrital zircon grains from pre-Cambrian sandstone in Australia, which suggests the erosion of granite-like crust older than the oldest known preserved crust.

Magnetostratigraphy

Radiometric dating paved the way for the principle of plate tectonics, perhaps the greatest conceptual breakthrough in the history of geology. Radiometric age determination of volcanic rocks on land and in the ocean basins, combined with measurements of the remnant magnetization in those same rocks, has shown that the earth's magnetic polarity has reversed itself frequently in the geologic past, that oceanic lithosphere has moved away from sea-floor-spreading centers, and that land masses have rotated and moved with respect to the poles. This discovery inaugurated a new era in human understanding of the earth and its history.

Figure 3-31 is a *geomagnetic time scale* for the last 160 million years. This scale shows *magnetopolarity* events indelibly imprinted in diverse assemblages of rocks: terrestrial lava flows, oceanic basaltic lithosphere, and even deep-sea sediments. Radiometric dating has provided an absolute time frame for the magnetopolarity events, whose time intervals are called *polarity chronologic units*. In deep-sea sediments the time value of the polarity chronologic units has also been verified in many places by their constant relationship to time-stratigraphic units based on fossils (chronozones). Thus

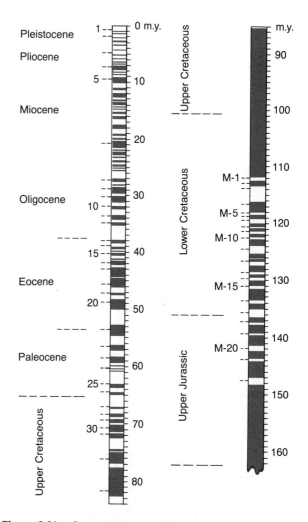

Figure 3-31. Geomagnetic reversal time scale from the present to the beginning of the Late Jurassic, 162 m.y. B.P. (From Robert L. Larsen, and Walter C. Pitman III, 1972, World-wide Correlation of Mesozoic Magnetic Anomalies, and Its Implications, Fig. 5, p. 3651: *Geol. Soc. America Bull., vol. 83.* Reprinted by permission of authors and Geological Society of America)

remnant magnetism is yet another property of rocks that shows time sequence and permits correlation. If particular magnetic events can be identified and related to other means of correlation, we have a powerful tool for correlating and dating deep-sea sediments on a world-wide scale (Fig. 3-32).

With today's sophisticted and highly sensitive magnetometers, remnant magnetism is being determined for many oceanic stratigraphic sequences. For example, a geomagnetic *stratotype* for the Cretaceous–Paleogene boundary has been designated near Gubbio, Italy (Fig. 3-33). Here an exposed section of deep marine limestone, representing continuous deposition and with excellent microfossil control and well-defined microfossil biozones, has yielded exceptionally good paleomagnetic data, facilitating correlation with sections in several ocean basins. Additional stratotypes utilizing polarity chronostratigraphy no doubt will be recognized for other parts of the Mesozoic and Cenozoic record.

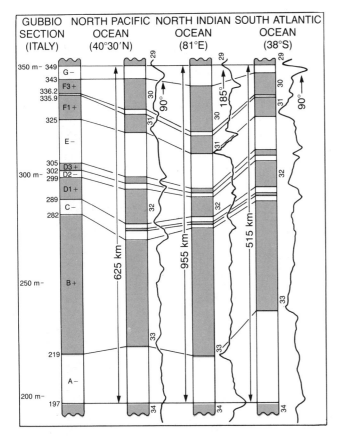

Figure 3-32. Comparison of the Late Cretaceous geomagnetic polarity sequence of Gubbio, Italy, with marine magnetic profiles and interpreted geomagnetic polarity sequences from three oceanic areas. Numbers represent normal polarity epochs.

(From William Lowrie, and Walter Alvarez, 1977, Upper Cretaceous-Paleocene Magnetic Stratigraphy at Gubbio, Italy, III. Upper Cretaceous Magnetic Stratigraphy, Fig. 2, p. 376: *Geol. Soc. America Bull.,* vol. 88. Reprinted by permission of the authors and Geological Society of America)

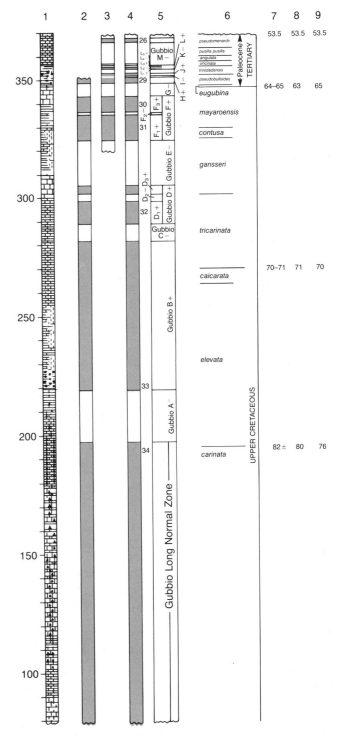

Figure 3-33. Results of stratigraphic studies at Gubbio, Italy. Column 1: Measured stratigraphic section showing lithostratigraphy. Columns 2 and 3: Magnetic results (black is normal polarity, white is reversed). Column 4: Combined magnetic results. Column 5: Designation of polarity zones (+, normal; −, reversed). Column 6: Planktonic foraminiferal (microfossil) biozones. Columns 7, 8, 9: Absolute age calibration according to proposed time scales.

(From Walter Alverez etal., 1977, Upper Cretaceous-Paleocene Magnetic Stratigraphy at Gubbio, Italy, V. Type Section for the Late Cretaceous-Paleocene Geomagnetic Reversal Time Scale, Fig. 2, p. 386–387. *Geol. Soc. America Bull.,* vol. 88. Reprinted by permission of authors and Geological Society of America)

SUMMARY

Although the science of historical geology has its roots in ancient Greece, the first stratigraphic principles were formulated in the last half of the seventeenth century by Nicolas Steno in Italy. The principles of original horizontality, lateral continuity, and particularly superposition paved the way for development of the geologic time scale. Earliest attempts at subdivision of the rocks of the earth's crust were motivated by the economic incentive of mining activity in western Europe. These subdivisions were based on superposition of stratified, younger looking sedimentary rocks and supposedly underlying, older looking, more complex igneous and metamorphic assemblages. Although such simplified subdivisions represented monumental achievements for the time and worked for local stratigraphic sequences, they were destined for ultimate failure as universal schemes for telling geologic time. The fallacy that all crystalline rocks formed at one time and all stratified rocks were formed at a later time continued to be perpetuated through the eighteenth century, mainly by Werner and his followers. The religious climate of the time was one of strict adherence to the notion of a 6000 year old earth—one that had been shaped by various catastrophes, such as the Noachian flood, wrought by a provident creator.

The legitimate father of modern geology was James Hutton, a Scotsman, who took a more realistic, albeit heretical, view of the earth by placing no time constraints on the age of the earth and by invoking natural laws rather than supernatural catastrophes to explain geologic history. "The present is the key to the past" became the slogan for the principle of uniformitarianism, which slowly displaced catastrophism as the guiding philosophy of the geological sciences.

Early in the nineteenth century, William Smith in England and George Cuvier in France independently articulated the principle of fossil succession. During the early decades of the nineteenth century in western Europe, application of fossil succession withstood the test of repeated observations and became the key that unlocked the door to development of a chronology of earth history.

The principle of fossil succession led gradually to the recognition of large-scale stratigraphic units containing unique aggregates of fossils upon which the stratigraphic systems—the major subdivisions of the modern time scale—were based. By the beginning of the twentieth century, the modern relative geologic time scale was firmly established as a scale based on a succession of fossiliferous units, each with its own stratotype and unique aggregate of fossils. The stratigraphic systems were subsequently lumped into three larger scale aggregates corresponding to three major chapters in the history of multicellular life: Paleozoic, Mesozoic, and Cenozoic Eras. More precise, refined stratigraphic correlations necessitated subdivision of the systems into smaller increments, including series, stages, and zones—all based on aggregates of fossils. Sir Charles Lyell, the father of modern historical geology, subdivided the Paleogene and Neogene Systems of the Cenozoic into series based on the relative percentages of fossil species still living today. He is also responsible for putting uniformitarianism on solid footing as the guiding principle of geology.

Twentieth-century refinements of the geologic time scale and stratigraphic principles and practice include, among others: stratigraphic classification, delineation of subsurface stratigraphy, facies analysis, and radiometric dating. Classification of formal stratigraphic units was necessary to avoid confusion and to clearly distinguish between material units—lithostratigraphic, biostratigraphic, and time-stratigraphic—in order to facilitate organization of the stratigraphic record and to establish consistency in terminology and communication.

Principles of stratigraphy have been applied to the subsurface as well as to outcrop sections. Exploration of subsurface sedimentary basins has been in response to a twentieth century economic incentive: the search for oil and gas. Deep drilling and seismic studies have delineated facies and formal stratigraphic units. Abandonment of strict Lyellian uniformitarianism has freed geologists to take a more realistic posture in the interpretation of geologic history. Actualism allows a more liberal application of "The present is the key to the past" by accepting changes in rate, magnitude, and intensity of geologic processes. It also accommodates the episodic event, which gives earth history a punctuated tempo.

The greatest twentieth century contribution to the geologic time scale is the radiometric scale, which has provided an understanding and appreciation of the antiquity and duration of the relative subdivisions and the age of the earth. The discovery of radioactive decay and its application to mineralogy launched the science of geochronology and soon demonstrated the age of the earth to be at least an order of magnitude greater than turn-of-the-century estimates of about 100 million years. Continued advances in age dating highlighted by the development of the mass spectrometer promulgated a vast network of uranium–lead, rubidium–strontium, and potassium–argon dates from throughout the world, resulting in tighter bracketing of the relative time-scale subdivision boundaries and assignment of absolute dates. Radiocarbon dating together with other techniques applicable to the late Neogene (last 3 million years) have completed the spanning of the geologic column with absolute dates that show greater reliability than thought possible.

Remnant magnetism has proved to be another property of rocks that shows time sequence and is amenable to global correlation. With the advent of plate tectonics theory, paleomagnetic studies on land and across the ocean basins have resulted in a reliable late Mesozoic and Cenozoic geomagnetic time scale whose magnetopolarity units can be plugged into the radiometric and relative times scales. Stratotypes for various parts of the upper Mesozoic and Cenozoic stratigraphic column combine magnetopolarity units with traditional chronostratigraphic units utilizing fossils to provide powerful correlation tools and refined land–land, ocean–ocean, and ocean–land global correlation for the last 100 million years of earth history.

SUGGESTIONS FOR FURTHER READING

Ager, Derek, 1981, *The Nature of the Stratigraphical Record,* 2nd ed.: John Wiley & Sons, New York.

Berry, W. B. N., 1968, *Growth of a Prehistoric Time Scale:* W. H. Freeman, San Francisco.

Eicher, D. L., 1976, *Geologic Time:* Foundations of Earth Science Series (paperback), Prentice-hall, New York.

Eisley, Loren, 1969, *Charles Lyell: Sci. American* Offprint No. 846, W. H. Freeman, San Francisco.

Faul, Henry, 1978, A History of Geologic Time: *Am. Scientist,* vol. 66, pp. 159–165.

Geikie, Archibald, 1905, *The Founders of Geology:* New Dover ed., 1962, Dover, New York. Unabridged and unaltered republication of 2nd. ed. (1905) of work first published by Macmillan, New York, in 1897.

Gillispie, Charles C., 1959, *Genesis and Geology:* Harper Torchbook ed., Harper, New York. Originally published as vol. LVIII of Harvard Historical Studies, 1951.

Gould, S. J., 1965, Is Uniformitarianism Necessary?: *Am. Journ. Science,* vol. 263, pp. 223–228.

4

DOCUMENTS OF LIFE: Preservation of Fossils and the Spectrum of Life

CONTENTS

KEY TERMS

Carbonaceous film
Permineralization
Recrystallization
Replacement
Mold
Cast
Trace fossil
Paleontology
Kingdom
Moneran
Protistan
Stromatolite
Foraminifer
Radiolarian
Plant
Diatom
Animal
Invertebrate
Vertebrate
Phyla
Sponge
Coral
Septa
Stromatoporoid
Archaeocyathid
Bryozoan
Brachiopod
Mollusc
Bivalve mollusc
Gastropod
Cephalopod
Suture
Arthropod
Trilobite
Echinoderm
Crinoid
Echinoid
Graptolite
Conodont

THE WAY IT WAS

It began, probably, as a practical joke. Students of Professor Johann Beringer were humorously impressed by the professor's dedication in combing the hills around the University of Würzburg, Germany, for fossils. Beringer lived during the early eighteenth century. At this time the argument of an organic origin versus an inorganic origin for fossils still was debated, a relic of the so-called "fossil controversy" that was mentioned in Chapter 2. Beringer was one of those who opposed the concepts of an organic explanation for fossils, favoring an elusive origin as "unique manifestations of nature." Anonymous students fabricated a number of artificial fossils by molding masses of clay into forms with various images (Fig. 4-1) and then proceeded to salt the hillsides where Beringer was likely to look for his specimens.*

Professor Beringer was enthused with his discoveries; he studied his fossilized finds and described them in detail. Unusual findings spurred him on to look further, and his fossils became more and more remarkable. Unusual markings were followed by discoveries of fish, insects, bees actually sucking honey from fossil plants, birds in flight, figures of astronomical bodies, and finally, unusual letters, some apparently Hebrew and some ancient Babylonian forms. Professor Beringer wrote a treatise in which he described his fossil discoveries, including one that actually had the name of the divine upon it. He illustrated his treatise with twenty-one folio plates.†

Prior to Beringer's publication of this work, some persons had expressed the opinion that his specimens were fakes, that they had been made artificially. In his treatise, Beringer devoted space in a specific chapter to denial of such a possibility. Then the day came when Beringer found on the hillside a fossil form with his name inscribed on it! His dismay was great. He attempted to buy back all the copies of his treatise publication and impoverished himself in the effort. Unfortunately, he did not destroy the copies that he had so diligently retrieved and they were found after his death. An opportunistic publisher purchased them, printed a new title page, and sold them in 1767 as a

second edition of Beringer's work. Professor Johann Beringer's name remains as part of the colorful background of the development of our concepts on the nature of fossils. He is remembered as the victim of a classic fossil hoax!

The word "fossil" originally referred to any discrete object dug from the earth. It did not, at first, have the organic limitations that are associated with it today. Fossils have been of interest to humans since ancient times. Prehistoric people buried them with their dead; writings of the ancients include references and drawings of fossil objects. Along with actual fossils of plants and animals, these objects include curious geometric forms and stones with markings that imagination is able to convert to clouds, landscapes, and castles.

Some ancient writers, especially among the Greeks, did recognize plant and animal fossils as being of organic derivation. Herodotus, journeying in Egypt between 460 and 443 B.C., observed fossil shells on the hillsides and straightforwardly called them shells. Writing much later in the eleventh century A.D., the Chinese scholar Shen Kua observed that in the cliffs of the Thai-Hang Shan mountain range there were layers of rock containing whelklike animals, oyster shells, and stones (fossil sea urchins) that he described as being like bird eggs. Shen Kua states that this locality, about 500 kilometers from the sea, must once have been marine. Unusual in his statements is Shen Kua's

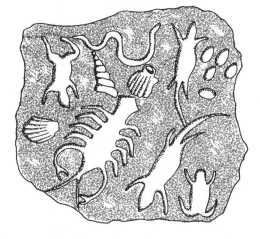

Figure 4-1. Examples of Beringer's "fossils." (From Johann Beringer, 1726, *Lithographiae Wirceburgensis Specimen Primum*)

---

*Written accounts of Beringer's "fossils" differ in whether it was a student(s) or another professor who made and scattered the clay images. Students are most commonly given responsibility for the deed.

†1726, *Lithographiae Wirceburgensis,* Specimen Primum: Würzburg.

recognition not only of the former marine nature of the area, but also that these fossils were shallow-water forms and that the area had probably been a shoreline.

More influential in the thinking of western cultures, however, were the ideas of Aristotle. As the founder of a highly influential center of learning, Aristotle championed the view that the earth had been shaped by observable processes that obeyed certain natural laws (see Chapter 3). However, Aristotle generally believed that fossils formed ("grew") directly in the rock, and, most often, under the influence of celestial bodies. During the Dark and Middle Ages, the early Christian church dogmatically accepted a great deal of Aristotelian thinking, thus strongly hindering acceptance of organic derivation of fossils.

We chuckle readily today in reading some of the explanations of fossils that come to us from the writing of the Middle Ages. To keep our perspective in balance, it should be remembered that these explanations were made in a historical period when it generally was accepted that "anything" could happen. Some persons wrote of fossils being formed as a result of the work of a spirit influence in nature, a "Virtu Divina." Imaginative pictures visible in rocks (such as forms of a virgin and child) were considered to be prophetic designs formed in the rocks at the time the earth itself was created and a direct prophecy from the creator. Still other writers suggested that fossils were the result of earlier ill-fated attempts by the creator—discarded abortive models, so to speak. Another opinion that was quite widespread during this time was that fossils were "jokes" of nature, humorous "sport" of a whimsical creator. More ominous, though, was the idea that fossils were the result of the forces of evil and that fossils were created to confound mankind, to terrify humans. One viewpoint was that fossils grew in the rocks because of the presence of vapors that were generated by a fermentationlike process. This particular concept went so far as to ascribe beautiful Etruscan vases and other pottery dug up in Rome to formation from these elusive vapors. These vases had grown in the rocks and did not involve human workmanship at all!

During the Renaissance (fifteenth to seventeenth centuries) scientific thinking changed (see Chapter 3). A number of individuals, such as Leonardo da Vinci, wrote of the more plausible organic derivation of fossils. This concept was still contrary to church dogma, however, and some persons expressing it were persecuted; one unfortunate victim, Giordano Bruno, was burned at the stake in 1600 by the Italian Inquisition.

Thus, even into the eighteenth century, the time of Johann Beringer, the "fossil controversy" was still evident. Early in that century it was expressed that fossils had been created to fulfill the role of "ornaments" for the inner portion of the earth, just as flowers had been formed to ornament the surface of the earth! As the eighteenth century progressed, the concept of the organic nature of fossils prevailed and became widely accepted. The term "fossil" came to mean what it means today: *Fossils are the remains or traces of ancient organisms preserved in rocks or sediments of the earth's lithosphere.*

BIAS OF THE FOSSIL RECORD
The Miracle of Preservation

The spectrum of life—from single-celled amoebas to the complex cellular structures of humans—is truly immense and impressive. Biologists have described and classified about 1.5 million living species of plants and animals. New species are being described each year and added to this figure, and the actual number of living species probably will be much higher; estimates have been made that 4.5 million species of living plants and animals eventually will be known. This figure is for only the *present* moment of time. It is not known how many species actually have existed throughout the history of life; the fossil record, as presently recognized, is composed of about 250,000 described and classified species of organisms. This is seemingly a large sample of past life, but in view of the total number of organisms that must have lived and died during several *billion* years, the number is undoubtedly small. Thus the fossil record is a biased record and does not give a true picture of life of the past.

When a plant or an animal dies, it is subject to bacterial decay or destruction by scavengers (Fig. 4-2). If the dead organism has hard parts, such as shell or skeletal material, these parts may remain after this first round of destruction. These hard parts, in turn, may be subject to breakage by abrasion or may be completely pulverized. However, strong shells and bones, or portions of them, may be buried under and within sediments before destruction and thus preserved as fossils. Common prerequisites for the formation of fossils, therefore, are the existence of hard parts and their relatively rapid burial. Even at this stage preservation of an organism is not assured.

The buried shells or skeletal parts later may be chemically destroyed by groundwater that percolates downward through the enclosing sediments and leaches

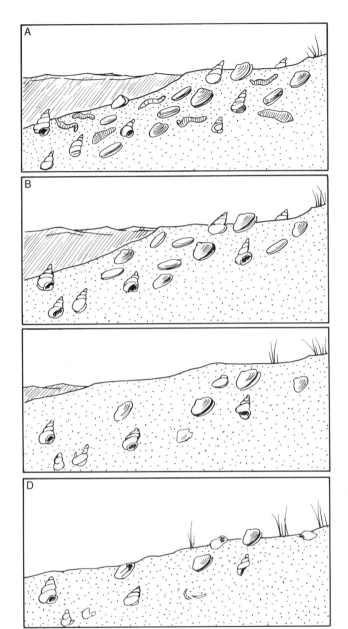

Figure 4-2. Forms of destruction that obscure the fossil record. A. Death assemblage. B. Bacteria decompose soft forms. C. Hard parts are buried and dissolution removes delicate shells. D. Erosion removes part of record, dissolution continues, and only a small part of the death assemblage remains.

the hard organic parts from the sedimentary material. Other destructive processes that may occur include metamorphism of the sedimentary material under conditions of deep burial and/or mountain-building events and destruction by uplift and subsequent erosion of the sedimentary rocks containing the fossils. Thus fossils can be preserved and can be destroyed by various aspects of the *geologic cycle.* An assemblage of dead organisms provides us with an exceedingly distorted glimpse of the variety and abundance in the original assemblage. For

example, about three-quarters of a million living species of insects are recognized, but only about 8000 fossil insect species have been classified. The fact that most insects live in terrestrial environments not conducive to rapid burial and that insects possess fragile exoskeletons explains why this group of animals is rarely fossilized. Even in more readily preservable groups of animals with shells (such as clams and snails), this bias and distortion of the picture of past life prevails. Estimates have been made that the known number of fossil species of preservable forms of life represents 2.5 to 13.5 percent of the total number of preservable species that lived since the beginning of the Cambrian. Whichever estimate is accepted, it is clear that the fossil record includes a *small* fraction of the total number of species that have lived in the past. In a sense, preservation is rather miraculous in light of all the events that may obliterate the record.

In spite of the bias of the fossil record, the significant interrelationships of rocks, fossils, and time have provided the basis for formulation of the concept of fossil succession and, subsequently, development of the geologic time scale (see Chapter 3). Fossils have become a major tool in our ability to look back over the abyss of geologic time. This chapter and the next present an overview of these "miracles" of preservation.

Preservation and Occurrence of Fossils

Most fossils are found in sedimentary rocks; it is the deposition of soft sediments (muds, silts, sands) enclosing the dead organisms that increases the possibility of preservation. Also, the accumulations of sediments throughout geologic history have been in those places (environments) where organisms have lived and died. Because the ocean is the "sink" to which most clastic debris is carried and dumped, there is a selective preservation of marine sedimentary rocks, which, in turn, exerts a direct influence on the fossil record. Because they represent areas of erosion more than accumulation, continental environments such as mountain and desert areas are less conducive to accumulation of sediments, a condition that has a profound negative effect on the fossil record. Nonetheless, unusual and outstanding occurrences of fossils have been found in some rocks formed in lake and stream environments.

Within the marine realm, the continental-shelf setting (the submerged margins of the continents) has contributed importantly to the sedimentary rock record, which, in turn, has had important implications for the fossil record. Continental shelves, such as those off the coast of the eastern United States and the Gulf of Mexico, are generally wide, shallow, submerged regions within the *photic zone* (the depth of sunlight penetration). The shelf environment supports abundant life and receives sediments from rivers emptying into the sea. Most of the fossil record consists of the hard-part remains of organisms that lived and died on the shallow marine bottoms of continental-shelf environments or other shallow seas during the *Phanerozoic*. Thus, the bias of the fossil record is influenced by preservation of organisms that secreted hard parts and lived in shallow marine environments (see paleoenvironments and facies, Chapters 5 and 6). The particular type of preservation is dependent upon the physical and chemical characteristics of the sedimentary environment and upon postdepositional events.

Metamorphism tends to destroy fossil remains through the high pressures and temperatures associated with metamorphic events. Occasionally fossils are found in low-grade metamorphic rocks; here the fossils commonly are deformed and partially altered, but not completely destroyed. An important occurrence of fossils in metamorphic rocks was located by Clarence King (later to become first director of the U.S. Geological Survey) in 1863 in the western metamorphic belt of the Sierra Nevada, California. The problem of the age of the slates associated with the gold deposits of the region had puzzled western geologists. For many days King had been searching for fossils in the slate beds, as others had previously; then, at the end of a wearying day's search, he picked up his hammer to return to camp, and, in his words, he "noticed in the rock an object about the size and shape of a small cigar. It was the fossil, the object for which science had searched and yearned and despaired! There he reclined comfortably upon his side . . . a plump pampered cephalopoda (if it is cephalopoda), whom the terrible ordeal of metamorphism had spared. . . . The age of the gold-belt was discovered!" The thrill of discovery is reflected in these expressive words from the 1860s.

Even more rare than fossils in metamorphic rocks are the occasional occurrences of fossils in igneous rocks. A striking example of the latter can be seen in Yellowstone National Park, where stream erosion has exposed a succession of twenty-seven forest levels encased in lava flows. The forests grew between periods of volcanism and then were buried in the fluid lavas (Fig. 4-3). Fossil plant-bearing sedimentary rocks are interstratified with the volcanic materials and so in this unusual geologic situation a fossil record is found in both sedimentary and igneous rocks. Because of the relative rarity of such occurrences, the discussion of the fossil record in this and succeeding chapters will concentrate on occurrences in sedimentary rocks.

The relative "miracle" of preservation is considered to be that because of the vulnerability of organisms to decay and destruction after death and the resultant loss

*Clarence King, 1872, *Mountaineering in the Sierra Nevada:* James R. Osgood, Boston.

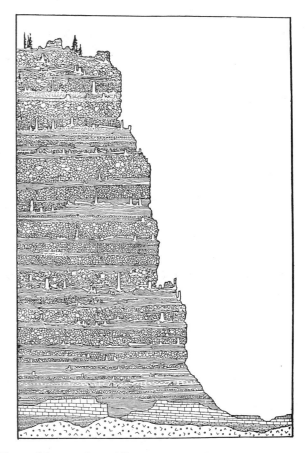

Figure 4-3. Amethyst Cliff, Yellowstone National Park, illustrates a succession of eighteen forests, each one killed when buried by volcanic flows of Paleogene age (section about 700 m thick). (Courtesy of U. S. Geological Survey)

of record of most individuals. When fossilization does occur, it may, in rare cases, involve soft organic parts, but hard-part preservation is much more common. Both soft and hard-part preservation can be accomplished by several processes (Table 4-1).

Unaltered soft remains of organisms are relatively rare and generally are confined to geologically young strata. The frozen mammoths and the other examples of unaltered preservation of soft parts listed in Table 4-1 are restricted essentially to deposits of Pleistocene age (see geologic time scale, Table 3-1). As time passes, of course, it is more likely that alteration will affect the organisms. Although exceedingly rare, these unaltered fossils are significant in that they present a picture of the complete animal. Unaltered hard parts and hard parts altered just by leaching also are restricted essentially to comparatively young strata that have *not* been mineralized or severely deformed.

Important and outstanding are the **carbonaceous films** of organisms found in dark shales deposited in oxygen-deficient environments. Plant material (Fig. 4-4) commonly is fossilized by this mode of preservation. As plant leaves, for example, were buried by accumulating

Table 4-1. Modes of fossil preservation

Preservation without alteration	
Soft parts (rare)	Freezing of organisms, such as the mammoths of Siberia
	Mummification of remains in dry climates
	Entrapment of organisms in resin or in oil seeps
Hard parts	Unaltered shells, bones, teeth, most commonly of calcium carbonate, calcium phosphate, silica, and chitin
Preservation with alteration	
Leaching	Chemical dissolution of the most soluble portions of the remains, commonly resulting in bleached and pitted shell and bone
Carbonization	Changing by chemical action of original plant or animal material to a thin film of carbon which outlines the shape of part or all of the organism
Permineralization	Deposition by underground solutions of mineral material, unlike original shell or bone composition, in the pore spaces of buried remains; most commonly calcium carbonate, silica, pyrite, dolomite
Recrystallization	Conversion into a more stable form (such as the calcite form of calcium carbonate) of less stable compounds (such as the aragonite form of calcium carbonate of some clams and snails)
Replacement	Complete (or nearly so) dissolution and replacement by new mineral matter of original organic material (such as shells or bones); common replacement minerals are calcite, dolomite, silica compounds, and iron compounds
Molds and casts	
Molds	Removal by dissolution of organic material buried in sediment; void left in the rock is a mold (e.g., an imprint); molds can be internal (expressing the shape of the inside of a shell or other feature) or external (expressing the shape of the outside of the object)
Casts	Filling of a mold (void) with sediment or mineral material, thus preserving the shape (internal or external) of the organic feature
Trace fossils	
Tracks and trails	Footprints of animals and birds
	Indications of movements by invertebrates
Burrows	Excavations made by worms and other animals as they tunnel into sediments
Borings	Round holes bored through shells by predator snails or other organisms, holes bored into rock by rock-boring organisms
Coprolites	Fossilized animal excrement; may give evidence of diet, animal size, and habitat

Figure 4-4. Plant fossilization by formation of a carbonaceous film (specimen 14 cm in width; Eocene, Green River Formation, Wyoming.)
(Specimen in Raymond Alf Museum, Webb School, Claremont CA. Photo by J. Streng)

Figure 4-5. Fossilization by permineralization. Fragile bird eggs are not commonly fossilized, but the eggs in this cluster have had mineral material deposited in the pore space of the organic matter, thus indurating them (Paleogene age).
(Specimens in Raymond Alf Museum, Webb School, Claremont CA. Photo by J. Streng)

muds, volatile materials (such as nitrogen and oxygen) were squeezed out and chemical action changed the tissues to a thin film of carbon. What remains is a residue forming an outline of a portion of the previously living leaves. If thick accumulations of plants derived from swampy coastal lagoons and deltas are carbonized more completely, coal deposits may develop. This important economic resource is a sedimentary rock derived from the chemical alteration (carbonization) of plant material that has progressed through a sequence of grades: *peat,* a fibrous mat of partially decayed plant material; *lignite,* a soft, brownish, lithified form of the original plant material; *bituminous* coal, black, so-called "soft coal" which typically leaves a tarry residue when burned; *anthracite* coal, black and glossy "hard coal" that gives much heat when burned and produces little smoke. Anthracite coal typically is found in areas where intense folding and low-grade metamorphism have occurred: The original plant material has been compacted and carbonized, then further altered by the heat and pressure of deformation (see Chapter 14).

Alteration by **permineralization** involves the deposition by underground solutions of mineral matter in the pore spaces of organic hard parts (Fig. 4-5). Calcium carbonate (lime), silica, or occasionally the mineral pyrite (an iron compound) may fill the pore spaces of shells or bones. This process has the effect of making the original organic material more indurated or rocklike. Hardening of the original shell or bone increases the possibility of preservation.

The **recrystallization** of stable forms of mineral material of shells—e.g., calcium carbonate in the form of *aragonite* to a more stable form such as calcium carbonate in the form of *calcite*—destroys the microstructure of the original shell. This form of alteration, which typically changes a shell to a mosaic of interlocking crystals, commonly intergrades with and is difficult to distinguish from replacement.

Alteration by **replacement** takes place by removal of the original hard part through dissolution and deposition of new compounds in its place. A familiar result of this process is petrified wood (Fig. 4-6); the woody cellulose material was removed and replaced by silica. This replacement may occur so precisely that even such fine detail as tree-ring structure is preserved.

A common mode of preservation is the formation of **molds** and **casts.** After burial of an organism, dissolution by groundwater may remove the original material and leave a void that expresses the external shape of the organic feature; this void is an external mold. If this mold is later filled with some type of sediment or mineral matter, an external cast is formed (Fig. 4-7). Molds and casts preserve the shape of an organic feature; either the inside (internal mold or cast) or the outside (external mold or cast) shape of the buried remains. In addition to being relatively common types of naturally formed fossils, molds and casts can be made artificially in the laboratory to provide specimens for student use, research, and for displays at museums and other institutions.

The modes of preservation that have been discussed and illustrated preserve the actual organisms or the shape of the original organic material; these fossils are referred to as *body fossils.* In addition, there is a group of fossils called **trace fossils,** in which the body parts of the organism are not preserved, but evidence of the former behavior and activity of an individual is. Trace fossils show evidence of activities such as walking, running, digging, resting, and feeding. Animal tracks (Fig. 4-8) in sedimentary rocks are examples of this mode of preservation. Other forms of trace fossils are listed in Table 4-1. The study of trace fossils is an important and

A   B   C

Figure 4-6. Preservation by replacement. A. Abundant slabs of petrified wood being exposed by erosion in the Petrified Forest National Park, Arizona. B. Preservation has been so detailed that some slabs have an outer barklike appearance, although no woody bark remains. C. Tree-ring structure is still visible in some petrified logs.
(Photos by J. Patterson)

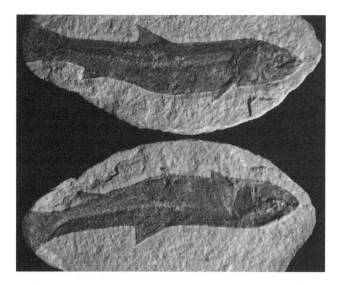

Figure 4-7. A fossil fish exposed by splitting a concretion illustrates an external cast (top) and an external mold or impression (bottom) (specimen 22 cm in length; Cretaceous age, from northeastern Brazil).
(Specimen in Raymond Alf Museum, Webb School, Claremont CA. Photo by J. Streng)

significant subscience (called *ichnology*) within the broad field of **paleontology,** the study of ancient life.

Each fossil, be it a delicate carbonaceous film, a permineralized oyster, the cast of a snail, or merely tracks indicating movement of an individual, is a special occurrence. Fossils contain a potential wealth of data that may provide *paleontologists* with information about where the organisms lived, how they lived, how they grew, and what they ate. Fossils are important tools for the study of the geologic past; they are the link between present life forms and the vast spectrum of life that has lived and died.

A PARADE OF CHARACTERS: UNICELLS TO PLANTS

Introduction

Recognition of the biases in the fossil record indicates that our knowledge of past life is incomplete. In the continuous parade of life forms that have lived and died during geologic history, we are generally limited

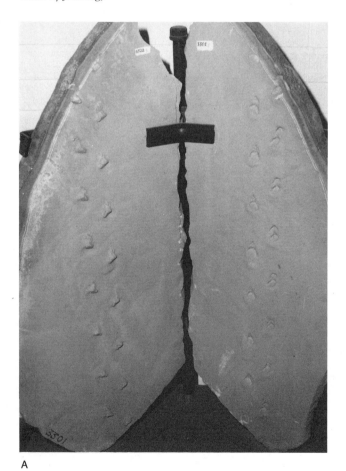

A

B

Figure 4-8. Animal tracks, a form of trace fossil. A. A "book" of reptile prints, with casts on the left side and molds on the right (Permian age Coconino Sandstone, Seligman, Arizona). B. A single dinosaur track (Jurassic age, Kanab, Utah).
(Specimens in Raymond Alf Museum, Webb School, Claremont CA. Photos by J. Streng)

in our observation and study to those organisms that secreted hard parts capable of being preserved. Even so, the available fossil record shows tremendous diversity. This chapter presents an overview of the life forms preserved as fossils; a more detailed classification is provided in Appendix B.

A number of life forms that have left a fossil record are familiar because they are presently common (Fig. 4-9). Other fossil groups are *extinct;* some of these have living relatives, others do not. In order to study this multitude of fossils, paleontologists group them into categories that are part of the present-day biological classification of organisms. The major subdivisions are called **kingdoms** (Table 4-2) and include the **monerans,** the **protistans,** and the more familiar plants and animals. Many biologists recognize a fifth kingdom, the fungi, but this group will not be included in the following discussion because it has left very little fossil record.

Table 4-2. Basic characteristics of the four kingdoms well represented in the fossil record.*

Kingdom	Characteristics
Monerae (monerans)	One-celled; nonnucleated cells
Protistae (protistans)	One-celled; nucleated cells
Plantae (plants)	Unicellular or multicellular; manufacture food from inorganic materials
Animalae (animals)	Multicellular; obtain nutrients from other organisms

*Synoptic classification, Appendix B.

Monerans

Monerans are single-celled organisms different from organisms of all other kingdoms in that the cells lack a distinct nucleus. Common examples of monerans are bacteria and cyanobacteria (blue-green algae), which occur as isolated cells or form colonies. Microscopic monerans are the oldest known forms of life (see Chapter 8). They are found in rocks of early Cryptozoic age, thus supporting the hypothesis that the origin of life goes back nearly 4000 million years.

Organosedimentary structures called **stromatolites** are especially noteworthy in rocks of Cryptozoic age because they are essentially the only *megafossils* (visible without magnification) found in rocks of this ancient time. Stromatolites are layered, mound-shaped structures (Fig. 4-10) formed by the calcium carbonate-binding activities of colonial cyanobacteria. These earliest stromatolites are associated with some of the first limestones and are important indicators of changing Cryptozoic environmental conditions (see Chapter 8).

Protistans

Protistans are one-celled organisms also, but, unlike the monerans, the individual cells have a nucleus. The evolutionary development of the nucleated cell indicates an order of complexity considerably advanced beyond the simplest monerans. The development of the cell nucleus is possibly one of the most important evolutionary steps in the history of life (see Chapters 9 and 10). The nucleus contains the heredity-controlling chromosomes and functions in cell reproduction as well as in regulating other activities of the cell.

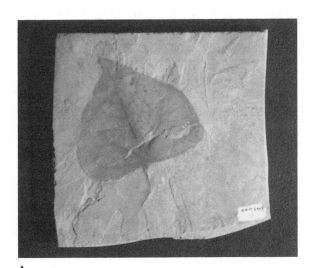

A

B

Figure 4-9. Fossils or organisms that are familiar forms. A. Plant leaf (specimen 12 cm in width; Eocene age from Green River Formation, Wyoming). B. Scallop shell (specimen 15 cm maximum width; Neogene age).
(Specimens in Raymond Alf Museum, Webb School, Claremont CA. Photos by J. Streng)

Figure 4-10. Wavy-type stromatolites (bottom) overgrown by mound-shaped stromatolites; note the laminated nature of the carbonate material. This specimen formed in a freshwater environment (specimen 4 cm in height; Pliocene age).
(Photo by J. Streng)

An important protistan group in the fossil record is the **Foraminifera** (Fig. 4-11). The variety of form in this abundant and diverse group is illustrated by the shape of the *test* (internal shell), which can be rod-shaped, globular, coiled, or spiral. Simpler varieties of Foraminifera have a single-chambered test, but a thin-section of a particular form, the *fusulinid* (Fig. 4-12), illustrates the multichambered construction of some more intricate tests. The fusulinids were marine bottom dwellers that became extinct at the end of the Paleozoic Era. They evolved rapidly during the late Paleozoic and their fossil remains are used for the biozonation of rocks of this age (see Chapter 11). In Mesozoic time floating varieties of foraminifers contributed their tests to the sediments that accumulated on the sea floor. This process occurs also in the modern ocean and millions of square kilometers of present sea floor are carpeted with foraminiferal *ooze*. Because most individuals of this fossil group are so small, they can be retrieved intact from well cuttings and have been used by petroleum companies in important subsurface stratigraphic studies.

Radiolarians are another group of protistans that have left an important fossil record. These microscopic organisms secrete siliceous tests that form a variety of complex shapes. Photographs of radiolarians taken through a *scanning electron microscope* (Fig. 4-13) show the intricacies of the finely detailed tests. The radiolarians are floating marine organisms that typically live in the photic zone of modern oceans and are distributed widely from polar to tropical regions. After

death, the tests, like those of some foraminifers, settle to the sea floor and contribute to the sedimentary ooze. Radiolarians are found in marine rocks that range in age from Cambrian to Holocene and are major components of *radiolarian cherts* that represent deep-sea-floor deposition (see Chapter 11).

Protistans form a diverse kingdom and have left an abundant fossil record. Most of the forms referred to in the paleontological literature as *microfossils* belong to this group of organisms, but many fossil groups from the other kingdoms have microscopic-sized individuals or microscopic-sized larval specimens or are represented as bits and pieces of individuals. The term "microfossil" refers specifically to size—those organic remains requiring use of a high-magnification microscope for study—and not to a kingdom or a portion thereof.

Plants

The **plant** kingdom is composed of life forms that are unicellular or multicellular, contain nucleated cells, and manufacture their food from inorganic materials through the process of photosynthesis. In comparison to animals, plants have left a less abundant fossil record, but it is, nonetheless, an interesting and important one (see Chapter 14). Plant fossils include such familiar forms as the fern imprint previously illustrated (Fig. 4-9A), preserved tree trunks and petrified wood (Fig. 4-6), and various other structures of multicellular and megas-

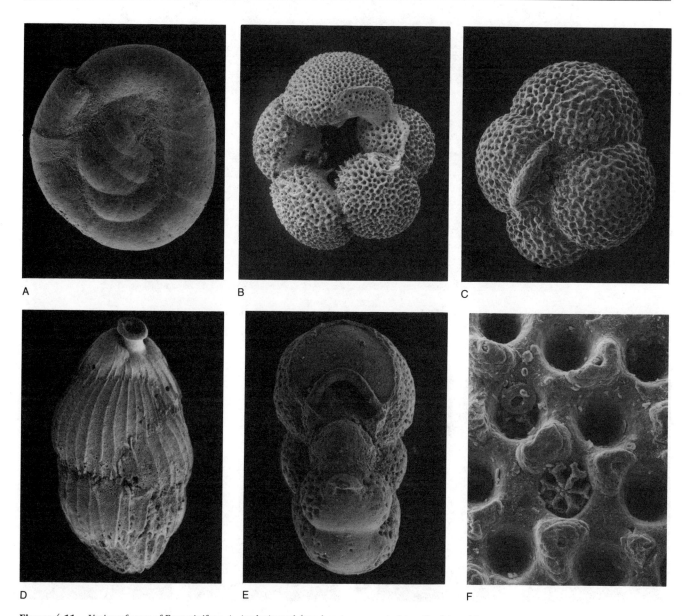

Figure 4-11. Various forms of Foraminifera. A. Agglutinated (sand grains cemented together) benthic foraminifer, ×180. B. Calcareous planktonic foraminifer, ×240. C. Calcareous planktonic foraminifer, ×435. D. Calcareous benthic foraminifer, ×195. E. Calcareous planktonic foraminifer, ×465. F. Calcareous nannoplankton (extremely small microfossils) in the pore spaces of a foraminifer, ×3000. (Scanning electron microscope photos courtesy of Union Oil Science and Technology Division, Breas CA)

copic land dwellers. Less familiar plant fossils include various types of unicellular algae such as the calcareous green and red forms, diatoms (golden algae), and microscopic *spores* and *pollen* of multicellular land plants (Fig. 4-14). The study of plant spores and pollen is a subdiscipline of paleontology called *palynology.* Like the foraminifers, spores and pollen have application in the exploration for petroleum (see Chapter 5).

Diatoms, a group of microscopic plants that have left an important fossil record, secrete siliceous bivalved structures called *frustules* that are commonly ornate

(Fig. 4-15). Diatoms appeared during Mesozoic time and thrived throughout the Cenozoic Era, producing an abundant fossil record. A sedimentary rock, *diatomite,* is composed primarily of these tiny frustules that accumulated as ooze on the floors of some ancient seas. Diatoms are abundant in the modern ocean, but several types have adapted to freshwater environments. These tiny plants are significant to marine life because they are among the most important *photosynthesizers.* They form the so-called "pasture of the sea," the base of many oceanic food chains (Chapter 5).

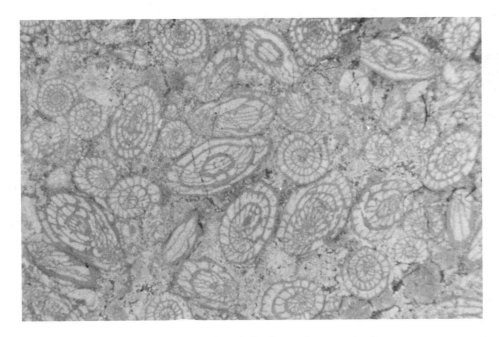

Figure 4-12. Thin section of a fusulinid (a type of Paleozoic Foraminifera); note the many chambers in the coiled form of one individual.
(Photo by J. D. Cooper)

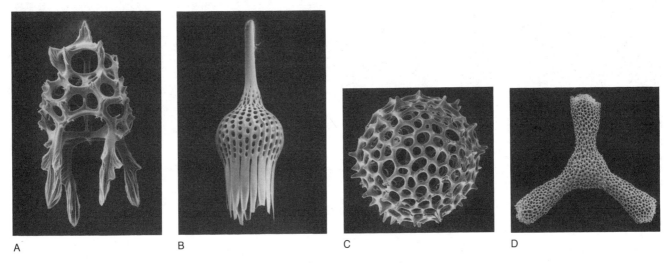

A B C D

Figure 4-13. Finely detailed tests of Radiolaria. A. ×540. B. ×280. C. ×530. D. ×435.
(Scanning electron microscope photos courtesy of Union Oil Science and Technology Division, Brea, CA)

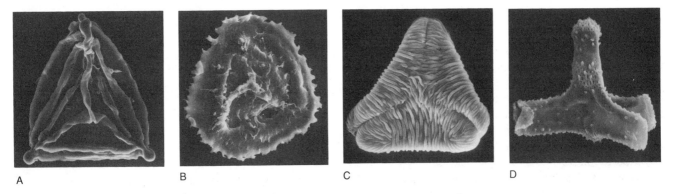

A B C D

Figure 4-14. A. Plant spores: A. ×1800. B. ×2600. Angiosperm pollen grains: C. ×2960. D. ×1600.
(Scanning electron microscope photos courtesy of Union Oil Science and Technology Division, Brea, CA)

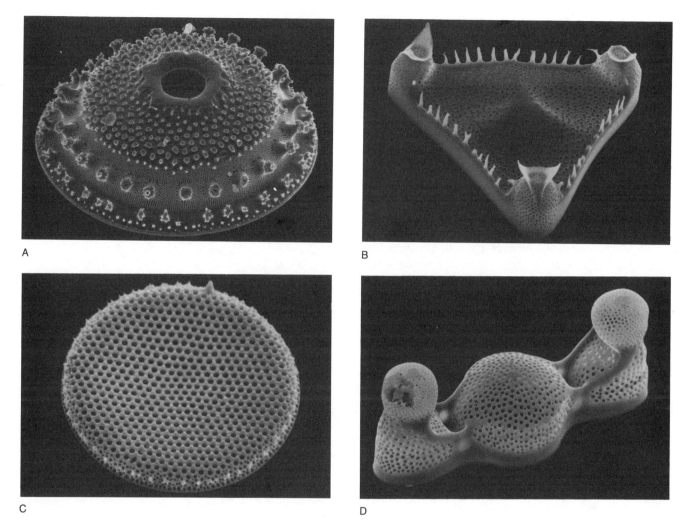

Figure 4-15. Ornate frustules of diatoms. A. ×1090. B. ×1090. C. ×1700. D. ×1450.
(Scanning electron microscope photos courtesy of Union Oil Science and Technology Division, Brea, CA)

THE PARADE CONTINUES: THE ANIMAL KINGDOM

The fourth kingdom represented in the fossil record is Animalae, the **animals.** These life forms possess nucleated cells and are multicellular organisms. They are unable to manufacture their own food from inorganic materials, and thus must obtain their nutrients from other organisms. Both **invertebrates,** animal forms without backbones, and **vertebrates,** animals with backbones, have left an abundant and diverse fossil record. The invertebrates, however, make up the most important part of this record and are emphasized in this discussion. It is the shallow-marine shelf-dwelling invertebrates that are most commonly preserved because they satisfy the prerequisites of fossilization: the possession of hard parts that were rapidly buried.

Marine invertebrates make up the most accessible part of the fossil record and are the kinds of fossils most likely to be found in the field. In the following discussion, the major **phyla** (singular, *phylum,* the major classification category below kingdom) of invertebrates that have left an important fossil record will be surveyed. A more formal and complete classification of organisms is presented in Appendix B.

Invertebrates first appear in rocks of earliest Phanerozoic time, approximately 700 million years ago. These early forms were relatively simple and lacked hard parts; the fossils occur as soft-bodied impressions in sandstone (see discussion of Ediacarian Period, Chapter 10). The fact that these soft-bodied creatures were preserved at all is amazing in light of their vulnerability to decay and destruction following death. The first appearance in the geologic column of an abundant and diversified shelly invertebrate record reflects the appearance (beginning of the Cambrian Period) of organisms with more easily fossilized hard parts composed of calcium carbonate, silica, or calcium phosphate. The development of such parts is responsible for the abundant and diverse fossil record of the invertebrates.

Sponges (Phylum Porifera)

Biologically, the **sponges** are the simplest invertebrates. Modern forms possess a saclike to globular body covered with pores. Water intake through the pores provides nutrients; water outflow passes through a larger opening, the osculum (Fig. 4.16A). Although barely multicellular, the sponge cells, nonetheless, are specialized for various functions such as food digestion, construction of skeletal elements, and reproduction. Yet each cell has remained so generalized that it is capable of respiration and excretion functions similar to those of the single-celled protistans. The biological relationship of sponges to other animals is not well understood and apparently no other animal group evolved from this

phylum. Modern sponges live attached to the sea floor and are found at all depths.

The skeletal elements of the sponges are either hard *spicules* (Fig. 4-16B) or a mass of organic fibers called *spongin.* The fossil record of the sponges is predominantly spicules, either isolated or in an interlocking mass. Spicules differ widely in form, from rod shaped to star shaped; some are composed of calcite and others are of silica. Siliceous sponge spicules have contributed to the formation of chert during various times in the geologic past. Several subdivisions of sponges are designated on the basis of the types of spicules present. The fossil record of the sponges is found in rocks of late Cryptozoic to Holocene age.

Corals (Phylum Coelenterata)

The phylum Coelenterata includes a diverse array of organisms (e.g., corals, sea anemones, jellyfish) of which the **corals** have left the most important fossil record. The corals (Fig. 4-17) possess relatively simple saclike bodies with a body wall that surrounds a digestive cavity, to which the mouth acts as both entrance and exit; there is no separate anus. Commonly the body is subdivided by radial, wall-like infoldings (mesenteries) that extend partway into the central digestive cavity. The small individual coral *polyp* secretes a calcareous external skeleton at its base and around the sides of its body. This skeleton reflects the internal body structure with the presence of radiating plates called **septa** (singular, *septum*) that alternate in position with the internal mesenteries. The septa of the coral skeleton are distinctive and identifying features of this animal group.

Some individual coral polyps (such as the horn coral, Fig. 4-17B) secrete a relatively large single shell; others aggregate with many individuals to form a colony and secrete a larger colonial skeletal structure. Both the single horn corals and the compound colonial corals are important elements of the fossil record: corals are known from Ordovician time to Holocene.

During several geologic periods corals were very abundant and, in association with other invertebrates and *algae,* formed massive *reef* structures similar to the present Great Barrier Reef of Australia. The discovery of one of these fossil reefs in sedimentary rocks provides a great deal of information about community life of the past.

An interesting type of fossil called a **stromatoporoid** is commonly included in the same phylum as the corals, but its biological affinity is uncertain. The stromatoporoids are an extinct group with a fossil record extending from Cambrian to Cretaceous. The fossils are typically laminated masses composed of calcium carbonate with vertical structures penetrating the horizontal

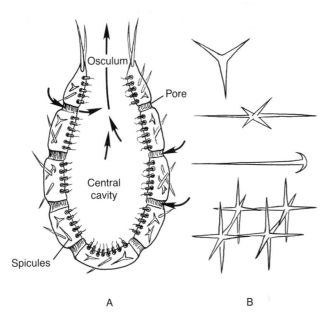

Figure 4-16. A. Basic anatomical features of the sponge include a perforated saclike body, water intake (nutrients) through the pores, and water outflow through the osculum. B. A variety of shapes of siliceous sponge spicules, the predominant feature found in the fossil record.
(After R. C. Moore, C. G. Lalicker, and A. G. Fischer, 1952, *Invertebrate Fossils,* Figs. 3-2(1) 3-3, pp. 82, 84: McGraw-Hill Book Co., New York. Reproduced by permission of publisher)

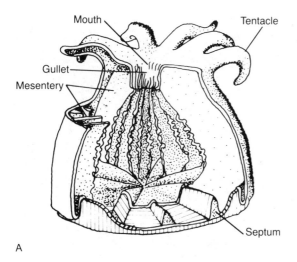

A

B

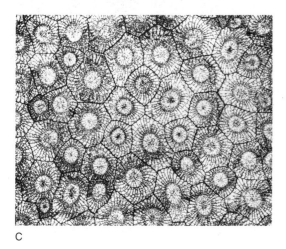

C

Figure 4-17. A. Basic anatomical structure of a coral polyp includes a saclike body with a mouth that is both entrance and exit for fluids, infoldings of the body (mesenteries) separated by radiating skeletal plates (septa) B. Solitary "horn" coral. Note the radial septa at the top of each specimen. C. Thin section through a colonial coral showing the radiating septa plus the thin dark walls separating individual corals from one another.
(A after R. C. Moore, C. G. Lalicker, and A. G. Fischer, 1952, *Invertebrate Paleontology*, Fig. 4-12(1), p. 117: McGraw-Hill Book Co., New York. Reproduced by permission of publisher. B and C, photos courtesy of National Museum of Natural History)

layers (Fig. 4-18). On the upper surface of the mass are star-shaped features that may have housed polyps. Stromatoporoids, in association with corals, were major contributors to reefs during the Silurian and Devonian Periods and they are found most commonly as large and irregular masses within the reef structure.

Archaeocyathids (Phylum Archaeocyatha)

Archaeocyathids are yet another fossil group of extinct organisms of uncertain biological affinity. Nothing is known of the soft-part anatomy of these enigmatic creatures, but the hard parts form two conelike structures, one inside the other (Fig. 4-19), connected by septa; both cones and septa are perforated.

Most archaeocyathid fossils are individual calcareous cones, but a few forms are colonial. Archaeocyathids are confined to marine rocks of Early and Middle Cambrian age. Because they possess some structural features similar to those of sponges, corals, and even calcareous green algae, archaeocyathids have been classified with each of these groups at various times. Archaeocyathids, like the stromatoporoids mentioned previously, are examples of extinct organisms that lack recognized living relatives. A "search for relatives" of these extinct groups is severely hampered by the lack of knowledge of soft-body anatomy, a factor that makes accurate biological classification difficult. Some other groups of extinct organisms also exemplify this problem, which focuses on yet another element of incompleteness of our knowledge of the fossil record.

Bryozoans (Phylum Bryozoa)

Bryozoans are colonial organisms whose calcareous encrusting masses contributed to limestones during the Paleozoic Era. Their plantlike appearance (Fig.

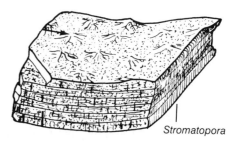

Stromatopora

Figure 4-18. *Stromatopora*, an extinct life form of unknown biological affinity. The basic structure is a laminated mass of calcium carbonate with star-shaped features on the upper surface.
(After R. C. Moore, C. G. Lalicker, and A. G. Fischer, 1952, *Invertebrate Paleontology*, Fig. 4-7(1), p. 108: McGraw-Hill Book Co., New York. Reproduced by permission of publisher)

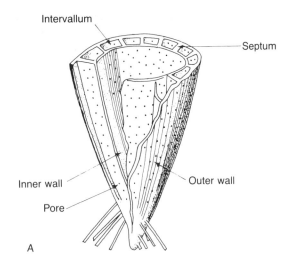

Figure 4-19. Archaeocyathids, extinct organisms of unknown biological affinity. A. Double wall structure of the cone-shaped organism. B. Cross section of this structure, with visible septas (partitions) is illustrated by the fossil specimen (specimen 2 cm in width; Lower Cambrian, Australia).
(A after R. C. Moore, C. G. Lalicker, and A. G. Fischer, 1952, *Invertebrate Paleontology,* Fig. 3.10, p. 94: McGraw-Hill Book Co., New York. Reproduced by permission of publisher. B, photo by J. Streng)

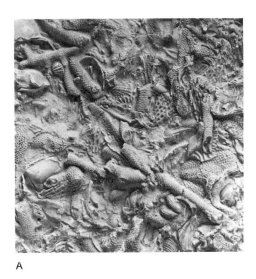

Figure 4-20. Diversity of form of bryozoans. A. Twig-shaped bryozoans in limestone slab (about natural size; Silurian). B. Distinctive spiral axes of *Archimedes* (Mississippian). Some of the lacy bryozoan fronds scattered on the limestone surface were attached to the spiral axes (slab about 15 cm across).
(Photos courtesy of National Museum of Natural History)

4-20) has been responsible for their common designation as "moss animals." The individual bryozoan is microscopic in size, but this animal always aggregates into colonies, commonly many centimeters in diameter.

The bryozoan soft-part anatomy is more complex than that of the previously discussed life forms. Bryozoa possess a U-shaped digestive tract (rather than the simple saclike cavity of the corals) with esophagus, stomach, and intestinal structures. Both a mouth and an anus are present.

Bryozoa are common in modern marine environments and commonly attach to rocks and seaweed along the shoreline. The significant bryozoan fossil record extends from Ordovician to Holocene.

Brachiopods (Phylum Brachiopoda)

Brachiopods also contributed to the formation of great volumes of limestone during the Paleozoic Era. Brachiopods were abundant animals in the warm, shallow Paleozoic seas, but in the modern marine realm they are rare in comparison to their earlier Phanerozoic abundance. Brachiopods have a stratigraphic range that extends from Cambrian to Holocene.

The anatomy of the brachiopod is more complex than that of the bryozoan. In addition to a digestive system, the brachiopod has a nervous system, reproductive organs, and well-developed sets of muscles (Fig. 4-21A). These soft parts are enclosed by two separate shells *(valves)* that protect the internal organs. Most shells are hinged at one end and open at the opposite end to allow water to circulate and to bring food to the animal. Protruding from the hinged end is a fleshy stalk, the *pedicle,* by which the animal is attached to the sea floor.

Brachiopods include two major groups, one of which secretes a calcium *phosphate* shell and the other a calcium *carbonate* shell. The shell has been very diverse through the Phanerozoic and has included smooth, ribbed, and spiny forms as well as those with short or long hinge lines and convex and concave shapes (Fig. 4-21B). Despite this shell diversity, brachiopods are readily recognized by the facts that the two enclosing valves are different in shape and size from one another (inequivalved) and each individual valve is *bilaterally symmetrical* (the two halves of a single valve are mirror images, Fig. 4-21C).

Molluscs (Phylum Mollusca)

Molluscs make up an animal group that contains many familiar modern examples such as clams, snails, octopi, and squids (Fig. 4-22). Other important members of the phylum, such as ammonoids (Fig. 4-26), are extinct. The molluscs are one of the most significant groups of fossils and constitute a *major* part of the marine, shell-bearing shelf fauna that is favored by the bias of the fossil record. It is a record that extends from Cambrian to Holocene.

Molluscs are so numerous and diverse that it is difficult to pick out a single (or even a few) representative that illustrates this important group. The soft anatomy of molluscs is more complex than that of the brachiopods and includes internal organs for the functions of digestion, circulation, and excretion, as well as better developed nervous, reproductive, and muscle systems. A characteristic molluscan feature is the *mantle,* which consists of two fleshy flaps that fold together to form a cavity (mantle cavity) within which the soft body mass is suspended. Body organization includes a fleshy mass that serves as a crawling foot for snails and clams; a well-defined head, with eyes, characterizes the *cephalopod*

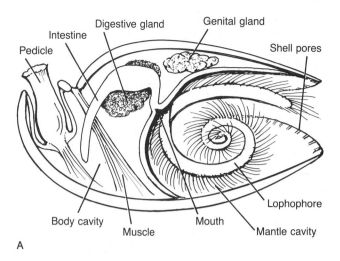

A

B

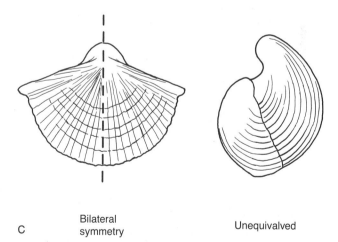

C   Bilateral symmetry   Unequivalved

Figure 4-21. A. The complex body of the brachiopod is enclosed between two valves which protect the soft parts. Much of the mantle cavity within the valves is occupied by the lophophore, the food-gathering structure. The lophophore consists of a pair of extendable "arms," the brachia. Note the position of the pedicle, the fleshy stalk by which the animal is attached to the sea floor. B. Brachiopods were diverse in size and shape (upper right specimen 5.5 cm in length). C. Brachiopods typically display bilateral symmetry (left), but the valves (right) are unequal in form (unequivalved).
(A after R. C. Moore, C. G. Lalicker, and A. G. Fischer, 1952, *Invertebrate Paleontology,* Fig. 6-3, p. 199: McGraw-Hill Book Co., New York. Reproduced by permission of publisher. B, photo by J. Streng)

Figure 4-22. The molluscs are an abundant and diverse group. Examples of molluscs include (left) two forms of coiled gastropods or snails; (center) the coiled *Nautilus,* cephalopod (specimen 6 cm in width); and (right) two forms of bivalves or clams (the lower specimen is a mussel).
(Photo by J. Streng)

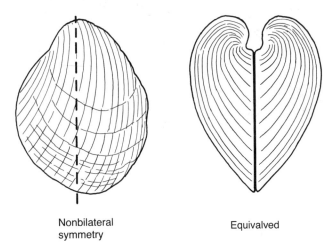

Nonbilateral   Equivalved
symmetry

Figure 4-23. Bivalves are characterized by nonbilateral symmetry (left), but valves are typically equal in form and size (right). Compare this symmetry with the opposite symmetry of the typical brachiopod (Fig. 4-21C).

Figure 4-24. Upper Cretaceous rudist bivalve illustrating nonsymmetry of some aberrant forms. Note the elongate lower valve and the lidlike upper valve (specimen 20 cm in length; Baja California).
(Photo by J. Streng)

molluscs (e.g., octopi, squids). Most molluscs share the characteristic of an external, calcareous shell that is secreted by the mantle, but some types of cephalopods, such as the octopus, have an internal shell. Modern shell-bearing molluscs are commonly referred to as shellfish.

Molluscs show a great range of environmental adaptations to a variety of marine, terrestrial, and freshwater habitats. Some molluscs live attached (e.g., oysters), some are active swimmers (e.g., squids, octopi), and many are crawlers (e.g., snails). Because of their tremendous abundance and diversity and their importance as fossils, three separate molluscan classes (classification below phylum, see Chapter 5) require closer examination.

Bivalve molluscs (also called *pelecypods*) are characterized by a body enclosed within a bivalved shell; common examples are clams, mussels, oysters, and scallops. Typically, the two valves are the same shape and size (equivalved), but individual valves are *not* bilaterally symmetrical (Fig. 4-23). Recall that the bivalved shells of the brachiopod have an opposite relationship: the individual valves are bilaterally symmetrical, but the two valves have different shapes and sizes (see Fig. 4-21). However, not *all* bivalve molluscs are symmetrical. Those that have adopted a sedentary life style, such as an attachment to the sea floor (e.g., oysters), most commonly have unequal, nonsymmetrical valves. This condition results from the mollusc resting or reclining on one valve or from one valve being cemented to rock or to another shell. An extreme example of this asymmetry is exhibited by the extinct *rudist* bivalves of the Jurassic and Cretaceous. These forms had a lower elongate corallike valve and an upper flattened lidlike valve (see Fig. 4-24). Bivalve mollusc shells are positioned on the left and right sides of the body and are hinged along the *dorsal* (back) margin. The ventral (bottom) unhinged margin of the valves opens to allow for protrusion of the large muscular foot with which many mobile types of bivalves crawl, as well as to allow nutrient waters to enter the shell.

There is great diversity in the mode of life within the class: some forms actively burrow into the bottom sediment; some bore into wood or rock; some forms are cemented to the bottom, while others recline; many are tethered to the bottom by organic threads called a *byssus;* and a few even propel themselves through the water by rapidly opening and closing the valves. Perhaps more than for any other group of invertebrates, this range of life habits among the bivalve molluscs is faithfully reflected in shell shape, thickness, and ornamentation. Bivalve mollusc fossils are found in rocks ranging in age from Cambrian to Holocene, but are most prevalent in strata of the Mesozoic and Cenozoic.

Gastropod molluscs (snails) are *univalved* forms whose bodies are enclosed by a single, spirally coiled shell. Gastropod shells exhibit a vast number of shapes and sizes including (Fig. 4-25) low-spired to high-spired forms, those with many whorls or only a few whorls, those ornamented with ribs, knobs, or spines, or those that are unornamented.

The gastropod shell contains only a single chamber into which the animal withdraws its body completely. The body may be protruded (in a manner similar to that of the bivalve) as a muscular foot on which the animal crawls. Anatomically, gastropods have a distinct head with a pair of eyes and a pair of feelers. They also possess a mouth with a strip of material holding filelike teeth (the *radula*). Carnivorous snails are able to drill with their radula directly through the shells of other organisms to gain access to the soft body parts within. Many Cenozoic fossil clams and snails have been found with a smooth circular hole penetrating the shell, a sure

A

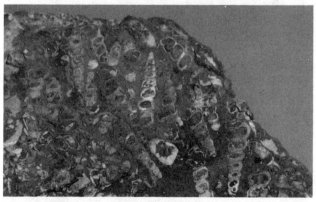

B

Figure 4-25. A. Variety of size and form of the modern gastropod shell; note high-spired form (left, specimen 9 cm in length) and low-spired cone shell (right, specimen 7 cm in length). The muscular foot of the gastropod protrudes from the aperture visible on several of the specimens. B. Upper Cretaceous fossil gastropods from southern California (specimens 2–3 cm in length).
(Photos by J. Streng)

sign of the cause of death! Gastropods have a Cambrian to Holocene fossil record, and like the bivalve molluscs, were most abundant and diverse during the Mesozoic and Cenozoic.

Cephalopod molluscs, unlike bivalves and gastropods, are swimming forms. These exclusively marine molluscs have the ability to jet propel themselves through the sea by the expulsion of water from a nozzlelike tube. Because of this ability to swim, cephalopods are agile and are active predators, rather than being confined to slow crawling on the sea floor or passively drifting with currents. The cephalopods possess several other distinctive features: highly developed eyes, tentacles commonly covered with sucker discs, and the ability to eject inky fluid as a defense or escape mechanism. Some modern squids have attained a length of 16 meters and are the largest known invertebrate animals.

Living cephalopods are relatively few in number (octopus, squid, chambered nautilus), but ancient cephalopods were abundant, diverse, and rapidly evolving and thus serve as important fossils for biozonation and correlation. Most of the important fossil forms had external shells similar to those of their modern-day distant cousin, the chambered nautilus (see Figs. 4-22, 4-26A). Most extinct *ammonoids* (Fig. 4-26B), for example, lived in a symmetrical flatly coiled shell that was divided into a series of chambers by *septa*. The animal secreted larger and larger septa as it grew and the adult body occupied only the large end (living chamber) of the shell. A fleshy tube, the *siphuncle,* extended back through the septal walls, allowing the living animal to maintain communication with the older unoccupied chambers. This communication involved the partial to complete filling of some or all of the unoccupied chambers with various gas–fluid mixtures that aided importantly in movement through the water.

In the cephalopod shell, distinct lines, called **sutures,** represent the contact between the septa and the inner wall of the shell. Cephalopod sutures range in form from a simple slightly curved line in the *nautiloids,* to various types of more complex crenulated designs in the *ammonoids*. This feature is significant in classification of the cephalopods below the rank of class. One group of extinct cephalopods, the squidlike *belemnites,* secreted internal, calcareous, cigar-shaped shells similar to the cuttlebone of the modern cuttlefish.

The geologic range of the cephalopod molluscs extends from Late Cambrian to Holocene, but the extinct ammonoids are confined to rocks of Devonian to Cretaceous age. The ability of the ammonoids to disperse over wide geographic areas, coupled with the evolutionary development of many distinctively sutured forms during the Mesozoic Era, makes them unsurpassed in their usefulness as guides to narrow time subdivisions

A   B

Figure 4-26. A. Interior of the modern chambered *Nautilus.* Living chamber is at top; note the septa that partition the shell's interior into the earlier used chambers; a fleshy tube, the siphuncle, extended from the living chamber back through the projection on each septum into the older chambers (maximum width of specimen 13 cm). B. Fossil cephalopods illustrating a simple suture (right) and a complex suture line (left; specimen 12 cm in diameter).
(Photos by J. Streng)

(chronozones) within rocks of this age (see Chapters 3 and 12).

Arthropods (Phylum Arthropoda)

Arthropods are another group that contains many familiar living forms: crabs, lobsters, shrimp, insects, and spiders. These familiar members, however, have not left as abundant or as significant a fossil record as an extinct class of arthropods, the **trilobites.** As biologically advanced invertebrates, the trilobites are especially important because they developed the ability to secrete a chitinous exoskeleton very early in the Phanerozoic. In fact, the first appearance in the rock record of trilobites previously had been used to define and recognize the boundary between rocks of Cryptozoic and Cambrian age in some areas. Trilobites are used to subdivide the Cambrian System into series and stages.

Arthropods, in a sense, are the most successful of all animals because they are the most abundant organisms on earth (remember this phylum includes insects!) and live in a greater variety of habitats than do the members of any other phylum. As the term arthropod ("jointed" + "leg") implies, the various subgroups in this diversified phylum are characterized by segmented bodies and jointed limbs. Arthropods also have a *chitinous* dorsal exoskeleton that covers the body and is molted as the individual organism grows. Many trilobite fossils probably represent discarded molted exoskeletons. Variation in the phylum is expressed by the nature of the segmentation, by specialization of appendages, by

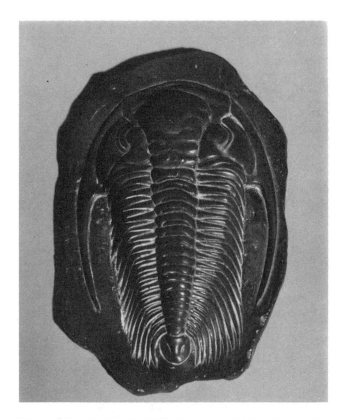

Figure 4-27. Model of a Middle Cambrian trilobite, *(Paradoxides harlani,* an important Paleozoic arthropod fossil. The distinct trilobed structure of the animal gives it its name. In addition, it has a three-part division into cephalon (head), thorax, and pygidium (tail). Note the crescent-shaped eyes located on either side of the central, globular part of the cephalon (specimen about 16 cm in length)
(Photo by J. Streng)

differences in nervous, respiratory (marine to air-breathing forms), and other systems, and by differences in locomotion (swimming to flying forms).

The trilobite body is divisible into three sections: *cephalon* (head), *thorax,* and *pygidium* (tail) (Fig. 4-27). In addition, the entire body is further divisible into three longitudinal lobes formed by grooves separating a central portion (axial lobe) from two lateral areas (pleural lobes); hence the derivation of the name trilobite.

Because trilobites are extinct and have no direct living descendants, our knowledge about the soft parts of the animal is incomplete. Sufficient information has been determined from fossilized remains to indicate the presence of well-developed eyes in some trilobites, to determine the placement of soft parts, and to detect the areas of muscle attachment to the exoskeleton. It is noteworthy that trilobites are complex and highly developed invertebrates and yet they are the major component of the oldest Paleozoic faunas. Many paleontologists believe the early appearance of biologically complex trilobites in the abundant Cambrian fossil record provides evidence of an "explosive" (dramatically rapid) evolutionary development of multicellular life forms prior to the beginning of Paleozoic time (see Chapters 9 and 10). Fossil arthropods occur in rocks of Cambrian to Holocene age, but the important trilobites are found only in rocks of Paleozoic age and are especially significant in strata of the Cambrian and Ordovician Systems.

Echinoderms (Phylum Echinodermata)

Echinoderms are a diversified, exclusively marine group that includes starfish, sand dollars, sea urchins (echinoids), and sea "lilies" (crinoids). The echinoderms are distinctive in several aspects: The body is commonly globular; most forms exhibit *pentameral* (five-rayed) *symmetry;* and the "shell" is actually an internally secreted series of calcareous plates that fit together like a tile mosaic. After the death of the organisms the plates commonly separate and become dispersed in the sediments, although many shells have been preserved intact.

Because the echinoderm phylum contains so many classes of organisms, it is divided into a number of *subphyla;* two are discussed here because of their importance in the fossil record. The subphylum Crinozoa includes forms that are most commonly attached to the sea floor *(sessile).* They have a structure typified by the **crinoids** (Fig. 4-28A), which have a stem, a cup-shaped body, and a series of branching arms extending upward from the body. The long flexible stems allow the crinoids to bend and wave in an almost plantlike manner (thus the designation sea "lily") in the ocean currents.

Although locally abundant, crinoids are not prominent members of the fauna of modern oceans; they are best known from an extensive fossil record. Some living cri-

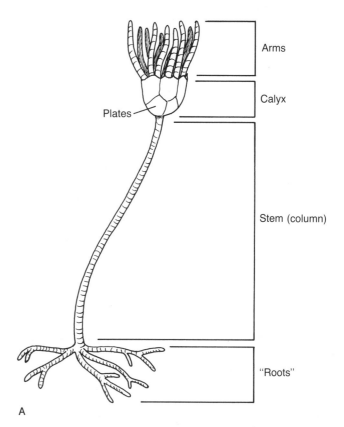

A

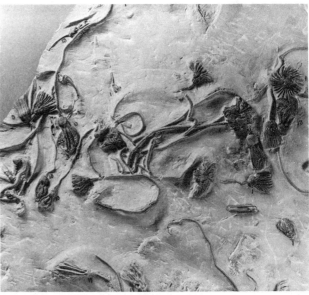

B

Figure 4-28. A. Crinoids illustrate the sessile (attached to the sea floor) form of echinoderm (subphylum Crinozoa). The crinoid body is enclosed in the cup-shaped calyx. B. Complete fossil crinoids on a slab of Mississippian limestone from LeGrand, Iowa (scale indicated by paper clip).
(B, photo courtesy of National Museum of Natural History.)

noids are brightly colored in hues of purple, red, yellow, and brown and have changed their mode of life from predominantly sessile during the Paleozoic and Mesozoic to *vagrant* in the Cenozoic. Although their fossil record extends from Ordovician to Holocene, crinoids were most abundant during Paleozoic time and formed abundant gardenlike masses on the warm, shallow sea floor (Fig. 4-28B). Some Paleozoic carbonate rocks are so full of crinoid fragments that they are referred to as crinoidal limestones.

Subphylum Echinozoa includes forms that generally are globular and without arms or a stem, as typified by **echinoids** (sea urchins and sand dollars, see Fig. 4-29). Within the subphylum there are some sessile forms, but most are mobile and move sluggishly along the sea floor; some burrow into the sediments. The pentameral symmetry characteristic of the echinoderms in general is well illustrated by the arrangement of plates on the sea urchins. Fossil echinoids are found in rocks of Ordovician to Holocene age, but are most diverse and abundant in strata of the Mesozoic and Cenozoic.

Graptolites (Phylum Hemichordata)

Graptolites are extinct, colonial, floating *(planktic)* organisms that lived during part of the Paleozoic Era (Cambrian–Mississippian). During the time that they lived they were geographically widespread and left a broadly dispersed, but rapidly changing fossil record that is useful in correlating rock units in various parts of the world. The graptolites (Fig. 4-30) are an excellent example of organisms preserved as carbonaceous films; they occur predominantly as carbonaceous impressions on the bedding surfaces of dark shales. Apparently the

organisms floated into an area, died, and settled into the dark muds in an *anerobic* (oxygen-poor) bottom environment; decay and scavenging were restricted under these conditions.

Graptolites also provide another good example of problems encountered in attempting to assign extinct organisms to their proper place in the scheme of biological classification; their biological affinities are not known with certainty. The structure of each graptolite colony was composed of many small cuplike features arranged along branches, each cup housing an individual animal. Living organisms most similar in structure to the extinct graptolites belong to a group known as *hem-*

A

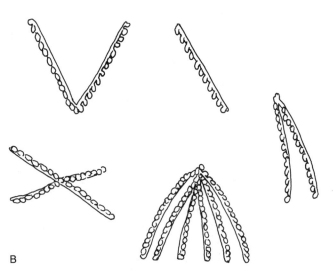

B

Figure 4-30. A. Graptolites are often preserved as carbonaceous films on a rock slab, as illustrated by these Middle Ordovician specimens from Canada. The rod-shaped notochord is an important element in the structure of these enigmatic organisms. On the notched notochord, each projection housed an individual; graptolites occur in colonies. Although the biological affinity of the graptolites is questionable, the group is excellent for dating and correlating some Paleozoic rocks. B. Some of the diverse forms that evolved in the graptolites.
(A, photo courtesy of National Museum of Natural History)

Figure 4-29. Modern echinoids illustrate the vagrant (mobile) form of echinoderm and, also, the radiating symmetry, often five-rayed, characteristically displayed by members of this phylum (note petal appearance on specimen on left; this specimen about 11 cm in width).
(Photo by J. Streng)

ichordates ("half chordate"), exemplified by the living *pterobranchs*. Because the hemichordates are characterized by the presence of a dorsal stiffened rod structure called a notochord, they have been classified as intermediate between the invertebrate and vertebrate condition. It is possible, therefore, that the graptolites represent an extinct group of these intermediate forms.

Conodonts (Phylum unknown)

Conodonts (Fig. 4-31) are another excellent example of an extinct group of organisms with uncertain biological affinities. These tiny toothlike structures have a phosphatic composition that is similar to that of vertebrate bone; however, no similar structure is known to exist in any living or fossil vertebrate. At various times conodonts have been linked to fishes, gastropods, worms, crustaceans, and cephalopods. In spite of the problem of the origin of the conodonts, they are important fossils for dating rocks of Middle to Upper Paleozoic age. The stratigraphic range of conodonts extends from Cambrian through Triassic.

Vertebrates (Phylum Chordata)

True vertebrates, or animals with backbones, include fish, amphibians, reptiles, birds, and mammals (Fig. 4-32). In addition to the backbone (vertebral column), which is a supporting structure composed of calcium phosphate, the internal vertebrate skeleton typically possesses paired limbs (fins, wings, legs) and a rib cage. Yet another important skeletal feature of the vertebrates is the skull, housing sense organs and a well-developed brain.

Vertebrates make their first known appearance in the fossil record as primitive jawless fish in rocks of Late Cambrian age (see Chapter 10) and they occur in the fossil record throughout the rest of the Phanerozoic. Vertebrates are probably the most dramatic of fossils and probably hold the most fascination for humans, perhaps because we are vertebrates also. Who has not been enthralled by a reconstructed dinosaur skeleton in a museum? Certainly such displays have kindled the imagination of writers and movie makers. Although dramatic and sometimes spectacular, the vertebrate fossils are, nonetheless, comparatively rare and not generally as complete as invertebrate fossils. Many vertebrates live in terrestrial and aerial environments that are not conducive to preservation, and the vertebrate skeleton easily disarticulates and becomes scattered following the death of an individual. The bias of the fossil record tells us that vertebrates have not been selectively preserved as readily as invertebrates.

Mammals, the biological group to which humans belong, are represented by many fossil forms in rocks of Cenozoic age. *Homo sapiens* is a very recent mammal

Figure 4-31. Conodont diversity illustrated by Devonian forms: conical form, upper left; bars with denticles, lower left and upper center; simple platform, lower center; complex platform, right. Conodonts, like graptolites, are of uncertain biological affinity, but are important fossils in dating and correlating Paleozoic rocks. In spite of great diversity of form, each conodont has a toothlike appearance. (Photo courtesy of Michael A. Murphy, University of California, Riverside)

Figure 4-32. Examples of vertebrates. Fish, amphibian (frog), reptile (alligator), mammal (man, dog), and bird all share the common structural feature of a vertebral column, the backbone. There is, as illustrated, *great* diversity of form, habitat, and life style within the phylum.

member of the animal kingdom although there is a fossil record extending back in time several million years. Human fossils are found most commonly in deposits of Pleistocene age in Africa. Some of the most exciting discoveries in paleontology since the 1950s have been geological and archaeological finds in the Olduvai Gorge of East Africa.

The foregoing has been intended as a brief overview of life forms that have left an important fossil record. The preservation of the spectrum of life forms, from the so-called lowly monerans to the highly developed vertebrates, has been accomplished in many different ways, dependent upon the physical, chemical, and biological conditions that prevailed in the death and burial environments. Now that some of the aspects of the preservation processes are understood and the types of organisms that are most readily fossilized have been described, how can we apply this information to achieve a better understanding of the geologic record? Historically, it has been shown that fossils were the basis of the development of the geologic time scale (see Chapter 3). But what about the use of fossils in dealing with current geologic problems? The next chapter discusses answers to these pertinent questions and presents reasons why fossils have present-day roles far beyond that of curious museum displays.

SUMMARY

The fossil record is not a complete picture of the past because only those organisms with hard parts (shells, bones) that become relatively rapidly buried tend to be preserved. These prerequisites for fossilization are most often fulfilled by animals in a shallow marine (continental-shelf) environment and so sedimentary rocks formed in this environment more often tend to be fossiliferous. The fossil record, therefore, is said to be a biased record.

Fossils are formed by a variety of methods: preservation without alteration; preservation with alteration, including leaching, carbonization, permineralization, recrystallization, and replacement; formation of molds and casts; and recording of organism activity as trace fossils.

To facilitate study, fossils are classified into categories that correspond to the present biological classification of organisms. In this chapter an overview of fossil forms belonging to four kingdoms was presented:

1. Kingdom Monerae
   Characteristics: one-celled; nonnucleated cells
   Example: stromatolites

2. Kingdom Protistae
   Characteristics: one-celled; nucleated cells
   Examples: foraminifers, radiolarians

3. Kingdom Plantae
   Characteristics: unicellular or multicellular; manufacture their own food (photosynthesis)
   Examples: diatoms, spores and pollen, trees

4. Kingdom Animalae
   Characteristics: multicellular; derive nutrients by consumption of other organisms
   Examples: sponges, corals, stromatoporoids, archaeocyathids, bryozoans, brachiopods, molluscs (bivalves, gastropods, cephalopods), arthropods (trilobites), echinoderms (crinoids, echinoids), graptolites, conodonts, vertebrates (fish, amphibians, reptiles, birds, mammals)

SUGGESTIONS FOR FURTHER READING

Case, Gerard R., 1982, *A Pictorial Guide to Fossils:* Van Nostrand Reinhold, Florence KY.

Halstead, L. B., 1982, *The Search for the Past:* Doubleday, Garden City, NY.

Lane, N. Gary, 1978, *Life of the Past:* Charles E. Merrill, Columbus, OH.

Levin, H. L., 1975, *Life Through Time:* Wm. C. Brown, San Francisco.

MacFall, Russell P., and Wollin, Jay C., 1983, *Fossils for Amateurs,* 2nd ed.: Van Nostrand Reinhold, New York.

McAlester, A. Lee, 1977, *The History of Life,* 2nd ed.: Prentice-Hall, Englewood Cliffs, NJ.

Simpson, George Gaylord, 1983, *Fossils and the History of Life:* Scientific American Books, New York.

Thompson, Ida, 1982, *The Audubon Society Field Guide to North American Fossils:* Alfred A. Knopf, New York.

5

ORGANIZING THE RECORD:
The Classification and Uses of Fossils

KEY TERMS

Taxonomy
Phylum (phyla)
Class
Order
Family
Genus (genera)
Species
Taxon
Biozone
Paleoecology
Habitat
Niche
Community
Ecosystem
Benthos
Epifauna
Infauna
Plankton
Nekton
Food chain
Producer
Consumer
Autotroph
Heterotroph
Biogeographic province
Endemic
Cosmopolitan
Fossil fuels

TWO "GIANTS"

Georges Cuvier in France and William Smith in England made major contributions to our understanding of the nature of fossils. The lives of these two men were very different and their paths did not cross, but in their individual ways each one utilized fossils in a significant manner.

Georges Cuvier (1769–1832), scientist and writer, politician and historian, made a major contribution to historical geology through his work on the stratigraphic succession of the Cenozoic units of the Paris Basin (Fig. 5-1). He recognized that distinctive fossils were confined to specific rock layers and he realized further that many of these fossils were of animals that are extinct. Extinction is a concept that was not widely accepted during Cuvier's lifetime. Cuvier believed that causes of extinction were sudden, catastrophic events and that successive levels of extinct fossils were explained by a series of catastrophic happenings (see discussion of catastrophism, Chapter 3).

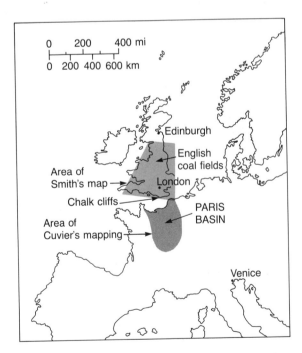

Figure 5-1. Index map of a portion of Europe, illustrating the geographic areas where Cuvier (Paris Basin) and Smith (England) did their late eighteenth century–early nineteenth century studies on fossils.
(After R. H. Dott, Jr., and R. L. Batten, 1981, *Evolution of the Earth,* 3rd ed., Fig. 2.8, p. 25: McGraw-Hill Book Co., New York. Reproduced by permission of publisher)

To accommodate early eighteenth century church dogma, Cuvier proposed a three-part history of the earth: (1) a Diluvian period, the time of the Noachian flood; (2) a post-Diluvian period that included all time *since* the flood; and (3) an ante-Diluvian period that included all the time *before* the flood. Cuvier accepted that the time of the flood and the time since the flood were interpreted accurately by the biblical scholars. He placed his extinct fossilized animals in the period of time prior to the flood—a time characterized by prevailing darkness and strange, now-extinct creatures. This period of time, Cuvier wrote, was separate from the two younger biblically related periods and was *not* governed by natural laws or amenable to reason or scientific method. In this way Cuvier could explain many catastrophic events and whole new assemblages of life forms without offending the position of the church.

Cuvier's compromising ideas were widely accepted because he was an influential person in both scientific and political circles. His academic background included studies in entomology and comparative anatomy. Cuvier was appointed as a professor at the College de France in 1800, and in 1803 he became Perpetual Secretary to the Division of Physical and Mathematical Science of the National Institute. In 1818 he was elected a member of the prestigious French Academy.

Cuvier lived during times of political unrest in Europe. Napoleon Bonaparte appointed him Counselor of State in 1814 because of Cuvier's stature in the world of science and his capabilities in administration. After Napoleon's downfall Cuvier was reappointed to the same position by Louis XVIII.

Both as Perpetual Secretary of the National Institute and later as Secretary to the French Academy, Cuvier was responsible for annual reports on work of members. He was well informed, as a consequence, of various aspects of science and especially on new avenues of research being pursued during these years. His annual reports are considered to be valuable contributions to the history of science. Cuvier's ability to write well created popular interest in fossils and his interpretation of the early history of the earth drew attention to the budding concepts of geology.

William Smith's (1769–1839) contributions to historical geology came about principally through his work as a civil engineer surveying and building canals (Fig. 5-1). He, like Cuvier, recognized an association of certain fossils with specific rock units. As canals were dug and layers of sedimentary rock exposed, Smith recognized the interrelationship of fossils and rock units. What a contrast, however, in the directions from which Smith and Cuvier came in their recognition of *fossil succession!* Cuvier was academically educated, was a member of the French court, and worked within its pomp and grandeur. Smith was born in a small village in England, was raised by an uncle in a more casual farm setting, and received no formal education beyond the local village school. His understanding of geology and the nature of fossils was self-acquired through his own countryside wanderings as a boy and later through his work on the canals.

As a young man of eighteen, Smith became an assistant to a surveyor; later he pursued this work on his own. At this time in England, prior to the invention of the steam locomotive, there was active interest in the construction of canals to be used for transportation of coal (one of the economic incentives for organizing the rock record). Smith traveled widely in England making survey lines for these canals and later was employed during construction of the waterways. Recognizing the importance of the rock–fossil associations he observed in the canal walls, Smith compared these associations in many locations in England. Not only did these observations lead him to formulate independently the principle of fossil succession, but the data he acquired had direct practical application for his work. His understanding of the order of the strata, based on the chronological succession of fossils, could be used in picking out the best canal routes (on the basis of the nature of the underlying bedrock), in selecting bridge foundations, and for supplying good construction stone. (A good knowledge of stratigraphy is likewise important to the modern-day engineering geologist.)

In essence, Smith was a professional geologist at a time when geology was virtually an unknown profession. He was in demand throughout the country, especially after successfully draining the Prisley Bog in 1801 and transforming it into prime agricultural land. It is estimated that he traveled as much as 17,000 kilometers a year in his work, a great distance

considering the time in which he lived. On his own, Smith traveled in Wales and southern Scotland also, always keeping complete notes and data based on his competent observations. One of his early manuscripts (1799), directly based on the data, was "Tabular View of the Order of Strata in the Vicinity of Bath with Their Respective Organic Remains." Later Smith turned over to the British Museum his extensive fossil collection of specimens from England and Wales.

The work of Smith's lifetime is remembered most lastingly for his great geological map of England, one of the first geological maps (plus cross sections and stratigraphic columns) ever published (1815).* The map consisted of fifteen sheets compiled at a scale of about three kilometers per centimeter; the completed map measured about two by two and one-half meters in size. Twenty colors were used on the map, plus various degrees of shading. It was a map without precedent, a great and original work, the result of over twenty years of geological observations in the field. The map shows the fossil succession in the rock strata; the mapped units are essentially biostratigraphic in nature and demonstrated that rock sequences could be subdivided on the basis of fossil content. As mentioned in Chapter 3, this application of the concept of fossil succession provided a foundation for the development of the geologic time scale during the nineteenth century.

William Smith was not a member of the academic-oriented Geological Society of London, but fortunately during his lifetime the significance of his work was recognized and he was honored by this society. Smith's imaginative vision saw beyond the local use of fossils to identify strata; he was able to grasp a greater significance for this inference: an application of fossil succession on a regional scale. During his later years when he was being honored for this original work, he was given the designation "Father of English Geology and Stratigraphy."

---

*Map of Strata of England and Wales with a Part of Scotland: Studio Press, Birmingham, England.

CLASSIFICATION OF ORGANISMS

Taxonomy

The spectrum of life preserved in the fossil record, as discussed in the overview of the previous chapter, may appear at first glance to present a confusing picture. However, this picture becomes clearer when the spectrum of living and fossil organisms is viewed within a framework of organization. Prerequisite to an ability to fully understand relationships among organisms and to communicate with others about these relationships is the development of a classification scheme. The science of classifying and naming organisms is **taxonomy.**

Living organisms as well as fossils are classified by a scheme that was developed by the Swedish naturalist, Carl Linné (better known as Linnaeus), who lived during the eighteenth century. Linnaeus established a hierarchical classification, officially published in the 10th edition of his book *Systemae Naturae* in 1758.* This was a system in which each individual organism was grouped with "like" individuals into various categories (Table 5-1). Progression upward through the hierarchical system is through a succession of categories that are more inclusive and more diverse. These categories within the classification scheme represent groupings based on the degree of similarity or difference among organisms. Especially important for assigning organisms to their proper categories in the hierarchical system are observable features such as the morphology of the external form (important in both paleontologic and biologic study) and of the internal anatomy (biologic study).

The most inclusive category in the Linnaean hierarchy is the *kingdom*. In the previous chapter, reference was made to five kingdoms, although only four are well represented in the fossil record: kingdoms Monerae, Protistae, Plantae, and Animalae. Linnaeus in the mid-eighteenth century differentiated only between plants and animals. The kingdom category is a very broad one and each kingdom contains a large spectrum of different-appearing, but related organisms. Each category below the level of kingdom becomes less inclusive and thereby more restrictive in the degree of similarity and biological relationship among the included organisms.

Thus each of the kingdoms is subdivided into various phyla (singular, **phylum**) and these phyla, in turn, are subdivided into **classes,** which, in turn, are subdivided into **orders.** Further subdivision of each category forms the kingdom-to-species classification scheme shown in Table 5.1. Viewed in the opposite direction (bottom to top) closely related **species** are grouped into **genera** (singular, **genus**), closely related genera are grouped into **families,** and so forth. In each situation, as dictated by the hierarchical system, organisms are grouped into broader, more inclusive categories.

The basic, fundamental unit of this classification scheme is the *species*. Biologically, a species is composed of related organisms that share a common *gene pool* (all the genes of a particular population). This means that a typical species is a group of organisms whose members are capable of interbreeding (exchanging genetic material) and producing viable offspring (see discussion of species concept, Chapter 7).

Linnaeus used the categories of kingdom, class, order, genus, species, and variety and listed about 4000 species in his book. Since the time of Linnaeus, however, the number of *known* species has increased greatly and with this increase the main categories of phylum and family have been added to the original classification. Further refinement of the Linnaean system has

---

Table 5-1. Linnaean and modern systems of classification of organisms

KINGDOM* (two: plants and animals)	KINGDOM
	PHYLUM
CLASS (six: worms, insects, fish, amphibians, birds, and mammals)	CLASS
ORDER	ORDER
	FAMILY
	GENUS
GENUS	SPECIES
SPECIES (about 4000)	
(variety)†	

*Linnaean system names in boldface.
†Variety indicates a distinctive group of individuals within a species population; not a formal subspecies.

---

introduced subcategories, some of the more common of which are *subphylum, subclass,* and *superfamily.* This further splitting of categories has increased the number of classification "cubbyholes" available for differentiating organisms. These refinements in the original Linnaean classification reflect increased knowledge, just as any classification reflects the state of knowledge of its subject matter. A classification can be viewed as an index to the level of understanding of a particular subject.

In addition to refinements of the classification of organisms, the assignment of organisms to appropriate categories is not static or inflexible, but instead may change with time as knowledge and understanding of evolutionary relationships increase. In paleontology this is particularly true for assignments below the rank of class, but for some enigmatic extinct groups (see examples, Chapter 4), correct placement in the proper phylum remains a problem.

Nomenclature

Using all seven major classification categories to name each organism would be cumbersome. One of the main aspects of the Linnaean classification is the *nomenclature,* whereby each species is identified by a two-part or *binomial* name that is latinized. Modern humans belong to a species designated by the binomial name *Homo sapiens* (Table 5-2). The first term, *Homo,* is the genus and, by custom, is capitalized. The second term, *sapiens,* is the *specific* name and is not capitalized.

Table 5-2. Classification of humans in the modern hierarchical system

KINGDOM	animalae
PHYLUM	Chordata (Vertebrata)
CLASS	Mammalia
ORDER	Primate
FAMILY	Hominidae
GENUS	*Homo*
SPECIES	*sapiens*

---

*Stockholm, Sweden.

Two species of early humans, both extinct, also are assigned to the genus *Homo: Homo habilis* and *Homo erectus*. The generic name, *Homo* in this example, can be used only once in naming organisms; once assigned it cannot be used again in any of the other kingdoms. The specific name, however, is a modifier and can be used any number of times in association with different generic names. The multiple use of a specific name is exemplified by *Placenticeras meeki* (Fig. 5-2), an ammonoid cephalopod, and *Dawsonella meeki,* a gastropod. In this example the modifying specific name was derived from the proper name of a paleontologist, F. B. Meek, who accompanied late nineteenth century exploration surveys of the western United States. Rules of *binomial nomenclature* (established by the International Commission on Zoological and Botanical Nomenclature and set forth in respective codes) require the names at every level of the classification to be latinized, thus avoiding the obvious problem of language barriers. It has been decreed that both generic and specific names be italicized when printed and underlined when written. A name, whether at the species, genus, family, order, or other level, is referred to as a **taxon** (plural, **taxa**).

Taxonomic Problems

Paleontologists commonly differ in opinion on the classification of some fossils. Should a fossil in question be "lumped" with a particular previously described taxon or should this fossil be assigned to another or perhaps even new taxon?

The assignment of fossils to particular taxa on the basis of their morphology has involved the judgment of paleontologists for more than two hundred years—since the time of Linnaeus. Linnaeus, however, was a creationist, assuming that each species was an immutable, nonevolving entity. We still use the Linnaean classification scheme, but the interpretation of the evolutionary thread of continuity that binds the various taxonomic levels (categories) is vastly different from the more simplified Linnaean concept of static categories differentiated solely on the basis of morphologic similarity.

A good meaningful classification of fossil organisms should attempt, to the full limits of knowledge available, to reflect the evolutionary biologic relationships among the organisms being classified. These evolutionary relationships are not indicated by morphology alone, but also are dependent upon stratigraphic position and geographic location of the specimens. When a fossil is classified, in effect its evolutionary relationships are designated; i.e., classification is a statement of how we think the evolution of that organism proceeded. There still remains, therefore, the problem of proper assignment of some extinct groups. Where do the extinct archaeocyathids fit into the evolutionary scheme of life? Archaeo-

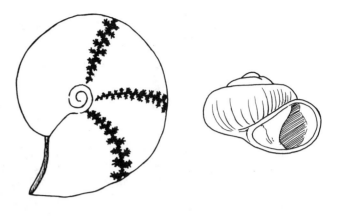

A *Placenticeras meeki* (×0.8)   B *Dawsonella meeki* (×8)

Figure 5-2. A generic name, such as *Placenticeras* or *Dawsonella,* may be used only *once* in naming organisms, but the species name, such as *meeki,* may be repeated. A. A Cretaceous ammonoid cephalopod. B. A Pennsylvanian gastropod.

cyathids were corallike in overall form, but not in detail. They were spongelike in some features, but do not have typical sponge spicules (see Chapter 4). At various times the Archaeocyatha have been placed in several different phyla, as well as in different classes within a single phylum. The question still persists as to whether they should be placed in the same phylum as the sponges, or the corals, or in a separate and distinct phylum, as is the present practice.

Likewise, the toothlike conodonts have been variously classified with fish, gastropods, worms, crustaceans, and cephalopods. As noted in Chapter 4, the usefulness of conodonts in dating Paleozoic rocks is not hampered by the question of their taxonomic affinity. Conodonts, as a reflection of the current state of knowledge, or lack of it, are presently placed in a utilitarian (for convenience) *order,* of an unknown class of an unknown phylum (see Appendix B).

Another important taxonomic problem in dealing with fossils is the existence of transitional forms. Where should mammallike reptiles be placed in the classification scheme? Other problems of this nature are found in the paleontologic literature. Regardless of differences of opinion (which are most often of degree rather than major schisms), the overall classification scheme described has been found to be a workable one. The basic scheme of the classification hierarchy has not changed during the two hundred years of its use, but refinements have involved detailed portions of it and made it more useful. Major changes in assignments of organisms to particular taxa within the classification scheme generally reflect more sound paleontologic judgment based on increased accumulation of data from the fossil record and better understanding of evolutionary relationships.

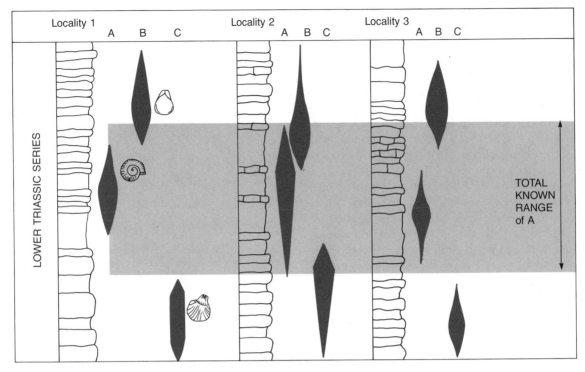

Figure 5-3. Use of several guide taxa to correlate Triassic rocks at three localities. The dark patterns of each taxon (A, B, C) indicate ranges and relative abundance, with the widest part of the pattern representing maximum development of the taxon. Use of several guide taxa in correlation is more accurate than use of a single species.
(After R. H. Dott, Jr., and R. L. Batten, 1981, *Evolution of the Earth,* 3rd ed., Fig. 3.16, p. 58: McGraw-Hill Book Co., New York. Reproduced by permission of publisher)

THE IMPORTANCE OF FOSSILS

Biostratigraphy and Correlation

Previous discussion (Chapter 3) indicated the relationship between time units (Cambrian Period) and time-stratigraphic units (Cambrian System) and explained the absence of considerations of time in the definition of rock-stratigraphic units such as the formation. A fourth category of stratigraphic classification is the biostratigraphic unit, which represents the division of strata on the basis of fossil content. Recall from previous discussion that the fundamental biostratigraphic unit is the **biozone.** Each biozone is named for a particular genus or species that occurs within it (although not necessarily *confined* to it). Basically, biozones are descriptive in nature; critical examination of the stratigraphic (vertical) ranges of the fossils is required before *interpretative* time-stratigraphic units (chronozones) can be defined (see Chapter 3).

Major evolutionary events can be recognized in the fossil record. For example, initial appearances and last occurrences (extinctions) are time significant and can be used in time-stratigraphic *correlation* of strata. Time equivalency of strata can be established in places by physical methods such as tracing the distribution of marker beds (volcanic ash beds, turbidites, and so forth;

see Chapter 2) or by recognizing distinctive paleomagnetic signatures (see Chapter 3). However, the most practical tools for determining time-stratigraphic equivalence on a regional to intercontinental (or interoceanic) scale are fossils. One of the most important uses of fossils, therefore, is the assignment of sedimentary strata to their proper position in the geologic time scale. To establish time correlation on the basis of fossils, it is necessary to establish biostratigraphic equivalency of the sections being studied. On a large regional to intercontinental scale this commonly can be accomplished by using guide taxa: fossil species or genera that are especially distinctive and serve as keys or guides to particular intervals of time (Fig. 5-3).

Guide taxa typically lived for relatively "short" spans of geologic time (thus have a short *stratigraphic* range, which is a reflection of rapid evolution). They also represent organisms that had widespread geographic distribution (which suggests a broad range of ecologic tolerance) and as fossils are easily recognized and abundant (Fig. 5-4). It should be noted, however, that refined time-significant correlation of widely separated regions (on an intercontinental scale) generally is much more complex a problem than can be resolved by simply matching one or a few guide taxa. As a common practice, *assemblages* (rather than individual taxa) of many

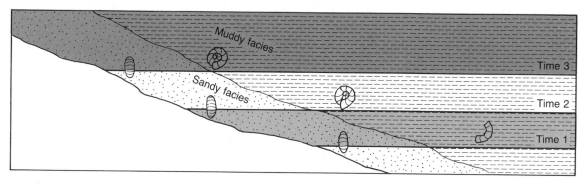

Figure 5-4. Usefulness of species as guide taxa. Brachiopod *Lingula* is a bottom dweller (benthic) and migrates with its shifting sandy environment. It has evolved slowly through time and the same form appears in many time spans; *Lingula* is *not* a useful guide taxon. The swimming cephalopod (nektic) is independent of the bottom environment and has evolved more rapidly, with distinctive forms representative of various time spans. The cephalopod also has a wide geographic distribution; it *is* a useful guide taxon.
(After R. H. Dott, Jr., and R. L. Batten, 1981, *Evolution of the Earth,* 3rd ed., Fig. 3.15, p. 57: McGraw-Hill Book Co., New York. Reproduced by permission of publisher)

diverse and perhaps dissimilar taxa are compared; this comparison may involve bringing in intermediate regions whose stratigraphic sections have taxa in common.

The use of *overlapping ranges* of taxa is an effective means of using biozones in determining time-stratigraphic correlation (Fig. 3-16B). This procedure involves careful collection of fossils and accurate plotting of the lower and upper occurrences of the taxa (Fig. 5-5) in each stratigraphic section. Time equivalency is indicated for the portions of the sections that have overlapping ranges (chronozones) of the taxa present. The most precise chronozone boundaries are based on the *first occurrences* of successive taxa.

It is essential to emphasize that the degree of accuracy of correlation using fossils is highly dependent upon accuracy of the identification of the fossil (proper taxonomic assignment), as well as on familiarity with ranges, evolutionary levels of development, relationships to living species, and other factors that enter into the total correlation procedure.

Documents of Organic Evolution

Evolutionary events, as previously indicated, serve as the basis for the distinction of paleontologically defined time-stratigraphic boundaries. Understanding of the scope of evolutionary processes that have operated through time has come largely from study of the fossil record. Paleontologists and biologists draw heavily on the fossil record for examples not only to support the concept of evolution, but also to interpret the rates, trends, tempo, and patterns of evolutionary change. The changes in assemblages of organisms through time (manifested by fossil succession) have been brought about through the interactions of environment and the *genetics of species populations*. The regulator, or impetus for change, is the environment, which exerts selective pressure on

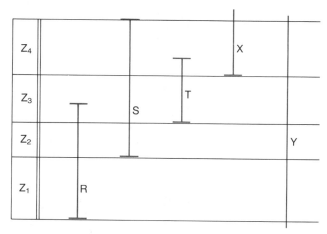

Figure 5-5. Overlapping range zones are determined by the concurrent appearance of an assemblage of species; species occurrence is represented by R, S, T, X, Y.

Zone 1 (Z_1) is characterized by fossils R and Y only.
Zone 2 (Z_2) is characterized by fossils R, S, and Y.
Zone 3 (Z_3) is characterized by fossils R, S, T, and Y.
Zone 4 (Z_4) is characterized by fossils S, T, X, and Y.

Note that each zone is characterized by a distinctive combination of fossils that is different from the assemblage in the zone above and the zone below it. Some species begin at the base of a zone, some end at the top of a zone, and some species occur throughout an entire zone or zones. Zones may differ in the length of time they represent; they are *not* defined on an established interval of a certain number of years.

individuals within the population and favors those that are most successfully adapted. (For more detailed discussion on evolutionary processes see Chapter 7.)

Although the fossil record is biased and incomplete, it is complete enough to reveal many good examples of evolutionary changes, for example, the development of the horse (see Chapter 15). The major features of evolution have been determined for a number of taxa, enriching our understanding of the development of life

Figure 5-6. A sequence of morphological changes is illustrated by the biostratigraphic succession of adult shells of cephalopod fossils in the Devonian Hunsrück Shale of Germany. Transitional forms indicate an apparent evolution of early ammonoids from the nautiloids. A–E are nautiloids; F–M are early ammonoids.
(From H. K. Erben, 1966, Über den Ursprung der Ammonoidea, Fig. 1, p. 643: *Biol. Reviews,* vol. 41. Reproduced by permission of Cambridge University Press)

through geologic time. An outstanding example of the evolutionary trend of organisms is shown by the biostratigraphic succession in the Devonian Hunsrück Shale of Germany. The fossils in the shale (Fig. 5-6) display evidence that the ammonoid cephalopods evolved from a stock of nautiloid cephalopods. Recall that Cuvier's interpretation of the succession of changes in the biostratigraphic record revolved around supposed catastrophic events that destroyed earlier life forms. In contrast to his concept, the transitional sequence from nautiloid to ammonoid illustrated in Figure 5-6 is evidence of the modern view, pioneered by Charles Darwin, that fossil succession (the biostratigraphic record) reflects *organic evolution* by *natural selection*. (See

also evolutionary trends of plants, Chapter 14; of reptiles, Chapter 12; and of mammals, Chapter 15.)

Paleoecology

Organisms are influenced in their stratigraphic distribution not only by evolutionary changes through time (e.g., cephalopods, Fig. 5-4), but also by environmental factors. Marine bottom-dwelling organisms such as *Lingula* in Figure 5.4 are limited in their geographic distribution by environmental influence (see Table 5-3). Swimming organisms such as fish and cephalopods, or floating forms such as graptolites (Chapter 4), are less affected by bottom environmental conditions. These an-

Table 5-3. Environmental parameters in a marine ecosystem

Physical parameters	Biologic parameters
Water temperature	Birth and death rates of organisms

Water composition
  Salinity
  Dissolved gases (oxygen, nitrogen)
  Trace elements present
	Size and number of organisms (biomass)

Water depth
  Light conditions
  Pressure
	Food supply and nutrients
	Mode of life (planktic, nektic, benthic, sessile, or vagrant)

Turbulence (energy level)	
Turbidity	
Upwelling currents	
Type of substrate (material forming the sea floor)	
Topography of the sea floor	

imals generally are more widely dispersed than bottom dwellers and consequently their fossils are more useful for time-stratigraphic correlation. Recall that the great bulk of the fossil record is composed of shell-bearing, shallow-marine, *bottom-dwelling* invertebrates; thus most fossils have the potential for providing information about ancient environments.

Paleoecology is that branch of paleontology that is specifically concerned with the study of relationships of fossil organisms to each other and to their ancient environments. It involves applications of principles of modern ecology, thus invoking the concept of *actualism,* whereby studies of living populations of organisms and their interactions with environments provide insight into ancient rocks and their contained fossils. Fundamental concepts in ecologic-paleoecologic study are the **habitat,** which is the specific environment ("address") occupied by a species; the **niche,** which is the specific role ("profession") or lifestyle of the species; and the **community,** which is an association of organisms living in close proximity and having a tendency to be found together. Habitats commonly are occupied by a number of species, each species having a distinct niche and each making a contribution to the interrelated activities of the community. Both the living species and the nonliving components of the environment interact with one another to form an **ecosystem.**

In a marine ecosystem such as that illustrated in Figure 5-7, three basic types of organisms typically are recognized. The **benthos** is composed of organisms *(benthic)* that live on or within the sea floor. Benthic forms living *on* the bottom *(substrate)* are the **epifauna,** some of which *(vagrant)* are able to move about and some of which *(sessile)* are attached to or recline on the substrate. Benthic forms that burrow into the substrate make up the **infauna.**

The **plankton** are organisms *(planktic)* that float freely in the water; planktic plants are *phytoplankton* and planktic protists and animals are *zooplankton.* The swimming organisms are referred to as the **nekton;** these organisms are able to control their own movements in the water. The relationships between the phytoplankton, the zooplankton, the nekton, and the benthos and the relationships between these groups and the physical environment form complex patterns.

Species existing in communities are fundamental links in **food chains,** which represent the feeding sequence from primary **producers** through various levels of **consumers.** An example of one simplified food chain existing in oceanic environments today is indicated in Figure 5-8. This example shows a food chain beginning with the phytoplankton, which utilize materials in the environment plus sunlight and by the process of *photosynthesis* produce their own food and energy. These producer organisms are called **autotrophs;** they have played a significant role in the history of the atmosphere and the biosphere (Chapter 8). Microscopic-sized autotrophs that float in the photic zone of the oceans are consumed by organisms on the next higher step of the chain, and these in turn are consumed by other organisms. The consumer organisms at each step are **heterotrophs** and must ingest their food. Among the marine invertebrates, for example, *filter* (suspension) *feeders* obtain food by filtering quantities of water across membranes to trap small organisms or organic particles; *deposit* (detritus) *feeders* obtain food by ingesting sediment and selectively removing the organic material; *grazers* ingest plants or scrape algae from hard surfaces; *decomposers* break down organic material. The particular mode of food gathering is a most important aspect of the adaptability of any particular species. Usually there is considerable interaction among the myriad

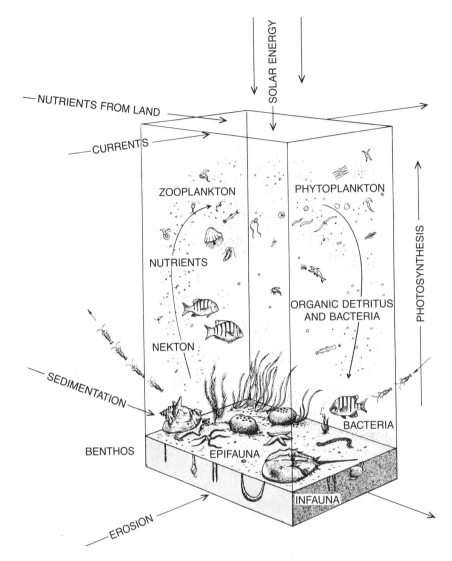

Figure 5-7. Elements of a marine ecosystem; see text for definition of terms.
(Based on graphic by T. L. Chase in R. L. Anstey and T. L. Chase, 1979, *Environments Through Time,* 2nd ed., Fig. 3.9, p. 27: Burgess Publishing Co., Minneapolis MN. Original source is J. W. Hedgpeth, Marine Ecology, in R. W. Fairbridge, ed., 1966, *The Encyclopedia of Oceanography.* Van Nostrand Reinhold Co. Inc., New York. Reproduced by permission of publisher)

species within food chains; this produces a much more complex association called a food web. A food web can be visualized as a complex energy-flow system progressing through successive levels.

A similar situation exists on land. Although the terrestrial food chain (Fig. 5-9) is more complex than that depicted for the ocean, both have been highly simplified for this discussion. Note that the base of this food chain is still composed of autotrophic species; there are many separate steps consisting of consumer heterotrophs representing higher and higher levels, eventually reaching the highest level. Where would humans fit into a food chain?

Another way of looking at communities of organisms is by determining the particular function (niche) of the various species. As indicated in Figure 5-10, most communities consist of basic nutrients and producer, consumer, and decomposer species. The nutrients include sunlight as well as organic and inorganic molecules; these are converted into food energy by the producers. This energy is used in the food chain as already described. After death, producers and consumers eventually are broken down by bacterial decomposition and thus may provide a source of nutrient molecules to help begin the cycle again. (Remember the geological cycles discussed in Chapter 1?)

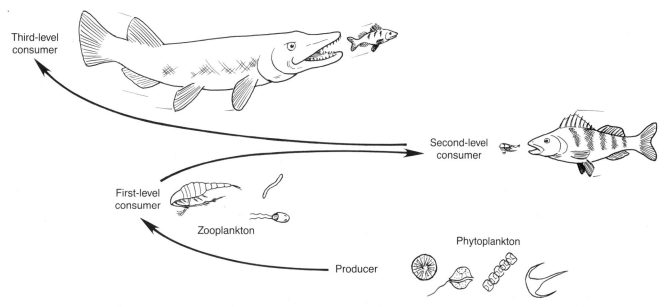

Third-level
consumer

Second-level
consumer

First-level
consumer

Zooplankton

Producer

Phytoplankton

Figure 5-8. Basic elements of a simplified marine food chain representing the feeding sequence from primary producers (phytoplankton) through various levels of consumers (zooplankton, fish). Consumption of the larger fish by humans would add a possible fourth level of consumer.

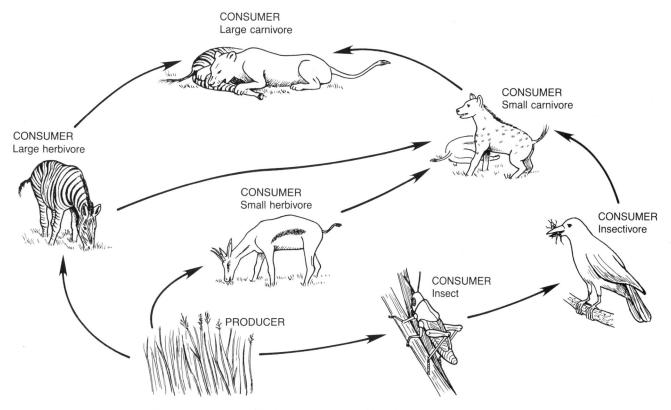

CONSUMER
Large carnivore

CONSUMER
Small carnivore

CONSUMER
Large herbivore

CONSUMER
Small herbivore

CONSUMER
Insectivore

PRODUCER

CONSUMER
Insect

Figure 5-9. Basic elements of simplified terrestrial food *chain* are illustrated in the sequence: producer = grasses, and consumers = antelope and hyena. A food *web* is also represented by the interrelationships of several food chains; i.e., there are several places where food chains connect with one another. As in the marine food chain, the base of a terrestrial chain is a producer (grasses) with various levels of consumers building upon this base. Humans play many roles in food chains and webs, but the most common role is that of a herbivore because grains and other plant materials compose a major proportion of the diet of most people.

By making a large jump in our scale, it is possible to consider the vast number of communities of organisms, which, when considered collectively, represent the earth's *biosphere*. Each community has been described in terms of food chains or by the niche of each member

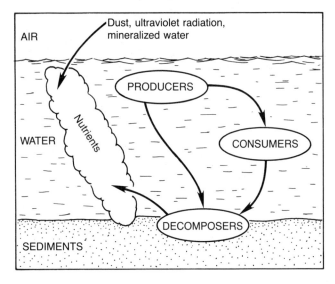

Figure 5-10. Interrelationships within a community based on the function (niche) of various organisms. Note the cyclic nature of the model, with the consumers and producers being decomposed into nutrient materials to be utilized by new organisms.

species, but individual communities are not isolated. They are somewhat interdependent and there is much interaction, one example being the food web. Different regions have communities composed of different species basically performing the same function (occupying the same niche), but acted upon by different environmental factors. Each organism in an ecosystem is controlled in its distribution by these environmental limiting factors (Table 5-3).

Paleontologists, working with fossil assemblages that usually represent only a part of the original community structure, attempt to reconstruct the ancient ecosystem in which the organisms lived. Reconstruction of ancient community structures involves consideration of the kinds and numbers of fossils present, the condition of the fossil material (for example, fragmented or whole fossils, fossils oriented in possible growth position, and so forth), mode(s) of preservation of fossils, and characteristics of the surrounding rock. All of these aspects of paleoecologic study are important in unraveling the life relationships and postdeath history of fossil organisms. Paleoecologic reconstructions, therefore, involve a synthesis of paleontologic and sedimentologic data (Fig. 5-11).

There is an important interrelationship between paleoecology (the relationships of organisms with ancient

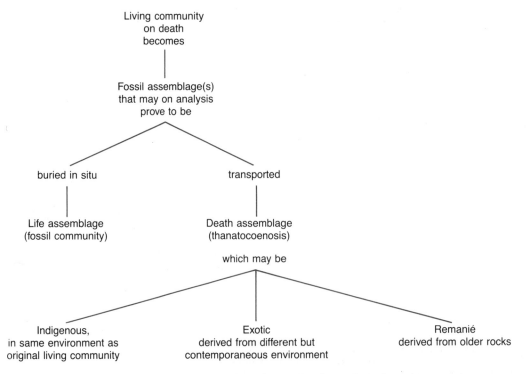

Figure 5-11. Status of a fossil assemblage. From a synthesis of paleoecologic and sedimentologic data it may be determined if a fossil assemblage represents an original living community or whether postmortem events have modified the record.
(From G. Y. Craig and A. Hallam, 1963, Size-Frequency and Growth-Ring Analyses of *Mytilus edulis* and *Cardium edule,* and Their Paleoecological Significance: *Paleontology,* vol. 6. Reproduced by permission of authors)

environments), biostratigraphy (the fossil record), and evolution (organic change through time). The ancient environments provided the driving force for evolutionary change, for the resulting variety of life forms, for distribution of organisms, and for resultant fossil preservation. Evolutionary changes were responsible for the *fossil succession* (biostratigraphic record) preserved in the stratigraphic record, and this, in turn, has provided data for subdivision into workable biostratigraphic units. These units provide the basis for the relative time scale and allow correlation of sedimentary sequences from local to global scales.

Synthesis of the data from this complexity of relationships, it is emphasized again, is based largely on the important concept of *actualism*. The use of studies of recent examples to provide interpretive comparisons is fundamental. This emphasis on modern-day analogues is a basic tool for geologic interpretation; it is an important part of the ticket that allows us to make this trip through time.

Paleobiogeography

Biologists recognize a basic relationship between the distribution of modern plants and animals and the geographic areas where they are dispersed. Areas characterized by a particular composition of organisms and distinctive climatic factors represent **biogeographic provinces.** Recognition of modern biogeographic provinces had its roots in the Renaissance voyages of exploration. As Europeans discovered new lands they also found new floras and faunas that were unlike those in their homelands. In the nineteenth century, naturalists journeyed widely to specifically study these different plants and animals and to map their distribution. It was from observations made on one such worldwide voyage that Charles Darwin amassed the documentation to support his concept of organic evolution (see Chapter 7). On the basis of these voyages, a classification hierarchy of areas was defined; on the continents various *realms, regions,* and *provinces* were recognized. Later this type of biogeographic division was applied to the oceans of the world.

Studies indicate that the latitudinal boundaries or limits of these biogeographic areas are determined primarily by climate, but also by other environmental factors such as topography, availability of food, salinity, and depth of water. An influential factor in the longitudinal separation of biogeographic provinces is the presence of geographic barriers such as mountains or even entire continents, as well as of bodies of water such as oceans. The existence of barriers is directly related to the geologic history of an area. Prior to Late Pliocene time, for example, the Isthmus of Panama did not exist and there was intermingling of the marine life of the Atlantic and Pacific Oceans. The Late Pliocene tectonic uplift of the isthmus created an emergent land barrier between these two oceanic realms, resulting in divergence of the marine faunas on either side. It is significant that the formation of this land barrier between the two oceans simultaneously formed a land connection between South America and North America and provided a bridge for terrestrial faunal mixing and convergence between these two continents. The biostratigraphic record of the areas adjacent to and including the isthmus helps to date the time of this tectonic event as Pliocene.

Several biogeographic provinces are recognized along the present marine shelf of the East Coast (Fig. 5-12). Because of climatic constraints, species characteristic of the present-day Nova Scotian Province are not likely to be found in the Carolinian or South Florida Provinces. Typically each province has **endemic** species that are confined to it and characterize the province and some **cosmopolitan** species that have a wider geographic distribution.

Very few, if any, taxa in the fossil record had actual worldwide distribution, but many of them were sufficiently widespread geographically that they provide good indexes for *continentwide* and *intercontinental* correlation. It is unmistakable that most fossil-bearing stratigraphic sequences are preserved as part of continental structure (even though perhaps originally marine). The modern-day ocean basins, remember, do not contain rocks recognized as being older than Jurassic. It is recognized that widespread cosmopolitan species are not the *exact* same age everywhere. Dispersal to regions beyond the area of origin must have taken some time. Studies of migrations of modern species, however, have shown that dispersal can occur rapidly once plants or animals have gained access to a new environmentally suitable territory. Migration histories differ considerably and are dependent upon the type of organism studied; migration of planktic or nektic forms, for example, occurs more rapidly than dispersal of benthic organisms. But even benthic organisms can achieve widespread dispersal because many groups have larval stages that are planktic. Research indicates that at least several thousand years are required for a species to reach all areas suitable for its propagation *once* migration routes are open. In terms of geologic time, several thousand or tens of thousands of years is a short interval and does not negate the apparent contemporaneous nature of cosmopolitan fossils.

The distribution of modern plants and animals defines present biogeographic provinces and the distribution of fossils defines *paleobiogeographic* provinces. A classic study by the British paleontologist J. W. Arkell of the biostratigraphic record of Jurassic ammonite cephalopods indicated a marked change in the provinciality (number of provinces) during the Early Jurassic com-

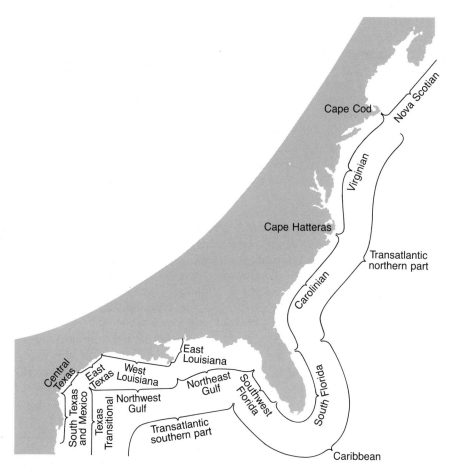

Figure 5-12. Present biogeographic provinces along the east coast of the United States. Climatic factors control the occurrence of species that are typical of each province. Species characteristic of the Nova Scotian Province, for example, would not be expected in the South Florida Province. (From J. W. Valentine, 1963, Biogeographic Units as Biostratigraphic Units, Fig. 1, p. 459: *Am. Assoc. Petroleum Geologists Bull.,* vol. 47)

pared to the remainder of the period. Early Jurassic ammonites were cosmopolitan: Species composition is similar in western North American and Europe and stratigraphic correlation is precise. Middle and Late Jurassic ammonites are less widespread, indicating greater provinciality. This led Arkell to define three faunal provinces (Fig. 5-13) for the Late Jurassic.

There is a direct relationship between provinciality and taxonomic diversity. During times of fewer distinct paleobiogeographic provinces (Early Jurassic, for example) organisms that migrated readily became cosmopolitan in distribution. Conversely, during times of greater provinciality (e.g., the Late Jurassic) organisms were less cosmopolitan, a greater number of endemic species existed, and there was comparatively greater total taxonomic diversity. The important paleontologic concept of paleobiogeography relates three aspects of the distribution of fossils: number of paleobiogeographic provinces, provincial versus cosmopolitan nature of the taxa, and taxonomic diversity. It also meshes as a fourth compo-

nent into the interrelationship involving evolution, paleoecology, and biostratigraphy. Furthermore, paleobiogeography provides important insight for understanding the timing of continental drift as well as the biologic consequences of moving continents. As continents have broken apart or moved together, greater or lesser numbers of paleobiogeographic provinces have formed, and thus the distribution of fossils in the biostratigraphic record, such as in the Jurassic example cited, is related to plate tectonics!

The Fossil Record and Plate Tectonics

The concept of moving continents has tenuous roots going back over two hundred years, long before Alfred Wegener's work early in the present century (see Chapter 1). In the mid-eighteenth century the French naturalist Comte de Buffon, on the basis of similarity of *modern* flora and fauna, speculated that North America and Europe had once been joined. Distribution of simi-

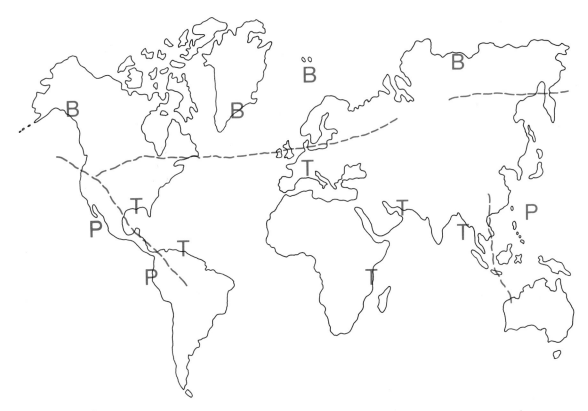

Figure 5-13. Late Jurassic marine realms based on occurrence of ammonite cephalopods. B = boreal; P = Pacific; T = Tethyan.
(From W. J. Arkell, 1956, *Jurassic Geology of the World:* McGraw-Hill, New York. Reproduced by permission of Oliver & Boyd, Edinburgh, Scotland)

lar *ancient* taxa in the biostratigraphic record of now-separated continents was one of Wegener's major lines of evidence for continental drift. Biostratigraphic correlations and paleobiogeographic distribution of fossils are important in both the former continental drift model and the present plate tectonic model and in this light represent a dynamic aspect of paleontologic interpretation.

A relationship between distribution of fossils and plate tectonics is illustrated by closure of a proto-Atlantic Ocean during the Devonian and Carboniferous Periods, with an accompanying decrease in floral provincialism on the colliding continents leading to the cosmopolitan *Glossopteris* floral *realm* of the Permian (see Chapter 11). During the Mesozoic Era the biostratigraphic record of terrestrial reptiles appears to indicate land connections during the Triassic, but provides less evidence of this during the later Mesozoic periods as Pangaea was breaking apart (see discussion, Chapter 12). The northward migration of Australia during the Cenozoic has been accompanied by the progressive adaptation of the flora. The vegetation has adapted to arid and semiarid climates in central Australia as that part of the continent moved into the dry climatic belt between 20° and 30° south latitude. If Australia's northward movement con-

tinues, the flora of central Australia will change to savannah vegetation and then to rain forest as the continent enters the equatorial belt.

Analyses have attempted to relate major evolutionary events such as rapid diversification or extinction of taxa to lithospheric plate motions. There is an interesting twofold relationship between fossils and plate tectonics. First, fossils in the biostratigraphic record provide evidence of moving continents and of the timing of opening and closing of ocean basins; second, plate motions through geologic history have been responsible for large-scale environmental changes that have influenced rates and patterns of organic evolution. Plate tectonics has produced not only a new model for interpretation of earth's tectonic structure, but also a new framework for paleontologic study.

Economic Importance of Fossils

Exploration for petroleum and natural gas provides a good example of the use of fossils in industry. Ancient organisms are the raw material from which the **fossil fuels** (petroleum, natural gas, coal) are derived. Burial and transformation of plant material into coal is relatively well understood, but the burial and transformation of predominantly marine phytoplankton and zooplankton into petroleum and natural gas is a more complex chemical process that is not understood as well. The buried organic debris disseminated in bottom sediments has been compacted and chemically changed to *hydrocarbons*. Following development of the hydrocarbon substances in the *source* rock, the processes of burial and compaction, expulsion of fluids into adjacent rock, migration through *permeable* beds, and final entrapment in *porous reservoir* rock result in accumulations of oil and/or gas (Figure 5-14).

It is the task of teams of petroleum geologists to locate subsurface hydrocarbon traps: anticlines, faults, and stratigraphic facies changes. Microfossils play an important role in the sedimentary-basin analysis that is required for systematically successful exploration ventures. One of the key tools in the exploration for hydrocarbons is the biostratigraphic correlation of rock units in the subsurface as discussed in Chapter 3. Exploration for oil and gas must be conducted in deposits buried beneath the earth's surface, commonly in areas that provide little or no surficial indication of the presence of hydrocarbons. Microfossils such as foraminifers, diatoms, and pollen grains are small enough to remain intact in the well cuttings derived from boreholes. These microfossils serve as valuable aids in correlating subsurface stratigraphic sequences which, in turn, enable geologists to conduct subsurface mapping and to locate geologic structures that are potential petroleum and natural gas traps. It is important to remember that the soft parts of ancient organisms have formed the hydrocarbons and their fossilized hard parts have made possible the biostratigraphic record aiding us in locating the resources.

Other mineral resources associated with sedimentary rocks—rock salt, gypsum, limestone (for manufacturing cement), phosphates (for making fertilizer), and uranium—are found in geologic settings whose characteristics are understood. Paleontologic studies may provide leads to these geologic settings and thus to the discovery of potentially valuable resources.

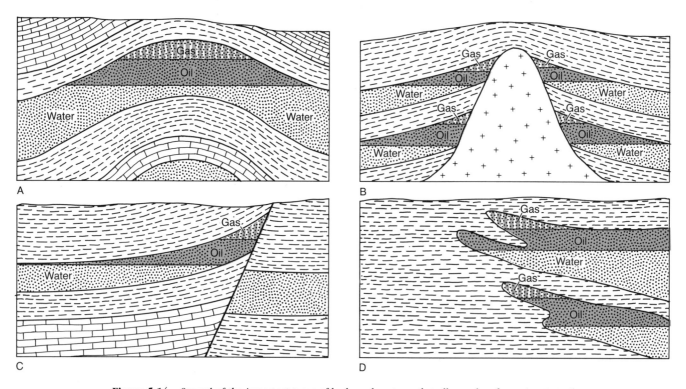

Figure 5-14. Several of the important types of hydrocarbon traps that allow subsurface migrating oil and gas to accumulate in sufficient quantity to form an economic resource. A. Anticlinal trap. B. Salt dome trap. C. Fault trap. D. Stratigraphic facies trap. Note that in each trap the fluids have migrated through a permeable bed (called reservoir rock; typically sandstone) until they reach impermeable rock units (typically shale) that act as barriers to further migration. Thus the hydrocarbons are entrapped.

SUMMARY

Fossils are classified on the basis of the Linnaean biological classification, a hierarchical system in which subdivisions become less inclusive as one progresses through the categories kingdom, phylum, class, order, family, genus, species. The fundamental unit in this system is the species: a group of related organisms that share a common gene pool and typically are capable of interbreeding and producing viable offspring. Each species is known by a binomial name; for example, modern humans are *Homo sapiens*. Problems of classification that arise—overlapping of terms or use of synonyms—and recognition of new species are controlled to some degree by an international commission. The science of classifying and naming organisms is taxonomy.

Fossils are more than museum curiosities; they are useful in providing

1. the basis of biostratigraphic study that led to the development of the relative geologic time scale;

2. an important tool in stratigraphic correlation (assemblages of fossils, biozones, guide taxa);

3. the record of evolutionary development of life forms;

4. the basis for paleoecologic studies (recognition of habitats, niches, and food chains within ecosystems);

5. the basis for paleobiogeographic studies (endemic versus cosmopolitan species and their association with biogeographic provinces);

6. a line of evidence in Wegener's original concept of continental drift and in the modern plate tectonics model; and

7. a tool for exploration for earth resources, especially hydrocarbons.

Fossils have long fascinated humans. These relics of former life have been articles of religious value and have been items of mere curiosity, but more significantly they are important tools in solving geologic problems.

SUGGESTIONS FOR FURTHER READING

Fortey, Richard, 1982, *Fossils: The Key to the Past:* Van Nostrand Reinhold, Florence KY.

Hallam, A., 1972, Continental Drift and the Fossil Record: *Sci. American,* vol. 227, no. 5, pp. 56–66.

Laporte, Leo F., 1978, Introductions, *in Evolution and the Fossil Record:* Readings from *Scientific American,* W. H. Freeman, San Francisco.

MacDonald, J. R., 1978, *The Fossil Collector's Handbook, A Paleontology Field Guide:* Prentice-Hall, Englewood Cliffs, NJ.

May, Robert M., 1978, The Evolution of Ecological Systems: *Sci. American,* vol. 239, no. 3, pp. 160–175.

McKerrow, W. S., ed., 1978, *The Ecology of Fossils: An Illustrated Guide:* Duckworth & Co., London.

Skinner, Brian J., ed., 1981, *Paleontology and Paleoenvironments:* Readings from *American Scientist,* Wm. Kaufman, Los Altos, CA.

Valentine, James W., and Moores, E. M., 1974, Plate Tectonics and the History of Life in the Oceans: *Sci. American,* vol. 230, no. 4, pp. 80–89.

6

SHIFTING SANDS AND MURKY MUDS: Ancient Sedimentary Environments and Facies

CONTENTS

KEY TERMS

Sedimentary facies
Graded bedding
Turbidity current
Turbidite
Ripple mark
Desiccation crack
Biofacies
Lithofacies
Facies tract
Walther's Law
Transgression
Regression
Onlap
Offlap
Sedimentary cycle
Reef
Back reef
Fore reef
Basin
Paleogeographic map

FROM PERMIAN SEA TO NATIONAL PARK

History has been described as a continuous stream of time—the origin and early beginning of the solar system flowing into and merging with the early geologic history of the earth, and this, in turn, flowing into and incorporating the origin and development of life, including the history of human existence. Guadalupe Mountains National Park in West Texas and southeastern New Mexico provides an excellent example of this stream of history. The park was created because of the natural beauty of the area, its geologic history (outstanding development of ancient sedimentary environments), its long period of associated human history, and more recent events involving the dedication of one person in particular who felt the area should be preserved for future generations to observe, understand, and appreciate.

The rocks of the Guadalupe Mountains are the exposed part of what has been called the largest fossil reef in the world. This feature, Capitan Reef, is a large barrier reef (carbonate buildup) that developed during Permian time. Geologically, the Capitan Limestone and associated rocks provide an outstanding example of the development of a spectrum of depositional environments: reef, back-reef lagoon, and deep basin seaward of the reef (Fig. 6-1). Each of these settings produced distinctive sediments and life forms. One of the principal attractions of this area, the world-famous Carlsbad Caverns, is developed in the deposits of this ancient reef complex. The rocks of this complex will serve as an illustrative example of sedimentary *facies* later in this chapter. Here, let us look briefly at the human history leading to the development of this area as a national park.

Indians were living in the Guadalupe Mountains as long as 12,000 years ago, as indicated by carbon-14 dating of charcoal from their fires. Evidence of their presence is found also in caves and in rock shelters where pictographs were painted on the walls. The first known written references to this area are in accounts from the 1700s of explorations of Spanish conquistadores. It is not known, however, if any of these explorers were within the present park boundaries. The name Guadalupe was first used for these mountains on a map dated 1828. The remoteness of the area, the aridity of the land, and the presence of the Mescalero Apache discouraged settlement.

Recognition of this area as an outstanding section of Permian rocks is reflected in the established names for subdivisions of the Permian System in the United States. The lower part of the Permian is divided into the Wolfcampian and Leonardian Series, both names derived from geographic sites (type areas) in the Glass Mountains of West Texas. The upper part of the Permian is divided into the Guadalupian and Ochoan Series, names also derived from this region. The stratigraphic sections from which these names are

A

B

Figure 6-1. A. El Capitan, limestone cliff in Guadalupe Mountains National Park. B. Rock units in the El Capitan section: 1. Capitan Limestone, massive geologic reef. 2. Delaware Mountain sandstones deposited in deeper water seaward of the reef. 3. Bone Spring Limestone also deposited in deeper water than the reef structure. The area between 1 and 2 is covered with Quaternary slope deposits. (Photo by J. Cooper)

derived are examples of *stratotypes* (see Chapter 3): type sections of regional time-stratigraphic units whose boundaries are clearly defined on the basis of paleontologic criteria. The Guadalupe Mountains region is the site of the standard American Permian stratigraphic section and is one of several regional reference sections for correlation and comparison of Permian rocks on a worldwide basis. The original type-Permian section was established by Roderick Murchison in 1821 in the province of Perm, Russia (U.S.S.R.). It is necessary to have a number of intermediate reference sections to enable worldwide correlation.

Wallace Pratt, a petroleum geologist, was intimately familiar with the Guadalupe Mountains and the impressive historical sequence they display. His enthusiasm for the area was not limited to wishful thinking that something should be done to preserve this rock record. Fortunately Pratt had considerable landholdings in the region; in 1961 he donated over 5000 acres to the federal government for the establishment of Guadalupe Mountains National Park. The donated area included McKittrick Canyon, whose north wall is 600 meters high and displays a graphic cross section of the various *carbonate rock* facies that are the lithified products of contemporaneous depositional environments of the ancient reef complex. Over 70,000 acres eventually were purchased by the government and included in the park. The park was opened officially to the public in 1970.

In 1976 Wallace Pratt, then in his 91st year, was interviewed and was asked why he had felt moved to donate this land. Pratt's reply expresses the depth of his feelings for this land and his concern for its preservation:

> The canyon exposes a precise cross-section of the Capitan Barrier Reef, which is unique in the Western Hemisphere. There are more than five hundred different fossils to be found here. But of greatest importance, the canyon clearly exposes the anatomy of the organic reef. By giving the land to the park service [I have the assurance] of preservation of a record of natural events over a period of two hundred million years.*

To be a "giant" of geology does not require being a nineteenth century founder of the science!

_____

*From interview printed in *Exxon USA,* 4th quarter, 1976.

ROCKS AND ENVIRONMENTS

The Grand Canyon Succession

The mighty cleft of the Grand Canyon (Fig. 6-2) arouses wonder in the minds of many of those who stand on its rim and look out over the imposing view. Questions arise as to how this "gash" in the earth's crust was formed: When did all of this begin? What is the significance of the different-colored rock layers? One generally is overcome, too, with a feeling of spaciousness

Figure 6-2. Grand Canyon view from the south rim; inner canyon in center of photograph is the course of Bright Angel Creek.
(Photo by J. Patterson)

and grandeur, even awe. The eighteenth century catastrophists would have invoked a great earth-splitting event to create the canyon, but *actualistically* the Grand Canyon, which exposes more than a billion years of earth history, is testimony to the long-term erosive downcutting of the Colorado River. It is a marvelous place to take a trip through time.

Joseph Wood Krutch, a well-known English professor and drama critic, reflected on the canyon view:

> As the river sawed slowly through the rising strata, its deepening walls exposed again to sight older and older formations going back more and more millions of years until, finally, they add up to more than a billion. . . . Seated at my point on the rim, I look up and down as well as east and west, and the vista is one of the most extensive ever vouchsafed to man. But I am also at a point in time as well as in space. The one vista is as grandiose as the other. I am small and alone in the middle of these great distances, vertical as well as horizontal. *But the gulf of time over which I am poised is inconceivably more vast and much more dizzying to peer into* (italics added).*

The present-day uplifted Colorado Plateau and deeply incised Grand Canyon are geologically young features. During the last billion years, the topography of this region and its paleoenvironmental conditions have changed many times. In the Grand Canyon's colorful Paleozoic sedimentary rocks alone, there is exposed a record of about 350 million years of change, of advancing Cambrian seas, of Carboniferous rivers and deltas, of Permian tidal flats and sand dunes. These changing paleoenvironments are recognized by reading the rock

---

*J. W. Krutch, 1957, *Grand Canyon: Today and All Its Yesterdays:* William Murrow, New York.

pages of this historical volume. Each rock layer in the vertical succession of Paleozoic formations in the walls of the Grand Canyon is a product of the paleoenvironment that existed in the district at the time the sediment was deposited.

The Meaning of Facies

Changes in paleoenvironments are not documented only in *vertical* sections, but also may be expressed by *lateral* variation, as exemplified by the Old Red Sandstone in Britain. This formation was recognized by early workers in southern Wales as a product of continental (nonmarine) deposition. The Old Red consists of thick sequences of red conglomerates, sandstones, and shales containing fossils of early land plants and amphibians and of freshwater fish. By tracing outcrops of the Old Red Sandstone southward through the British Isles, Sedgwick and Murchison determined that the nonmarine beds interfinger with and grade laterally into marine carbonate rocks of the *type Devonian* in northern Devon, England. The regional intertonguing of these units indicates that during Devonian time both marine and nonmarine paleoenvironments existed in different parts of what is now the British Isles and that each paleoenvironmental area was the site of accumulation of distinctively different sediments (Fig. 6-3). This lateral change from nonmarine to marine rocks of equivalent age is an example of a facies change. A **sedimentary facies** is *lateral* variation in sedimentary rock units that are partly or wholly equivalent in age. The lateral variation is produced by deposition in different, but laterally adjacent, environments. Most commonly, changes between sedimentary facies are gradational transitions, rather than the sharply defined tongues of marine and

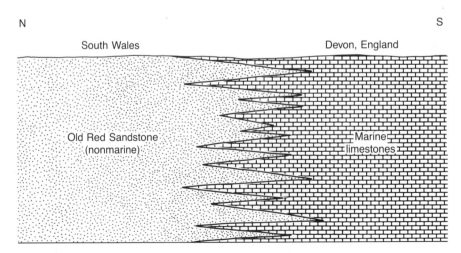

Figure 6-3. Sedimentary facies change in Devonian rocks of Great Britain. To the north, in Wales, deposition consisted of nonmarine sandstones and conglomerates; southward, in England, marine limestones were deposited. The lateral equivalency of the units is diagrammatically illustrated by the interfingering relationship between the two lithologies.

nonmarine rocks graphically depicted in Figure 6-3.

The original work leading directly to the recognition of sedimentary facies relationships was done in the 1830s by a Swiss geologist, Amanz Gressly. Gressly traced Jurassic sedimentary rock units in the Jura Mountains and found that the lithology of the rocks varied laterally; contemporaneous limestone and shale lithologies were the result of differences in the paleoenvironments in which the original sediments accumulated. Gressly was the scientist who first used the name *facies* for the distinctive units in this sort of lateral relationship.

THE RELATIONSHIP BETWEEN FACIES AND ENVIRONMENTS

Introduction

Sedimentary facies represent products of different environments of deposition. Facies analysis involves study of the internal characteristics of facies as well as of distribution patterns and relationships and is critical for correct interpretation of ancient environments (paleoenvironments). One of the keys to recognition of ancient sedimentary environments is understanding present environments; this is an example of the use of modern analogues. On the earth's surface today there

are many different sedimentary environments (Fig. 6-4): river channels and floodplains, deltas, coastal lagoons and beaches, continental-shelf and offshore areas, and various other settings. Each of these environmental settings is an area where sediment accumulates and where organisms live and die. Each of these environmental areas produces a representative sedimentary rock type or assemblage, dependent upon the physical, chemical, and biological processes operating. It is necessary to understand present-day processes and environments in order to identify and interpret their ancient counterparts (see Table 6-1). Many criteria useful in the recognition of ancient sedimentary environments are determined from study of present-day processes; the most important are lithology, geometric shape of the rock unit, sedimentary structures, and fossils.

Lithology

Lithology refers to the type of sedimentary rock, determined by *composition* and *texture*. These physical characteristics result from several factors related to the rock cycle (see Chapter 1). Some of the factors are the type of source rocks that are weathered to provide clastic debris, the method of transport of particles, the conditions prevalent in the depositional environment, and

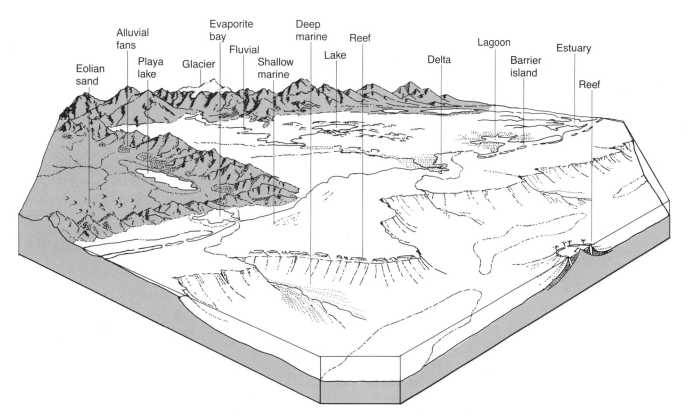

Figure 6-4. Overview of modern sedimentary environments.
(Based on W. K. Hamblin, and J. E. Howard, 1980, *Exercises in Physical Geology,* 5th ed., Fig. 98, p. 49: Burgess Publishing Co., Minneapolis, MN)

Table 6-1. Physical characteristics of some sedimentary environments

Environment	Common sediments (texture)	Sedimentary features
NONMARINE (terrestrial) Fluvial (stream)	Gravel, sand, mud	Poorly sorted material Rounded pebbles Lenses of sandstone Cross-beds
Lacustrine (lake)	Silt, mud, lime*	Laminated beds Freshwater fauna
Swamp	Silt, mud, organic debris	Decayed vegetation, peat
Desert Dune	Sand	Well-sorted material Frosted, rounded grains Large-scale cross-beds
Playa lake	Mud, evaporite salts	Laminated beds Desiccation cracks
Alluvial fan and debris flow	Boulders, gravel, sand	Poorly sorted material Beds of limited extent Lenticular or wedge-shaped units Cross-beds
Glacial Ice deposits	Morainal debris of all sizes	Unsorted debris Striated boulders
Glacial lake Fluvial (meltwater streams)	Silt, mud	Varved (annual) layers
TRANSITIONAL Delta	Complex intermingling of many subenvironments such as channels, swamps, freshwater lakes (see text discussion)	
Coastal lagoons, bays, estuaries	Silt, mud	Thin beds Brackish-water fauna
MARINE Supratidal (typically broad flats occasionally inundated by wind tide)	Evaporites, dolomite, silt, mud	Laminated beds Desiccation cracks Stromatolites
Littoral (intertidal)	Cobbles, sand, mud	Laminated beds Mud cracks Trace fossils Stromatolites (carbonate) Well-sorted sands
Sublittoral Shelf	Sand, silt, mud	Wide geographic extent Thin to massive bedding Cross-beds Diverse fauna
Carbonate platform	Calcareous sediments (lack of land-derived clastic debris)	Oolites Skeletal sands Lime mud Intraformational conglomerates

    Also includes the reef environment (see text discussion)

*Carbonate deposition (lime) may occur in a wide spectrum of environments; wherever carbonate material is in solution and there is relatively little influx of clastic debris.

Table 6-1. Physical characteristics of some sedimentary environments (continued)

Environment	Common sediments (texture)	Sedimentary features
Bathyal depths (100+–4000 m)	Mud Sand and gravel in submarine fans	Graded bedding (turbidites) Thick, coarse beds Sole marks Deep-water fossils
Abyssal depths (more than 4000 m)	Mud, ooze (biogenic)	Relatively thin and even bedding Deep-water fossils

the processes of lithification that convert loose sediment particles into indurated sedimentary rock. Terrigenous clastic rocks (rocks derived from the accumulation of erosional debris of preexisting rocks) reflect the ancient environment of deposition mainly in their textures. However, because the clastic debris is *extrabasinal* in origin, the composition of the framework particles is controlled mainly by the geology of the source area. *Chemical* sedimentary rocks, such as limestone and evaporites, are *intrabasinal* (within the depositional basin). Their lithologies commonly reflect paleoenvironmental conditions in that these rocks are deposited within their source area and are not composed of solids derived from preexisting rocks.

Table 6-1 shows that identification of ancient environments on the basis of lithology alone may be inconclusive. For example, conglomerates and sandstones (gravel and sand) can form in fluvial, alluvial fan, glacial, littoral, and offshore submarine fan environments, among others. Lithology commonly is a good index to the *kinds of conditions* that operated in the depositional environment. Grain size, sorting, degree of rounding of particles, and color furnish clues about turbulence, current strengths, water depth, and geochemical aspects, but seldom do these features serve as fingerprints of specific environments. The lithology of a sedimentary unit is one of the easiest characteristics to observe directly, but to be useful in paleoenvironmental interpretation, it must be studied *in combination with other parameters*.

Geometric Shape of Rock Bodies

The overall three-dimensional shape of a rock unit is the result of the architecture and topography of the depositional area, of compaction during lithification, and, possibly, of postdepositional erosion processes. A long, narrow, *linear unit* of sandstone may represent an ancient stream channel (Fig. 6-5) or an offshore bar, among other possibilities. A pattern of *radiating* sandstone units might represent deposits in a series of dis-

tributary channels on an ancient delta. An extensive *sheetlike* unit of sandstone may represent deposition along the migrating shoreline of an ancient sea or perhaps may be the product of a laterally meandering stream complex. A thick wedge of sediment may result from accumulation in a slowly subsiding basin.

Unfortunately postdepositional erosion may destroy the original geometry of a rock unit. Sedimentary sequences deposited in terrestrial environments are particularly vulnerable to erosion, whereas the marine setting is more favorable for preservation. (Recall from Chapter 4 how this influences bias in the fossil record.)

Where abundant rock outcrops (exposures) are available for direct study, the geometry of a rock unit commonly can be determined. In the subsurface, the geometry must be determined by indirect methods such as mapping the unit from borehole data or from modern geophysical seismic surveys (see Chapter 3). The latter method can trace the outlines of a reef or of a submarine fan unit, for example, even though the unit may

Figure 6-5. Stream-channel sands and conglomerates of Pleistocene age. Note the linear nature of the sands. (Photo by J. Patterson)

be buried beneath thousands of feet of overlying sediments. This technique is exceedingly useful in the determination of the geometry of units that are potential petroleum traps.

Sedimentary Structures

Primary sedimentary structures, which formed at the time the sediment was accumulating, provide significant clues to the original conditions of deposition and, therefore, are valuable for interpretation of paleoenvironments. Reference to Table 6-1 indicates that one such primary structure, stratification, is a helpful feature that has many forms. For example, the laminated strata characteristic of a lacustrine (lake) environment are distinctly different from the large-scale cross-stratification typically found in sand dunes (Fig. 6-6). Note, however, that cross-bedding on a smaller scale can be displayed in sediments deposited in a lacustrine environment, as

well as in deposits of streams, beaches, deltas, tidal flats, and other environments. Like sediment textures, sedimentary structures overlap environmental categories and generally cannot be used as a single, diagnostic criterion in paleoenvironmental interpretation. Also, like textures, sedimentary structures are signatures of paleoenvironmental conditions, such as current flow. Variations in certain kinds of sedimentary structures, such as cross-bedding, indicate changes in the depositional medium; for example, changes in the energy level of running water or wind in transporting sedimentary particles. Remember throughout that *analysis requires the utilization of multiple lines of evidence,* rather than simply honing in on one feature.

Graded bedding is an *internal* sedimentary structure that typically is formed in sediment deposited by a **turbidity current,** a dense sediment-laden current that moves rapidly down a submarine canyon or basin slope. At the mouth of the canyon or base of the slope the

A

B

Figure 6-6. Sand dune cross-bedding on a massive scale (section 100 m thick) in the Permian Coconino Sandstone, Grand Canyon National Park.
(Photo by J. Patterson)

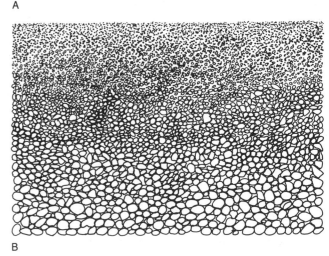

A

B

Figure 6-7. Turbidite graded bedding illustrating the upward-increasing fineness of the sediment formed as clastic debris settled out of the water.
(Photo courtesy of Critter Creations, Gary Peterson collection, no. 4)

velocity of the turbidity current is checked; the coarsest particles of the sediment load settle out of the water first, followed by finer and finer debris, resulting in the bottom-to-top, coarse-to-fine bedding shown in Figure 6-7. Graded bedding occurs in other kinds of deposits, but it is a characteristic feature of **turbidities** deposited in deep-sea fan settings.

Ripple marks are *external* sedimentary structures that indicate direction of movement of the transporting medium and are helpful features in paleoenvironmental interpretation. *Oscillation ripple marks* are symmetrical in cross section (Fig. 6-8) and indicate a back-and-forth (swash) movement of water, such as the uprush and backwash motions of waves on a beach face.

Current ripple marks are asymmetrical in cross section (Fig. 6-9) and indicate a prevailing direction of current flow either of water (such as in a stream channel or on a submarine surface) or of wind (such as over a dune surface). Detailed studies of ripple marks forming in modern sedimentary environments have resulted in the establishment of criteria that facilitate distinction between water and wind currents, and these criteria can be applied to ancient ripple marks. (Ripple marks are *external* sedimentary structures that form at the sediment–air or sediment–water interface. In rocks they are seen on the bedding surfaces and in cross section they are expressed as various kinds of cross-stratification; thus external ripple marks and internal cross-strata of all scales are genetically related.)

Desiccation cracks are good environmental indicators because they generally indicate some subaerial exposure (Fig. 6-10). They imply a sedimentary setting of alternating intervals of wetting and drying of surfaces such as river floodplains, desert playas, and shorelines.

Some of the sedimentary structures used as examples in this section, plus others such as various kinds of *sole markings* (e.g., flute casts), have been examined previously in discussion of determining the stratigraphic

A

A

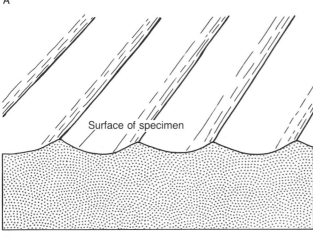

B

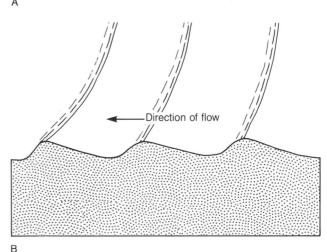

B

Figure 6-8. Oscillation ripple marks. In the cross-section graphic, note the symmetrical form of this type of ripple mark, indicative of water moving back and forth as on a beach face (specimen 33 cm in length).
(Photo by J. Streng)

Figure 6-9. Current ripple marks. In the cross-section graphic, note the asymmetrical form of this type of ripple mark, indicative of current flow from right to left (specimen 23 cm in length).
(Photo by J. Streng)

A

B

Figure 6-10. A. Desiccation cracks (mudcracks) on the edge of a modern lake bed. B. Desiccation cracks expressed in a rock surface by ridges of sediment that filled in the cracks (specimen 38 cm in length).
(A, photo by J. Patterson. B, photo by J. Streng)

top of a bed (see Chapter 2). Sedimentary structures are significant, therefore, in determining superposition as well as in the interpretation of ancient depositional environments.

Fossils

Just as living organisms reflect the environments in which they dwell, fossils are one of the best indicators of ancient sedimentary environments (see the discussion of the use of fossils in paleoecologic studies, Chapter 5). The concept of facies applies to fossils as well as to physical, or lithologic, aspects of sedimentary rocks. Lateral changes in fossil content within contemporaneous strata are called **biofacies,** as distinguished from the facies defined on lithology, the **lithofacies.** Commonly these two kinds of facies coincide because *benthic* (bottom-dwelling) organisms tend to be controlled in their distribution by bottom sediment type; thus, benthic fossils commonly are facies-controlled in their occurrences.

The distinction between marine fossils such as oysters and corals and nonmarine fossils such as dinosaur bones is an obvious clue to the depositional setting of rocks containing the fossils. The use of fossils as paleoenvironmental indicators, however, does necessitate caution. Fossils are useful in paleoenvironmental interpretation if the paleoecologic organism–sediment relationship can be worked out either through comparisons with modern counterparts or by deducing the adaptive characteristics of the fossils. Fossils commonly occur in sedimentary rocks that represent *not* where the organisms lived, but where they were transported after death and eventually buried and preserved. A good example

of this situation is the wood of Petrified Forest National Park, discussed in Chapter 14. The trees (conifers) lived in an upland area, died, were broken up in transport (probably during flood stage), and eventually accumulated (displaying current orientation) mostly as fragments on floodplains and in stream channels in a lowland area. The site of burial and preservation does not represent the original upland environment where these conifer trees were growing during Triassic time. Marine forms likewise can be subject to postmortem transport.

In some instances, biofacies can be interpreted to represent parts of ancient benthic communities, whose members have undergone comparatively little postmortem transportation from their original life habitats. As an example, three benthic paleocommunities are recognized in upper Ordovician rocks of the Appalachian region. The nearshore paleocommunity, including both infaunal and epifaunal forms, lived on a sandy bottom and was characterized by inarticulate brachiopods, gastropods, and bivalves (Fig. 6-11). This type of assemblage appears to reflect coastal, brackish paleoecological conditions.

An intermediate paleocommunity, living on a silty substrate, consisted of a more diverse association of thin, smooth-shelled and round-shaped brachiopods, some gastropods, trilobites, infaunal bivalve molluscs, and bryozoans (Fig. 6-11). This community existed in shallow, nearshore shelf environments of normal salinity. The offshore paleocommunity (Fig. 6-11) lived on a muddy bottom and was characterized by a lack of infauna, but included a diverse assemblage of epifauna composed of abundant branching bryozoans and several types of brachiopods. Some crinoids, gastropods, and bivalves also were present. This paleocommunity re-

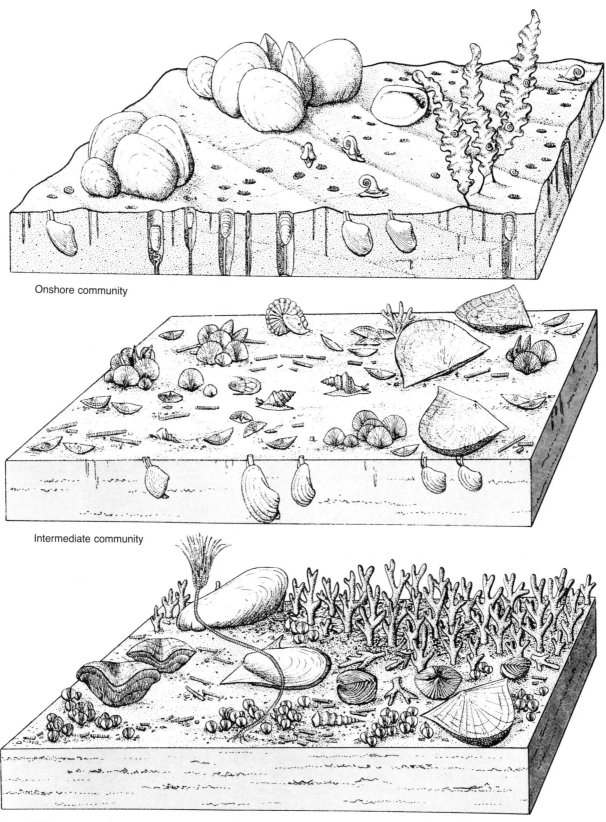

Onshore community

Intermediate community

Offshore community

Figure 6-11. Three benthic paleocommunities recognized in upper Ordovician rocks of the Appalachian area. Analysis of fossil data defines an onshore community of infauna plus epifauna with sturdy hard parts to withstand water turbulence; an intermediate community with forms of greater diversity reflecting a less rigorous environment; and an offshore community including some delicate life forms that require quiet water.
(Based on graphics by T. L. Chase in R. L. Anstey and T. L. Chase, 1979, *Environments Through Time,* 2nd ed., Fig. 6.1, p. 54: Burgess Publishing Co., Minneapolis, MN. Original source is P. W. Bretsky, 1969, Central Appalachian Late Ordovician Communities: *Geol. Soc. America Bull.,* vol. 80)

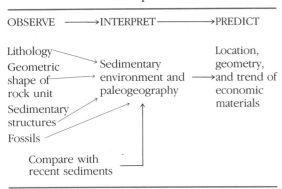

Table 6-2. Basic approach to determine how sediment was deposited

OBSERVE ——————→INTERPRET——————→PREDICT

Lithology
Geometric
shape of
rock unit   → Sedimentary
Sedimentary   environment and
structures   paleogeography
Fossils

Location,
geometry,
and trend of
economic
materials

Compare with
recent sediments

From Richard C. Selley: *Ancient Sedimentary Environments,* 2nd edition, © 1970, 1978 by Richard C. Selley. Used by permission of the publisher, Cornell University Press.

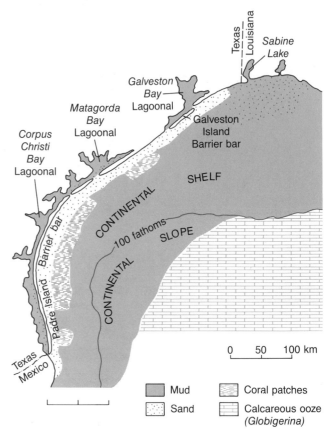

Figure 6-12. Examples of some modern sedimentary environments in the western Gulf of Mexico.

flects deeper and quieter water, probably below wave base, on the outer part of the shelf.

It has been emphasized that the use of lithology, geometry of the rock units, sedimentary structures, and fossils, as *single* parameters to interpret ancient sedimentary environments would be shortsighted and the results inconclusive. A synthesis of information integrating several or all parameters is much more meaningful and definitive (Table 6-2).

A PRESENT-DAY MODEL

The Delta

Depositional environments along the coast of the western Gulf of Mexico include delta, beach, bay, lagoon, barrier bar, and shallow-marine shelf (Fig. 6-12). Each of these environments has a set of physical and chemical factors such as water energy (turbulence), circulation patterns, temperature, depth, and salinity that influence the type of sediment that accumulates and the kinds of organisms that are adapted. Stratigraphers and sedimentologists from all over the world have used the Gulf Coast region as an instructive actualistic model for observing sedimentary processes and the characteristics of depositional environments.

An examination of one particular environment along the Gulf Coast, the delta (Fig. 6-13), reveals a complex association of a number of small subenvironments such as swamps, sand bars, and stream channels. Deltas are important environments of deposition because they represent the building out of land areas into bodies of water. Because they form at the termini of many major rivers, such as the Mississippi, they also are instrumental in introducing terrigenous clastic sediment into the marine setting. The principle of actualism tells us that ancient deltas likewise were significant as local areas for the feeding of sediments onto marine shelves. Their recognition in the ancient rock record is important for re-

construction of geologic history as well as for exploration for fossil fuels.

Deltas consist of a mosaic of subenvironments that interact with one another and undergo continuous shifting and changing of position. Rivers flow more slowly as they enter the sea; because they no longer have the capacity to transport their load, they become choked with sediment and their channels are divided. The *subaerial* parts of deltas include *deltaic plains,* which are characterized by a system of bifurcating *distributary channels* that become filled with silts and sands; *natural levees,* which form as ridges of interlaminated sand and mud deposited adjacent to the channels during overbank spreading of flood waters; and *back-swamp* areas where muds transported away from the channels by high water are deposited (they may become sites of luxuriant vegetation, the possible precursor to coal deposits). The *subaqueous* delta subenvironments include the distributary mouth bars, which are sand bodies deposited as distributaries enter the sea, and the *prodelta,* which is the foreslope of the delta and the site of clay and silt deposition (Fig. 6-13). Prodelta deposits (facies) merge gradually with the muds of the marine shelf. Modern and ancient deltas show a complex association of facies that represent the sediment-response products of the shifting subenvironments of deposition.

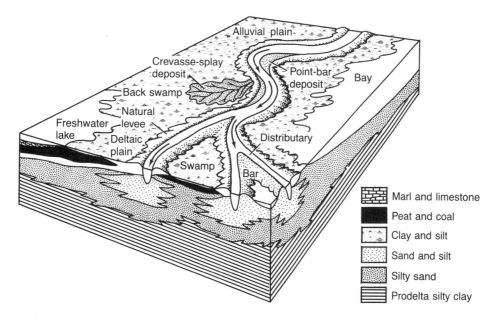

Figure 6-13. Subenvironments associated with a delta. Note the various sediment types associated with the different depositional subenvironments. Changes in position of these subenvironments produce a sedimentary record that illustrates Walther's Law. In the locality represented by stratigraphic section lines, prodelta silty clays are overlain by channel silts and sands, and these, in turn, are overlain by natural levee and then deltaic plain deposits. In the vertical succession of sediments in this section, we observe the record of subenvironments that once were lateral to one another but now are superimposed due to shifting of the delta subenvironments.
(Based on graphic by T. L. Chase in R. L. Anstey and T. L. Chase, 1979, *Environments Through Time,* 2nd ed., Fig. 14.2, p. 104: Burgess Publishing Co., Minneapolis, MN)

Lateral and Vertical Facies Relationships

The Gulf Coast delta model with its mosaic of co-existing subenvironments provides a two-dimensional map view of lateral facies changes; this is a **facies tract,** a series of laterally adjacent facies, different but genetically related. As environments shift through time, the migration of facies over the tops of one another produces a superposition of different facies vertically and results in a complex three-dimensional intertonguing of these facies. If a continuous core could be examined from a well borehole drilled at one particular location on the delta, a vertical succession of facies might include the products of back-swamp, natural-levee, freshwater-lake, distributary-mouth-bar, and prodelta environments. All of these facies at any one time were laterally adjacent to one another and formed a facies tract. This relationship between lateral and vertical sedimentary facies was expressed at the beginning of this century by the German geologist Johannes Walther (1860–1937), who travelled extensively and studied modern sedimentary environments. Present-day depositional processes, he believed, were the keys to interpretation of paleoenvironments. Walther observed that in a continuous sequence (not punctuated by unconformities) facies that precede or succeed one another in a vertical section were laterally adjacent to one another at any single mo-

ment in time (comprising a facies tract). This concept is known as **Walther's Law** of the correlation of facies. At a single locality, therefore, one is able to determine what the lateral paleoenvironmental relationships were by observing the nature of the overlying and underlying facies (Fig. 6-13). The interpretation of the geologic history of an area commonly involves the recognition of lateral facies that were contemporaneous and of vertical succession of facies produced through time. The vertical and lateral facies relationships together provide a three-dimensional stratigraphic view within which to interpret the paleoenvironmental implications of individual facies.

Facies relationships are illustrated by numerous examples in the stratigraphic record of ancient sedimentary rocks. The delta model, for example, has been recognized in rocks of several Paleozoic systems in the Appalachian and midcontinent regions and in rocks of Cretaceous age in the Rocky Mountain region (see discussion of the Devonian and Pennsylvanian, Chapter 11, and the Cretaceous, Chapter 12).

THE FLUCTUATING SHORELINE
Onlap and Offlap

Sedimentary facies patterns have been influenced by changes in position of the shoreline; seas have advanced upon and receded from the continents many

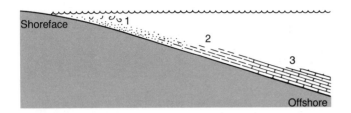

Figure 6-14. Idealized model of marine deposition. Sand is deposited (1) where an influx of clastic debris is available and there is some water turbulence. Farther seaward (2) water is less turbulent and muds settle to the sea bottom. Calcareous sediment (3) is deposited offshore where terrigenous clastic debris is not available. In a shoreward-to-seaward direction, the sediment texture and composition may change.

times during the geologic past. Migration of shallow-marine environments, evidenced by sedimentary facies changes, accompanied these **transgressions** and **regressions,** respectively.

In a simplified, *idealized model* (Fig. 6-14), sand accumulates in a shoreline or beach environment with finer muds deposited offshore in less turbulent water. These terrigenous muds, in turn, grade laterally farther seaward into calcareous mud as the influx of terrigenous clastic material decreases. This example illustrates a facies tract—a lateral change of facies from coarse to fine terrigenous sediment to calcareous sediment—deposited in a shoreface-to-offshore direction.

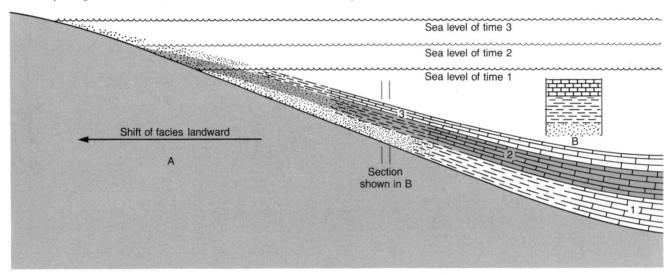

Figure 6-15. Marine transgression sequence. A. As sea level rises, the sedimentary facies shift in a landward direction. B. A sedimentary rock sequence of marine sandstone overlain by marine shale and limestone reflects this transgression and is called an *onlap* succession (not to scale of A).

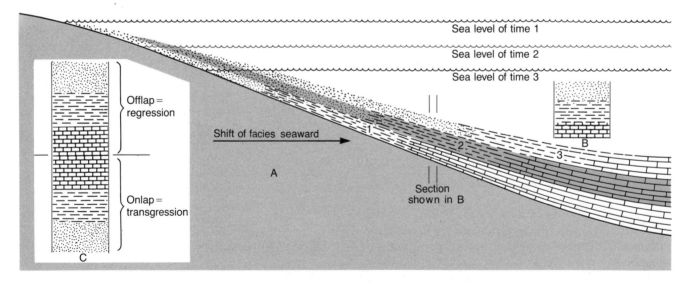

Figure 6-16. Marine regression sequence. A. As sea level falls, the sedimentary facies shift in a seaward direction. B. A sedimentary rock sequence of marine limestone overlain by marine shale and sandstone reflects this regression and is called an *offlap* succession (not to scale of A). C. An onlap sequence followed by an offlap sequence is called a *sedimentary cycle.*

If the sea transgresses the land, each of these sedimentary environments will shift landward and, in time, a vertical succession of sand, mud, and calcareous mud will be produced (Fig. 6-15). A stratigraphic pattern of this type, which shows progressively more seaward facies *overlying* more landward facies, is called an **onlap** succession. *Stratigraphic onlap is the result of marine transgression.* Note that this vertical succession is an illustration of Walther's Law; the facies in the stratigraphic column (Fig. 6-16) were positioned adjacent to one another at a single moment in time.

If the sea regresses in our same hypothetical facies tract model, an opposite pattern will be produced (Fig. 6-16), with each of the sedimentary environments shifting seaward; in time, a vertical succession of calcareous mud, terrigenous mud, and sand will be produced. A stratigraphic pattern of this type, which shows progressively more landward facies *overlying* more seaward facies, is called an **offlap** succession. *Stratigraphic offlap is the result of marine regression.* In both situations, onlap and offlap, the vertical succession of rock units (lithofacies) at a specific location reflects the shifting or migration of depositional environments through time. Both onlap and offlap stratigraphic sequences are expressions of Walther's Law.

A **sedimentary cycle** is a repetition of facies or sequence of facies (Fig. 6-16C) in vertical section and represents both onlap and offlap deposits and the consequent repetition of depositional environments in the area.

Recognition of onlap-offlap patterns on a regional scale is enhanced by the correlation of a number of stratigraphic sections. This provides a three-dimensional time and spatial framework for understanding facies relationships. The lithofacies patterns make possible interpretation of shoreline fluctuations in a geographic region through time. It should be emphasized that the foregoing discussion has presented an idealized situation for purposes of explanation and that actual examples typically are more complex. Also, the lithofacies changes that result from the transgressive-regressive cycle are likely to correspond to *biofacies* changes because each of the shifting environments carries its own particular assemblage of benthic organisms.

In the midwestern United States the St. Peter Sandstone is a distinctive, widespread rock unit of Ordovician age that crops out in several states. The St. Peter is a well-sorted, mature, cross-bedded quartz sandstone that was deposited as a blanketlike lithofacies in a shallow-marine environment. It represents the deposits of the leading edge of the transgressing Ordovician sea. As such, the lithofacies transgresses time; its outcrops in Illinois, for example, are older than those in Iowa and Wisconsin (Fig. 6-17). In interpretation of the depositional history of an area, it is important to remember

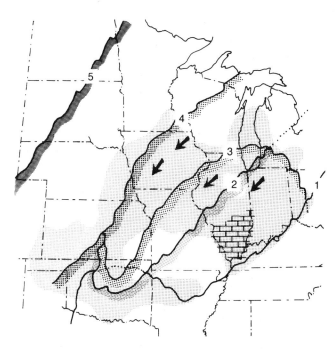

Figure 6-17. Shorelines of the St. Peter Sandstone. The wavy lines (1–4) indicate shorelines during transgression of the sea in which the sediment was deposited. Line 5 represents the possible westernmost extent of the sea. The stippled area indicates areas of outcrop of the sandstone. The St. Peter Sandstone was deposited over a span of time and is *not* everywhere the same age.
(After E. W. Spencer, 1962, *Basic Concepts of Historical Geology,* Fig. 13-6, p. 228: Thomas Y. Crowell, New York. Reproduced by permission of author)

that sedimentary facies are lithologic units that may have been deposited, as was the St. Peter Sandstone, over an interval of geologic time. Their boundaries generally do *not* coincide with time-stratigraphic boundaries (see discussion of Cambrian transgression in the Grand Canyon region, Chapter 10).

Causes of Transgression and Regression

Geologists have long recognized that changing sea–land relationships caused by transgression and regression have produced onlap and offlap stratigraphic patterns. Actual sea-level changes include *eustatic* (worldwide) fluctuations, as well as smaller local variations. The largest eustatic changes generally have been attributed to cycles of continental glaciation, whereby buildup and advance of ice sheets have resulted in lowered sea-level stands, and melting and retreat of major glaciers have resulted in raised sea-level stands (see Chapter 16). Smaller scale fluctuations in position of the shoreline may result from tectonic activity or isostatic adjustments involving uplift or subsidence of the lithosphere. The smallest fluctuations generally result from changing sedimentation patterns. Rapid influx of sediment, such as in delta building, can cause local regression of the sea as the land builds seaward. If, however, sedimentation stops, the delta is no longer nourished and subsidence allows marine waters to transgress its surface.

The present model of plate tectonics accounts for major eustatic sea-level changes in the past, for example, encroachment of shallow seas into continental interiors during Paleozoic and Mesozoic times. As continents have welded together by collision or fragmented by rifting, changes in the cubic holding capacity of the ocean basins have occurred. For example, it has been suggested that fragmentation of continents, accompanied by decreased volume of ocean basins, has caused extensive marine transgression; major regressions have resulted from increases in holding capacity of ocean basins, with corresponding withdrawal of marine waters from continental platforms (see discussion of cratonic sequences, Chapter 11). Transgression and regression have been important influences in forming the stratigraphic record from which we read geologic history.

A PERMIAN FACIES MODEL
Guadalupe Mountains National Park

The beginning of this chapter featured some human history of Guadalupe Mountains National Park, with a reference to the geologic features for which this area was set aside as a unit of the national park system. The park exposes rocks of a Permian reef complex that provides an excellent model for illustrating facies relationships, relationships between facies and paleoenvironments, onlap-offlap sedimentary cycles, distinctions between rock-stratigraphic and time-stratigraphic units, and transition from shallow platform to deep basin. These rocks are developed as four major facies that represent deposition in adjacent **reef, back-reef, fore-reef,** and **basin** paleoenvironments (Fig. 6-18).

Reef Facies

The Capitan Limestone, named for its exposure in the massive limestone face of El Capitan Peak at the south end of the Guadalupe Mountains, is a gigantic "fossilized" reef. The Capitan reef was a major carbonate buildup that formed a 600-kilometer loop around the margin of the Delaware Basin during much of Permian time (Fig. 6-19). The east flank of the Guadalupe Mountains exposes just the northwestern edge of this reef structure. The impressive mountain front consists predominantly of massive limestone composed of the exoskeletons of reef-building organisms, as well as other kinds of carbonate debris. Much of the reef complex is buried in the subsurface beneath younger sedimentary rock and can be located only by drilling or seismic profiling.

Geologists working in this area several decades ago interpreted the facies within the Capitan as various parts

of a huge barrier reef similar to the present-day barrier reef of Australia. More recent paleontologic and lithologic studies suggest that the Capitan reef was not so much a structure composed of frame-building organisms as it was a carbonate buildup composed predominantly of fragmented exoskeletons and other carbonate particles. But whether it was an ecologic reef in the strict sense or a buildup of fragmental material, the important consideration is that it was a wave-resistant, moundlike feature that grew near the edge of a shallow-marine platform and profoundly influenced the development of facies both seaward and landward of it.

Back-Reef Facies

Back-reef facies sediments were deposited in the relatively quiet waters of a platform shelf behind (landward of) the reef. In this back-reef area (Fig. 6-18) the water was shallower and less turbulent than on the seaward side of the reef; the sediments that accumulated were predominantly lime muds and skeletal limestones near the reef and more saline deposits landward. Within this setting there existed a lateral gradation of subenvironments that produced several subfacies. In a north-to-south direction across the northwestern shelf (seaward direction, see Fig. 6-19) the spectrum of paleoenvironments included supratidal and intertidal flats and shelf lagoon. The supratidal area, covered by a film of shallow water only during the highest tides, was the site of deposition of reddish terrigenous silt and sand and of evaporites such as dolostone and laminated gypsum. Abundant desiccation features attest to dry, exposed depositional conditions. The intertidal flats were sites of accumulation of algal stromatolites and dolostone, and in the shelf lagoon shelly limestones and lime mudstone accumulated.

Fore-Reef Facies

The fore-reef facies includes the apron of debris eroded from the seaward face of the reef. These limestone beds had original depositional dips of as much as 30° seaward into the deep basin. The fore-reef limestones are composed of bits and pieces of broken skeletal debris from the reef and variously sized clasts and blocks of limestone (Fig. 6-20). Sliding and slumping apparently occurred on this fore-reef slope between the shelf margin (reef) and the basin floor, and dense mixtures of mud and rock fragments accumulated with the broken shells to form a reef *talus*.

Basin Facies

The basin facies was developed seaward from the reef in moderately deep waters of the restricted Delaware Basin (Fig. 6-19). This facies is divided into two interbedded subfacies. One is composed of black, well-laminated claystones and siltstones. This subfacies was deposited in water well below wave base under *anaerobic* conditions. The second subfacies consists of fine-grained sandstones with some graded bedding. These sandstones have been carefully mapped because they are hydrocarbon reservoirs in the subsurface. The sand-stones are developed as elongate pods that are oriented perpendicular to the margin of the basin and have been interpreted as turbidite channel and fan deposits.

Biofacies

The depositional environments that produced the distinctive lithofacies also produced characteristic biofa-

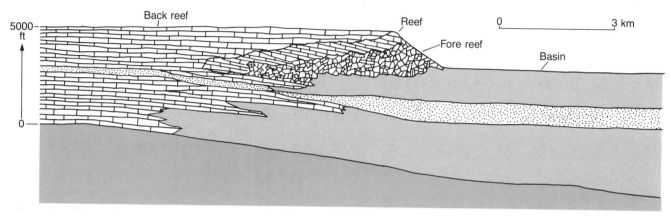

Figure 6-18. Permian reef complex of Guadalupe Mountains area, West Texas. The back-reef, reef, fore-reef, and basin subenvironments are well expressed in the rocks of this area and form a model for illustrating facies relationships.
(After N. D. Newell, *et al.,* 1953, *The Permian Reef Complex of the Guadalupe Mountains Region, Texas and New Mexico,* Fig. 5D, p. 13: W. H. Freeman, San Francisco. Reproduced by permission of author)

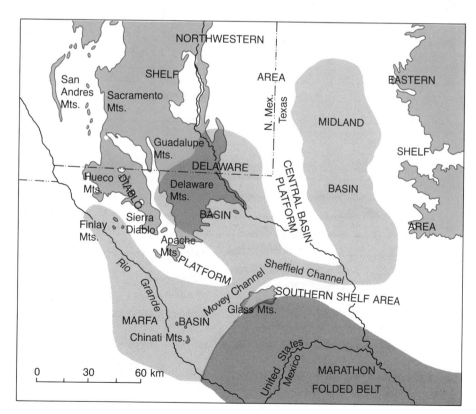

Figure 6-19. Permian age tectonic setting in West Texas and southeastern New Mexico.
(After P. B. King, 1948, Geology of the Southern Guadalupe Mountains, Texas, Fig. 3, p. 25: *U.S. Geol. Survey Prof. Paper,* no. 215)

cies. The reef limestones are the most fossiliferous of the facies (Fig. 6-21), containing both the greatest abundance and the greatest diversity of organisms. The reef environment was a well-oxygenated, sunlit, resource-rich setting that was able to support a rich community of sponges, bryozoans, crinoids, gastropods, calcareous algae, and rare corals. Because of poor circulation and low oxygen levels on the sea floor of the basinal environment, the rocks of the basin facies contain an impoverished fauna of sponge spicules, radiolarians, and ammonoids. The back-reef deposits carry a moderately diverse assemblage of fusulinids, gastropods, brachiopods, and algae. Particularly significant elements of the biofacies are the fusulinid foraminifers (Chapter 4), which are amenable to biostratigraphic zonation. The sequence of fusulinid biozones has provided time lines

that enable time-stratigraphic subdivision and correlation. The lithofacies boundaries cut across the biozonal boundaries, illustrating the fundamental difference between lithostratigraphic, biostratigraphic, and time-stratigraphic units.

The Permian Reef Complex and Walther's Law

The analysis of the described lithofacies and biofacies has defined a paleoenvironmental picture for Late Permian time in West Texas and southeastern New Mexico. The spatial relationships of the various facies show that their environments of deposition shifted through time as fluctuations of sea level occurred. Near the end of the Permian Period the sea gradually retreated from the region. This regression is expressed by the shifting of the marine paleoenvironments to the south and the gradual replacement of the marine setting by nonmarine conditions (Fig. 6-22). The depositional pattern resulting from this regression is an offlap sequence. An important aspect of this offlap sequence was the deposition of hundreds of meters of evaporites as the original deep-water basin shoaled, became restricted, and was eventually replaced by a coastal *sabkha* (low, salt-encrusted coastal plain) containing evaporative playa lakes.

Deposits consisting of alternating layers of anhydrite and gypsum accumulated in this setting (Fig. 6-23). As waters became even more saline (about ten times normal salinity) thick deposits of halite (rock salt) formed;

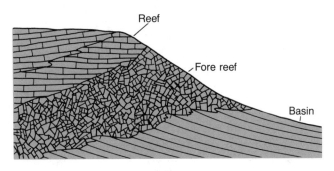

Figure 6-20. Steeply dipping reef rubble (breccia) that forms the dominant lithology of the fore-reef facies.

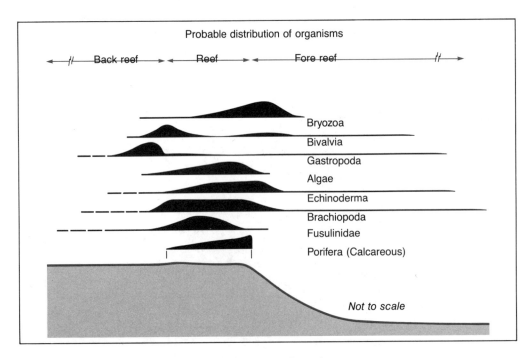

Figure 6-21. Components of the biofacies associated with the Permian reef complex.
(From N. D. Newell, *et al.,* 1953, *The Permian Reef Complex of the Guadalupe Mountains Region, Texas and New Mexico,* Fig. 79, p. 203: W. H. Freeman, San Francisco. Reproduced by permission of author)

these presently are mined as a natural resource. As the seas continued to retreat from the region, the evaporites, in turn, were *offlapped* by a relatively thin cover of reddish nonmarine sandstones and siltstones. Thus in vertical stratigraphic columns in the Delaware Basin, rocks of the basin facies are overlain successively by

rocks of the fore-reef, reef, back-reef lagoon, and tidal-flat facies, and finally by coastal sabkha and terrestrial facies, a complete offlap cycle that reflects the southward regression of the sea. In accordance with Walther's Law, facies that formed beside one another in a lateral relationship also lie on top of one another. Particularly

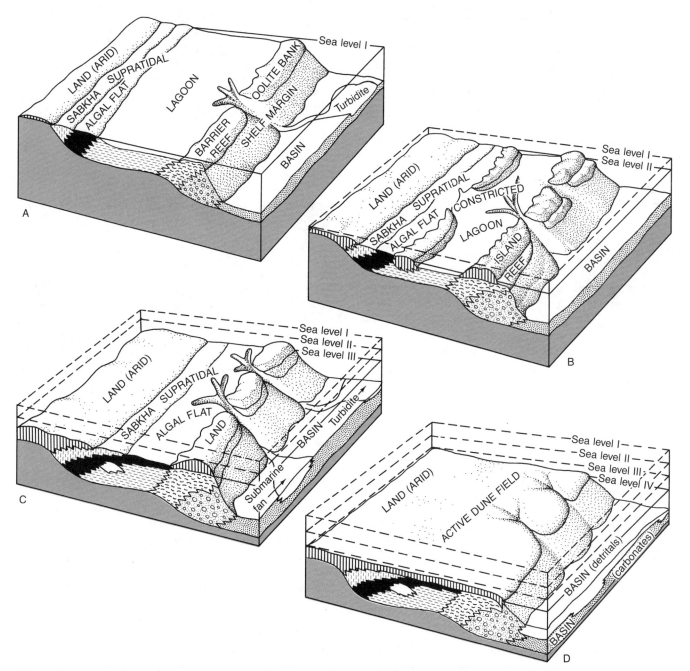

Figure 6-22. Regression of the Permian sea southward (A–D) resulted in replacement of conditions that formed the Permian reef complex by nonmarine conditions. The reef environment and the nonmarine coastal plain environment existed simultaneously in a lateral relationship and these two environments also existed at the same location at different times. The illustrated facies relationships are an example of Walther's Law.
(After B. A. Silver and R. G. Todd, 1969, Permian Cyclic Strata, Northern Midland and Delaware Basins, West Texas and Southeastern New Mexico, Figs. 4–7, pp. 2227–2230: *Am. Assoc. Petroleum Geologists Bull.,* vol. 53)

Figure 6-23. Gypsum beds of the Castille Formation, Guadalupe Mountains National Park, West Texas. The evaporites were deposited in a playa lake on a low coastal plain as the Permian seas regressed to the south and an offlap sequence was developed in the area of the Permian reef complex.
(Photo by J. Cooper)

noteworthy in this regard is the southward migration of the evaporites and red beds from the landward side of the shelf out into the Delaware Basin.

On the north wall of McKittrick Canyon in Guadalupe Mountains National Park, the verticolateral facies relationships among the reef, back-reef, fore-reef, and basin facies are clearly displayed (Fig. 6-24). This important location was included in the land donated by Wallace Pratt to the National Park Service (see beginning of this chapter).

Another significant area where part of the Permian reef complex can be observed is Carlsbad Caverns. The caverns were formed by dissolution of the carbonate rocks long after the rocks were initially deposited. The entrance to the caverns and the upper level of the caves have developed in the reef core (Capitan Limestone), but parts of the cavern complex, such as the Big Room, have formed in the fore-reef talus. Thus a walk through Carlsbad Caverns provides a walk through three distinctive sedimentary facies—the back reef, the reef, and into the fore reef.

END PRODUCTS

Paleogeography

Both lithofacies and biofacies are the products of their ancient sedimentary environments. As mentioned previously, these paleoenvironments were areas that had distinctive sets of physical, chemical, and biological characteristics that left their impress on the sediments that accumulated. Reconstruction of these paleoenvironments through facies analysis of the rock record has enabled geologists to fashion an end product, a **paleogeo-**

Figure 6-24. A portion of the north wall of McKittrick Canyon, Guadalupe Mountains National Park, West Texas. Exposed rocks are the Permian Capitan Limestone (Pc). The massive limestone along the skyline (Pc/m) is part of the reef facies and the slopes below are part of the dipping reef rubble of the fore-reef facies.
(Photo by J. Cooper)

graphic map (Chapters 11 and 12). Such a map is based on outcrops and/or subsurface sections of rocks of the same age, whose paleoenvironments have been interpreted. The ideal paleogeographic map is constructed for a single time-stratigraphic level such as a chronozone boundary; thus the map surface is a time plane. More commonly, however, the degree of refinement possible constrains these maps to various narrow time-stratigraphic intervals such as the duration of a chronozone or stage.

Paleogeographic maps depict the location and distribution of ancient seas, shorelines, basins, rivers, and other paleogeographic features such as mountain belts. This is not purely an academic exercise; this understanding of the geologic past is important in the exploration for mineral resources (Table 6-2). In the Permian model of West Texas, the subsurface mapping of the basinal turbidite sandstones and the reef carbonates is important for hydrocarbon exploration because these are possible reservoir facies.

With the advent of plate tectonics, *world* paleogeography during successive time intervals of the Phanerozoic Eon has become an exciting and fruitful area of research. The compiling of global paleogeographic maps involves synthesis of many diverse kinds of data from various fields of geology and geophysics: paleontology, sedimentology, paleomagnetism, paleoclimatology, and structural geology.

As pointed out in Chapter 1, we know that continental movement, the result of plate tectonic activity, has caused the geography of the earth to change slowly and intermittently during the last several billion years. The initial phase of reconstructing scenes of this ever-changing world of the past involves identification of regions that functioned as separate paleocontinents. This is accomplished by integrating a wealth of geologic data. At the heart of world paleogeographic reconstructions is the positioning of these paleocontinents in their correct orientation. The basis for this positioning is paleomagnetic information, which enables determination of paleolatitude (see Chapter 1).

The compiling of data indicative of ancient geographic features and paleoclimatic conditions contributes importantly to these reconstructions. In this regard, igneous intrusions, andesite volcanics, metamorphic complexes, folded and thrust-faulted rocks, and unconformities provide evidence for the existence of ancient mountain belts and lithospheric plate boundaries, and paleontologic and radiometric data provide timetables for various continental collisions and separations. Climatically sensitive sedimentary deposits such as thick sections of terrigenous clastics, red beds, tillites, evaporites, organic reefs, fossils, and coal deposits are signatures of various paleoclimatic conditions. Finally, interpretation of the distribution of paleoenvironmental conditions on each paleocontinent completes the picture. In addition to allowing identification of emergent areas and/or tectonic lands, determination of the character of paleoenvironments, as revealed by environmentally sensitive rock types and fossils, characterizes the particular continent.

For global paleogeographic reconstructions to be most meaningful, they should represent configurations of ancient continents and ocean basins during comparatively narrow intervals of geologic time. In this regard, the smallest time-stratigraphic subdivision that commonly can be recognized worldwide is the *stage* (see Chapter 3). The Phanerozoic record can be subdivided into about seventy-five stages, which correspond to *time ages* of about 5 to 15 million years duration. By assuming rates of 2 to 8 cm/year for lithospheric plate motions during the Phanerozoic, the total amount of change in

paleocontinent configuration is sufficiently contained during these narrow time intervals to be meaningfully portrayed; thus the record of a single *stage* provides a useful basis for a single paleogeographic reconstruction. Some of these reconstructions will be discussed for various parts of the Paleozoic, Mesozoic, and Cenozoic in the chapters on the evolution of North America. The continent of North America will provide the primary model for our trip through time, but appreciating ancient North America in its more global context provides an exciting perspective for historical geology.

An Economic Incentive

As noted previously, some economic mineral deposits such as coal, oil, and natural gas are associated with certain kinds of sedimentary rocks that are products of specific depositional environments. Most of the world's petroleum reserves have been found in sandstones that represent ancient delta and barrier beach complexes or in carbonates closely associated with ancient reefs. One of the reasons the Capitan reef tract has been studied so intensively is that around the edge of the Midland Basin (Fig. 6-21) the subsurface reef complex contains important oil reservoirs. Because of the economic value of the reef structures, the surface outcrops have been studied to aid interpretation of facies in the subsurface units. In the Texas Permian reef complex, oil is produced mainly from lagoonal calcareous sands. Oil produced from a reef complex of Devonian age in the subsurface beneath the plains of Canada, however, comes directly from a reef-core facies. Much petroleum has been located through paleoenvironmental analysis of facies and prediction of favorable reservoir rocks in the subsurface.

Paleoenvironmental facies analysis is important, too, in locating metalliferous deposits. In the Mississippi Valley region, various limestone facies are laced with ore deposits of lead and zinc that are commercially valuable and are mined. Many of the identified uranium occurrences in the United States are in Mesozoic freshwater sediments of the Colorado Plateau region. The uranium minerals were derived from weathering of uranium-bearing igneous rocks, were transported as detrital grains, and were delivered to fluvial and deltaic depositional environments, where they became incorporated in the sediments. Low-grade uranium resources are contained in marine organic shales in the same geographic region, but these are not mined at present because strip mining of these shales would produce great quantities of rock waste and consequent deterioration of environmental quality.

The recognition of ancient sedimentary environments and sedimentary facies gives us a better appreciation and understanding of the dynamics behind sedimentation, of the relationships of bodies of rock and stratigraphic patterns, and of earth history—most commonly in answer to an economic incentive. The Guadalupe Mountains are a part of this story and the national park there was created specifically to preserve a portion of the record of the ancient sedimentary environments of the Permian reef complex there.

SUMMARY

The physical, chemical, and biologic characteristics of various depositional environments produce different types of sediments and, after lithification, sedimentary rocks. At a single point in time, distinctive environments exist side by side—for example, a beach environment adjacent to a shallow-marine environment—and so different rock types may form in contemporaneous and adjacent units. These lateral variations are called *sedimentary facies*. An important part of the study of sedimentary rocks is determination of ancient depositional environments and of interrelationships of facies.

Sedimentary facies analysis involves study of a number of characteristics of rock units: lithology (composition and texture of the rocks); geometric shape of rock bodies, such as the narrow and linear shape of a stream-channel sandstone; sedimentary structures, including variations in bedding, ripple marks, and desiccation cracks; and fossils. Facies defined primarily on physical features (lithology) are called *lithofacies;* those defined primarily on fossils are called *biofacies.*

A river delta is an excellent example of a complex interrelationship of subenvironments (channel, swamp, lake, levee) and, therefore, of sedimentary facies. Study of the modern Mississippi River delta has provided a basis for recognition of ancient deltas preserved in the rock record.

Facies change position through time—sea level may rise and produce a *transgression* upon the land or fall and produce a *regression* of the sea from the land. A transgressive event may be recorded in the rock record by a shifting of a sandy beach facies landward—the St. Peter Sandstone in the midcontinent area. A shallow-marine environment then comes into existence where the beach had originally been located, typically depositing finer sedimentary material upon the beach sand. The change of sedimentation from coarse-grained (beach) to fine-grained (shallow-marine) is an *onlap* succession.

A regressive event is exemplified by the rock record of the Permian reef complex of West Texas and southeastern New Mexico. In Permian time a series of adjoining environments formed back-reef, reef, fore-reef, and basin facies. As time progressed, the sea regressed southward and these environments shifted in their locations. In the rock record, consequently, the reef facies is overlain by back-reef facies and, eventually, nonmarine coastal facies rocks, an *offlap* succession. A stratigraphic section at one location exposes vertically a succession of sedimentary facies that were at one time laterally adjacent to one another, an expression of *Walther's Law.*

The cause of specific transgression and regression events is often difficult to determine, but it has been suggested that, in general, fragmentation of continents may be accompanied by decreased volume of the ocean basins and subsequent transgression. Conversely, suturing of continents may be accompanied by increased volume of ocean basins and subsequent regression.

An end product of facies/paleoenvironmental study is the drawing of a paleogeographic map of an area. These maps reconstruct ancient shorelines, deltas, and rivers, for example, and are important tools in the quest for earth-related resources.

SUGGESTIONS FOR FURTHER READING

King, P. B., 1948, Geology of the Southern Guadalupe Mountains, Texas: *U. S. Geol. Survey Prof. Paper* 215.

Laporte, Leo, 1979, *Ancient Environments,* 2nd ed: Prentice-Hall, Englewood Cliffs, NJ.

Newell, N. D., 1972, The Evolution of Reefs: *Sci. American,* vol. 227, no. 6, pp. 54–65.

Newell, N. D., et al., 1972, *The Permian Reef Complex of the Guadalupe Mountains Region, West Texas and New Mexico.* Hafner, New York. Original publication 1953 by W. H. Freeman, San Francisco.

Rigby, J. K., and Hamblin, W. K., ed., 1972, Recognition of Ancient Sedimentary Environments: *Soc. Econ. Paleontologists and Mineralogists Special Publ.,* no. 16.

Selley, R. C., 1978, *Ancient Sedimentary Environments,* 2nd ed.: Cornell University Press, Ithaca, NY.

7

THE PERVASIVENESS OF CHANGE:
Evolution and Extinction

CONTENTS

KEY TERMS

Evolutionary theory
Natural selection
Heredity
Chromosomes
DNA molecule
Genes
Genotype
Phenotype
Mutation
Gene pool
Mitosis
Meiosis
Reproductive isolation
Divergence
Ontogeny
Phylogeny
Lineage
Speciation
Allopatric speciation
Punctuated equilibrium
Phyletic gradualism
Adaptive radiation
Extinction
Food chains
Niche
Stenotopic
Eurytopic
Biogeographic province
Food webs

TOO MUCH MONKEY BUSINESS

Tennessee, 1925, Chapter 17, House Bill 185. An act prohibiting the teaching of the Evolutionary Theory in all Universities, Normals and all other public schools of Tennessee, which are supported in whole or in part by the public school funds of the State, and to provide penalties for the violations thereof.

The preceding paragraph is the introduction to the Butler Act, which was passed by the General Assembly of the State of Tennessee in 1925. It, and the many other acts subsequently introduced into state legislatures and occasionally passed into law, represented political attempts to control or prohibit the teaching of the theory of evolution in public schools or universities. Evidently by the early 1900s the concept of evolution had transcended the realms of education and science and entered the world of politics.

As we will see in this chapter, modern **evolutionary theory** provides an explanation for the variety and changes in biological taxa through geological time. The presently understood theory has its own evolutionary history as it developed from bits and pieces and occasional large chunks of knowledge contributed by many scientists. One of the first scientific explanations of evolution was proposed by Jean Baptiste Lamarck in the early 1800s. However, Lamarckian concepts were completely overshadowed following the publication in 1859 of *On The Origin Of Species,* by Charles Darwin. Darwin's concepts of evolution represent one of the major milestones in biology and had a profound impact on science and society in general; thus the following sketch of his background provides an interesting backdrop to the history of evolutionary theory.

Charles Darwin grew up in England in a relatively well-to-do family, studied medicine at the University of Edinburgh, and then prepared for the ministry at Cambridge University. However, his main interest was natural history, and it was this interest that led him to embark as shipboard naturalist on an inauspicious voyage that began in December 1831. Darwin's voyage on the *H.M.S. Beagle* provided a vast amount of data that were later synthesized into his concepts of evolution. One of the most significant areas he visited was the Galapagos Islands, approximately 1000 km west of the coast of South America and lying astride the Equator. On these isolated islands, formed by volcanic activity, Darwin marveled at large marine iguanas, giant tortoises, the great variability of finches, and an array of unusual vegetation.

Upon his return to England in 1835, Darwin moved to the county of Kent and spent more than twenty years carefully analyzing data collected on his voyage and making further studies. Although he planned to write a major treatise on his interpretations and confided many of his ideas to geologist Charles Lyell and botanist Joseph Hooker, it is possible that Darwin never would have published his ideas but for the manuscript sent to him for review by Alfred Russel Wallace in 1858. To Darwin's surprise, Wallace had independently come up with similar concepts. With this stimulus a joint paper by Darwin and Wallace was presented to the Linnaean Society of London. Darwin then rapidly completed his major manuscript, and it was published in the fall of 1859. The first edition was sold out in its first day. Reaction to the book in scientific circles and by the public ranged from very favorable to violently negative. The well-known biologist Thomas Huxley commented upon reading the *Origin:* "How extremely stupid not to have thought of that."

Darwin's book was the outgrowth of decades of research and certainly had a profound influence on human thought; it described an elaborately researched and carefully reasoned concept and was based on a large volume of data. For the biological sciences the concept was the beginning of a revolution that led to many new ideas and developments. Scientists and the public in general were presented with a new interpretation that provided a coherent and comprehensive explanation for the richness of life on earth. The major emphasis of the book was its elaboration of **natural selection,** whereby the environment controls the probability of survival for each individual in a population of organisms. Those organisms best adapted to environmental conditions were viewed as having the best chance of survival and by reproducing offspring they would influence the characteristics of succeeding generations. The appearance of new species occurred by survival and reproduction of successful variants in original populations.

Darwin explained to the satisfaction of many, but by no means all, that life evolves continually and is always being modified through natural selection. To Darwin, changes by natural selection explained why diverse organisms could have strikingly similar kinds of shells or arrangements of bones and organs, how plants and animals can be well adapted to particular environments, and the underlying causes responsible for the previously recognized succession of different faunas and floras observed in the fossil record.

Because of the far-reaching significance of the theory of evolution, modern scientists such as the evolutionary biologist Ernst Mayr have attempted to recognize how various scientific and social trends influenced Darwin's thinking and what steps were involved in the gradual formulation of his concepts. These studies provide an intriguing glimpse into the thought process operating within the human mind. Evident from these studies is that ideas as momentous as Darwin's hypothesis often involve the assimilation of considerable data from widely diverse sources; thus the ideas are greatly dependent upon previous contributions to total human knowledge.

Since Darwin's time much new information has been added to our understanding of evolutionary processes. For example, Darwin knew nothing of genes or of the science of genetics; the modern concepts of this branch of biology have been formulated only in the twentieth century. Today we have extensive knowledge of biochemistry, population biology, and microbiology, again areas of science not known to Darwin. Our concepts of evolution have, therefore, undergone considerable refinement since Darwin's time, but still can be traced to his original ideas.

Modern concepts of organic evolution have surpassed the hypothesis stage. Evolutionary theory consists of a considerable body of supporting information from many scientific disciplines.

Surprisingly, we find that a very vocal minority still is convinced that the theory is totally false. Consider the opening paragraph of this chapter, which represents a law passed sixty years after most scientists were no longer concerned as to whether evolution occurs, but rather how it operates.

A result of the Tennessee antievolution law was the 1925 trial of high school biology teacher John Scopes in Dayton, Tennessee. The Scopes trial, now known as the "Great Monkey Trial," pitted William Jennings Bryan against Clarence Darrow. Bryan, a staunch antievolutionist, thundered and preached the state's case, fighting the cutting wit of Darrow. The result of the trial, which became one of the most famous of this century, was a victory for the state. However, in the process antievolutionists, and especially Bryan, were ridiculed. A number of state legislatures introduced antievolution bills during the middle 1920s, but almost all were defeated and by 1930 most of the outward controversy had died away. However, even today there are still significant protests against evolutionary theory. These have taken the form of censuring biology textbooks and lobbying in the legislatures to prohibit use of the books by some school districts. A recent tactic used by many fundamentalist groups is legal suit to demand equal time. These suits ask that religious creation be given coverage equal to evolution in biology textbooks, ignoring that creation is not a scientific concept nor is it open to modeling or experimentation. Numerous leaflets, pamphlets, and even an antievolution journal are published to discredit the theory. Let us consider this theory of evolution and determine its effects on the history of life on earth.

MECHANISM OF ORGANIC EVOLUTION

Organic Change Through Time

Fossils preserved in the lithosphere have provided us with a nearly continuous but often fragmentary record of the life that has existed for approximately 3500 million years of the earth's history. One of the most interesting discoveries made by studying this history of life is that there have been continuous changes in the kinds of organisms existing throughout this immense interval of time. Studies of the fossil record, especially that part representing what we now term the Phanerozoic Eon, had provided documented evidence of these changes to paleontologists by the late 1700s (see Chapters 2 and 3). One example of these changes is provided by the history of giant reptiles so familiar to us from museum displays but not recognized by eighteenth century paleontologists. These reptiles first appeared during Triassic time, were abundant and diverse in the Jurassic, and disappeared at the end of Cretaceous time. Progressing one step farther we find that all fossil taxa exhibit similar patterns; that is, they appear and then disappear, although some may last considerably longer than others. This paleontological pattern provides strong evidence that evolution occurs, but by itself does not explain the mechanisms controlling the process.

Accumulation and analyses of data from the fossil record raised perplexing questions related to the fossils themselves. A major question of this type, recognized as early as the late 1700s, was how to explain appearances and disappearances of taxa recorded throughout the stratigraphic record. The French naturalist Cuvier cham-

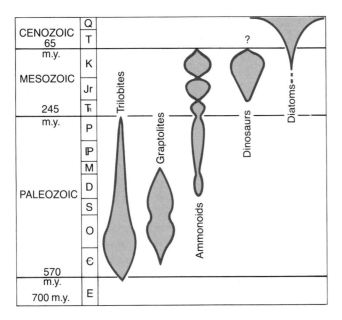

Figure 7-1. Origin and extinction of taxa including trilobites, graptolites, ammonoids, dinosaurs, and diatoms. The width of each group represents approximate diversity; wide spacing indicates large numbers of species (high diversity; diversity among various groups is not to scale). Note that each taxon has a distinctly different evolutionary history.

pioned the idea of catastrophism, whereby the earth's history was interpreted as having been punctuated by episodes of worldwide revolution. Catastrophism, as applied to the fossil record, visualized a succession of annihilations and subsequent repopulations by new kinds of organisms; this concept provided the crux of Cuvier's interpretation of the principle of fossil succession. Recall from Chapter 3 than an alternative explanation was the concept of uniformitarianism, proposed by Charles Lyell in the early 1800s. This idea recognized no worldwide catastrophic events, but emphasized that small, more local changes, taking place over the great length of geologic time, were sufficient to produce immense change. Given the impetus supplied in the last half of the 1800s by Darwin's book, the concept of evolution was applied to the fossil record and provides a logical if not completely understood explanation for the history of various taxonomic groups (Fig. 7-1). The subsequent discussion provides an outline of the development of tenets important to evolutionary theory.

Historical Development: Natural Selection, Genetics, and the DNA Molecule

Three historically important discoveries provide a basis for our discussion of evolution. The first was Darwin's concept of the process of evolution, which provided an explanation for changes in species and which

Darwin described as transformation. As he interpreted the process, changes in organisms are brought about by selective survival and reproduction of individuals whose characteristics are transmitted to successive generations. The second discovery was the mechanism of **heredity,** first described by Gregor Mendel, who published his results in 1866. His conclusions were used in the early 1900s to provide a fundamental impetus for the developing science of genetics. The concept of heredity, when coupled with subsequent discoveries of **chromosomes** and **genes,** provides a powerful tool for explaining the mechanisms by which characteristics are transmitted from one generation of organisms to another. The third discovery involved unravelling the extremely complex chemical nature of genetic material, namely, the chemistry and structure of the **DNA molecule,** described in 1953 by biologists James Watson and Francis Crick of Cambridge University.

The life of an individual organism and the evolutionary history of a species are controlled by many events and conditions. These can be divided into two categories: The first represents environmental conditions and the second represents genetic characteristics. Darwin recognized that environmental conditions influenced populations of organisms by effectively determining which, if any, individuals of a species survive and, even more importantly, which members reproduce successive generations, thereby perpetuating their genetic characteristics. This process of selective survival and reproduction he termed natural selection—popularly, but somewhat misleadingly, known as "survival of the fittest." Natural selection operates so that successive generations of the species will include offspring possessing characteristics similar to those individuals of the previous generation that were best adapted to the environment and thereby were able not only to survive but also—of the utmost importance—to reproduce. Thus each generation represents a segment of an evolutionary continuum. Significantly, over many generations the general character of the species can and may change. Darwin recognized these changes on the basis of morphological variations, especially in the populations of distinctive birds and reptiles he saw on the Galapagos Islands during his voyage on the *Beagle.* However, his concept suffered from a lack of information, and he could not explain which mechanisms within the organisms themselves could produce such changes; indeed this difficulty was a source of frustration for him.

This example illustrates a parallel development between the concepts of evolution and continental drift: Both for many years were hypotheses in seach of a viable mechanism. Just as sea-floor spreading vindicated early explanations of continental drift, so did recognition of genetics and the mechanism of inheritance put evolution on a solid footing.

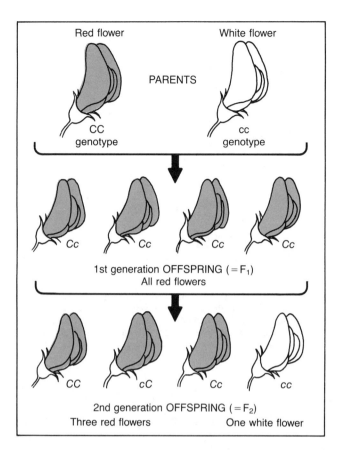

Red flower   White flower

PARENTS

CC genotype   cc genotype

Cc   Cc   Cc   Cc

1st generation OFFSPRING (=F₁)
All red flowers

CC   cC   Cc   cc

2nd generation OFFSPRING (=F₂)
Three red flowers   One white flower

Figure 7-2. Results of Gregor Mendel's genetic experiments illustrating the mechanisms of heredity and ratios of morphologic features (color of flowers) in offspring from successive generations of sweet pea plants. The genes responsible for the color are represented by C (red) and c (white). This experiment was performed with hundreds of offspring to produce the statistical ratio of 3:1.
(Adapted from LIFE, AN INTRODUCTION TO BIOLOGY, Second Edition by G. G. Simpson and W. S. Beck, copyright © 1965 by Harcourt Brace Jovanovich, Inc. Reproduced by permission of publisher)

The mechanisms that Darwin sought in vain were discovered by Gregor Mendel, ironically, at about the same time that the *Origin* was being published. Mendel was an Augustinian monk and taught natural science at a monastery in what is now Brno, Czechoslovakia. After developing an interest in the process of heredity, he experimented over many years with the breeding of sweet pea plants. From these experiments he developed a hypothesis that today represents one of the fundamental principles of genetics (Fig. 7-2). Mendel made the following assumptions based on his experimental results: (1) Each plant contained a pair of hereditary factors that controlled flower colors. (2) Both of these factors were derived from the parents and in each parent the factors separate during formation of germ cells. (3) The red or white color of the flowers is an alternate form of the same factor, although red was found to be dominant over white. Thus Mendel recognized that variability of traits was transmitted to successive generations of his

pea plants and that the ratios of color (red and white) in each generation could be mathematically predicted.

Mendel published his results in an obscure journal and did no further research on the subject. His results were overlooked by biologists for about thirty-five years and then were rediscovered in the early 1900s when the modern science of genetics was in its infancy. This brings to mind that often in the history of science individual scientists have been ignored and their ideas buried in scientific literature because they were "ahead of their time" or failed to publish in appropriate journals. At a later date the ideas may be rediscovered and recognized as historical milestones. You might speculate on the number of ideas or discoveries that may have been published but are overlooked in the vast scientific literature.

From the work of Watson and Crick, as well as from contributions by many other scientists, we recognize that genetic information in organisms is contained within comparatively large, complex nucleic acid (DNA) molecules. The DNA molecules form genes that are located along strands called chromosomes. These complex molecules control those critical chemical reactions within cells that play a major role in determining the physical characteristics and responses to environmental changes of individual organisms. The genetic makeup of an individual organism is described as its **genotype.** The resulting physical characteristics of the organism are due to a combination of genotype and environmental conditions and are collectively described as **phenotype** (Fig. 7-3). Variation in genes is produced by alterations of the internal chemistry of the DNA molecule, a process called **mutation.** Genetic variation can occur also by structural rearrangement or reordering of the sequence of genes along chromosomes; these processes are known as recombination and crossing over.

Modern Synthesis: Population and Reproduction

The collective genetic material of any population constitutes the **gene pool** of that population. Evolutionary changes in gene pools can occur only through successive generations of populations of a species and do not occur by changes in individual organisms (Fig. 7-4). The key link in this change is reproduction, and sexual reproduction provides the most effective method for recombining genetic information.

Reproduction involves replication (duplication) of a multitude of organic chemical molecules; it can be accomplished through **mitosis** or **meiosis.** Mitosis is characteristic of prokaryotic organisms of the kingdom Monera (see Chapter 5 and Appendix B). These organisms replicate their DNA and then divide into two smaller cells that contain genetic information that is

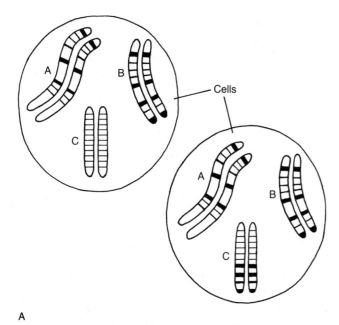

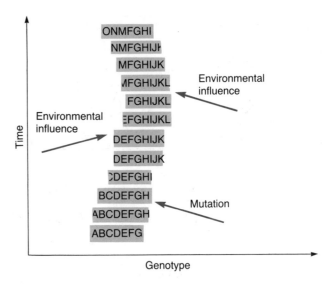

Figure 7-4. Hypothetical example of changing gene pools in successive populations over time, as represented by the changes in letters (= genes). Environmental conditions and other mechanisms such as mutation produce irreversible changes. One difficulty is to determine the point in this continuum of populations where biological speciation has occurred.

Figure 7-3. Difference between genotype and phenotype. A. Note the differences in genes (genotypes) in chromosome C of two single-celled organisms. B. Example of specimens of gastropods. Do these represent a single species with considerable morphological variation or two species?

identical to that of the parent cell (Fig. 7-5); the various groups of prokaryotic bacteria and cyanobacteria illustrate a number of variations of this mitotic process.

Much more complicated is the process of meiosis, or reduction division, which is characteristic of eukaryotic organisms that represent the other four kingdoms of organisms. In meiosis there are two successive divisions of the replicated DNA strands (chromosomes); the results are known as gametes (Fig. 7-6). Each gamete contains only half the genetic information of the parent;

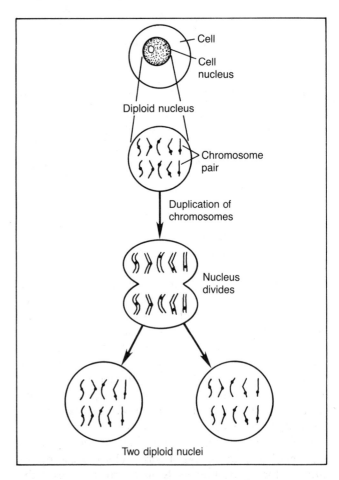

Figure 7-5. Model of reproduction in organisms illustrating the process of mitosis. This is a basic type of cell division; it occurs in all kingdoms of organisms, but is the only type of reproduction in monerans.

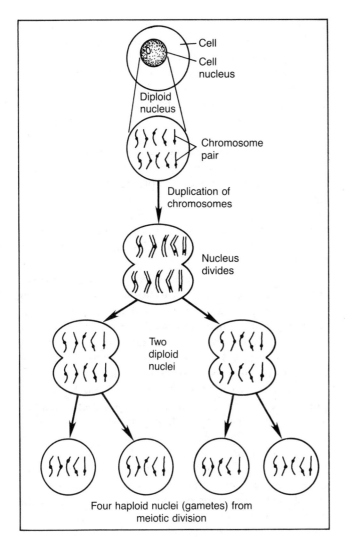

Figure 7-6. Example of meiosis, which is characteristic of sexual reproduction in nucleated organisms. Mitotic cell division, including nonsexual reproduction and replication of cells during growth, also occurs in nucleated organisms.

offspring are produced by combining gametes from two different parents; thus each offspring is genetically somewhat different from either parent.

Sexual reproduction by meiosis has played an enormous role in the evolutionary diversity of life. This can be illustrated by the following example, provided by the work of paleobiologist J. William Schopf at the University of California, Los Angeles. In a hypothetical asexually reproducing population consisting originally of one genotype, ten mutations will result in eleven different genetic combinations: the original plus the ten mutations. In contrast, ten mutations within a single genotype in the simplest sexually reproducing population could result in over 59,000 different genetic combinations! This large number dramatically expresses the potential for recombining genetic material through sexual reproduction. Evidence of this immense potential variation is

provided by the multitude of living species and fossil taxa.

One of the major features of the fossil record is that it provides good documentation of evolutionary changes through the long history of the earth. An example that serves to document evolution is provided by the history of horses. As discussed in Chapter 15, a very large collection of fossil horses has been painstakingly assembled from Europe, Asia, and North America by many paleontologists. These remains illustrate the many species that have existed through the 50-million-year history of the group. Collections of other fossils also provide evidence of speciation and lineages that have existed during the Phanerozoic Eon.

Evolutionary changes from one species to another take a considerable length of time, much greater than that of a single human life span. Nonetheless it is possible, within existing populations, to document the smaller genetic changes that may lead to the emergence of new species. A very well-known example that illustrates changes in gene frequency is provided by the color changes in a species of British moth over the last 120 years. The extensive burning of coal for energy in the late 1800s and early 1900s produced considerable pollution that gradually killed the lichens growing on the outer bark of trees, darkening the trunks. As the trees became darker, light-colored moths became more visible and were easily preyed upon by birds. A strain of darker colored moths (melanistic variants) gradually became more abundant because they were better camouflaged on the darker trees, were not as intensively preyed upon by birds, and thus were able to produce more offspring than the lighter moths. The melanistic moths were selected for in this environment and over successive generations became the most abundant strain in the populations. This classic example, known as industrial melanism, provides a good illustration of the effect of environment and reproductive patterns on population gene pools and is an example of natural selection. Interestingly, the installation of antipollution devices has allowed the lichens to grow again, and lighter moths are becoming more abundant, apparently reversing the earlier trend. This would suggest that the environmental change had influenced the selection of darker variants in the moth populations, but was not of sufficient duration to produce a permanent genetic shift in the moth gene pool. Thus, this example illustrates processes that may lead to speciation.

MAJOR FEATURES OF EVOLUTION

The Species Problem

In living organisms species often can be distinguished from one another on the basis of differences in phenotype, and this influences our classification of them. However, the underlying basis of our biological definition of a species is the genotype: the potential ability of individuals of a single species to interbreed and produce viable offspring. Conversely, members of different species cannot interbreed. As defined by Harvard University biologist Ernst Mayr, species are "groups of actually or potentially interbreeding natural populations, which are reproductively isolated from such groups."* The key test for distinguishing different living species, especially if they closely resemble each other in morphology, is the determination of their potential for interbreeding. That is, different species should not be capable of interbreeding and producing fertile offspring. This isolation among species is known as **reproductive isolation,** but may not be complete in all species. For example, horses and donkeys can interbreed and produce mules, but mules are sterile and, because they cannot reproduce, are not viable offspring. This reproductive test indicates that horses and donkeys are genetically somewhat similar and that the evolutionary process of **divergence** of species has not yet become complete in this case. Naming and recognizing most modern species on the basis of morphology and reproduction are relatively easy; the recognition of species is somewhat more difficult for paleontologists working with fossils.

The scientist who collects, studies, and identifies fossils obviously cannot determine the ability of the specimens to interbreed! Paleontologists, therefore, rely heavily on the phenotype of the individuals, but also

---

*1965, Animal Species and Evolution, p. 19: Belknap Press of Harvard University Press, Cambridge, MA.

consider geograhic proximity and stratigraphic position of the specimens and the range of variation within the entire population undergoing study. We previously made the observation that morphology is influenced to a great extent by genetic control; thus morphological differences or similarities among fossils provide a guide, although occasionally imperfect, to their evolutionary relationships. For example, comparison of the morphology of an individual specimen with the total variation in morphology observed in a collection of fossils judged to represent a single species can provide a strong clue for proper identification of the specimen (Fig. 7-7). The stratigraphic position in the geologic column also may provide a clue to species identification of the individual specimens (Fig. 7-8), but by itself is a poor tool for identification.

A further feature related to identification is **ontogeny,** which is growth of an individual organism during its lifetime. In many organisms there is considerable phenotype difference expressed during ontogeny. For example, amphibians such as the frog have a young stage represented by a larval fishlike tadpole. After metamorphosis and growth of legs and lungs, loss of tail, and other changes, an adult frog has a totally different phenotype but has not undergone any genetic changes. Many other organisms, including most benthic marine invertebrates, also have a larval stage. Even in the absence of obvious metamorphosis the young and old of the same species may be considerably different in size, ornamentation, and other morphologic features. These morphologic variations present major problems for the paleontologist who attempts to recognize and identify species. Still another difficulty for identification is pre-

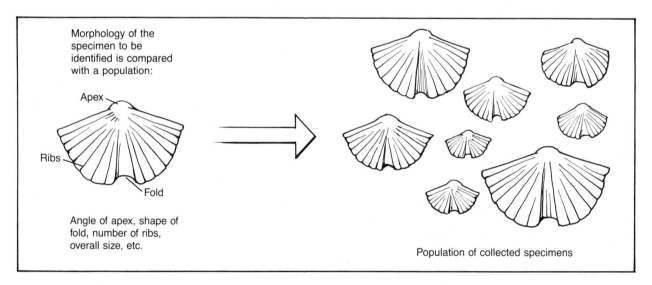

Figure 7-7. Comparison of a single brachiopod specimen with a large population of individuals judged to represent a single species. If the morphology (phenotype) of the individual falls within the range of the large population most paleontologists would consider that the individual belongs to the same species.

sented by the differences in morphology between males and females of the same species, a characteristic known as sexual dimorphism.

Evident from this discussion is that recognition and identification of species of fossils is a more subjective process than recognition of living species. Paleontologists attempt to distinguish species on the basis of significant morphologic differences. Traditionally these kinds of judgments have been made from visual, qualitative comparisons; more recently, sophisticated mathematical approaches have been used for more objective, if less personal, discrimination of species. Nonetheless, the overall experience and judgment of the paleontologist who has studied a particular group of fossils play an important role in the recognition of species.

Models and Patterns of Evolution

We have emphasized the genetic and morphologic variability within living species and the problems encountered in identification of fossil species. The next step is to consider the changes of species and higher taxa through time. Just as an individual organism undergoes ontogenetic changes during its life history, species, genera, families, or higher taxa to which the individual belongs likewise undergo changes during their geologic history. The record of these changes is called **phylogeny,** which refers to the evolutionary relationships of species and supraspecific taxa. Phylogenetic relationships are illustrated most commonly by the branching so-called tree of life (Fig. 7-9), which is a graphic expression of what is known about the phylogenetic re-

lationships of the taxa involved. Remember from our discussion of classification in Chapter 5 that organisms in the fossil record are assigned to their appropriate taxonomic categories in the classification scheme on the basis of known phylogenetic relationships.

Evolutionary changes within the geologic history of genetically related species constitute a **lineage.** Changes within a lineage are called phyletic changes and occur by a process termed **speciation.** A major problem in recognizing the various species within lineages is the difficulty of determining the rates and processes by which speciation occurs. Two models have been proposed to explain speciation. In one model, which is called **allopatric speciation** (Figs. 7-9A, 7-10), species are envisioned as forming from small parts of large populations that become geographically isolated. Because of factors such as genetic drift and natural selection, these isolated populations branch and eventually become reproductively isolated from the main population. After a succession of many generations, they

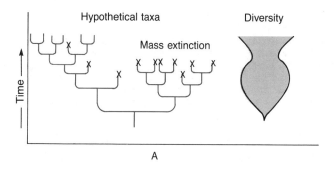

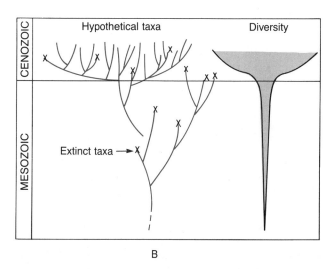

Figure 7-9. Hypothetical relationships of various related taxa (species, genera, and so forth). Each branch represents a separate taxon and the various groupings represent supraspecific taxa. This "family tree" can be expressed by the diversity diagram to the right and represents the allopatric model where evolutionary change occurs between species. B. Another way of expressing phylogenetic relationships, evolutionary history, and diversity is by phyletic gradualism, where evolutionary change occurs within lineages.

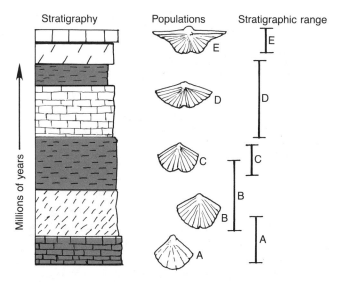

Figure 7-8. The stratigraphic position of a specimen indicated by asterisk may provide an important clue to its identification if the stratigraphy and associated fossilized taxa for the region have been worked out in detail. Note that the total amount of geologic time represented by the rocks is too long for these taxa (A–E) to represent variations of interbreeding populations.

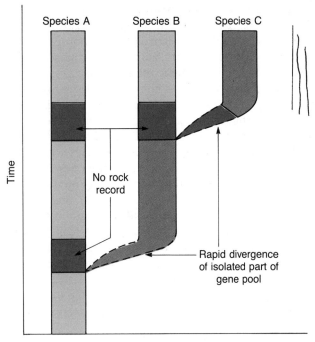

Figure 7-10. Simplified model of allopatric speciation. The dashed lines represent divergence, which occurs rapidly in a small isolated part of the total population. This small gene pool may undergo rapid changes leading to genetic and reproductive isolation and thus the appearance of a new species. Note that for the major length of a species existence there is no major genetic or morphologic change.

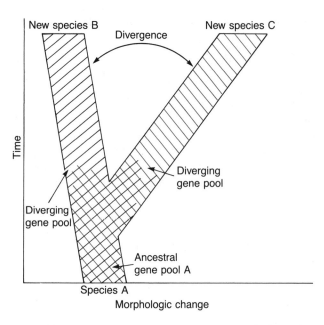

Figure 7-11. Example of phyletic speciation, or gradualism, where gradual divergence in the genetic composition of two populations leads to reproductive isolation and evolution of new species. Note that each species undergoes genetic and morphologic change through time.

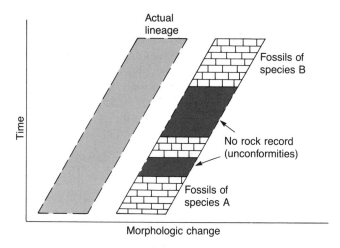

Figure 7-12. Sedimentary gaps in the rock record create breaks in the preserved record of a taxonomic lineage; these gaps are used by paleontologists to help define new species. In this example species B is morphologically and presumably distinct from species A.

become genetically distinct and thus biologically new species. Because this process occurs rapidly in small isolated populations, the intermediate steps are rarely, if ever, preserved. This may explain the abrupt appearance of species in the rock record with no apparent intermediate forms. In this allopatric model all evolutionary changes occur between species. Within each individual lineage few or no evolutionary changes occur. This example, representing rapid speciation events and longer lived unchanging lineages, has been termed **punctuated equilibrium,** or sometimes the punctuational model.

Another model of speciation is called **phyletic gradualism,** whereby one species of a lineage gradually changes into another through time (Figs. 7-9B, 7-11). In this model gradual evolutionary transition from one end-member species to another renders it difficult to recognize exactly where the two lineages separate. This decision, however, commonly is simplified for the paleontologist because of gaps in the fossil and rock record (Fig. 7-12). In this sense, nature has solved the species problem to some degree, but at the expense of other geologic information.

Evolutionary biologists and paleontologists currently are engaged in a controversy over which of the two evolutionary models best explains the history of life.

As we have seen, the rock and fossil records are incomplete and interpretations of evolutionary events suffer from many biases. However, the fossil record is sufficiently complete to provide the data necessary for evaluation of evolutionary patterns; the models developed from these patterns are based on how the record is interpreted and how the evolutionary process is viewed. What is important is the fact that speciation by evolutionary processes occurs and can be documented by the fossil record; the method or methods by which speciation occurs are more difficult to interpret.

The geologically rapid proliferation of new species and higher taxa from one or a few ancestral stocks is called **adaptive radiation.** Examples of adaptive radiation are numerous in the fossil record and seemingly are related to the allopatric model of speciation; however, this is by no means firmly established. Such evolutionary patterns have occurred in many groups. Some spectacular examples are the abrupt appearances of shelled invertebrates in the early Paleozoic (Chapter 11), the evolution of ruling reptiles in the Mesozoic (Chapter 12), and the evolution of mammals in the Cenozoic (Chapter 15).

Emergence of Higher Taxa

So far our discussion has focused on the species, but how do we explain the origin of higher taxa? As we saw earlier in this chapter, species may originate rapidly by allopatric speciation or more slowly by phyletic speciation. The myriad of species that have existed during geologic time all have interacted with the environment with varying degrees of success. In our discussion of the classification scheme for organisms we recognized that taxonomic categories above species are artificially created by taxonomists; however, these ranks still represent evolutionary relationships among organisms.

At various times in the geologic past, significant environmental changes coupled with adaptive innovations or breakthroughs have produced broad new adaptive patterns. As a consequence of these far-reaching evolutionary opportunities, major adaptive radiations have occurred, highlighted by the appearance of new higher taxa—phyla, classes, and orders—in response to the accessibility of new ways of life. Lower taxa within these groups—families, genera, and particularly species—diversified most rapidly at later times within these adaptive radiations and in the process essentially partitioned the broad adaptive zones invaded earlier. A few examples include the evolution of hard parts in the early Paleozoic, paving the way for the radiation of invertebrate phyla; the evolution of limbs and lungs in a group of fish that later invaded the land as the class Amphibia; and the evolution of temperature regulation and feathers that allowed the development of flight, characteristic of birds.

CRISES IN THE HISTORY OF LIFE

Introduction

We have briefly considered some of the processes of evolution that control the appearance, diversity, and distribution of organisms, but there is yet another major feature to consider, and that is the phenomenon of **extinction.** Evidence from our previous discussion of models of speciation suggests that extinction is a major

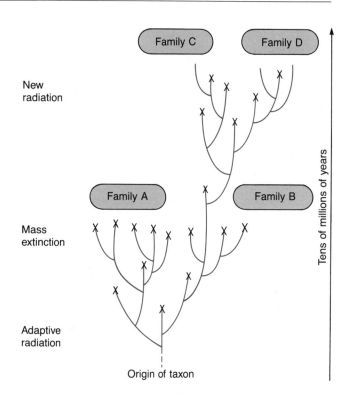

Figure 7-13. Origin and subsequent radiation of taxa classified into distinct families. Families A and B are affected by an episode of mass extinction. Only one lineage survives and it serves as the ancestor for a new adaptive radiation. Although this is a hypothetical example, the geologic record of ammonoids in the Mesozoic followed a similar pattern (see Chapter 12).

feature of evolution and furthermore is the ultimate fate of all species. However, study of the distribution of extinctions of taxa through time indicates that at certain intervals of time relatively severe waves of extinction affected large numbers of species and higher taxa on a worldwide scale. As paleontologist Norman Newell has aptly pointed out, these times of mass extinctions were "crises in the history of life."* Mass extinction, in a sense, is the opposite of adaptive radiation; we could consider extinction an expression of environmental foreclosure, whereas adaptive radiation reflects environmental opportunity (Fig. 7-13).

Although extinctions occur continuously throughout geologic time, our focus will be on those that have affected a large number of taxa within a relatively short span of time. The most interesting major episodes of extinction were those that occurred near the ends of the Cambrian, Ordovician, Devonian, Permian, Triassic, and Cretaceous Periods. As we mentioned in Chapter 3, such events in the fossil record have been used to define and recognize these and other chronostratigraphic boundaries. We will consider as examples mass extinctions at the ends of Permian and Cretaceous time.

---

*1963, Crises in the History of Life: *Sci. American,* Feb.

The Permian and Cretaceous Extinctions

A tabulation of the fossil record made by Newell indicates that approximately 55 percent of the families of invertebrates became extinct within a span of only 5 million years near the end of the Permian. A similar tabulation indicates that about 30 percent of the families of invertebrates became extinct at the end of the Cretaceous. Although this is a lower percentage than for the Permian, the total number of families was greater in the Cretaceous because the total diversity apparently was greater. At both times many land and marine-dwelling vertebrate taxa also become extinct (Fig. 7-14).

During the Permian crisis some groups, such as trilobites and fusulinid Foraminifera, completely disappeared. Many other taxa were severely affected; some of these were brachiopods, crinoids, corals, and terrestrial amphibians and mammallike reptiles. Interestingly, this episode of extinction did not greatly affect terrestrial plants. In general, similar kinds of organisms became extinct at the end of the Cretaceous. These included marine invertebrates, such as ammonites and many kinds of bivalves, and planktic Foraminifera. Many vertebrates also became extinct, especially the large reptiles that lived on land and in the oceans. Again, plant groups were not, for the most part, severely affected.

By comparing these two episodes of extinction we find that a wide variety of land- and water-dwelling vertebrates and invertebrates become extinct. The fact that plant groups do not seem to show a similar decline in numbers presents an interesting problem. The fossil record indicates that major extinctions and replacement of plant groups occurred slightly before the Permian and Cretaceous animal extinctions. What, if any, are the relationships between these plant extinctions and the subsequent animal extinctions? This is discussed later in this chapter.

Perhaps one of the most nagging questions in paleontology is what caused extinctions of a wide variety of taxonomic groups within relatively short intervals of time. Why are some taxa affected and some not at all? What are the ideas and hypotheses that have been proposed to explain extinctions?

Major physical and biological phenomena known from study of the earth's history and from study of the earth today might have caused or contributed to major episodes of extinction. Figure 7-15 and the following list indicate some of the events.

1. Major volcanic eruptions

2. Major mountain building (orogeny)

3. Heating or cooling of global climate

4. Trace-element changes in oceans

5. Magnetic reversals and solar radiation

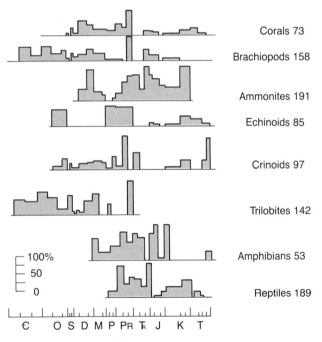

Figure 7-14. Selected representatives of major taxonomic groups that became extinct at the end of the Permian and Cretaceous Periods. Note percentages of families.
(Adapted from N. D. Newell, 1962, Paleontological Gaps and Geochronology, Fig. 6, p. 599: *Journ. Paleontology,* vol. 36, no. 3. Reproduced by permission of Society of Economic Paleontologists and Mineralogists)

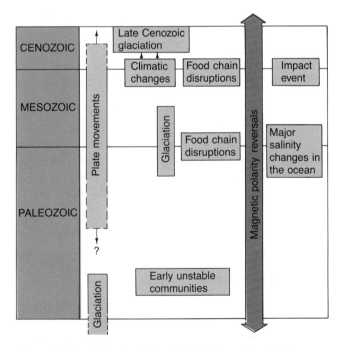

Figure 7-15. Some physical and biological events which have occurred in the last 700 million years and which might have triggered episodes of mass extinction.

6. Meteorite impacts

7. Community disruptions

8. Species competition

9. Plate tectonics

10. Changes in atmospheric composition

One of the most interesting as well as frustrating features of hypotheses based on these events is that for the most part such hypotheses cannot be tested experimentally; very likely it will never be possible to determine which, if any, is correct.

The explanations for extinction can be divided into two main categories: those proposing a physical cause and those proposing a biological cause. Perhaps a combination of these explanations is involved, or perhaps an event or events not yet recognized may have contributed.

Extinctions: Physical Hypotheses

Two physical causes for extinction will be considered here: movements of lithospheric plates and catastrophic impact events. Both represent physical conditions that could have affected the entire planet, thus had the potential for producing complete extinction of one or more groups of organisms.

Movement of continents and opening and closing of ocean basins could have played a major role in extinctions. This idea, related to the concept of plate tectonics, has been developed only since the late 1960s. Let us look at some factors involved and determine their possible consequences for organisms. Shifting of the lithospheric plates would be expected to produce a number of physical changes in the environment. First, we know that movements of the continents as parts of the lithospheric plates have influenced mountain-building and volcanic activity, especially at convergent plate boundaries. This has resulted in tectonic uplift of mountain belts and thereby changed the relative position of sea level. Second, any northward or southward components of movement would result in a change in position of the continents relative to east–west distribution of climatic zones. Finally, the opening and closing of ocean basins, along with changing climate, would create environmental stresses by changing the number as well as the characteristics of habitats to which organisms had become adapted. For example, a clustering of continents would produce more severe climatic conditions on land by eliminating any buffer effect of large intervening bodies of water. Such clustering would also decrease the number of shallow-marine environments by eliminating many square kilometers of coastline and shelf area. Could such environmental changes, caused by

plate tectonic activity, have affected habitats and lowered rates of speciation enough to produce worldwide extinctions of many diverse taxa? Let us consider this possibility further.

Paleontologic, stratigraphic, and paleomagnetic evidence all suggest that the continents were joined together to form Pangaea in late Paleozoic time (Fig. 11-68). The effects of this movement and joining were (1) uplift of land areas and regression of the shallow Paleozoic seas, (2) reduction in areas of shallow-shelf seas that bordered the continents, and (3) accentuation of climatic extremes because of the clustering of the continents. It is quite reasonable to assume that a combination of these events would have placed great stress on existing organisms, primarily because of climatic changes on land and reduction in habitats and changes in circulation patterns in the oceans. However, the basic question still remains: Were these factors sufficient to produce the extinctions noted in the record?

Moving ahead in time about 150 million years we find that by mid-Cretaceous time there were considerable changes in positions of continents and ocean basins compared with positions in the Paleozoic (Chapters 11 and 12). In particular, opening of the south Atlantic and partial opening of the north Atlantic had occurred, and Australia and India had moved northward. Corresponding with these changes was the establishment of warm climate, possibly related to formation of widespread, shallow, continental seas reminiscent of those existing during the middle Paleozoic before the formation of Pangaea. Thus environmental conditions were considerably different from those that had existed on Pangaea in late Paleozoic to early Mesozoic time. Study of the fossil record indicates that Mesozoic marine and terrestrial environments were filled with a great diversity of organisms. However, in rocks representing the last 30 million years of the Cretaceous Period, the fossil record changes dramatically and provides evidence of another mass extinction.

Major environmental changes that occurred from Medial to Late Cretaceous time were a result of the fragmentation of Pangaea and ensuing plate motions. Increased tectonic activity associated with subduction of oceanic lithosphere and withdrawal of the extensive shallow seas from the continents may have influenced development of drier and cooler climatic zones. These changes would have reduced the number of terrestrial habitats and conceivably could have played a major role in the Cretaceous extinctions.

The present positions of the continents and ocean basins indicate that considerable movement has occurred during the 65 million years since the end of the Cretaceous Period. Today continents are still moving; average elevation of land areas is relatively high compared with elevations in the past, and climate is relatively cool.

Furthermore, we recognize that there have been considerable changes in kinds of organisms and their geographic distributions in the last few million years of the Cenozoic Era. These changes may reflect continent positions. Overall, the movements and positions of lithospheric plates through time appear to play an important role in the evolutionary history of taxa and may indeed have been a major cause of mass extinction. Could there be other mechanisms significant enough to cause mass extinction?

The other extinction hypothesis we will discuss applies only to the Cretaceous event and was first proposed in early 1980. Study of the sedimentary rocks at the stratigraphic position of the Cretaceous–Tertiary boundary provided evidence of an unusually high concentration of iridium and other rare elements. Subsequent research has confirmed that in some areas this element occurs in significantly higher concentrations at the boundary than in other strata. This anomaly suggests an extraterrestrial event: a meteorite burning through the atmosphere and crashing on earth, either on land or in the oceans. Disintegration of this meteorite, which is calculated to have been 6 to 10 kilometers in diameter, and the force of the impact event would have produced iridium.

More controversial than the impact itself is the hypothesis that the Cretaceous extinctions were directly related to this event. Current ideas are very tentative and are being revised almost on a monthly basis. An impact on land would have produced many cubic kilometers of dust, whereas an impact in water would have produced ice crystals. These materials literally would have been blown into the upper atmosphere and would have caused a severe reduction in solar radiation reaching the earth's surface. As presently calculated, the resulting decrease in sunlight would have lasted a few months—long enough to produce massive extinctions in marine phytoplankton and thus disruption of the entire oceanic food structure. Significantly, this length of time would have been insufficient to drastically affect terrestrial plants; this is in accord with the record of plants.

Current studies have focused on determining the iridium concentrations in strata bracketing the Cretaceous–Tertiary boundary. Cores from deep-sea drilling and outcrops in areas such as the Rocky Mountain region are currently being studied. Along with determination of the iridium content, comparisons of distribution and changes in taxa associated with iridium-rich layers are being made. Relationship of these layers to the established position of the boundary will, perhaps, indicate the validity of this hypothesis (Fig. 7-16).

It is evident that both these physical hypotheses for extinction operate by producing major alterations of the environment, thereby affecting communities of organisms. Possibly the changes triggered within and among

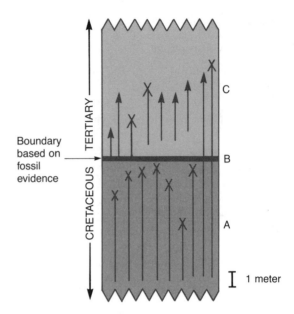

Figure 7-16. Stratigraphic distribution of taxa used to recognize the Cretaceous–Tertiary boundary. X represents an extinction, arrow indicates continued range of taxon. A, B, and C represent potential locations of iridium-rich layers with respect to the boundary. Current research may provide an indication of the stratigraphic position of this layer with respect to the boundary in many geographic locations.

communities themselves would be important contributions to extinction. This possibility is considered by examining biological hypotheses for extinction.

Extinctions: Biological Hypotheses

Biological hypotheses used to explain extinctions invoke mechanisms operating within or among communities or organisms rather than those involving changes in the physical environment. Some hypotheses within this category are species competition, alteration of community relationships, and changes of **food chains** among interdependent species. As you will see, these hypotheses are not isolated from one another but tend to overlap and often have to be considered together. In the following discussion we will combine these ideas.

Although each species occupies a **niche** that is generally distinct from that of other species, requirements for various resources create considerable competition among individuals of a single species and among the species themselves. Physical changes in the environment can alleviate or intensify this competition. Just as individuals have different abilities to cope with competition and environmental stress, so do species differ in their ability to survive environmental changes. In general, each species is physiologically adapted to the conditions in which it lives, but some species can survive and reproduce in a much greater range of conditions than others. In oceanic environments those with narrow tolerances are considered **stenotopic** and those having wide tolerances are considered **eurytopic.** Stenotopic species are specialized taxa more susceptible to changes; eurytopic species are generalists and are able to survive by adapting to changes. Stenotopic species would be more likely to become extinct and have shorter phylogenetic histories than eurytopic species.

Modern organisms exist in complex communities that may consist of tens or hundreds of species and a vast number of individuals. During times of environmental stability, well-established communities consisting of a large number of interdependent stenotopic and eurytopic species develop. Each species exists within a niche, but the boundaries of niches commonly overlap; this creates competition for living space, sunlight, food, oxygen, and other resources. This competition occurs in all communities and promotes strong selective pressure on organisms. Competing species exist in complex systems of interacting communities within major **biogeographic provinces.** The species, together with the general characteristics of communities in each province, can and do change through time. Perhaps extinctions are explained more easily at the community level of organization than at the individual species level. Let us examine some ancient communities.

Paleontologist Helen Tappan Loeblich at the University of California, Los Angeles, has suggested that mass extinctions at the end of the Paleozoic were caused in part by a breakdown in the stable community structures that had developed earlier in the Paleozoic. This breakdown may have been precipitated by lithospheric plate movements that produced major changes in distribution of land and ocean-basin positions, thereby triggering catastrophic changes in primary producer organisms and in feeding patterns in oceanic **food webs.**

Careful analysis of organisms preserved in the fossil record shows that there has been a trend toward greater efficiency in food gathering through time. The ability of individual organisms and species to obtain food in competition with others determines in large measure which survive and, most importantly, which reproduce. Thus natural selection will favor the most efficient food gatherers. For example, the bivalve molluscs, many of which are efficient filter feeders, rapidly increased in diversity in the Mesozoic and replaced brachiopods, which had been abundant during most of Paleozoic time but had been greatly affected by extinction in the Permian Period.

The significance of food chains and food webs in development and maintenance of communities is well documented for living organisms. In modern communities, disruptions of food chains or food webs have disastrous consequences for the structure of the community and the survival of its species. For example, changes

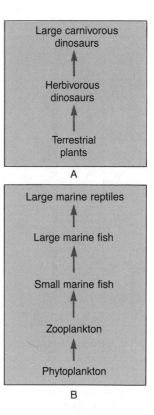

Figure 7-17. A. Simplified terrestrial food chain for Mesozoic time, which had primary producers (land plants) and a variety of consumers up to the level of carnivorous dinosaurs. B. Simplified oceanic food chain for Mesozoic time to illustrate the interdependence of species from the producers (phytoplankton) to the top-level consumers (marine reptiles).

in climate on land will affect the type of vegetation. Because land plants are the base of terrestrial food webs, changes in these taxa greatly alter the types of consumer species living in a region. In the oceans, phytoplankton blooms such as red tides greatly affect the other organisms, although these changes may be only temporary. We have noted a possible connection between changes in producers and mass extinctions at the end of the Permian. Does this hypothesis of food-chain disruption as a triggering mechanism for extinctions hold for the Cretaceous also?

Reconstruction of a simplified terrestrial food chain for Mesozoic time indicates that it consisted of seed plants such as gymnosperms as the basic producers and herbivorous and then carnivorous dinosaurs as the consumer groups (Fig. 7-17A). Study of the fossil record provides evidence that during the latter part of Mesozoic time many taxa of gymnosperms became extinct and were replaced by flowering plants (angiosperms). This is particularly evident in rocks of the Cretaceous Period. The dinosaurs also became extinct at the end of the Cretaceous. It is tempting to assume a direct cause-and-effect relationship between changes in gymnosperms and

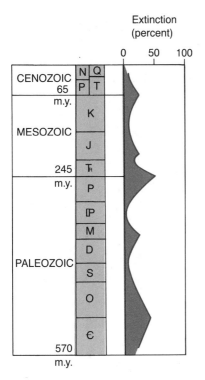

Extinction
(percent)

Figure 7-18. Rates of extinction of major taxa during the Phanerozoic, indicating the percentage of each becoming extinct at various intervals of time.

subsequent extinction of dinosaurs. However, plant replacement took place primarily during the early part of the Creataceous and major extinction of dinosaurs occurred only near the end of this time, many millions of years later. Were the dinosaurs decreasing in abundance throughout this interval, and, if so, were the changes in plants a significant contributing factor? As yet, we do not know the answers to these questions. Also, this mechanism does not explain the simultaneous extinctions of many marine-dwelling species at the end of the Cretaceous.

A food chain can be reconstructed for Mesozoic marine communities (Fig. 7-17B). This food chain has two more steps than the terrestrial example and involves more species. The producers consisted of phytoplankton such as diatoms. These were consumed by various plankton such as *Foraminifera* and *Radiolaria* and eventually the sequence ended at large marine reptiles such as plesiosaurs and ichthyosaurs. The fossil record indicates that significant changes in phytoplankton occurred near the end of the Mesozoic. We find that the first major diversification of diatoms and other microscopic algal groups occurred during the Cretaceous Period. Drastic changes affected planktic *Foraminifera*, algal groups, ammonite cephalopods, and marine reptiles at the end of Cretaceous time; perhaps there is a con-

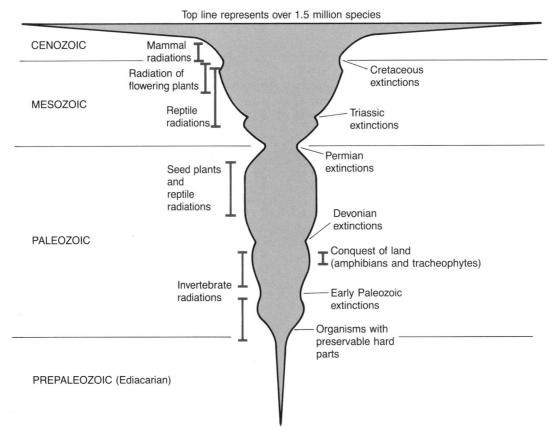

Figure 7-19. Model of diversity changes through the Phanerozoic; based on the known fossil and modern record. Note important events and the extremely rapid increase that has occurred in the last million years.

nection among all of these taxa which explains their similar phylogenetic histories.

After considering these biological explanations as the cause of mass extinction and recognizing that geologists currently lack sufficient evidence to provide proof for any of these mechanisms, we can suggest an area of research. This would involve detailed study to determine feeding characteristics of the various taxa affected by mass extinction and then an attempt to reconstruct their community relationships. Perhaps analyses of feeding types would indicate that certain kinds of organisms such as filter feeders were more affected during mass extinctions than organisms having different feeding habits. Evidence that extinction selectively affected taxa on the basis of their feeding type and community organization would provide strong support for the hypothesis that biological events play an important role in mass extinctions.

Armed with these ideas on extinction we can look at a representation of the relative extinction rates of families of eleven major taxa of vertebrates and invertebrates (Fig. 7-18). Based on a scale of 1 to 100 percent it is evident that mass extinction occurred at a number of fairly regularly spaced time intervals within the last 600 million years. Especially evident are the extinctions that occurred at the end of the Cambrian and Permian;

slightly less dramatic are those that occurred at the end of Devonian, Triassic, and Cretaceous times. We have considered those events that occurred in the Permian and Cretaceous, but it is evident that our explanations must account for these other episodes of mass extinction. At the present state of paleontological knowledge, we have not yet been able to detect a common feature relating all of these episodes to a single event or a combination of events. This is obviously an area of geological research that will continue to present a challenge for a long time.

Other Ideas on Diversity and Extinction

Perhaps what are seen to be significant episodic mass extinctions are in part a bias of the fossil record. Discussion of the diversity of taxa through the Phanerozoic Eon provides an illustration of this idea. Many tabulations of the number of species in the fossil record of the Phanerozoic indicate that a very rapid, spectacular increase in diversity occurred in the late Mesozoic and in the Cenozoic (Fig. 7-19). However, other interpretations of the same evidence suggest this geologically recent apparent increase may instead represent a bias in the older rock and fossil record (Fig. 7-20). For exam-

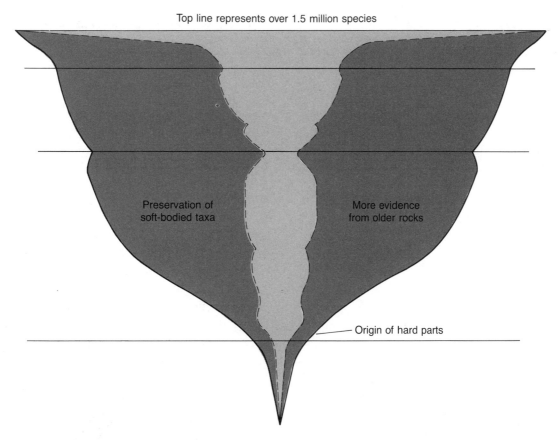

Top line represents over 1.5 million species

Preservation of soft-bodied taxa

More evidence from older rocks

Origin of hard parts

Figure 7-20. Alternative model of diversity changes through the Phanerozoic; based on extrapolation of organisms not preserved. This model and the one in Figure 7-19 represent "end members" in concepts of diversity increase. Perhaps the correct explanation lies between the two.

ple, younger sedimentary rocks are more abundant on the earth's surface, commonly contain more abundant fossils, are less altered than older rocks, and also are more intensively studied by paleontologists. Furthermore, the fossil record consists primarily of the remains of organisms having hard parts and tends to reflect only a small portion of the true diversity existing during any particular interval of time. Consider the large number of modern soft-bodied taxa, few of which would be preserved. The alternative explanation suggests that the major increase in diversity occurred in the early Phanerozoic, triggered by the appearance of sexual reproduction and aided by the evolution of hard parts and the conquest of land. Support for this idea comes from very recent studies which note much higher diversities of soft-bodied fossils in rocks of Ediacarian and Cambrian age than had previously been recognized.

These models as well as other explanations of evolution and extinction considered in this chapter serve to illustrate the many problems of unravelling the complex nature of the history of life. The models also touch on a fascinating aspect of geology: the existence of alternative answers, none of which may be proved correct and any of which may be evaluated or reinterpreted upon acquisition of new data. Indeed, these examples illustrate the scientific method and underscore the fundamental differences between scientific and religious explanations for evolution.

SUMMARY

Evolutionary theory has had a stormy history since publication of Charles Darwin's *Origin of Species* in 1859. Evidence of natural selection operating on populations, the experiments of Gregor Mendel and later development of the science of genetics, and the unravelling of DNA structure by Watson and Crick provided three important support pillars for the theory. We recognize that environmental influences coupled to mutation provide mechanisms for changes which occur through sexual reproduction, or meiosis, over successive generations. These changes in gene pools of populations eventually lead to reproductive isolation, and such reproductively isolated populations represent biologically distinct species. Speciation events through geologic time, if preserved in the rock record, represent what we call evolutionary lineages. Study of such lineages has provided evidence of immense changes in the biosphere over the last 3500 million years. Two models have been proposed to explain speciation. One is known as phyletic gradualism and the other is allopatric speciation. Evidence has been put forth to support both models, and possibly both have served as pathways for speciation.

A phenomenon recognized in study of higher taxonomic ranks is adaptive radiation, characterized by appearance of many new taxa within a relatively short interval of geologic time. An opposing phenomenon is extinction: the disappearance of taxa from the rock record. Episodes of mass extinction have occurred at a few intervals during geologic time, in particular at the end of the Permian and Cretaceous Periods. Explanations to account for these apparently catastrophic events have called upon a wide variety of physical and biological events. Plate movements, with corresponding changes in position of ocean basins and land areas and in climate, or possible impacts of large extraterrestrial objects are currently the most favored physical hypotheses. Changes in food chains and disruptions of major community structures have been suggested as biologically related hypotheses.

SUGGESTIONS FOR FURTHER READING

Darwin, C. R., 1859. *On the Origin Of Species:* John Murray, London.

deCamp, L. S., 1968. *The Great Monkey Trial:* Doubleday, New York.

Laporte, L. F., 1978. *Evolution and the Fossil Record:* Readings from *Scientific American,* W. H. Freeman, San Francisco.

Note especially the articles by Dobzhansky (1950), L. C. Eisley (1956) and N. D. Newell (1963).

Mayr, E., 1978. Evolution: *Sci. American,* vol. 239, no. 3, pp. 46–55.

Newell, N. D., 1962. Paleontological gaps and geochronology: *Journ. Paleontology,* vol. 36, no. 3, pp. 592–610.

Raup, D. M., 1977. Probabilistic models in evolutionary paleobiology: *Am. Scientist,* vol. 65, no. 1, pp. 50–57.

Raup, D. M., and S. M. Stanley, 1978. *Principles of Paleontology,* 2nd ed.: W. H. Freeman, San Francisco.

Watson, J. D., 1968. *The Double Helix:* Signet, New York.

8

ORIGINS OF EARTH AND ITS SPHERES

KEY TERMS

Coacervate droplets
Lithosphere
Hydrosphere
Atmosphere
Biosphere
Protoplanets
Jovian planets
Terrestrial planets
Nova
Supernova
Planetesimals
Astroblemes
Outgassing
Anaerobic
Ozone
Red beds
Banded iron formations
Cells
Fermentation
Photosynthesis
Respiration
Organic macromolecules
pH

ORIGIN OF LIFE

Perhaps the most fascinating phenomenon of planet Earth is the almost bewildering array of organisms that exist in virtually every nook and cranny. As far as we presently know, neither these organisms nor any others are known to exist on any other planet within our solar system. Investigation of other solar systems has not even begun; in fact, evidence for existence of other solar systems is very sparse. The possibility that life is not unique to Earth is the subject of a number of proposed studies (Fig. 8-1) but little information is presently available. Our understanding of life on Earth, however, has progressed tremendously since the mid-1800s. New techniques have given us much information about the way life works and about its history.

Because of its significance to us, many people have contemplated the meaning of life and how it may have originated. These speculators have diverse backgrounds: philosophy, psychology, biology, geology, chemistry, and religion. Each ponders the questions and apparent mysteries of life and develops explanations and interpretations that are colored by his or her training and temperament. The antiquity of life and its history is a provocative and often perplexing field of study.

The fossil record provides an important account of the history of life. With the tremendous increase in our knowledge of fossils in the last 200 years has come recognition of amazing changes in types of organisms that have populated the earth. Many scientists have observed evidence of the sudden appearance of hard-part bearing fossils and the disappearance of such creatures as trilobites and dinosaurs. How are these events explained on a scientific basis? Since publication of Darwin's momentous *Origin of Species* various scientific hypotheses used to explain the observed changes and to account for the origin of life have focused on evolutionary theory. In recent years, teams of scientists have attempted to determine the history of the earliest life on our planet, to search for evidence of life on other planets, and to reconstruct models of the ancient Earth on which life originated. The complexities of these models require synthesis of evidence from many disciplines, such as astronomy, biology, and geology, and such cross-fertilization has provided added insights.

Most scientists agree that life originated by a sequence of chemical reactions that produced aggregates of organic molecules under environmental conditions significantly different from those existing on the planet today—conditions, in fact, that would be lethal for most presently living organisms. This model provides a logical complement to the theory of organic evolution in that it suggests that many of the same mechanisms—competition, replication of molecules, natural selection—that control populations today influenced development of nonliving, or abiologic, organic molecules that preceded the first living cells.

In 1870, the English biologist T. H. Huxley presented a stimulating presidential address to the British Association for the Advancement of Science in which he discussed the perplexing and controversial idea that biological systems evolved from abiological molecules. At the time of his speech most scientists had become convinced from many previously reported experiments that life could come only from previous life. As Huxley carefully pointed out, the ancient concept of spontaneous generation, where life supposedly developed from decaying organic matter, had been disproved, but only through a variety of experiments performed at various times for hundreds of years. Huxley lauded the invention of the microscope and its gradual improvements through the eighteenth century. Better resolution and magnification aided in the performance of many of these experiments in that they allowed studies of tiny single-celled organisms. Among many others, the tests by Spallanzani in the 1700s and Pasteur in the 1800s were noted as simple but significant experiments that demonstrated clearly that any organisms appearing on decaying organic matter were brought in by other organisms or were carried in by air currents. Similar experiments with nonliving organic material sealed off from the atmosphere illustrated that no new organisms developed.

Having made his point, Huxley then strongly hinted he believed life had in fact originated from nonliving chemicals. He was at a loss to explain the method by which this may have happened, but he felt that were he present on primitive Earth he "should expect to be a witness of the evolution of living protoplasm from not living matter."*

Perhaps because of a lack of models for primitive Earth, studies on the origin of life languished in the late 1800s and early 1900s. Beginning in the late 1920s and continuing in the 1930s however, the work of two

---

*1896, *Discourses Biological and Geological:* D. Appleton, New York.

170

biologists provided stimulus for more detailed investigations into the origin of life. A. I. Oparin, a Russian, and J. B. S. Haldane, an Englishman, published thought-provoking ideas that supported hypotheses that life began on Earth as a result of chemical reactions in oxygen-deficient environments. As Oparin envisioned the process, a transitional step between nonliving chemicals and living cells may have resembled what are today known as **coacervate droplets.** These microscopic-sized droplets consist of large organic, but nonliving, molecules that remain suspended in water. Under appropriate conditions these droplets behave like living cells. They can be used to simulate the reactions that may have occurred in prebiologic

molecules, and therefore they provide clues for understanding how the first cells may have developed.

Haldane was a pivotal figure in promoting subsequent research into the origin of life. In a paper published in 1954 he considered four possible explanations to account for the appearance of life and judged as most reasonable a process in which life arose by chemical evolution in oxygen-poor paleoenvironments. His ideas stimulated a number of laboratory experiments in which attempts were made to duplicate conditions thought to exist on primitive Earth.

A key experiment was performed in the 1950s in the laboratory of Harold C. Urey at the University of

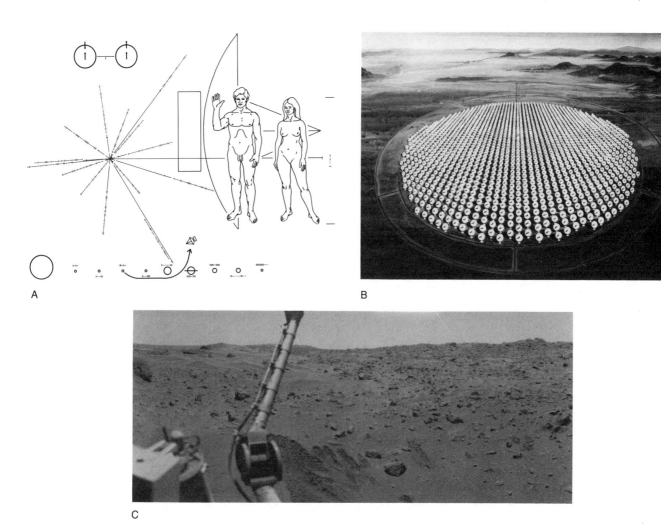

Figure 8-1. Attempts to search for and communicate with extraterrestrial life. A. Aluminum plaque (15 × 23 cm) placed on the *Pioneer* spacecraft, which was launched in 1972 and is heading to the fringes of our solar system and beyond. The plaque, designed by Carl and Linda Sagan, is much like the proverbial message in a bottle. Omission of female genitalia caused a minor flap and drew both positive and negative responses. B. Large array of radio telescopes, such as that of the proposed *Project Cyclops,* will be used to search for broadcast signals from space, presumably from other civilizations. C. Experiments to test for evidence of organic activity on Mars were part of the *Viking* landings of 1976; no evidence of Earth-type organic molecules or cells was found.
(A, Reproduced by permission of Carl Sagan; B and C, courtesy of National Aeronautics and Space Administration)

Chicago. Earlier work had suggested that in the early stages of Earth's history, its atmosphere lacked oxygen gas, instead consisting of ammonia and methane, and therefore was reducing. One of his graduate students, Stanley L. Miller, constructed an apparatus containing these gases and water and energized the mixture with electrical discharges (Fig. 8-2). Within a short time, a variety of organic molecules had formed, many types that occur in living organisms. This experiment provided strong evidence for a method by which complex organic molecules and possibly living cells could develop from simple chemicals in appropriate environments.

Since these pivotal experiments in the 1950s, studies have suggested that the early atmosphere on Earth was probably not composed of ammonia and methane—compounds that had been vaporized early in the evolution of the planet—but rather was a mixture of nitrogen, carbon dioxide, and water vapor. Subsequent workers have agreed that there was a virtual absence of free oxygen gas. Further laboratory experiments using these newer ideas have also been successful and have produced a variety of complex organic molecules such as amino acids. These molecules represent the major building blocks for cells. Although a complete living system has yet to be produced, these experiments have clearly demonstrated a process by which life could have evolved. Thus, nearly a hundred years later, we find

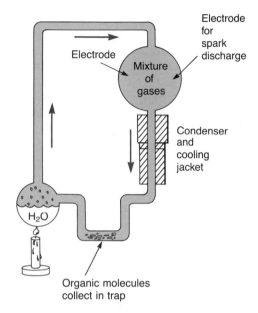

Figure 8-2. Type of apparatus used by S. L. Miller at the University of Chicago. Methane, ammonia, water vapor, and hydrogen gas were circulated in a system sealed off from the atmosphere and subjected to electric discharge. Amino acids formed in a relatively short time and collected in the trap.
(Adapted from S. L. Miller, 1953, A Production of Amino Acids under Possible Primitive Earth Conditions, Fig. 1, p. 528: *Science,* vol. 117. Copyright 1953 by the American Association for the Advancement of Science.

strong support for Huxley's concepts at the time of his presidential address in the 1870s.

BEGINNINGS

Earth's Spheres

This chapter is concerned with origins. We will consider ideas and models for the origins of Earth and of its lithosphere, atmosphere, hydrosphere, and biosphere. Accumulated evidence indicates these spheres developed early in Earth's history: It appears that all were present at least 3800 million years ago, near the beginning of Archean time. During and since their formation each has been affected by the others and a change in one will eventually be reflected by changes in one or all of the others. This interconnection of spheres has implications for us. Developing awareness of what has come to be thought of as spaceship Earth has suggested to many people that our rapid development, exploitation, and alteration of these spheres may initiate long-term changes unfavorable to us, but which we only dimly perceive at the present time.

In Chapter 1 we discussed Earth's spheres and it will be helpful to briefly review them here. The **lithosphere** represents the outermost rock layer (Fig. 8-3).

It consists of a variety of rocks that make up the continents and floors of the ocean basins and is divided into about a dozen large plates that have a maximum thickness of about 75 km. Surrounding the lithosphere are the **hydrosphere** and the **atmosphere.** We recognize the hydrosphere as including water in oceans, freshwater lakes, streams, and rivers, along with groundwater

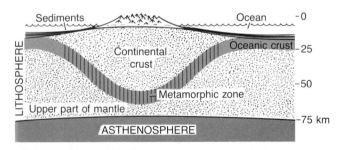

Figure 8-3. Earth's lithosphere (maximum thickness of about 75 km) includes mantle rock, oceanic crust, and in some areas continental crust. The lithosphere is thickest where the continental crust has been deformed into mountains; examples are the Rocky Mountains and the Himalayas.

and water vapor in the atmosphere. Part of this sphere is locked up in solid form as glacial ice. The atmosphere is the gaseous envelope surrounding Earth and consists mainly of nitrogen and oxygen gases. The final sphere, known as the **biosphere,** represents life on Earth. Living organisms are generally restricted to areas containing water.

Origin of the Solar System

Earth is but one planet in the solar system. To consider its origin we must look to the initial formation of the sun and all the planets. Our solar system is believed to have formed from a swirling, contracting cloud of cosmic gas and dust. As gravitational contraction of this cloud began, its speed of rotation increased so that the mass gathered into a disclike form with a dense central zone. This central globular core ultimately became the

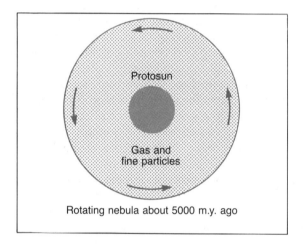

Rotating nebula about 5000 m.y. ago

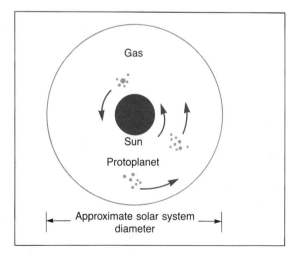

Approximate solar system diameter

Figure 8-4. Hypothetical succession of events in the formation of our solar system. A. Condensation of primordial sun and beginning of formation of the disc. B. Central portion has collapsed and heated to the point of thermonuclear reactions; remaining particles collide and form protoplanets which have fixed orbits around the sun.

sun, which represents over 99 percent of the mass of our solar system. Remaining dust particles and gas of the primeval solar disc separated into belts. Within these belts, accretion of adjacent dust and gas particles through gravitational attraction produced swarms of larger masses that rotated about their own centers and revolved around the globular core of the cloud (Fig. 8-4). These orbiting masses, called **protoplanets,** gradually condensed into solid bodies that were precursors of the present-day planets. A time framework would put this formational stage around 4600 to 5000 million years ago.

This hypothesis actually represents a synthesis of various earlier ideas. It represents an attempt to account for various physical characteristics of our solar system: mass and angular momentum of the sun and planets, distinct differences in size and composition of the inner planets and outer planets, and distances and rotational directions of the planets relative to each other and to the sun. Earlier hypotheses, some of which date back to the mid-1600s, suggested possible collisions between a sun and a passing comet; a condensing gas cloud; a near-miss between two suns; and an exploding sun in a double star system.

The next step is to interpret what events may have led from protoplanets to planets. Since the late 1960s much information about our solar system has come from flybys and landings of spacecraft such as *Luna, Ranger, Mariner, Apollo, Pioneer, Viking,* and *Voyager.* This information has been combined with that available from Russian spacecraft and other evidence collected from years of observations by earthbound astronomers. Coupled with our new understanding of plate tectonics, new perspectives of the nature and evolution of the solar system have emerged.

What kind of history can we envision for the origin of the elements and early evolution of the solar system on the basis of observable planetary characteristics? The large-size, outer planets, or **Jovian planets,** have a low specific gravity and are composed largely of hydrogen and its simple compounds, a condition that most likely reflects the primitive composition of the solar cloud and protoplanets. By contrast, the inner, rocky planets, or **terrestrial planets,** such as Earth, are smaller, have high specific gravities (Table 8-1) and relatively little hydrogen. They are composed primarily of elements heavier than hydrogen: oxygen, silicon, aluminum, sodium, calcium, and iron. These elements are rare in the universe as a whole. They probably were produced at the centers of giant stars under extremely high temperatures by the fusion of helium from hydrogen and the addition of neutrons to the nuclei of helium. We hypothesize that most of the heavier elements formed in giant stars by processes known as neutron capture and decay. Very heavy elements such as uranium may fuse

Table 8-1.   Comparison of terrestrial and Jovian planets*

	Diameter (km)	Mass (Earth = 1)	Density (Water = 1) (g/cm^3)	Gravity (Earth = 1)
Terrestrial				
Mercury	4,835	0.06	5.6	0.38
Venus	12,194	0.82	5.1	0.89
Earth	12,756	1.00	5.5	1.00
Mars	6,760	0.11	3.9	0.38
Jovian				
Jupiter	141,600	318.0	1.2	2.64
Saturn	120,800	95.1	0.6	1.17
Uranus	47,100	14.5	1.6	1.03
Neptune	44,600	17.0	2.2	1.50

*Data from various sources may vary by 10 percent.

only at very high rates of neutron capture and decay. This may occur when a star implodes as a **nova** or **supernova.** Such activity not only produces the heavier elements, but literally blasts them into space where they become the "stuff" of cosmic nebulae for the later building of other stars or planets. In a sense, each new generation of stars and planets builds from the ashes of the last, and the population of elements in any solar system is both a product and a record of its prior history. The abundance and range of heavy elements and isotopes on Earth suggest that matter of our solar system has been recycled through at least one supernova. Our sun is probably a second or third generation star associated with other stars in the Milky Way galaxy, whose origin and evolution preceded that of our solar system.

Stellar evolution of our sun may have produced the radiation necessary to heat the inner protoplanets to such an extent that their hydrogen and helium were released into space and solid hydrogen compounds such as ammonia and methane became vaporized. The outer protoplanets retained most of their original light gases and solid compounds because they were not heated as much as the protoplanets between Mars and the sun. Satellites separated from these protoplanets in much the same fashion as the protoplanets themselves had separated from the solar disc. Eventually the protoplanets became condensed into solid masses as particles became more concentrated through gravitational attraction.

Protoplanet to Planet

Our present-day interpretive model of Earth's interior structure and composition has been developed from seismological evidence, composition of meteorites, studies of the magnetic field, and direct evidence of rocks from Earth and its moon. What implications does this model have for interpreting thermal history and chemical and physical evolution from protoplanet to dif-

ferentiated planet? Nearly all of the elements that occur naturally on Earth were formed prior to the protoplanet stage, probably through a series of complex thermonuclear fusion reactions that represented an evolution from the element hydrogen. A few have formed within the planet as byproducts of radioactive fission or by decay from originally incorporated unstable isotopes. Initial heating of the protoplanet by solar radiation and gravitational contraction, followed by radioactive decay and to some extent meteoric impact, probably raised the temperature at the center to several thousand degrees Centigrade and caused partial or complete melting.

Presumably most of the denser elements such as iron and nickel sank into the protoplanet's deep interior and formed a core, probably around 4700 million years ago. Convection and radiation of heat within this early hot interior may have facilitated separation of lighter elements and formation of silicate minerals, which then accumulated into a layer called the mesosphere. Persistence of a liquid outer core (Fig. 1-2) beneath a solid mesosphere is explained by temperatures in that area above the melting points of nickel and iron. However, the melting points of the magnesium- and iron-rich silicate minerals typical of the deep mesosphere are higher than that of the outer core; the mesosphere, therefore, remains in a solid state. Is Earth still getting hotter or is it becoming cooler? Differentiation of the interior clearly suggests that Earth was once largely molten, but it is not known whether it ever completely melted or if the heating is still increasing, is decreasing, or is remaining steady.

Radiometric dating of meteorites and rocks from the moon's surface, along with information from lead isotope ratios and lead evolution curves (Chapter 2), indicates that Earth evolved from its large, low-density, heterogeneous protoplanet stage to a smaller, dense, internally differentiated planet, with an organization of distinct mineral phases, by approximately 4600 million years ago. Developmental events occurring between

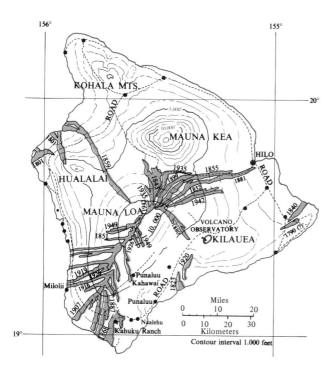

Figure 8-5. Geologic map of Hawaii showing the gradual buildup of a basaltic shield volcano.
(Adapted from G. A. Macdonald and A. T. Abbott, 1970, *Volcanoes in the Sea,* Fig. 45, p. 52: University of Hawaii Press, Honolulu)

about 4600 and 3800 million years ago have left no tangible rock record, so much of our interpretation has come from the study of other planets and of meteorites.

As you might expect, ideas about the origin of the lithosphere, atmosphere, and hydrosphere fit into our picture of overall segregation of Earth and are particularly related to separation of lighter mineral phases from original mesosphere material. The protolithosphere very likely was forged through volcanic activity, perhaps similar in style and composition to that occurring on the island of Hawaii today (Fig. 8-5). Eruption of gases such as carbon dioxide, nitrogen, and steam can readily account for the atmosphere and hydrosphere, and these two spheres must also have evolved early, perhaps before 3800 million years ago.

THE LITHOSPHERE

Evidence from the Moon and the Planets

Although various lines of evidence point to an age of about 4600 million years for Earth, the oldest remaining rocks are only about 3800 million years old. So the rock record for the first 800 million years of our planet's history appears to have been obliterated, probably by a combination of subduction, metamorphism, weathering, and erosion—processes of the geological and rock cy-

cles. It is also possible that for part of this time interval the lithosphere, in which these oldest rocks are preserved, had not developed to its present thickness and composition. In fact, the origin and early evolution of the lithosphere is poorly understood and is the focus of considerable research. In part this research is multifaceted because it must also consider characteristics and changes in the atmosphere and hydrosphere, both of which would have significantly influenced and been affected by the evolving lithosphere.

What kinds of activity were taking place within the lithosphere and on its surface during the first 800 million years of Earth's history? Study of the moon provides a window through which we are able to catch a glimpse of this enigmatic formative period. Exploration of the moon during the *Apollo* program provided on-site investigations and sample recovery that inaugurated studies of the petrologic and geochemical history of a planetary body other than Earth. This information together with extralunar photography, has demonstrated that, in contrast to Earth, the moon's early history is fairly well preserved. Rock debris as old as 4500 million years has been identified from the lunar surface. Cratered highland areas, crater ejecta, and lava flows forming basins (low-lying dark regions) make up a composite of rocks and structures (Fig. 8-6) whose stratigraphy has been interpreted by study of superposition and cross-cutting relationships. We may attribute this preservation of older features to the absence of a significant geologic cycle. In particular, plate tectonics has not been a factor and there has been no extensive weathering or erosion.

The moon has a thick, solid crust that formed early in its history from global melting and cooling. The cratered highlands, which dominate the lunar surface, are exposed surface remnants of this early crust. Alteration of the crust occurred primarily by impact of extralunar material, much of which may have been small **planetesimals** remaining from the condensation of the nebula. This cratering was at first extensive but began to decline about 3000 million years ago. Extremely large impacts excavated gigantic depressions and spread ejecta over extensive areas as evidenced by radially textured deposits; lunar mountains are piles of debris ejected by meteoric impact. The next stage of lunar history was dominated by emplacement of dark mare plains that are relatively thick layers of basalt. This major outpouring of lava occurred from 3900 to 3200 million years ago; there has probably been no extensive igneous activity on the surface during the last 3000 million years. The moon's crust has changed but little since then; and evolution of the surface was essentially complete by 2500 million years ago or earlier. In the absence of an atmosphere or hydrosphere to povide the impetus for weathering, meteoric impact has been the primary process in modifying surface features.

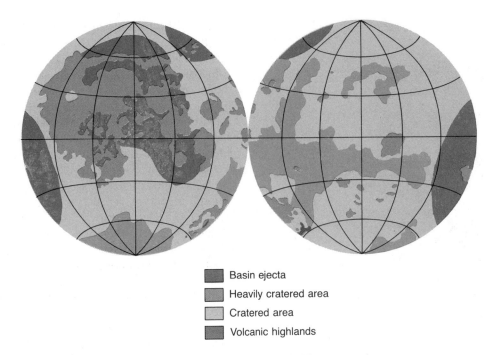

Basin ejecta

Heavily cratered area

Cratered area

Volcanic highlands

Figure 8-6. Geologic map of our moon showing volcanic stratigraphy. Compiled from many photographs and very limited sampling by the *Apollo* missions.
Adapted from D. E. Wilhelms and J. F. McCauley, 1971, U. S. Geol. Survey Map I-703, and K. A. Howard, D. E. Wilhelms, and D. H. Scott, 1974, Lunar Basin Formation and Highland Stratigraphy, Fig. 14, *Rev. Geophysics and Space Physics,* vol. 12. Copyright by the American Geophysical Union.

In contrast to the moon, 98 percent of Earth's surface is less than 2500 million years old, and 90 percent is less than 600 million years old. Compared with the moon, which has essentially been dead for the last 3000 million years, the earth's lithosphere has been rearranged constantly through plate tectonics activity and has been effaced by the weathering and erosion processes of a dynamic atmosphere. During the last few thousand million years or so, the geologic cycle has been dramatically more active on Earth than on the moon.

The key point of this present discussion, however, is that our understanding of the early history of the moon suggests an analogous history for Earth. The pregeologic history of Earth's surface and near surface probably involved basaltic volcanism and meteoric impact, perhaps even as violent as that on the moon. About 4000 million years ago the surface of Earth probably looked similar to today's lunar surface, which is intensively cratered (Fig. 8-7). Since that time, when infall of meteorites from the solar cloud was much greater than now, comparatively few have hit the earth. Craters that have survived are called **astroblemes** and are recognized as circular depressions (Fig. 8-8) displaying impact structures and containing materials of meteorite composition in a few places.

Our knowledge of lunar history and processes is only part of the story. We know that meteoric impact has left its mark also on the surfaces of Mercury and Mars. Mars also shows evidence of extensive and varied vol-

canism, large-scale linear tectonic features, and a complete lexicon of surface features produced by running water (Fig. 8-9). But from our accumulating knowledge, impact cratering and volcanism are the two processes that have dominated the surface histories of the terrestrial planets in terms of areal coverage, volume, and time duration. Thus it would seem that the early history of the solar system, whose signature has undergone destruction on dynamic Earth, has been laid before our very eyes on the moon and Mars. The early history of our planet can be interpreted from studies of rocks and events preserved on the other terrestrial planets, because they all shared a common early history.

Formation of the Earth's Lithosphere

Our preceding discussion supports a hypothesis that Earth's earliest lithosphere, like that of other inner planets and of the moon, was thin and composed largely of basalt rich in iron- and magnesium-bearing silicate minerals. However, presence on Earth of an atmosphere and hydrosphere capable of weathering rocks at the surface began to modify this basaltic layer. By 3800 million years years ago, at the beginning of what we term Archean time, part of the basaltic layer had been altered and probably resembled the continental or granitic part of the lithosphere of today (Fig. 8-3). This continental or granitic crust of the lithosphere is relatively rich in silicon and aluminum. By what processes did it form?

Figure 8-7. The moon's visible surface, illustrating the undestroyed evidence of craters.
(Reproduced by permission of Lick Observatory)

Figure 8-8. Meteor Crater, Arizona, an astrobleme. The crater is approximately 1.3 km in diameter
and 100 m deep and was made by a meteorite.
(Reproduced by permission of Yerkes Observatory)

It is likely that the basaltic crust must have solidified
as Earth cooled from its early molten state. This crustal
material would have been exposed to the primitive atmosphere and hydrosphere and as a result would have
been subjected to chemical and physical weathering and
to erosion of the rock cycle. Exposed areas of basalt

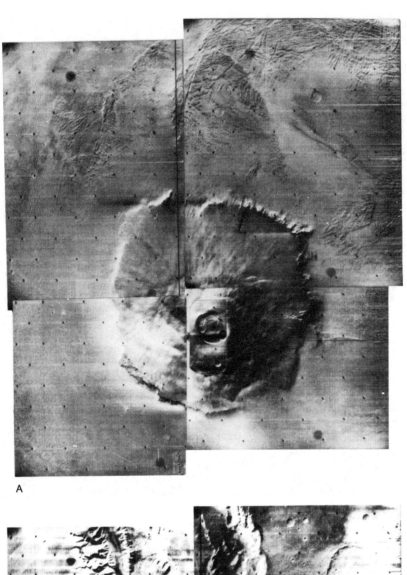

A

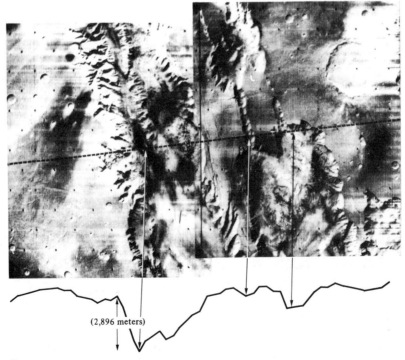

(2,896 meters)

B

C

Figure 8-9. Mars. A. Olympus Mons volcano, 27 km above the plains and 600 km across at the base. How does this compare in size to volcanoes on Earth? B. Evidence of water erosion and mass movements in very large canyons. C. Surface photograph from *Viking* 1 showing a dune field. (Photos courtesy of National Aeronautics and Space Administration)

would have become chemically altered because various silicate minerals making up these rocks break down at different rates. This process produced clastic sediments enriched in potassium feldspars and quartz relative to basaltic rocks; these sediments would have been transported and deposited in the primitve seas. Burial, followed by metamorphism and possibly melting of the sediments resulted in formation of metamorphic and igneous rocks of intermediate rather than basaltic composition. Repetition of these events would eventually form granitic rocks, which represented early continental masses. Although this presents a sketchy outline of the formation of the continental crust of the lithosphere, it fits nicely with our understanding of the rock cycle, hydrologic cycle, and geologic cycle.

Although much evidence indicates that Earth is about 4600 million years old, the oldest remaining continental crust so far discovered is an area of 3800-million-year-old rocks in southern Greenland. Their age indicates that formation of continental crust of the lithosphere had begun before 3800 million years ago and also indirectly indicates that erosion and sedimentation were operating on Earth's surface at an early date. From these beginnings the rock record indicates that continental crust has grown thicker and increased in volume throughout Earth's history. Because their specific gravity is less than that of the basaltic layers from

which they were ultimately derived, continental rocks have remained topographically high, in accordance with the principle of isostasy.

THE ATMOSPHERE AND THE HYDROSPHERE

As primitive Earth heated up and underwent the internal differentiation that produced its core, mesosphere, asthenosphere, and lithosphere, volcanic activity must have been common at the surface, as it has continued to be to the present time. Along with the pyroclastic debris and lava of these eruptions, it is likely that abundant gases similar to those associated with modern volcanism, were produced. This phenomenon is known as **outgassing** and provides an explanation for the origin of the atmosphere and hydrosphere (Fig. 8-10). The gravitational field of the earth was sufficient to prevent these gases from escaping into space and a major portion remained to form an atmosphere rich in nitrogen and carbon dioxide and a hydrosphere composed of condensed water vapor. As currently known, no other planet has a dense gaseous atmosphere and a fluid hydrosphere; it is probably no coincidence that life is known only on Earth.

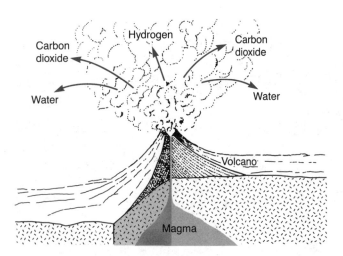

Figure 8-10. Volcanic outgassing of carbon dioxide, hydrogen, nitrogen, and a few other minor gases. Most of the water enters the hydrosphere, nitrogen and carbon dioxide are added to the atmosphere, and hydrogen gas generally escapes into space.

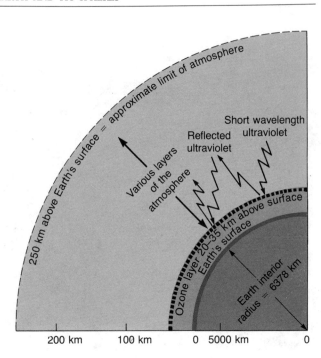

Figure 8-11. Formation of the ozone (O_3) layer (approximately 15 km thick) in the atmosphere serves to screen out a significant amount of solar radiation, especially in the ultraviolet range.

Neither the atmosphere nor the hydrosphere contained oxygen gas because it is not produced by volcanic activity. Therefore, pre-Archean and Archean paleoenvironments were **anaerobic,** or reducing. An absence of oxygen gas also indicates that the atmosphere lacked an **ozone** (O_3) layer. Ozone, formed by bombardment of O_2 molecules by ultraviolet radiation in the atmosphere, serves to screen out much short-wavelength ultraviolet radiation from the surface (Fig. 8-11). Such radiation is lethal to exposed cells: High levels of ultraviolet radiation would have provided a barrier to any developing life near or at the surface. As we will see, the absence of oxygen and ozone and the presence of high levels of ultraviolet radiation at Earth's surface provided important environmental conditions that controlled development and survival or prebiologic molecules and later primitive cells.

Characteristics of early Archean sedimentary and metasedimentary rocks preserve some evidence of the conditions in which they were deposited. For example, although younger sedimentary rocks commonly contain oxidized iron particles and have a characteristic reddish hue, there are no **red beds** of Archean age. This implies that iron particles were not oxidized but rather remained in a reduced state as ferrous iron even through atmospherically controlled surface processes of weathering, erosion, and deposition. Evidence of locally and periodically abundant oxygen gas in the hydrosphere is provided by layers of chert alternating with layers containing oxidized ferric iron in unusual **banded iron formations** which are known only within age limits of 3400 to 1800 million years ago (Table 9-4).

These rocks also provide evidence of the chemical characteristics of the Archean oceans. Abundance of terrigenous rocks and a lack of carbonates can be explained by existence of an acidic hydrosphere. Carbon dioxide-rich and oxygen-deficient waters would favor the formation of bicarbonate ions ($HCO_3^=$). Such a system would maintain hydrogen ion concentrations, or **pH,** of less than 7.0 and therefore would be acidic. In such acidic conditions calcium carbonate is soluble and would not be deposited.

Composition and characteristics of the early atmosphere and hydrosphere are in distinct contrast with those existing today. We can assume that weathering, erosion, and deposition were occurring as they do today, but because of the different conditions under which they were operating, rates and products were different from those of today's Earth. However, conditions in this early interval favored the formation of a wide spectrum of chemical molecules that eventually produced a living cell. As we will see, this event had far-reaching consequences, in particular by altering the chemistry of the hydrosphere and the atmosphere and by affecting rates of weathering and types of sediments produced.

THE BIOSPHERE

Definition and Characteristics of Life

In the 1950s Stanley Tyler and Elso Barghoorn discovered microfossils in rocks previously considered unfossiliferous. At the same time radiometric dating techniques became more precise. This evidence indicated

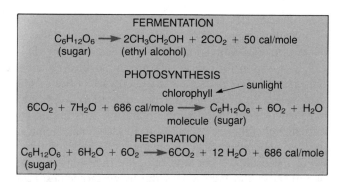

Figure 8-12. Important metabolic reactions in cells. Fermentation is an anaerobic process that produces alcohol and releases about 50 calories of heat energy. In contrast, photosynthesis and respirations are aerobic processes and provide a much larger amount of heat energy.

that life first evolved on Earth during early Archean time. Further studies have provided a fascinating glimpse of what the earliest life was like and have considerably altered our perceptions of the biosphere and how it originated.

The boundaries of the biosphere are best defined as corresponding to those areas in which water occurs in a liquid state, as it does in all parts of the hydrologic cycle (Chapter 1). A variety of physical and chemical properties of water are also vital for support of living organisms. One important property of water is that it expands on freezing; thus ice is less dense and floats on water. It may act as an insulator, preventing bodies of water from freezing solid and killing organisms. Because of its bipolar structure, water is also an excellent solvent; it makes up a major portion of organisms and serves as the transporting and reaction medium for a large variety of chemicals. Even a casual glance at a chemistry book will provide a list of many important characteristics of this elixir of life.

Other environmental conditions also influence the ability of organisms to exist in the biosphere; some of these include the amount of solar radiation reaching Earth's surface, the temperature of the land surface, atmosphere, and hydrosphere, and the chemical composition of the atmosphere and hydrosphere. Furthermore, as mentioned in Chapter 7, considerable evidence indicates that environmental changes have affected diversity and abundance of organisms and that variations in Earth's biota throughout geologic time have been controlled by a variety of physical and biological events. We should expect that similar, though perhaps not identical, conditions influenced prebiologic molecules and the development of life.

Living systems are recognized by particular diagnostic characteristics and essentially represent chemical factories that consist primarily of four elements—carbon, hydrogen, nitrogen, and oxygen—that combine to form

organic molecules. These elements combine also with lesser amounts of about twenty other elements to provide a large number of complex molecules. We can consider in more details some other characteristics of life: (1) Organisms consist of one or more **cells** that are capable of reproducing themselves by chemical replication of a wide variety of complex organic molecules. (2) Individual organisms undergo many other chemically related activities such as metabolism, excretion, and growth; important examples of metabolic reactions are **fermentation, photosynthesis,** and **respiration,** which involve carbon dioxide, water, and oxygen along with more complex molecules such as sugars and alcohol (Fig. 8-12). (3) Organisms are able to move or cause motion in water or air; most single-celled and multi-celled organisms have appendages such as cilia, tentacles, or other structures capable of such movements. (4) Individuals, whether unicellular or multicellular, have a boundary layer such as a cell membrane, cell wall, or skin that permits selective exchanges of gases and liquids with the environment. (5) All life as we know it has the characteristic ability to adapt to environmental changes, whether by physiological response to small-scale seasonal changes or by genetic change of successive generations through a longer interval of time (Chapter 7). Living organisms should possess all of these characteristics. Paleontologists recognize that fossils are the remains or traces of once-living organisms, so we may assume that such fossils also possessed these characteristics when they were living.

The Origin of Life: A Working Model

Excluding various religious explanations of creation, which cannot be subjected to scientific methods, we find that most scientists subscribe to a hypothesis that suggests life on Earth originated from a series of abiological evolutionary processes during pre-Archean or early Archean time. As currently envisioned, this evolutionary process consisted of a number of progressively more complex chemical reactions that involved prebiologic molecules within the early hydrosphere. The reactions involved the four main chemical building blocks of carbon, hydrogen, nitrogen, and oxygen and led to the formation and subsequent recombination of organic compounds into increasingly more complex **organic macromolecules.** These macromolecules responded to environmental changes and may have resembled the coacervate droplets of Oparin. Eventually a combination of macromolecules formed that possessed a combination of characteristics that fit our definition of life, and so the simplest living system, the cell, appeared on Earth (Fig. 8-13).

A working model for the origin of life combines information obtained from laboratory experiments and

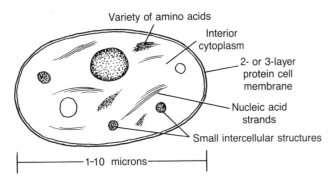

Figure 8-13. The cell is the smallest living system. Procaryotic, or nonnucleated, cells are the simplest type and consist of various organic macromolecules distributed within membrane-bound cytoplasm.

from studies of remaining early Archean rocks. This evidence suggests that living organisms developed (1) by a sequence of successively more complex chemical reactions, (2) during an unknown length of geologic time after formation of the lithosphere, atmosphere, and hydrosphere, and (3) within an anaerobic hydrosphere and atmosphere, lacking an ozone layer, that allowed high levels of ultraviolet radiation to reach the surface. According to this model, life originally formed from chemicals that underwent a succession of many different reactions. We assume that these reactions occurred in the primitive hydrosphere because the various properties of water would have provided the best environment for necessary chemical reactions and also would have provided some protection from ultraviolet radiation. These chemical combinations and recombinations may have occurred over a relatively short interval of time.

Within the global primitive hydrosphere a wide variety of temperatures and chemical and physical energy conditions would have existed. Such different conditions would naturally affect the kinds of chemical reactions and types of molecules produced, much as formation of chemical compounds and their subsequent reactions can be altered by changing conditions in beakers with chemistry lab experiments. These chemical changes occurring within the primitive hydrosphere represented a form of natural selection, whereby certain organic molecules were favored because of their ability to form and exist in a particular environment. Once formed, these organic molecules, which were adapted to particular conditions within the hydrosphere, could undergo changes in composition in response to changes in environmental conditions. This description suggests that the primitive oceans may have, as the biochemist J.

B. S. Haldane described them, "reached the consistency of hot dilute soup."* Within this soup and over millions of years of recombinations of molecules, it is not difficult to envision formation of some molecules or combinations of molecules having the characteristics of life that we set forth earlier. By our definition we would have to consider them living organisms. It is important to realize that there was and is no fundamental chemical difference between large nonliving molecules and large living molecules. What was basically different was the organization and behavior of the living molecules. Can we document this hypothesis with evidence from the rock record?

The Rock Record

As we previously discussed, indirect evidence for a lack of free oxygen in Earth's early atmosphere is preserved in sedimentary and metasedimentary rocks of Archean age. These rocks contain no oxidized iron minerals such as occur in geologically younger red beds. Also the rocks are composed predominantly of detrital particles; nonterrigenous rocks such as limestone and dolostone are very rare. This scarcity of nonterrigenous clastic (chemical) rocks is explained by the existence of low pH values in the primitive oceans. Furthermore, many of these old rocks contain unstable minerals, such as uraninite and pyrite, that could not have survived in the presence of oxygen (Chapter 9).

Within these rocks are micron-sized structures that are interpreted to represent carbonized organic remains of simple cells (Fig. 9-32). These structures have received considerable study and have been photographed with electron microscopes and light microscopes. Recognized only since the 1950s, they have been considered to represent bacteria and consequently classified in the kingdom Monera. These fossils indicate that life had originated by 3500 million years ago. As you will see in Chapter 9, younger Archean-age rocks provide evidence of other microscopic and macroscopic monerans, including bacteria and various types of cyanobacteria. The subsequent geologic history of the biosphere is closely tied to changes in the hydrosphere and atmosphere. In turn such changes influenced the lithosphere, within which we find the preserved records of events. The history of these changes provides the subjects for the following chapters.

---

*1954, *New Biology,* vol. 16: Penguin, New York.

SUMMARY

Study of the origin and early history of Earth and its spheres—lithosphere, atmosphere, hydrosphere, and biosphere—is challenging. Much of the evidence has been obliterated by dynamic processes such as the hydrologic and geologic cycles. However, evidence of the early history of other, less active, planets has been preserved and has been studied by a variety of remote-sensing techniques and actual samplng. This information has been applied to understanding of Earth's history. The sun and planets appear to have originated at least 4600 million years ago from a condensing gas nebula. Our planet underwent a partial or complete molten state that led to formation of various layers: core, mantle, asthenosphere, and lithosphere. Formation of the lithosphere probably began very early, but the oldest remaining rocks are between 3800 and 3900 million years old.

A process known as outgassing was responsible for our hydrosphere and atmosphere, but as we know from studying modern volcanic eruptions, no oxygen gas was produced. The reducing nature of these spheres is reflected in the absence of red beds and carbonates of Archean age.

The biosphere, representing life on Earth, originated from chemical reactions involving carbon, hydrogen, nitrogen, and oxygen in the form of complex molecules. These molecules existed in a reducing hydrosphere and were influenced by a variety of environmental conditions. About 3400 million years ago some of these molecules attained characteristics that we use to define living cells. These first organisms were probably much like modern bacteria. Some specimens are preserved in Archean-age rocks.

SUGGESTIONS FOR FURTHER READING

Barghoorn, E. S., 1971, The oldest fossils: *Sci. American,* vol. 231, no. 5, pp. 30–42.

Barghoorn, E. S., and S. A. Tyler, 1965, Microorganisms from the Gunflint Chert: *Science,* vol. 147, no. 3658, pp. 563–577.

Cloud, P. E., 1970, *Adventures in Earth History:* W. H. Freeman, San Francisco.

Cloud, P. E., 1978, *Cosmos, Earth and Man:* Yale University Press, New Haven.

Dickerson, R. E., 1978, Chemical evolution and the origin of life: *Sci. American,* vol. 239, no. 1, 70–109.

Head, J. W., Wood, C. A., and Mutch, T. A., 1977, Geologic evolution of the terrestrial planets: *Am. Scientist,* vol. 65, no. 1, pp. 21–29.

Hoyle, F., 1975, *Astronomy and Cosmology:* W. H. Freeman, San Francisco.

Press, F., and Siever, R., eds., 1975, Planet Earth: Readings from *Scientific American,* W. H. Freeman, San Francisco.

Sagan, C., 1973, *The Cosmic Connection:* Anchor, New York.

Sagan, C., 1980, *Cosmos:* Random House, New York.

9

CRYPTOZOIC HISTORY:
The Age of Microscopic Life

CONTENTS

KEY TERMS

Cyanobacteria
Prokaryote
Stromatolite
Cryptozoic
Phanerozoic
Precambrian
Archean
Proterozoic
Geochronometric unit
Canadian shield
Orogenic front
Kenoran orogeny
Greenstone belt
Banded iron formation
Red beds
Diamictite
Geosyncline
Methanogen
Eukaryote

LIFE BEFORE TRILOBITES

Until about thirty years ago, one of the great mysteries of geology had been that definitive evidence of life prior to the beginning of the Cambrian Period had not been discovered. The Cambrian fauna, dominated by such complex organisms as trilobites (Fig. 9-1), appeared to come into existence abruptly and without known predecessors. How could life have begun with organisms as complex as trilobites? In *On the Origin of Species,* Charles Darwin wrote: "To the question why we do not find rich fossiliferous deposits belonging to . . . periods prior to the Cambrian system, I can give no satisfactory answer. . . . The case at present must remain inexplicable."* During the last several decades, however, an answer has been found; a long pre-Cambrian history of life has been discovered and it extends back through geologic time almost 3000 million years before the beginning of the Cambrian.

In the early 1950s, the late Stanley Tyler, an economic geologist at the University of Wisconsin, made a startling discovery that was to inaugurate a new dimension in our understanding of early life on this planet. Tyler was involved in a petrographic study of iron-bearing chert deposits in the approximately 2000-million-year-old Gunflint Formation, exposed around the shores of Lake Superior. For decades, this iron-rich formation and the Biwabik Iron Formation, a geologically correlative unit in Minnesota, had provided the ore for the nation's iron and steel industry. Tyler was looking for clues to the precise origin of the minerals. In several thin sections, Tyler observed some puzzling spherical and elongate microstructures that he believed might be organic in nature. He sent them to Elso Barghoorn, an expert on fossil microorganisms, at Harvard University. Barghoorn was convinced that the structures, beautifully preserved in three-dimensional detail in chert, were **cyanobacteria** (formerly referred to as blue-green algae). This discovery from rocks about 2000 million years old represented the first definitive evidence of life before the age of visible animal life (Fig. 9-2). Tyler and Barghoorn published a report on the discovery in 1954. Continued work on the Gunflint microbiota has identified a diverse paleocommunity of **prokaryotic** microorganisms, including different kinds of bacteria and cyanobacteria.

Prior to Tyler's discovery, the only previously published reports on pre-Cambrian fossils involved

---

On the Origin of Species, p. 309. Mentor paperback ed., 1963.

Figure 9–1. Cambrian trilobite. Until recently, the first appearance in the stratigraphic record of complex organisms such as trilobites represented a paradox of major proportion—the sudden appearance of complex organisms with little evidence of prior evolution. (Photo by R. Miller)

peculiar megascopic pillarlike and mound-shaped laminated structures called **stromatolites** (Fig. 9-3A). American paleontologist Charles D. Walcott had hypothesized that stromatolites were fossilized reefs and had been formed by various kinds of algae. However, few people accepted the biologic origin of stromatolites. Barghoorn and Tyler's examination of the Gunflint showed that many of their microfossils were from silicified stromatolite structures. This substantiated Walcott's idea and, coupled with discoveries of algal stromatolites in a few modern-day environments (Fig. 9-3B), gave credence to the interpretation that stromatolites are organosedimentary structures, whose laminations are formed by the carbonate sediment-trapping and binding activities of matlike communities of cyanobacteria.

As an undergraduate geology major at Oberlin College in Ohio, J. William Schopf was enthralled by the mystery of scant evidence of life in the pre-

A

Figure 9–2. Prokaryotic microorganisms from the 2000-million-year-old Gunflint Formation. The discovery of the Gunflint microbiota by Stanley Tyler and its description by Tyler and Elso Barghoorn inaugurated an exciting chapter in the recognition of life before trilobites.
(Photo courtesy of J. William Schopf)

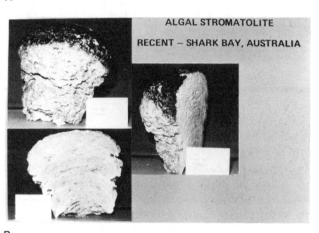

B

Figure 9–3. Stromatolites. A. Calcareous stromatolites from the approximately 3000-million-year-old Bulawayan System of Rhodesia. B. Recent calcareous stromatolites from Hamelin Pool, Shark Bay, Western Australia. The discovery of modern stromatolites and their biologic origin shed important light on similar-looking structures in the pre-Phanerozoic rock record.
(Photos courtesy of J. William Schopf)

Cambrian record and the sudden appearance of complex, diverse life forms in the Cambrian. Intrigued by publications on the Gunflint microbiota by Barghoorn and Tyler, he decided to embark on a quest—to learn as much as he could in his lifetime about life before the Cambrian. From Oberlin, Schopf went to Harvard for a program of graduate studies under the supervision of Barghoorn. In 1965 Barghoorn collected specimens from the Fig Tree Formation in the Barberton Mountain Region of South Africa. At that time the Fig Tree rocks were estimated to be slightly in excess of 3000 million years old and were regarded as the oldest known unmetamorphosed sedimentary succession. Schopf joined Barghoorn in study of the Fig Tree specimens, and, from petrographic and electron microscope examination, they were able to identify a number of rod-shaped and threadlike filaments, as well as larger spheroidal forms. These finds, identified as certain kinds of bacteria and cyanobacteria, were interpreted as the inhabitants of the earliest known community of organisms.

Schopf and Barghoorn also studied chert samples from the Bitter Springs Formation in the Northern Territory of Australia. From these rocks, estimated to be about 1000 million years old, they found examples of unicellular eukaryotes: microorganisms with a cell nucleus (Fig. 9-4). This discovery was of great importance for a clearer understanding of pre-Cambrian evolution prior to the dawn of multicellular animal life. Thus, during the 1950s and 1960s, major discoveries not only provided evidence of life before the Cambrian but also shed light on the crossing of three major evolutionary thresholds: (1) The Fig Tree microfossils were evidence that the transition from

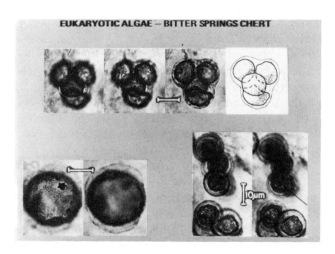

Figure 9–4. Nucleated(?) unicells from the Bitter Springs Formation (approximately 900 million years old), central Australia.
(Photo courtesy J. William Schopf)

chemical evolution to organic evolution had been crossed prior to 3000 million years ago; (2) the Gunflint microbiota was evidence of the crossing of the threshold of diversity at least 2000 million years ago; and (3) the Bitter Springs eukaryotes were evidence that the greatest of all thresholds had been crossed—that of the evolution of the nucleated cell. Once the nucleated cell evolved, the chromosomes and the DNA and RNA molecules could be confined and organized, paving the way for sexual reproduction and, eventually, the emergence of multicellular life.

After leaving Harvard, Schopf joined the faculty at UCLA, where he has continued to conduct research on pre-Cambrian life and evolution and to make important contributions to our understanding of the age of microscopic life. Preston Cloud, at the University of California at Santa Barbara, has made many contributions to the study of the early relationships of lithosphere, atmosphere, hydrosphere, and biosphere. Among his many contributions has been a critical examination of the evidence that alleged pre-Cambrian microstructures are in fact organic in nature. This is by no means a simple task. According to Cloud, "To suggest a biological origin for a given assemblage of non-living microstructures is permissible only if they are demonstrably carbonaceous, reasonably abundant, show a narrow or approximately polymodal (bell-shaped) size distribution, and have a morphology that is consistent with the proposed origin."*

In this chapter we will examine more closely the evidence of life in the pre-Cambrian rock record and the close interconnection of the evolution of biosphere and atmosphere.

---

*1976, Beginnings of Biospheric Evolution and Their Biochemical Consequences, p. 355: *Paleobiology*, vol. 2.

TERMINOLOGY

In 1930, G. H. Chadwick proposed the term **Cryptozoic** to include all of geologic history prior to the earliest evidence of visible animal life. He likewise introduced the term **Phanerozoic** to include the geologic record characterized by conspicuous animal life. The term **Precambrian** became popularized as geologists became aware that the stratigraphic systems, recognized by their fossil content (see Chapter 3), subdivided only that part of the geologic column in which remains of organisms were commonly preserved. Beneath the Cambrian, the lowest of these paleontologically defined systems, they recognized a vast sequence bearing no fossils and not amenable to subdivision into interpretive time-stratigraphic units. It was the seemingly bottomless aspect of this "primitive" rock complex that inspired James Hutton to write "no vestige of a beginning." Thus the designation Precambrian has long been the popular inclusive term for all of geologic time prior to the beginning of the Cambrian. Because of this, Cryptozoic has had only limited use; Phanerozoic, embracing the Paleozoic, Mesozoic, and Cenozoic Eras, has enjoyed wide acceptance (Table 9-1).

Until recently, the base of the Cambrian has been considered to mark the earliest evidence of animal life and has been viewed as synonymous with the base of the Paleozoic and Phanerozoic. In this context, Precambrian, representing the long preceding time interval that had yielded no convincing record of visible life, was an acceptable time-scale name. During the past few decades, however, a widespread soft-bodied fauna, representing the world's earliest known animals, has been documented and clearly demonstrated as pre-Cambrian in age. This important Ediacara fauna (discussed more fully in Chapter 10) provides the substantive basis for the Ediacarian Period and System (Table 9-1) and represents the earliest convincing record of visible animal life. The designation Precambrian is, therefore, not suitable. Also, if we are to be consistent in defining the Phanerozoic as the age of visible life, the base of the Phanerozoic and Paleozoic should be extended downward to the base of the Ediacarian. Again, in this context, the term Precambrian loses its original meaning.

We realize that the deeply entrenched term Precambrian is probably here to stay and that semantic problems make unambiguous discussion difficult. However, in an attempt to use terminology that best reflects geologic history as we know it, we are abandoning the term Precambrian and resurrecting the designation Cryptozoic to refer to that phase of geologic history preceding the first appearance of *animal* life on earth. Where appropriate, we also employ the informal terms pre-Cambrian or pre-Phanerozoic—for example, when discussing crystalline basement rocks nonconformable beneath Cambrian or younger strata. The first seven-eighths of geologic time deserves a formal name of its own—Cryptozoic, instead of pre-something else.

GEOCHRONOLOGIC SUBDIVISIONS

Prior to the widespread use of radiometric dating, the magnitude of the Cryptozoic was not fully appreciated.

Table 9-1. Comparison of conventional time scale and time scale used in this text*

Conventional Time Scale			Time Scale Used in this Test			Chapter Treated in Text
Eon	Era	Period		Eon	Era	Period	
PHANEROZOIC	CENOZOIC	Quaternary	Neogene	PHANEROZOIC	CENOZOIC	Neogene	13
		Tertiary	Paleogene			Paleogene	
	MESOZOIC	Cretaceous			MESOZOIC	Cretaceous	12
		Jurassic				Jurassic	
		Triassic				Triassic	
	PALEOZOIC	Permian			PALEOZOIC	Permian	11
		Pennsylvanian				Pennsylvanian	
		Mississippian				Mississippian	
		Devonian				Devonian	
		Silurian				Silurian	
		Ordovician				Ordovician	
		Cambrian				Cambrian	10
						Ediacarian	
PRECAMBRIAN	PROTEROZOIC	Late Proterozoic		CRYPTOZOIC	PROTEROZOIC	Late Proterozoic	9
		Middle Proterozoic				Middle Proterozoic	
		Early Proterozoic				Early Proterozoic	
	ARCHEAN	Late Archean			ARCHEAN	Late Archean	
		Middle Archean				Middle Archean	
		Early Archean				Early Archean	

*Note that the main difference is in abandonment of the term Precambrian and inclusion of Ediacarian (named by P. E. Cloud, Jr., and M. F. Glaessner; see chapter 10) in the Paleozoic.

In the early days of geology, the complex of granites, schists, and gneisses upon which the younger fossiliferous strata rested was regarded as merely a foundation or basement. These sub-Cambrian rocks showed no clear sequence of events, in marked contrast with the more orderly arrangement of rocks of the Paleozoic, Mesozoic, and Cenozoic. Remember from Chapter 3 that at the turn of the century the oldest accepted estimates of the age of the earth did not exceed 100 million years. Most of this time was believed to be required for the Cambrian and younger record, thus allowing relatively little time for the "Precambrian."

Perhaps the greatest contribution of radiometric dating has been its applications to the Cryptozoic rock record. Many uranium–lead, potassium–argon, and rubidium–strontium dates from all over the world have provided a totally different perspective on the vastness of the Cryptozoic—a time span of some 3000 million years. Because of radiometric dating, there has been a ground swell of renewed interest in Cryptozoic rocks and history. Extreme complexity of the rocks in many places still hampers accurate work. Today, however, we have a considerable body of information on the Cryptozoic.

Table 9-2. Classification schemes for Cryptozoic chronology

Age in m.y. B.P.	Canada (Stockwell, 1964)	North America		
		(U.S. Geol. Survey, 1971)	(Harrison and Peterman, 1982)	(Scheme adopted in this text)
500	pALEOZOIC Cambrian	pALEOZOIC Cambrian	pALEOZOIC Cambrian	pALEOZOIC Cambrian
	570	570	570	
700	Hadrynian	Precambrian Z	Late Proterozoic	Ediacarian
				700
		800		Late Proterozoic
900	880		900	900
1100				
1300	Helikian	Precambrian Y	Middle Proterozoic	Middle Proterozoic
1500	PROTEROZOIC			
	1640	1600	1600	1600
1700			PROTEROZOIC	PROTEROZOIC
1900				
2100	Aphebian	Precambrian X	Early Proterozoic	Early Proterozoic
2300				
	2390			
2500		2500	2500	2500
2700			Late Archean	Late Archean
2900				
			3000	3000
3100	ARCHEAN	Precambrian W	Middle Archean	Middle Archean
3300				
			3400	3400
3500			ARCHEAN	ARCHEAN
			Early Archean	Early Archean
3700				

Table 9-2. Classification schemes for Cryptozoic chronology—continued

Age in m.y. B.P.	Canada	North America		
	(Stockwell, 1964)	(U.S. Geol. Survey, 1971)	(Harrison and Peterman, 1982)	(Scheme adopted in this text)
3900			(pre-Archean)	Pregeologic history of the earth
4100				
4300				
4500				

Data from C. H. Stockwell, 1976, Geologic and Economic Minerals of Canada: *Geol. Survey Canada Econ. Rept.* 1; H. L. James, 1972, Subdivision of Precambrian: An Interim Scheme to Be Used by the U.S. Geological Survey: *Am. Assoc. Petroleum Geologists Bull.,* vol. 56, pp. 1128–1133; J. E. Harrison and Z. E. Peterman, 1982, Adoption of Geochronometric Units for Divisions of Precambrian, Fig. 1, p. 802: North American Commission on Stratigraphic Nomenclature Rept. 9.

Because superposition is difficult to work out in many pre-Cambrian terranes, Cryptozoic history has not been subdivided according to a refined scheme such as we have for the Phanerozoic. This condition has been imposed by the complexity of the rocks coupled with the fact that Cryptozoic sedimentary rocks do not contain the kind of fossil record that is amenable to subdivision.

As discussed in Chapter 3, the birth and development of the Phanerozoic time scale, based on fossil succession, took place in western Europe. However, the important development of a pre-Phanerozoic chronology occurred primarily in the Great Lakes region of North America in the southern part of the Canadian Shield. Early attempts to subdivide the complex history were met with frustration, but with time and effort a number of sequences were described and two very generalized subdivisions emerged: **Archean** and **Proterozoic** (Table 9-2). The terms Archean and Proterozoic, although representing an early attempt at subdivision of the Cryptozoic record, are still used. Prior to the availability of radiometric dating, there was a tendency toward equating Archean with the older looking pre-Cambrian rocks (plutonic and metamorphic) and Proterozoic with the younger looking rocks (mostly volcanic and sedimentary). We have already seen (Chapter 3) how this rationale of judging relative age on physical appearance met with failure in early attempts at fashioning a time scale; it is no wonder that this Wernerian approach plagued attempts to organize the more obscure record of the Cryptozoic.

Various regional subdivision schemes have been proposed, but there has been difficulty in correlating Cryptozoic rocks and events from place to place. Until recently there has been little success in putting together

a worldwide chronology of the Cryptozoic. In 1972 the U. S. Geological Survey adopted an interim scheme for subdividing the "Precambrian" (Table 9-2) on the basis of significant breaks in Cryptozoic history pinpointed by good control of radiometric dates. The U. S. Geological Survey 1974 edition of the geologic map of the United States uses this scheme and Precambrian W corresponds closely to Archean. In 1982, the North American Commission on Stratigraphic Nomenclature suggested a classification of **geochronometric units** for divisions of pre-Cambrian time. This scheme employs the terms Archean and Proterozoic as eons, which, in turn, are subdivided into early, middle, and late eras (Table 9-2).

The North American Stratigraphic Code and the International Stratigraphic Guide customarily recognize time-stratigraphic (chronostratigraphic) units and boundaries defined at type sections (stratotypes) as the basis for corresponding geologic time units and boundaries. For the Cryptozoic, however, this procedure has proved impractical owing to the fact that much of the Cryptozoic record is developed as unfossiliferous nonstratified rocks, not amenable to applications of superposition and fossil succession—the two fundamental bases for the Phanerozoic time scale. As defined in the 1983 North American Stratigraphic Code, geochronometric units are direct divisions of geologic time and do not necessarily have a corresponding rock sequence to which they are referred; thus for the Cryptozoic, they are more suitable than standard chronostratigraphic subdivisions. In this scheme the geochronometric units have been selected on the basis of the temporal positions of the more important geologic events in major regions of the earth. The boundaries have been selected so as to interrupt as few as possible major sequences of sedimentation, igneous intrusion and extrusion, and

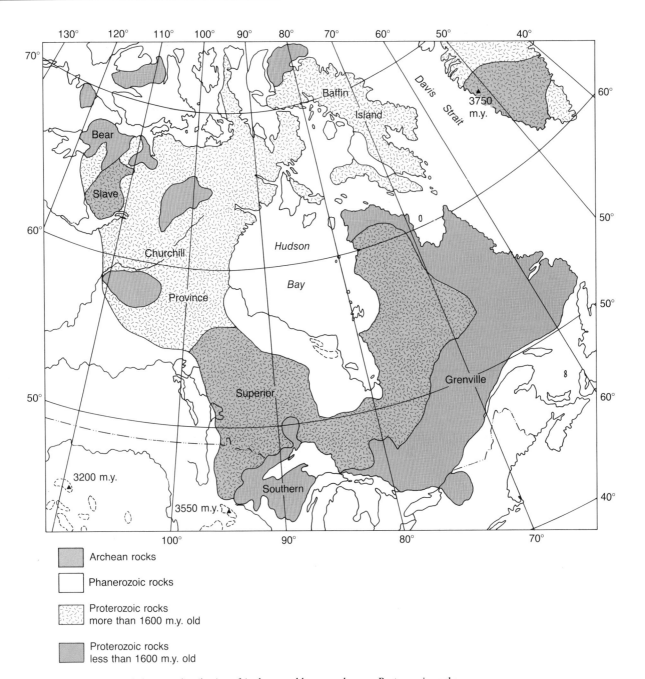

Figure 9–5. Canadian Shield showing distribution of Archean and lower and upper Proterozoic rocks. (Compiled from C.H. Stockwell, 1965, Tectonic Map of the Canadian Shield: *Geol. Survey of Canada; Econ. Geology Rept.* No. 1: P.B. King, 1969, Tectonic Map of North America: U.S. Geological Survey)

orogeny. They are defined by chronometric age so as to provide a common basis for region-to-region and continent-to-continent application. In this text we will use a slightly modified version of the North American Commission on Stratigraphic Nomenclature scheme by referring to subdivisions of Cryptozoic rather than Precambrian and by placing the Proterozoic–Phanerozoic boundary at the base of the presently defined Ediacarian System, the lowest time-stratigraphic subdivision of the Paleozoic Erathem (Table 9-1).

CANADIAN SHIELD

Sequence of Events

We shall begin our examination of the Cryptozoic history of North America with the **Canadian Shield,** a vast lowland rimming Hudson Bay and occupying the eastern two-thirds of Canada, the United States margins of Lake Superior, and most of Greenland (Fig. 9-5). Each of the continents of the world has a nucleus or core of predominantly Cryptozoic basement rocks, either ex-

Figure 9–6. Permian continental configuration showing pre-Phanerozoic basement nuclei of continental blocks, including shield areas, flanked by Phanerozoic mobile belts.
(From Brian F. Windley, THE EVOLVING CONTINENTS, Fig. 1, p. 1. Copyright © 1977, John Wiley & Sons, Ltd. Reprinted by permission of John Wiley & Sons, Ltd.)

posed as a shield region or covered with a veneer of younger platform sediments (Fig. 9-6). The Canadian Shield contains the oldest and most expansive surface exposures of Cryptozoic rock on the North American continent. Granite and granite gneiss are the most abundant rock types, but the shield is a complex patchwork of various kinds of metamorphic rocks, intrusive and extrusive igneous rocks, and sedimentary rocks. As discussed in Chapter 1, the shield represents the

exposed part of the nucleus of the continent and is flanked by younger stratified rocks of Paleozoic and Mesozoic age.

William Logan, who established the Geological Survey of Canada in 1842, was responsible for the pioneering effort in working out the geologic history of the Canadian Shield. It was a herculean task for Logan and his field parties to penetrate the harsh terrain on horseback or foot or by canoe. Excellent field mapping continued to the twentieth century, stimulated in large part by the search for metallic ore deposits (particularly copper, gold, nickel, and platinum), and paved the way for development of a workable pre-Cambrian chronology (Table 9-3). This was a remarkable achievement and speaks highly of the Canadian Survey and its team of hardy, perceptive geologists.

This and subsequent work has shown that in the southern Canadian Shield there are major subdivisions of sedimentary and volcanic rocks, separated by unconformities and intrusive igneous and metamorphic complexes that bear evidence of at least four major orogenies. This sequence of events is set in a general relative time frame of Archean (early Cryptozoic) and Proterozoic (late Cryptozoic). Since the early days of geologic investigations of the complicated rocks and structures of the shield, geologists have recognized a major break—marked not only by a widespread unconformity but also by a change in lithology and style of deformation—within the Cryptozoic complex. This boundary between the Archean and Proterozoic has a radiometric age of about 2500 million years. The Cryptozoic relative chronology of the Canadian Shield was pieced together by use of fundamental principles of historical geology: superposition—layered sequences or "packages" of

Table 9-3. Cryptozoic time-stratigraphic classification in relation to orogenies of the Canadian Shield

Eon	Era	Sub-era	Orogeny mean K-Ar mica Age, m.y. B.P.
PROTEROZOIC	Hadrynian		
	Helikian	Neohelikian	Grenvillian (955)
		Paleohelikian	Elsonian (1370)
	Aphebian		Hudsonian (1735)
ARCHEAN			Kenoran (2480)

From C. H. Stockwell, 1976, Geologic and Economic Minerals of Canada Table VI-1, p. 51: *Geol. Survey of Canada Econ. Geology Rept.* Vo. 1. Reprinted by permission of Geological Survey of Canada.

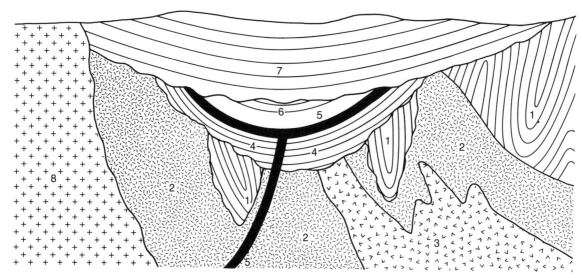

Figure 9–7. Complex superpositional and cross-cutting relationships of rocks in southern Canadian Shield. Numbers indicate oldest (1) to youngest (8).
(Information from C.H. Stockwell, 1976, Geologic and Economic Minerals of Canada: *Geol. Survey Canada, Econ. Geology Rept.* No. 1)

rock—and cross-cutting relationships—faults, igneous intrusions, and unconformities (Fig. 9-7). Once the sequence of events was worked out, it took the application of radiometric dating to place shield rocks in a regional time frame (Table 9-3).

One of the most significant discoveries from the application of radiometric dating is that the shield can be subdivided into regions or provinces on the basis of isotopic age and rock types. Radiometric dates clustering around 2500, 1700, 1400, and 950 million years B.P. indicate four major orogenic episodes and major times of regional resetting of radiometric clocks. These orogenies define seven major structural provinces (Fig. 9-8), each of which, in addition to characteristic isotopic age and rock types, has a style of deformation and a structural orientation different from those of adjacent regions. The structural-isotopic age provinces are separated by rather abrupt metamorphic or fault boundaries called **orogenic fronts.**

Archean Rocks: Greenstone Belts and Granite Gneiss

Rocks of Archean age occur mainly in the Superior and Slave Provinces (Fig. 9-8). As shown in Table 9-3, the orogenic event that brought the Archean to a close is called the **Kenoran** and occurred approximately 2500 million years ago (Fig. 9-9). Characteristically unique rock suites of Archean terranes are the **greenstone belts,** which occur as giant pods in great elongated downwarps (Fig. 9-10). These belts are thought to be remnants of ancient volcanic-sedimentary basins. The rock successions in the greenstones include ultramafic and mafic volcanic rocks (see Fig. 1-9B) in the lowest

parts, overlain by and interstratified with immature graywacke-type sediments rich in mafic volcanic rock fragments; these sediments in turn are topped by andesitic to felsic volcanic rocks (see Fig. 1-9B) and by interstratified pyroclastic rocks and more silicic sedimentary rocks. The name greenstone refers to the basaltic rocks in the lower part of the pile; these rocks have been altered by metamorphism to produce green minerals such as epidote, chlorite, and serpentine. Greenstone belts are found in all the Cryptozoic Shield terranes of the world and are developed in essentially the same fashion.

On the Canadian Shield greenstone belts are concentrated in large tracts several hundred kilometers long and several tens of kilometers across. Between these greenstone belt tracts lie vast regions of granite and granitoid gneiss and other associated high-grade metamorphic rocks. The relationship of the greenstone belts to the surrounding granitoid terranes is not clearly understood. Some geologists believe that the more ultramafic and mafic lower parts of the greenstone belts are remnants of originally more extensive volcanic-sedimentary successions representing earlier oceanic crust (Fig. 9-11). In this context, these remnants are the deepest downfolds left after massive intrusion of granites (and their close derivatives) and deep erosion. Another school of thought considers the greenstone belts as having been formed along downwarp and fracture zones in an early thin, brittle granitic crust (Fig. 9-12). In this context they represent more localized volcanism and orogeny. Logan originally suggested that the granite and granite gneiss are older than the greenstone belts and are part of an ancient crust; this idea has been revived in recent years. Still another idea proposes that some

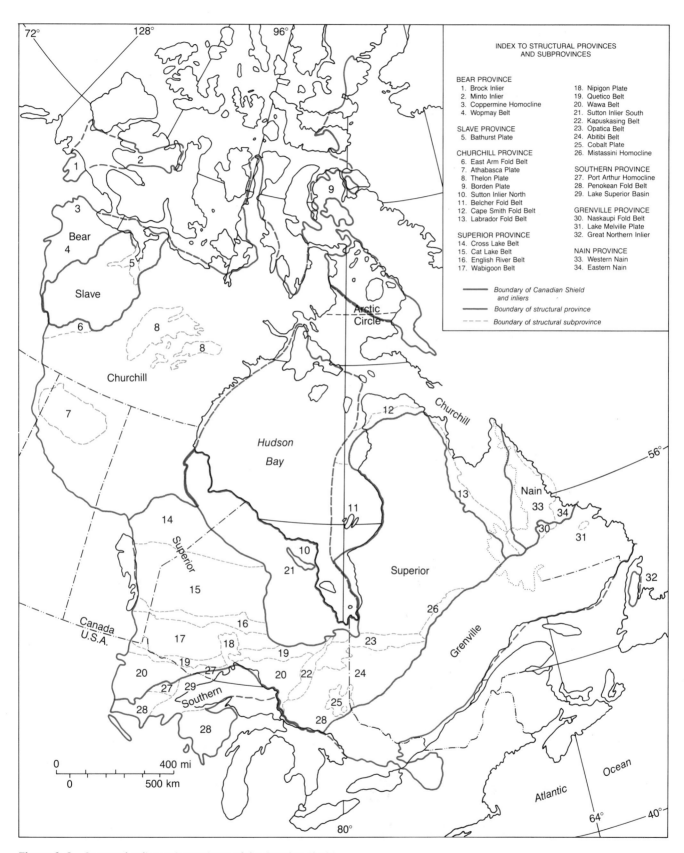

Figure 9–8. Structural-radiometric provinces of the Canadian Shield.
(From C.H. Stockwell, 1976, Geologic and Economic Minerals of Canada, Fig. IV–1, p. 46: *Geol. Survey Canada, Econ. Geology Rept.* No. 1. Reprinted by permission of Geological Survey of Canada)

195

greenstone belts developed initially as terrestrial equivalents of lunar maria (Fig. 9-13). Certainly the correct model would have important implications for the nature of the Early to Middle Archean crust in terms of how much of it was granitic. It is noteworthy that ultramatic volcanic rocks are known only in Archean terranes.

Their presence indicates processes were operating then that no longer occur on the earth.

Chemical analyses of the highly altered volcanic rocks of the greenstone belts indicate rocks ranging from peridotites and basalts through andesites to rhyolites and their pyroclastic equivalents (see Fig. 1-8). Pil-

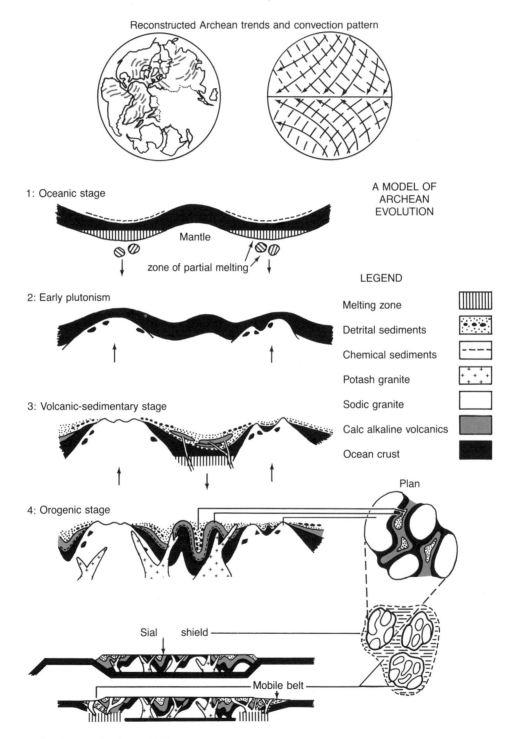

Figure 9–9. Model of evolution of Archean shields.
(From A.Y. Glickson, 1972, Early Precambrian Evidence of a Primitive Ocean Crust and Island Nuclei of Sodic Granite; Fig. 12, p. 3338: *Geol. Soc. America Bull.,* vol 83, no. 11. Reprinted by permission of author)

low structures in many of the volcanic rocks show that extrusion was under water. The sedimentary rocks include graywacke sandstones, slates, and conglomerates as well as peculiar rock sequences called **banded iron formations.** Many of the sandstone beds are graded and have the characteristics of turbidites. Some of the sedimentary rocks in the upper parts of the greenstone belt successions are arkoses and granite pebble conglomerates, indicating derivation from granitic rocks that were present before the belts were formed or at least before they were completed. The greenstone belt sequences are generally about 10,000 to 30,000+ m thick, although seldom are their bases observed. Ages range from less than 3800 to a little less than 3000 million years.

We know very little about the composition and structure of the earth's early Archean crust. The Kenoran orogeny that brought Archean history to a close marks a major change in the development of the earth's lithosphere. Present estimates put the amount of granitic crust during the Archean at about 10 to 40 percent of the present figure. It was during the widespread Kenoran event that tremendous volumes of granitic lithosphere formed and transformed what had been comparatively thin slabs of granitic crust into the present thickness of about 40 km. The granites that now surround the greenstone belts comprise some of the thickest and most stable parts of the continental crust. Unfortunately it was the emplacement of so much granite that has obscured the original greenstone belt–granitoid crustal relationship and reset the radiometric clocks. As the Kenoran granites formed, the greenstone belts, being denser pods, progressively sank into synclinal configurations (Fig. 9-14). When the Proterozoic Eon commenced, after a considerable period of erosion, lithospheric conditions were much different than they had been during the Archean.

Proterozoic Rocks: Continental Lithosphere on a Grand Scale

Proterozoic rocks of the shield are quite diverse and have been subdivided into three erathems separated by two major orogenies: The Hudsonian orogeny (about 1800 million years ago) and a later Grenville orogeny (about 1000 million years ago). The erathems are called the Aphebian, the Helikian, and the Hadrynian (Table 9-3). Aphebian rocks include the first widespread quartz-rich sandstones and carbonate rocks, as well as the peculiar but important banded iron formations. Rocks in the Southern Province south and west of Lake Superior contain banded iron formations that have served as North America's main source of iron ore for more than a century. The Gowganda Formation in southern Ontario consists of unsorted and unbedded muddy conglomerates that are believed to represent tillites (Fig. 9-15). Many of the pebbles, cobbles, and boulders in the conglomerates have striated surfaces and in

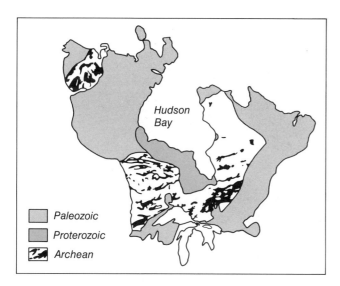

Figure 9–10. Canadian Shield, showing Archean greenstone belts (black).
(From W.R.A. Barager, and F.C. McGlynn, 1976, Early Archean Basement in the Canadian Shield: A Review of the Evidence: *Geol. Survey of Canada Paper* 76–14. Reprinted by permission of Geological Survey of Canada)

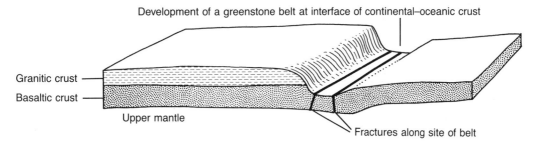

Figure 9–11. Possible development of a greenstone depository in accordance with a more uniformitarian view.
(After Carl R. Anhaeusser et al., 1969, Reappraisal of Some Aspects of Precambrian Shield Geology; Fig. 3, p. 2179: *Geol. Soc. America Bull.,* vol. 80, no. 11. Reprinted by permission of senior author)

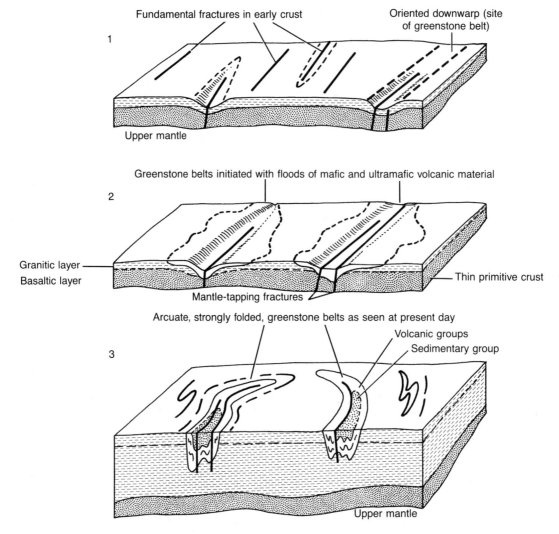

Figure 9–12. Suggested evolution of Archean greenstone belts.
(After Carl R. Anhaeusser et al., 1969, Reappraisal of Some Aspects of Precambrian Shield Geology; Fig. 4,
p. 2180: *Geol Soc. America Bull.,* vol. 80, no. 11. Reprinted by permission of senior author)

places the conglomerates rest on striated and polished older basement rock. Parts of the Gowganda are developed as isolated outsized cobbles and boulders in a muddy matrix, giving the impression that the large stones were dropped into muddy bottom aquatic environments. These features all point to a major episode of continental glaciation during the early Proterozoic. Aphebian rocks were folded and intruded by granites during the Hudsonian orogeny (Table 9-3).

Helikian rocks, for the most part, have been severely deformed and metamorphosed by the Grenville orogeny. A mid-Helikian disturbance called the Elsonian orogeny defines a lower Paleo-Helikian and an upper Neo-Helikian interval. Thick successions of volcanic rocks, quartzites, marbles, and red beds occur mainly in elongate troughlike areas flanked by granite and gneiss.

The Keweenawan Group of the Lake Superior area consists of several thousand meters of basalt flows, arkosic and quartzose sandstones, and shales and contains metallic copper deposits. Hadrynian rocks are not well represented in the shield proper, but accumulated to great thicknesses in marginal areas of the Appalachian and Cordilleran regions and will be discussed later.

CRYPTOZOIC ROCKS OUTSIDE THE SHIELD

Other Provinces

Reference to Figure 9-16 reveals, in addition to the Canadian Shield, scattered smaller exposures of Cryptozoic rocks in different parts of North America. All of the

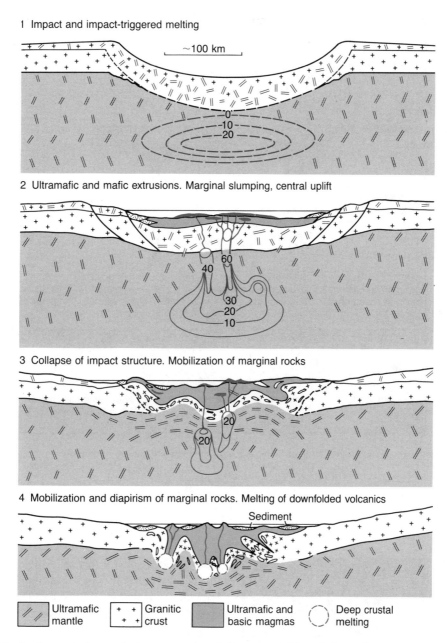

1 Impact and impact-triggered melting

~100 km

2 Ultramafic and mafic extrusions. Marginal slumping, central uplift

3 Collapse of impact structure. Mobilization of marginal rocks

4 Mobilization and diapirism of marginal rocks. Melting of downfolded volcanics

Sediment

| Ultramafic mantle | Granitic crust | Ultramafic and basic magmas | Deep crustal melting |

Figure 9–13. Possible development of some Archean greenstone belts as terrestrial equivalents of lunar maria.
(From D.H. Green, 1972, Archean Greenstone Belts May Include Terrestrial Equivalents of Lunar Maria; Fig. 2, p. 267: *Earth and Planetary Science Letters,* vol. 15. Reprinted by permission of Elsevier Publishing Co.)

Cryptozoic crystalline rocks of these regions are parts of the core or nucleus of the continent generally referred to as the basement complex of North America. The Canadian Shield is that part of the basement complex that has been extensively exposed by erosion during the last 1000 million years. The rocks of the shield extend toward the margins of North America beneath the cover of Paleozoic rocks on the platform interior. Radiometric age data from outcrop areas beyond the shield as well

as from the subsurface through drilled core samples show a continuation of the pattern illustrated in Figure 9-8. Structural trends in the basement complex have been mapped by geophysical methods.

Figure 9-17 shows the present state of our knowledge of the isotopic age-structural provinces beneath the continental platform outside the Canadian Shield. Much of the middle part of the United States is underlain by the Central Province, which has a comparatively

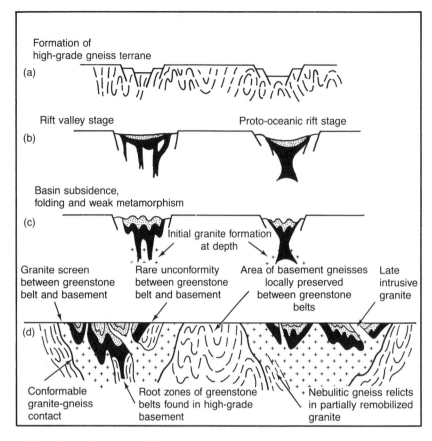

Figure 9–14. Stages in evolution of Archean greenstone belts and gneissic terranes. (From Brian F. Windley, 1973, Crustal Development in the Precambrian; Fig. 3, p. 330: *Philos. Trans. Royal Soc. London,* A 273. Reprinted by permission of author)

Figure 9–15. Gowganda tillite; Proterozoic. (Photo by J. Cooper)

small outcrop area on the shield itself. The relatively small Wyoming Province is composed of Archean rocks metamorphosed in the Kenoran orogeny. Examples of rocks representative of the various provinces are shown in Figure 9-18.

The oldest rocks are in the Lake Superior region in Minnesota (approximately 3700 million years old) and in western Greenland (approximately 3800 million years old). The age of deformation (plutonism, uplife, and metamorphism) shows a crude concentric pattern of progressively younger, more outward bands, suggesting that the North American continent has grown by lateral accretion—the welding of successive organic belts.

Grand Canyon

The spectacular Grand Canyon and adjacent parts of the Colorado Plateau provide a panoramic view of a significant part of earth history. This region will serve as a model of reference as we continue our discussion of the geologic history of North America. Well exposed in the

Figure 9–16. North America, showing areas of exposure of Cryptozoic rocks outside the Canadian Shield. 1. Appalachian region; 2. cores of domes in midcontinent; 3. cores of ranges in Rocky Mountains; 4. southeastern California and Arizona; 5. Canadian Rockies; 6. Alaska. (From P. B. King, 1979, Tectonic Map of North America: U.S. Geological Survey)

narrow, inner gorge (Fig. 9-19) of the Grand Canyon are schists, gneisses, and granites that comprise a complex crystalline assemblage that formed about 1500 to 1300 million years ago during a major orogenic disturbance roughly equivalent to the Elsonian orogeny of the Canadian Shield. Metamorphic rocks such as the Vishnu (Fig. 9-20) and Brahma Schists represent former marine mudstones and sandstones that were subsequently uplifted, deformed, metamorphosed, and intruded by

granite during a major orogeny. These rocks later were eroded to a rather subdued landscape. In many respects, these crystalline basement rocks in the Grand Canyon are similar in style to many other Cryptozoic complexes. They tell a story of sedimentation, volcanism, folding and faulting, metamorphism, igneous intrusion, and erosion—the birth and death of mountain belts.

All of this happened before a succession of younger pre-Cambrian stratified rocks, the Grand Canyon Super-

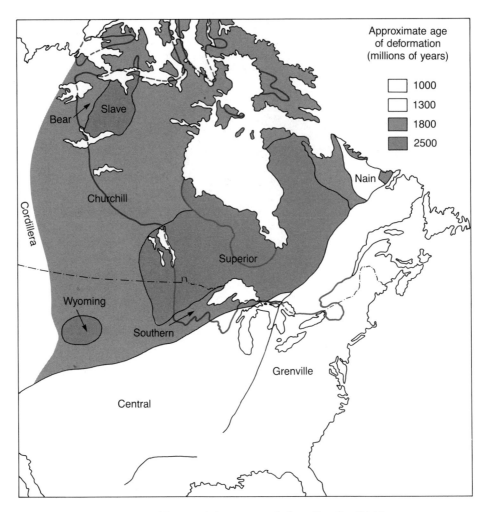

Figure 9–17. North America, showing extension of Cryptozoic basement rocks from Canadian Shield into United States, including structural-radiometric age provinces.
(From C. W., Stearn, R. L. Carroll, and T. H. Clark, GEOLOGICAL EVOLUTION OF NORTH AMERICA, 3rd ed., Fig. 9–14, p. 164. Copyright © 1976, John Wiley & Sons, Inc., New York. Reprinted by permission of John Wiley & Sons, Inc.)

group, was deposited. These rocks consist of a thick sequence of quartzites, argillites (compact siltstones and mudstones), **red beds** (Fig. 9-21A), limestones, and some lava flows nonconformably overlying the basement complex and underlying a Paleozoic succession in a spectacular angular unconformity (Fig. 9-21B). In addition to illustrating classic stratigraphic relationships, the pre-Cambrian rocks of the Grand Canyon also provide an example of how lithology is *not* a good guide to age. The old-looking crystalline basement rocks were originally assigned to the Archean; however, radiometric dating has demonstrated their Middle Proterozoic age.

Sediments of Late Proterozoic Continental Margin

The Grand Canyon Supergroup is only a part of a wedge-shaped 300-km-wide and 3500-km-long belt of

sandstone, shale, and minor limestone that was deposited along the western margin of the Middle and Late Proterozoic continent (Fig. 9-22). Rocks of this belt are well exposed in the Death Valley area of California, the Uinta Mountains in Utah, and the northern Rocky Mountains from Glacier National Park, Montana, northward into Alberta and British Columbia. At Glacier Park they are called the Belt Supergroup. The Belt stratigraphic succession consists of argillite, limestone, and some interstratified basaltic lava flows and sills (Fig. 9-23). These rocks have a composite thickness of up to 8000 m and comprise an essentially unmetamorphosed section that was deposited in a northwest–southeast-oriented marine trough about 1200 to 800 million years ago. Some of the rocks bear ripple marks and desiccation cracks (Fig. 9-23)—clues of shallow-water deposition; others show graded bedding and dark colors—evidence of deeper water deposition—and appear to be turbidites. These

Figure 9–18. Cryptozoic basement rock. A. Gneiss with vertical foliation, Grenville Province, southern Canadian Shield. B. Gneiss in basement rock of Grand Tetons, Wyoming Province. C. Granite and gneiss, Mount Rushmore, Black Hills, Southern Province. D. Gneiss in Royal Gorge of the Arkansas River, Colorado, Central Province. E. Gneiss in Blue Ridge Mountains, Virginia, Grenville Province. (Photos by J. Cooper)

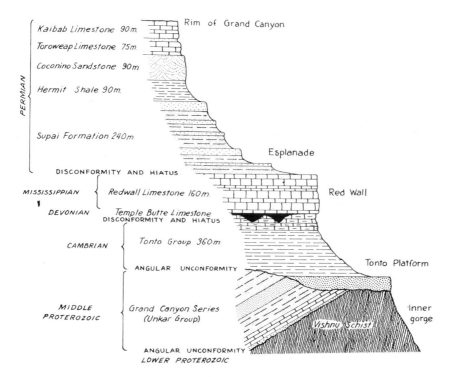

Rim of Grand Canyon

Kaibab Limestone 90m.
Toroweap Limestone 75m.
Coconino Sandstone 90m.
Hermit Shale 90m.

PERMIAN

Supai Formation 240m.

Esplanade

DISCONFORMITY AND HIATUS

MISSISSIPPIAN { Redwall Limestone 160m.

Red Wall

DEVONIAN Temple Butte Limestone
DISCONFORMITY AND HIATUS

CAMBRIAN { Tonto Group 360m.

Tonto Platform

ANGULAR UNCONFORMITY

MIDDLE
PROTEROZOIC
{ Grand Canyon Series
(Unkar Group)

Inner
gorge

Vishnu Schist

ANGULAR UNCONFORMITY
LOWER PROTEROZOIC

Figure 9–19. Wall of Grand Canyon showing pre-Paleozoic basement rocks (Vishnu Schist); pre-Paleozoic stratified sequence (Grand Canyon Series, which is now called Grand Canyon Supergroup because the term series is a time-stratigraphic term, the subdivision of a system). Unconformably overlying basement; and Paleozoic succession, unconformably overlying pre-Paleozoic rocks.
(After Philip B. King, *The Evolution of North America.* Copyright © 1959, rev. ed. © 1977 by Princeton University Press. Fig. 63, p. 105, reprinted by permission of Princeton University Press and author)

Figure 9–20. Vishnu Schist, inner Gorge of Grand Canyon.
(Photo by J. Cooper)

A

B

Figure 9–21. A. Red beds of Hakatai Shale, Grand Canyon Supergroup, Grand Canyon, cut by basaltic dike. B. Angular unconformity between Cambrian Tapeats Sandstone and strata of Proterozoic Grand Canyon Supergroup, Grand Canyon, Arizona.
(Photos by J. Cooper)

rocks of the western Cordillera as well as coeval conglomerates, sandstones, and siltstones of the Great Smoky Mountains in the Appalachian belt are the early

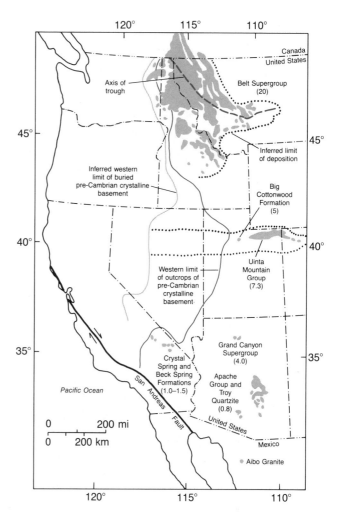

Figure 9–22. Distribution and inferred basins of deposition of Middle and Late Proterozoic (900?–1450? m.y. B.P.) rocks of the Belt Supergroup and related strata and known and inferred western limit of Cryptozoic crystalline basement rocks (1450?–2400 m.y. B.P.), western United States. Numbers are maximum thicknesses in kilometers.
(From J. H. Stewart, 1982, Regional Relations of Proterozoic Z and Lower Cambrian Rocks in the Western United States and Northern Mexico, Fig. 2, p. 172: Geol. Soc. America Cordilleran Section Guidebook)

deposits of rifted continental margins as proto-North America was emerging as a separate continental mass (Fig. 9-24).

The rifted continental margins evolved into long, linear, subsiding basins called **geosynclines,** in which accumulated thick sequences of sediment that made up a continental shelf–slope–rise wedge. During the initial stages of rifting, three rift arms developed about 120° apart. Two of these evolved into an incipient rift ocean basin to inaugurate the *Wilson cycle* of ocean-basin opening and closing. The third arm of the rift system, called an *aulacogen,* developed as a linear, fault-bounded basin trending into the interior of the continental block (Fig. 9-24B). The aulacogens represent the failed or aborted attempt at complete rifting. These

A

B

C

Figure 9–23. Proterozoic sedimentary sequences. A. Belt Supergroup, Glacier National Park, Montana. B. Uinta Mountain Group, Uinta Mountains, Utah. C. Stirling Quartzite, southern Death Valley region, California.
(Photos by J. D. Cooper)

failed arms eventually became filled with sedimentary deposits and were incorporated into the continental structure. The geosynclines continued to evolve in a setting marginal to the rift ocean basin.

In a number of localities (Fig. 9-25) peculiar pebbly mudstones called **diamictites** of Late Proterozoic age have been interpreted by many workers as being of glacial origin. The scattered pebbles, embedded in the muddy matrix of some argillite beds, are composed of a great variety of rock types representing a variety of source terranes and appear to have been dropped onto the site of deposition from above. Similar dropstones occur in the muddy deposits of modern polar seas where floating icebergs have rafted coarse glacial debris away from land areas and delivered it to the ocean floor. Late Proterozoic glaciomarine deposits as well as tillites have been recognized on all the continents, attesting to a time of global refrigeration and large-scale glaciation as a giant supercontinent was beginning to split apart.

EVOLUTION OF THE LITHOSPHERE

Evidence from the Archean

The progressive increase in granitic lithosphere during the Archean is reflected in the petrologic character of detrital sediments. Most Archean sedimentary sequences typically consist of compositionally and texturally immature rocks rich in plagioclase feldspar, mafic minerals, and mafic rock fragments. Graywackes typically contain abundant basalt fragments. In upper Archean sedimentary rocks volcanic rock fragments and plagioclase are still abundant, but appreciable amounts of quartz and potassium feldspar are also present. The presence of these last two minerals signals the existence of significant amounts of granitic rocks and bears evidence that large tracts of continental lithosphere were being eroded by this time. Poorly sorted and angular detrital grains in Archean sedimentary rocks as well as associated volcaniclastic rocks suggest rapid erosion and deposition in volcanically active, structurally mobile basins. These rock suites, generally associated with the greenstone belts described earlier, indicate crustal instability and are similar to present-day island-arc sediments. The comparatively small volumes of quartzite and carbonate rocks suggest that continental shelves were at best poorly developed during most of Archean time. During the Archean the crust was thinner than at present and microcontinental masses gradually combined to form larger continents. Large-scale formation of granite during the Kenoran orogeny produced an estimated one-half to two-thirds of the present continental lithosphere and the first thick, stable continental platforms.

Evidence from the Proterozoic

Proterozoic sandstones, in general, consist primarily of quartz-rich graywackes, arkoses, and pure quartz sandstones. The mature quartzose sandstones display

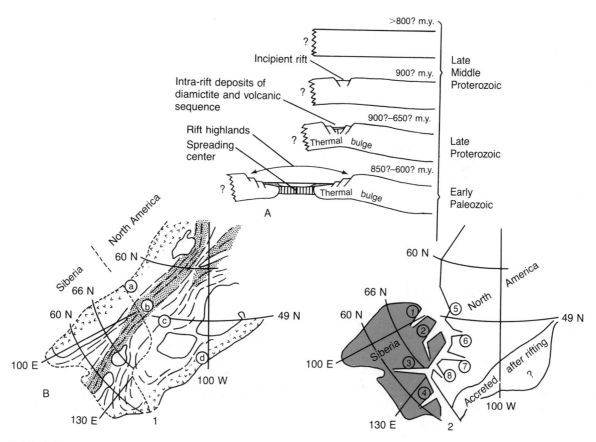

Figure 9–24 A. Tectonic model of Late Proterozoic development of continental margin of western United States: 1. part of supercontinent including Siberia and western North America; 2. initial rifting; 3. continued rifting to form incipient ocean basin separating Siberian and North American continental blocks; 4. rift ocean with marginal geosynclines. B. 1. Pre-rifting correlations between tectonic subdivisions of the Siberian and North American pre-Cambrian cratons; a, b, c, d represent volanic-plutonic and orogenic belts. 2. Siberian-North American pre-Cambrian rift system and related fault-bounded basins and aulacogens; 6, 7, 8 represent the Belt, Uinta, and Death Valley aulacogens, respectively. (A, from J. H. Stewart, 1982, Regional Relations of Proterozoic Z and Lower Cambrian Rocks in the Western United States and Northern Mexico, Fig. 11, p. 183: Geol. Soc. America Cordilleran Section Guidebook. B, from J. W. Sears and R. A. Price, 1978, The Siberian Connection: A Case for Precambrian Separation of the North American and Siberian Cratons, Fig. 2, p. 269: *Geology,* vol. 6, no. 5. Reprinted by permission of authors and Geological Society of America)

abundant ripple marks and cross-stratification; such sedimentary structures together with the high degree of textural maturity of the sediments suggest a greater volumetric importance of depositional environments where reworking, sorting, and rounding of grains occurred. The high percentage of quartz is an indication of successive cycles of weathering, erosion, and concentration during which there was progressive enrichment in quartz at the expense of less stable constituents such as feldspar, mafic minerals, and rock fragments. The Proterozoic sedimentary record thus provides evidence of rapid growth of continental lithosphere, much of it relatively stable and deeply eroded. Associated with the quartz-rich rocks were the first significant amounts of limestone and dolomite and the unique banded iron formations. Abundant carbonate rocks indicate deposition in tectonically stable areas away from the rapid influx of terrigenous clastic detritus. Thus Proterozoic sed-

iments (Fig. 9-26), in sharp contrast to their Archean forebears, indicate a greater preponderance of more stable environmental conditions such as we see today on continental shelves. Much Proterozoic sedimentation occurred on broad, relatively flat, slowly subsiding shelves and basins marginal to continental platforms. Early Proterozoic rocks of the Slave Province (Fig. 9-8) represent the oldest well-documented *geosyncline,* a long, linear subsiding basin that developed marginal to a continental platform and was the site of accumulation of several thousand meters of sediments. Much of the section is of shallow-marine character (Fig. 9-27).

Continental growth and stability that were attained by the Early Proterozoic resulted in the development of continental margins and interiors basically similar to what continents have exhibited during later geologic history to the present. This is in marked contrast to the situation during most of the Archean, when continental

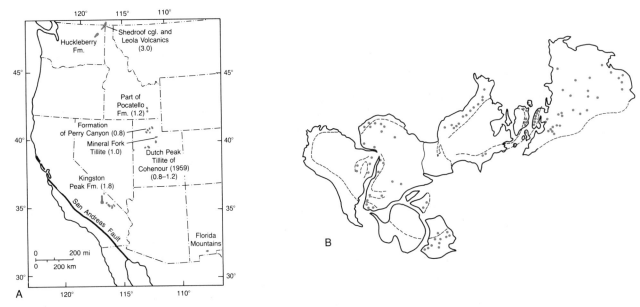

Figure 9–25. A. Distribution and maximum thickness (km) of diamictite (pebbly mudstones of possible glaciogenic origin) and volcanic sequences (650?–900? m.y. B.P.), western United States. B. Distribution of Late Proterozoic diamictites plotted on the Late Proterozoic supercontinent reconstruction of Piper (1976). Also shown are boundaries of late pre-Cambrian geosynclines.
(A, from J. H. Stewart, 1982, Regional Relations of Proterozoic Z and Lower Cambrian Rocks in the Western United States and Northern Mexico, Fig. 3, p. 173: Geol. Soc. America Cordilleran Section Guidebook. B, from B.F. Windley, *THE EVOLVING CONTINENTS*, Fig. 7–3, p. 121. Copyright © 1977, John Wiley & Sons, Ltd. Reproduced by permission of John Wiley & Sons, Ltd.)

masses were small and continental crust was thin. Plate tectonic activity as we understand it probably was not an established phenomenon prior to about 2500 million years ago. By Early Proterozoic time, however, with increased rigidity of the lithosphere—permanently thick granitic continental lithosphere and formation of relatively stable continental interiors—plate margins became more sharply defined and continent-to-continent, ocean basin-to-continent, and ocean basin-to-ocean basin interactions and environments began to characterize the global plate tectonic style that has dominated to the present time. Proterozoic orogenies most certainly are expressions of various plate interactions, including suturing of continental blocks.

EVOLUTION OF THE BIOSPHERE AND THE ATMOSPHERE

As pointed out in Chapter 8 there are several lines of evidence that strongly suggest the earth's early atmosphere was devoid of free oxygen. Many laboratory experiments have demonstrated the synthesis of organic compounds under conditions similar to those proposed for the primitive earth. Such syntheses are inhibited by even very small concentrations of molecular oxygen.

These experiments leave us with the distinct impression that life probably would not have developed at all had the early atmosphere been oxygen rich. Yet if we look ahead from the origin of life to the first appearance of multicellular life, we see organisms with strong oxygen requirements. A paradox you say? Not at all. The observation simply underscores the close relationship between evolution of the biosphere and of the atmosphere during the Cryptozoic. What were the evolutionary highlights and their preconditions from the time when the world was anoxic to the time of oxygen dependency? We will begin our examination of this question by looking at three momentous discoveries made during the 1970s. All shed important light on the earliest life.

Fig Tree Revisited

Until recently the oldest conclusive evidence of life has come from the Fig Tree Group of South Africa. Recall from the beginning of the chapter that Barghoorn and Schopf in the late 1960s reported cyanobacteria and bacterialike organisms from the Fig Tree. However, with continued work on the material, clouds of doubt arose concerning the true organic nature of the microstructures. Many were of disturbingly large size for prokaryotes and were spheroidal (Fig. 9-28). Spheres—for example, bubbles—can be produced by a host of inorganic processes. Ten years later Barghoorn and associate Andrew Knoll proposed new evidence of life from the Fig Tree and, on the basis of tighter radiometric control, a new age of 3400 million years. The new

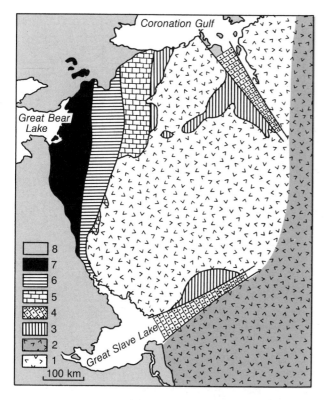

Figure 9–26. Tectonic subdivisions of Early Proterozoic rocks of the northwestern Canadian Shield: 1, exhumed Archean basement rocks; 2, Archean basement rocks subjected to Early Proterozoic metamorphism; 3, thin Early Proterozoic supracrustal rocks of the continental platform; 4, thick Early Proterozoic supracrustal rocks deposited in aulacogens; 5, thick Early Proterozoic supracrustal rocks of continental shelf and clastic wedge of Coronation geosyncline (see Fig. 9–27); 6, metamorphosed Early Proterozoic supracrustal rocks of continental rise and clastic wedge of Coronation geosyncline intruded by 7, batholith; 8, Middle Proterozoic and younger cover rocks.
(From Paul Hoffman, 1973, Evolution of an Early Proterozoic Continental Margin: The Coronation Geosyncline and Associated Aulacogens of the Northwestern Canadian Shield; Fig. 2, p. 551: *Philos. Trans. Royal Society of London,* A273. Reprinted by permission of the author)

evidence from the Fig Tree cherts consisted primarily of nonspheroidal microstructures far more convincing than the previous discoveries. Interpretation of these microstructures as organic rests upon five principal arguments:

1. The new microstructures are within the size range of modern prokaryotes; many of the earlier structures were disturbingly large.

2. The new microstructures form a bell-shaped or polymodal size distribution with a limited range of 1 to 4 microns and have a diagnostic size like that of modern prokaryotes.

3. The new microstructures exhibit a variety of shapes: flattened, wrinkled, and folded, as well as spherical (Fig. 9-29).

4. Some of the microstructures are preserved in what appear to be various stages of cell division.

5. The carbon-12:carbon 13 ratios are too high for inorganic origin and even suggest carbon fixed by the process of photosynthesis.

These five lines of evidence, considered together, satisfy the requirements set by Preston Cloud to distinguish true fossils from pseudofossils or dubiofossils. So then, if prokaryotes were well established 3400 million years ago, how much further back in time shall we seek the origin of life itself? Certainly one contraint on the evidence is posed by the age of the oldest rocks yet discovered on earth—3800 million years for the Isua supracrustal terrane of southwest Greenland. The oldest part of this complex is granite gneiss, but from a metamorphosed banded iron formation higher in the succession, Cloud has reported traces of probable organic carbon; no microstructures were found, however. Could it be that life originated prior to 3800 million years ago?

An Unlikely Place Called North Pole

Yet another major discovery at a place called North Pole in Western Australia has produced reasonably convincing microorganisms from a chert-bearing sedimentary sequence perhaps as old as 3500 million years (Fig. 9-30). This succession, the Warrawoona Group, is part of an Archean greenstone belt assemblage and contains abundant volcanic rocks in the form of pillow basalts and pyroclastics. These interstratified volcanic rocks are amenable to radiometric dating and abundant isotopic analyses support the approximate age of 3500 million years.

Considering the great antiquity of the North Pole sequence, its degree of preservation is remarkable. As described by geologists D. L. Groves, J. S. R. Dunlop, and Roger Buick at the University of Western Australia, pervasive silicification has preserved primary sedimentary structures such as cross-bedding, ripple marks, intraformational flat-pebble breccias, desiccation features, and structures that look like stromatolites (Fig. 9-31). Beds of evaporative sulfates in the form of barite are also present. Collectively, the rock types and structures suggest an ancient tidal flat and lagoonal environment, with waters warmed by periodic volcanic emanations, for this paleoenvironment accommodating perhaps the oldest known biotic community (Fig. 9-32).

Evidence of life in several of the chert beds consists of microstructures in the form of several micron-sized microspheroids and elongate forms resembling filamentous bacteria and cyanobacteria. Stromatolite-looking structures (Fig. 9-33A), some of which are similar in form to modern types growing in a hypersaline lagoon in Sharks Bay, Western Australia (Fig. 9-33B–D), also provide evidence of life. Additionally, carbon and sulfur isotope ratios suggest not only the presence of organic

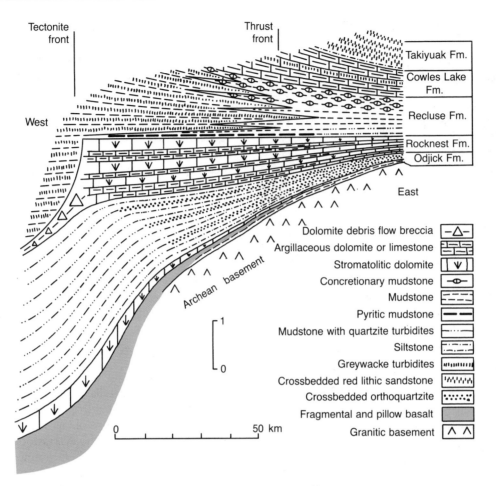

West

East

Dolomite debris flow breccia —△—
Argillaceous dolomite or limestone
Stromatolitic dolomite
Concretionary mudstone
Mudstone
Pyritic mudstone
Mudstone with quartzite turbidites
Siltstone
Greywacke turbidites
Crossbedded red lithic sandstone
Crossbedded orthoquartzite
Fragmental and pillow basalt
Granitic basement ∧ ∧

0   50 km

Figure 9–27. The Coronation geosyncline, Epworth Basin, northwest Canadian Shield: 1, pre-quartzite depositional phase; 2, quartzite phase; 3, dolomite phase; 4, pre-flysch phase; 5, flysch phase; 6, calc-flysch phase; 7, molasse phase.
(From Paul Hoffman, 1973, Evolution of an Early Proterozoic Continental Margin: The Coronation Geosyncline and Associated Aulacogens of the Northwestern Canadian Shield, Fig. 5, p. 554: *Philos. Trans. Royal Society of London,* A273. Reprinted by permission of the author)

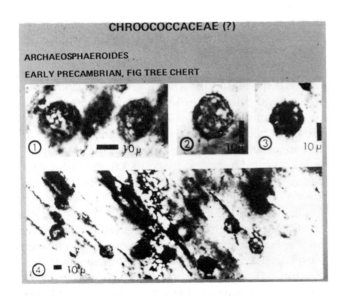

Figure 9–28. Spheroidal prokaryotic microfossils in chert from the Fig Tree Group, South Africa.
(Photo courtesy of J. William Schopf)

carbon, but perhaps also that photosynthesis was going on.

The feature that originally lured researchers to closely investigate these Warrawoona rocks was bedded barite ($BaSO_4$). It was previously thought that such sedimentary sulfate minerals occurred only in much younger rocks, deposited when the atmosphere had attained a significant oxygen concentration. It was also assumed that when the Warrawoona deposits formed, the atmosphere and the oceans were essentially anoxic. The barite appears to have replaced and retained the original crystal form of gypsum ($CaSO_4 \cdot H_2O$), the primary evaporite mineral. Because sulfates are rich in oxygen, the presence of these unusual sulfates at North Pole suggests that local conditions there were more oxidizing than was presumed normal for the time. One possibility is that oxygen was being released by oxygen-generating organisms indulging in photosynthesis.

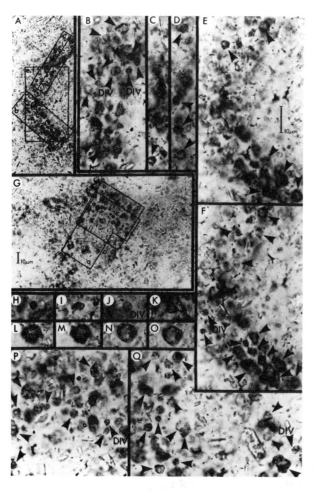

Figure 9–29. Spheroidal, prokaryotic microfossils (arrows) from chert of Kromberg Formation, Onverwacht Group (approximately same age as Fig Tree), South Africa. Scale in G is for A and G; magnification of all other parts indicated by scale in DIV refers to apparently paired spheroids resembling algae in the process of dividing. A. Small domed arc of spheroids, oriented perpendicular to the bedding laminations of the chert boxes indicate areas shown at higher magnification in B–F. G. Globular, unstructured cluster (irregular colony?) of spheroids; boxes indicate areas shown at higher magnification in P and Q. H–K. Small carbonaceous spheroids. L, M. Relatively large spheroid, shown at two different focal depths. N, O. Relatively large spheroid, shown at two different focal depths. (Photo courtesy of J. William Schopf)

The Methanogens: A Link to the Earliest Forms of Life?

The third major discovery of the 1970s involved biochemical studies of certain modern prokaryotes. Evolutionary paleobiologist Stephen Jay Gould of Harvard University has pointed out that in retracing the course of Cryptozoic evolution it is not necessary to rely solely on the fossil record. An entirely independent signature is preserved in the metabolism and biochemical pathways of living cells. Vestiges of Cryptozoic biochemistries have been retained in certain living organisms. By studying their modern distributions, it is occasionally

possible to deduce when certain biochemical capabilities first appeared.

In 1977, a University of Illinois research team announced that a group of bacteria that grow by oxidizing hydrogen and reducing CO_2 to methane are in fact not bacteria at all, but may represent a distinctly different type of prokaryotic life. These microorganisms, called **methanogens,** are anaerobic; they die in the presence of oxygen. In the modern world they are confined to unusual oxygen-free environments such as deep hot springs and fetid mud bottoms of stagnant ponds. Methanogens have been known for some time, but the pronouncement that they are different from other prokaryotes came on the heels of discovery that they have a unique RNA nucleotide sequencing. In fact, the research team that made the discovery believes that RNA sequencing to be so significantly different between methanogens and other prokaryotes (bacteria and cyanobacteria) that they are ready to place methanogens in a separate kingdom, distinct from the Monera! According to Gould, however, the inference of this biochemical difference is that the prokaryotes, including methanogens, bacteria, and cyanobacteria, must have had a common ancestor. Thus all prokaryotes presumably had the same RNA sequence at one point in their past and any current measurable differences arose by divergence from this common ancestor. If bona fide monerans (bacteria and possibly even cyanobacteria) had already evolved by the time of deposition of the Fig Tree Group, 3400 million years ago, then the common ancestor of methanogens and monerans must have existed even earlier.

We have good reason to believe that the earth's original atmosphere was devoid of oxygen and rich in CO_2, the very conditions under which methanogens thrive and under which the earth's original life might have evolved. Two workers, W. B. N. Berry and P. Wilde at the University of California, Berkeley, have postulated that the world's first biota might have evolved prior to 3900 million years ago—when geothermal energy was more readily available than photic energy—at Archean analogues of modern oceanic hydrothermal vents. Could methanogens be a remnant of the world's first biota—a biota that evolved to match the earth's primordial conditions but one which now is restricted, because of the rise in oxygen levels, to a few marginal environments? Just as the moon provides us with information about the pregeologic (prior to 3800 million years ago) history of the earth (see Chapter 8), so might the modern methanogens give us some insight regarding the world's earliest life.

If Gould's interpretation is correct, the Fig Tree organisms, and perhaps also those at North Pole, Australia, were monerans already indulging in photosynthesis. This is direct evidence of life in the oldest rocks that

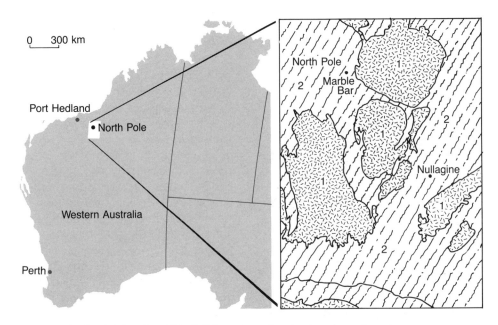

Figure 9–30. Western Australia showing location of North Pole region and regional geology of granite basement (1) and greenstone belts (2).
(Data from D. L. Groves, J. S. R. Dunlap, and Roger Buick, 1981, An Early Habitat of Life: *Sci. American,* vol. 245, no. 4)

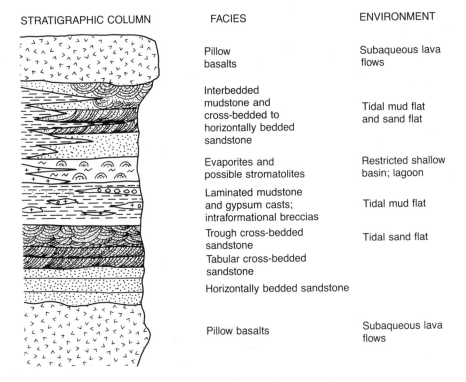

Figure 9–31. North Pole sedimentary rocks and volcanics showing facies and interpreted environments.
(Data from D. L. Groves, J. S. R. Dunlop, and Roger Buick, 1981, An Early Habitat of Life: *Sci. American,* vol. 245, no. 4)

reasonably could contain it. By reasonably strong inference there is reason to believe that a major radiation of methanogens predated the Fig Tree and Warawoona monerans. As expressed by Gould, Life probably arose on this planet as soon as the lithosphere cooled enough to support it. Given the anoxic atmosphere, the earliest

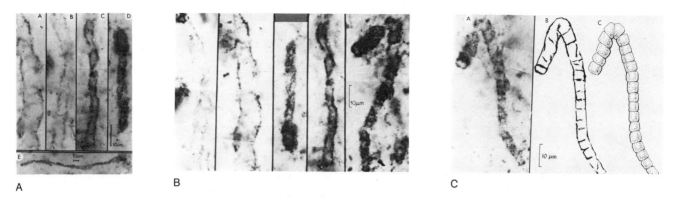

Figure 9–32. Microfossils from carbonaceous chert, early Archean Warrawoona Group, North Pole dome region of the Philbara block, Western Australia. A. Tubular or partially flattened bacterial or cyanobacterial sheaths. B. Elongate rod-shaped apparently nonseptate fossil bacteria. C. Unbranched, septate, apparently somewhat tapering filamentous fossil prokaryote shown in petrographic thin section (A) and reconstructions (B, C).
(Photos courtesy of J. William Schopf)

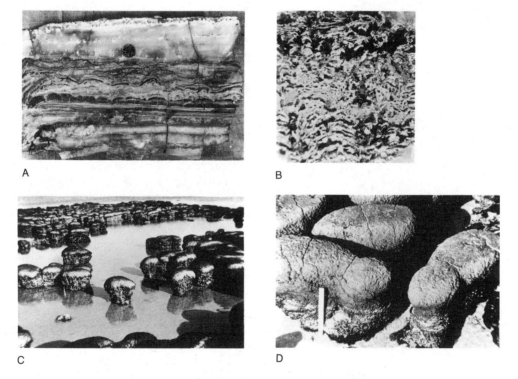

Figure 9–33. A. Possible stromatolite from Warrawoona Group, North Pole dome region of Philbara block, Western Australia. B. Modern-day stromatolite from Shark Bay, Western Australia. Modern forms are much like their fossil counterparts. C, D. Domal stromatolites from intertidal zone, Hamelin Pool, Shark Bay, Western Australia. Such scenes were commonplace during the Proterozoic; today, however, well-developed stromatolites are confined to a few hypersaline environments like Hamelin Pool in Shark Bay. The forms are rare or absent from more normally saline marine environments because of grazing by invertebrates, absent during the Late Proterozoic.
(A, B, photos courtesy of J. William Schopf; C, D, photos courtesy of John B. Dunham, Union Oil Co.)

hydrosphere, a lithosphere that could hold bodies of water, and a period of chemical evolution, "the origin of life might have been as inevitable as quartz or feldspar."*

_____
*_The Panda's Thumb_, p. 218.

The Oxygen Revolution: Evidence from Sedimentary Rocks

One important line of evidence that our early atmosphere was devoid of oxygen is found in the study of certain sedimentary minerals from Archean greenstone belts. Workers such as Cloud have investigated the possibility that certain very unstable minerals might reflect the concentration of free oxygen at the time they were deposited. One particular mineral of significance is uraninite. In the presence of oxygen, grains of uraninite are readily oxidized and dissolved. Uraninite probably could not have accumulated in Archean deposits if there had been an appreciable (as much as 1 percent of present atmosphere level) concentration of atmospheric oxygen. Significant amounts of unoxidized pyrite also support the notion of anoxic conditions. But what is our evidence in the rock record for the timing of the transition from an oxygenless to an oxygen-rich atmosphere?

Perhaps our best evidence comes from iron-rich sedimentary deposits. The banded iron formations contain iron oxides embedded in chert. Banded iron formations occur sporadically throughout the Archean. They are particularly abundant in stratigraphic sequences of Early Proterozoic age between about 2400 and 2000 million years old. The rich prokaryotic microbiota of the Gunflint Formation occurs in a banded iron formation approximately 2000 million years old. Cloud and others believe the banded iron formations represent the deposits of "oxygen sinks": aquatic environments where oxygen released by photosynthesizing cyanobacteria was quickly combined with free ferrous iron. Interestingly, the time interval of most abundant banded iron formation development coincides with a major development of carbonate rocks and stromatolites in the geologic record.

Stromatolites are known from rocks 3100 and 2800 million years old in South Africa and Rhodesia, and perhaps from rocks as old as 3500 million years at North Pole, Western Australia (Table 9-4). However, the Early Proterozoic represents a tremendous explosion in stromatolite development and, with it, a major buildup of oxygen produced by the photosynthetic activity of abundant and widespread mats of cyanobacteria. Thus the record of major oxygen release is provided by the

DAYLIGHT
Upward growth (*S. calcicola*)
and sediment trapping

DARKNESS
Horizontal growth (*O. submembranacea*)
and sediment binding

Figure 9–34. Day–night accretion cycle in stromatolites.
(From C.D. Gebelein, 1969, Distribution, Morphology, and Accretion Rate of Recent Subtidal Algal Stromatolites, Bermuda; Fig. 14, p. 60: *Jour. Sedimentary Petrology,* vol. 39, No. 1. Reprinted by permission of Society of Economic Paleontologists and Mineralogists)

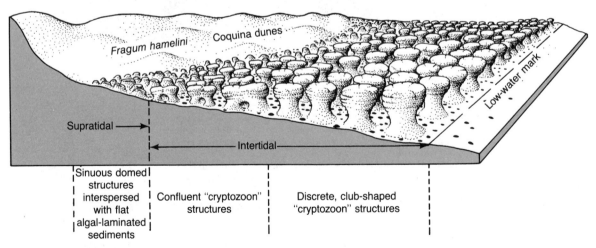

Figure 9–35.   Idealized form zonation of stromatolites at Flint Cliff, Hamelin Pool, Shark Bay, Western Australia.
(From B. W. Logan, 1961, Cryptozoan and Associated Stromatolites from the Recent, Shark Bay, Western Australia: Fig. 3, p. 528: *Jour. Geology,* vol. 69. Reprinted by permission of University of Chicago Press)

banded iron formations and the record of the oxygen-releasing agent is provided by the stromatolites.

A final note on Proterozoic stromatolites concerns their use in biostratigraphic studies. Some success has been achieved in a few places by use of stromatolites for biostratigraphic subdivision (biozonation) and correlation. Such applications have been limited in scope because of the difficulty in distinguishing between the effects of evolution and the effects of paleoecology on morphology of stromatolites. We know that stromatolite form is a sensitive indicator of environment (Figs. 9-34, 35, 36), but we do not know much about the evolutionary component—the change in morphology through time. Certainly stromatolite biozones are much more difficult to define than those based on metazoan fossils in the Phanerozoic.

Another kind of iron-rich deposit called red beds includes detrital sediments whose particles are coated with iron oxides, mostly the mineral hematite (Fe_2O_3). Red beds are known in sedimentary sequences younger than about 2000 million years, but *not* in older ones. Curiously, banded iron formations are virtually unknown in rocks younger than about 2000 to 1800 million years old. Is this temporal pattern of change in the character of iron-rich sedimentary rocks fortuitous or real?

Table 9-4. Major evolutionary events during Archean and earliest Proterozoic

Age in m.y. B.P.	Event	Evidence
2000	**PROTEROZOIC** — 7 Advent of significant O_2 in atmosphere (Oxic atmosphere)	7 Earliest major development of red beds
2200–2400	6 Major development of aerobic photosynthesis; rapid evolution of cyanobacteria	6 Major development of banded iron formations (B.I.F.) and widespread, diverse stromatolites
2600–3000	5 Beginning of aerobic photosynthesis; early evolution of cyanobacteria	5 Oldest definitive stromatolites
3200–3600	**ARCHEAN** — 4 Early diversification of anaerobic bacteria; first primitive anaerobic photosynthesis; simple prokaryotes (Anoxic atmosphere)	4 Oldest definitive evidence of life on earth: moderately diverse prokaryotes in Warrawoona Group, Western Australia; Fig Tree and Onverwacht Groups in South Africa
3800–4200	3 Formation of primitive lithosphere, atmosphere, hydrosphere, biosphere	3 Oldest dated rocks on earth (Isua supracrustal terrane of southwest Greenland, ~3800 m.y. B.P.); indicates time of metamorphism of sedimentary sequences and plutonic igneous rocks; presence of possible organic carbon and B.I.F.
4400	2 Major volcanism and meteoric impact	2 Lunar craters and rocks
4600	1 Formation of the earth	1 Age of meteorites; Pb-evolution curves

Information from P. E. Cloud, Jr., 1974, Evolution of Ecosystems, in *Paleontology and Environment: Readings from American Scientist*; 1976, Beginnings of Biospheric Evolution and Their Biochemical Consequences: *Paleobiology*, vol. 2, pp. 351–387; 1983, The Biosphere: *Sci. American*, vol. 249, no. 3, pp. 176–189; J. William Schopf, 1978, The Evolution of the Earliest Cells: *Scientific American* Offprint no. 1402, W. H. Freeman, San Francisco; 1983, *Earth's Earliest Biosphere: Its Origin and Evolution*: Princeton University Press, Princeton NJ.

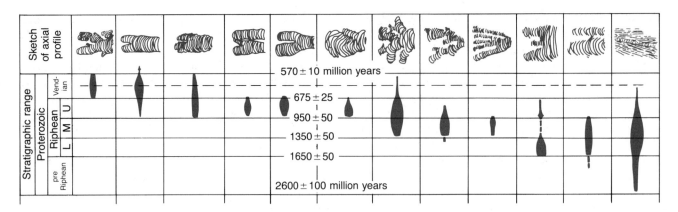

Figure 9–36. Stratigraphic biozonation of stromatolites, Proterozoic of Russia.
(From P. E. Cloud Jr., and M. A. Semikhatov, 1969, Proterozoic Stromatolite Zonation, Fig. 3, pp. 1026–27:
Am. Jour. Science, vol. 267. Reprinted by permission of *American Journal of Science* and senior author)

Pervasively iron-stained red sediments are conventionally attributed to oxidation of iron by free oxygen in the atmosphere and are most common in nonmarine (subaerially exposed) environments. Could it be that the onset of red beds (and the disappearance of banded iron formations) in the geologic record signals the initial buildup of oxygen in the atmosphere after the oxygen sinks were filled up—perhaps around 2000 million years ago (Table 9-4)?

Photosynthesis

Essentially all of the free oxygen in our atmosphere is and always has been the product of organic aerobic photosynthesis. Aerobic photosynthesis is a process whereby the energy of sunlight is employed to make carbohydrates from water and carbon dioxide, and molecular oxygen is released as a byproduct. Aerobic photosynthesis is performed by cyanobacteria, algae, and higher plants. Essentially all of the oxygen produced during the early stages of the oxygen revolution was by stromatolitic communities of cyanobacteria. Modern-day bacteria that photosynthesize are of minor importance but do perhaps provide us with insight regarding the origin of photosynthesis. Bacterial photosynthesis today is entirely anaerobic. Such photosynthesis does not release oxygen as a byproduct and cannot proceed in the presence of oxygen. Anaerobic photosynthesizers employ hydrogen sulfide (H_2S) instead of water and release sulfur instead of oxygen.

Anaerobic bacterial photosynthesis likely evolved to meet the world's first global energy crisis. The first living cells were probably small spheroidal anaerobes that received their energy by fermenting organic molecules formed nonbiologically in an anoxic environment. But as the "organic soup" became depleted by these original anaerobic heterotrophs, there arose a need for an alter-

native energy source. The importance of such ready-made nutrients diminished, however, when the first photosynthetic organisms evolved. Anaerobic photosynthesis probably evolved among primitive bacteria or methanogens early in the Cryptozoic when the oceans and atmosphere were essentially anoxic and H_2S or reduced sulfur from volcanic vent outgassing persisted due to high solubility of H_2S. According to the Berry and Wilde hypothesis, increased competition for available sulfur probably resulted in more dependency on photic energy. These early photic autotrophs might have lived in matlike communities in shallow water but under conditions of comparatively low light intensity, which was controlled by the cloudy, CO_2-rich Venuslike atmosphere. Somewhat later, as the quantity and range of light reaching the ocean surface improved and water splitting accelerated oxidation, primitive prokaryotes probably gave rise, as a mutant strain, to the first organisms capable of aerobic photosynthesis; these organisms were the ancestors of modern cyanobacteria. This new brand of photosynthesis was more efficient as a means of energy and nourishment and it was selected for; however, the molecular oxygen released was a toxin that no doubt poisoned many kinds of anaerobic organisms. Schopf suggests one result was that the new aerobic photosynthesizers were able to replace the anaerobic forms in the upper portions of the mat communities. The anaerobic forms adapted to the underparts of the mats—a place of less light penetration, but also of little oxygen. Many photosynthetic bacteria occupy such habitats today. As recounted earlier, anaerobic photosynthesis was very likely going on during deposition of the Fig Tree and Warawoona beds 3400 to 3500 million years ago.

The cyanobacteria spread rapidly and dominated virtually all accessible habitats by the beginning of the Proterozoic. And with them, the rise of aerobic photosynthesis about 2000 million years ago introduced a change in the global environment that was to profoundly influence all subsequent evolution. Oxygen was here to stay.

The Eukaryotes: The Greatest Evolutionary Step

Since the 1960s it has become ever more apparent that the greatest division among living organisms is not between plants and animals, but rather between organisms whose cells have nuclei and those that lack a nucleus. Methanogens, bacteria, and cyanobacteria are the principal types of nonnucleated cells and belong to the superkingdom Prokaryota. Organisms whose cells have nuclei are called eukaryotes. Table 9-5 shows the important fundamental differences between prokaryotes and eukaryotes. One particular difference between the two is of great importance in the study of their evolution—namely, the extent to which they tolerate oxygen. As we have already seen, oxygen requirements are quite different among the prokaryotes. Some are totally anaerobic; others can tolerate small amounts of oxygen, and some are fully aerobic. In contrast to this range of adaptations, eukaryotes exhibit a pattern of great consistency; with but a few rare exceptions, eukaryotes have an absolute

requirement for oxygen. Comparison of the metabolism and biochemistry of prokaryotes and eukaryotes provides strong evidence that the latter group arose only after a significant quantity of oxygen had accumulated in the atmosphere (Table 9-5). Thus it is appropriate to ask when eukaryotic cells first appeared.

Table 9-6 lists the main stratigraphic occurrences of relatively complex microfossils from Proterozoic strata. The oldest definitive eukaryotes come from the Beck Springs Dolomite of southeastern California—most reasonably dated at about 1200 to 1300 million years old. These are highly branched filaments of comparatively large diameter and with rare crosswalls and are similar in some respects to certain green or golden-green eukaryotic algae. Several occurrences of likely eukaryotes have been reported from rocks approximately 1400 million years old in Northern Territory Australia. Perhaps the most productive stratigraphic unit for early eukaryotes is the approximately 900-million-year-old Bitter Springs Formation in central Australia. Schopf has documented a diverse microbiota of nucleated unicells (Fig. 9-37), some of which may actually show stages of meiotic cell division. If this is the case, simple sexual reproduction had evolved by perhaps 1000 m.y. B.P. Studies of both the morphology and size (Table 9-6) of unicellular microfossils suggest that in rocks older than

Table 9-5. Features distinguishing eukaryotes from prokaryotes

Distinguishing characteristics	Unicellular organisms	
	Prokaryotes	Eukaryotes
Cell size	Very small; generally 1–10 microns	Larger; generally 20–100 microns
Cell organization Nucleus Organelles Genetic organization	Poor Absent Absent Loop of DNA in cytoplasm	Organized Present Mitochondria and chloroplasts DNA organized in chromosomes within nucleus
Oxygen requirements Metabolism Energy production	Intolerance to tolerance Aneorobic or aerobic Fermentation or respiration	Oxygen required Aerobic Respiration
Reproduction	Binary fission	Mitosis or meiosis
Organisms represented	Bacteria, methanogens, cyanobacteria	Protists

Data from J. William Schopf, 1978, The Evolution of the Earliest Cells: *Scientific American* offprint no. 1402, W. H. Freeman, San Francisco.

Table 9-6. Eukaryotic microfossils in proterozoic rocks*

Stratigraphic unit	Location	Bracketed radiometric age m.y. B.P.	Occurrence of eukaryotes
Olkhin Formation	Siberia	725 (680–800)	Branched filaments composed of cells with distinct crosswalls; resembling modern, fungi or green algae
Kwagunt Formation	Eastern Grand Canyon, U. S.	800 (650–1150)	Complex, flask-shaped microfossils
Bitter Springs Formation	Central Australia	850 (740–950)	Unicellular algae containing intracellular membranes and organellelike bodies; four sporelike cells in a tetragonal configuration representing mitosis or meiosis
Unnamed shales	Siberia	950 750–1050)	Spiny cells or algal cysts several hundred microns in diameter
Beck Spring Dolomite	Eastern California	1300 (1200–1400)	Highly branched filaments of large diameter with rare crosswalls similar to those of some green and golden-green algae
Skillogalee Dolomite	South Australia	850 (290–867)	Similar to those of Beck Spring Dolomite; spheroidal microfossils exhibiting two-layered walls and with medial splits on surface, possibly representing encystment stage of eukaryotic algae; tetrahedral group of four small cells resembling spores produced by mitotic cell division of some green algae; well-preserved unicellular fossils containing small membrane-bounded structures
McMinn Formation	Northern Territory Australia	1400 (1280–1450)	
Amelia Dolomite	Northern Territory Australia	1500 (1390–1575)	
Bungle-Bungle Dolomite	Same region and about same age as Amelia Dolomite		

*The task of identifying microscopic, single-celled organisms as eukaryotic is not simple. Also, precise age dating is limited because ages of the fossiliferous sedimentary deposits are interpolated between the ages of the nearest overlying (youngest bracketing age) and underlying (oldest bracketing age) datable rock units such as lava flows. From the information available, the oldest definitive eukaryotic microfossils are about 1500 million years old. Numerous microfossils from older sediments, such as the well-studied Gunflint Chert (about 2000 million years old), appear to be exclusively prokaryotic.
(Data from J. William Schopf, 1978, The Evolution of the Earliest Cells: *Scientific American* Offprint no. 1402, W.H. Freeman, San Francisco)

1500 million years eukaryotic cells are rare or absent; in rocks younger than that, such cells become increasingly more abundant. Remember that the moderately diverse Gunflint microbiota, about 2000 million years old, is exclusively prokaryotic; thus it would seem that eukaryotes first evolved sometime between about 2000 and 1500 million years ago; 1500 million years ago is the time of abundant occurrence of red beds also. Perhaps just prior to 1500 million years ago, an oxygen level in the atmosphere of about one to several percent present atmospheric level (PAL) was attained, providing an evolutionary opportunity for life to evolve beyond prokar-

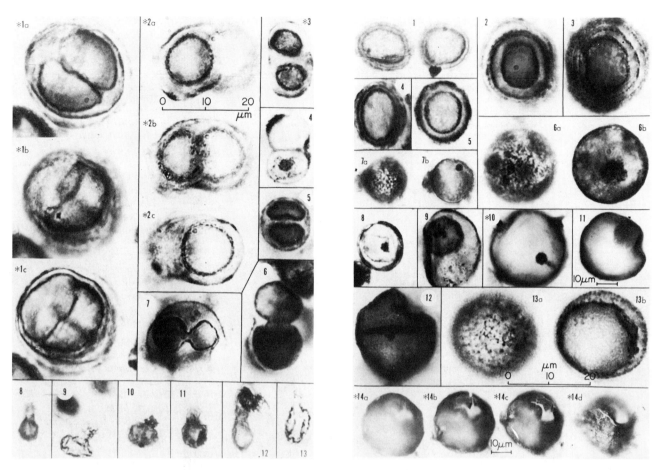

Figure 9–37. Spheroidal eukaryotic algae from bedded, black, stromatolitic chert of the Bitter Springs Formation, Late Proterozoic (about 859 m.y. B.P.), central Australia. (Photos courtesy of J. William Schopf)

yotes. The prokaryotes evolved mainly during the time when environmental oxygen concentration in the atmosphere changed from zero to slight. However, by the time the eukaryotes appeared, the oxygen levels in the atmosphere were significant and conducive to respiration, the central metabolic process of organisms with nucleated cells.

It may seem astonishing that for at least the first 2000 million years in the history of life (more than half!), only the simple prokaryotic level was involved. It is important to remember, however, that major evolutionary breakthroughs, such as the emergence of a new super-kingdom (Eukaryota), do not automatically happen with the passage of time. The main prerequisites for evolutionary change are appropriate antecedent biological systems and evolutionary opportunities. Why the "delay" in the evolution of life beyond simple prokaryotes? The answer probably lies in the oxygen revolution. The origin of the nucleated cell was the most important, far-reaching step in the history of life. According to Cloud, the evolutionary biogeochemical consequences of this important innovation included, among others: increases in atmo-

spheric O_2 and CO_2, increase in carbonate rocks, formation and stabilization of the *ozone* layer, and rise of sexual reproduction. Sexuality coupled with increasing atmospheric oxygen levels (to perhaps 6 to 10 percent PAL) may have provided the critical evolutionary triggers that promoted Late Proterozoic eukaryotic diversification and ultimately the evolutionary emergence of multicellular organisms about 700 million years ago (Table 9-7).

Evolution of life during the Cryptozoic was under conditions that differed greatly from those prevailing today, but the mechanisms of evolution were the same. Genetic variations made some individual unicells better suited to survive and reproduce in a given environment and pass on their heritable traits to succeeding generations. The emergence of new forms of life through this principle of natural selection resulted in great changes in turn in the physical and chemical environment, thereby altering the conditions and directions of evolution. Life as we know it could not have arisen in the presence of oxygen, but it also could not have evolved beyond bacteria without it.

Table 9-7. Major evolutionary events during latest Archean and Proterozoic

Age in m.y. B.P.	Event				Evidence
600	Cambrian	12	Evolution of exoskeletons		12. Calcareous and calcareophosphatic shells
800	Ediacarian	11	Origin and early diversification of multicellular organisms	6–10 PAL O_2	11. Ediacara—soft-bodied metazoan fauna
1000		10	Diversification of eukaryotes; evolution of sexual reproduction	5% PAL O_2	10. Abundant eukaryotes in Bitter Springs Formation; evidence of meiosis
1200					9. Increase in diversity of microfossils; increase in size of spheroidal microfossils
1400	PROTEROZOIC	9	Origin of eukaryotes		
		8	Diversification of aerobic prokaryotes		8. Gunflint and Fortescu diverse microbiotas
1600					
1800					7. Earliest major development of red beds; youngest detrital uraninites
2000		7	Advent of significant O_2 in atmosphere	1% PAL O_2	
2200		6	Major development of aerobic photosynthesis and respiration; rapid evolution of cyanobacteria		6. Major development of banded iron formations and widespread, diverse stromatolites; abundant microfossils from stromatolites
2400					
2600					
2800		5	Early diversification of cyanobacteria; aerobic photosynthesis		5. Oldest definitive stromatolites

Oxygenated atmosphere · Anoxic(?) atmosphere · ARCHEAN

Sources as in Table 9-4

SUMMARY

Prephanerozoic history is subdivided into Archean and Proterozoic Eons (Table 9-8), whose joint boundary has a radiometric age of about 2500 million years before present. Archean and Proterozoic are further subdivided into Early, Medial, and Late Eras, which are chronometric units not based on stratotypes. Cryptozoic, not Precambrian, is the preferred term for pre-Phanerozoic history.

The Canadian Shield contains the oldest and most expansive surface exposure of Cryptozoic rocks on the North American continent. The shield has been subdivided into eight major isotopic age-structural provinces separated from one another by abrupt orogenic fronts. Archean rocks of the shield are found mainly in the Su-

perior and Slave Provinces and consist of long, narrow greenstone belts within wider tracts of granite, granitoid gneisses, and other metamorphics. The Kenoran orogeny that brought Archean history to a close was largely responsible for transforming what had been comparatively thin slabs of granitic crust into larger volumes of granite that approached the present average continental thickness of about 40 km. By the beginning of Proterozoic time, general lithospheric conditions were much different than during Archean time. Proterozoic sedimentary rocks include quartz-rich deposits, banded iron formations, glacial tillites, true red beds, and carbonates.

Beyond the Canadian Shield, exposures of Cryptozoic basement rocks are present in the cores of numerous mountain ranges such as the Rockies, in structural domes like the Adirondacks, and in deep canyons like

Table 9-8. Summary of major events during the Cryptozoic

Age in m.y. B.P.	Event	Environment	Era/Period
600	Origin and early evolution of metazoans	6–10% PAL O_2	Ediacarian
800	Diversification of eukaryotes and evolution of sexual reproduction	Glaciation	Late Proterozoic
1000		5% PAL O_2	
1200		Oxic Environment	Middle Proterozoic
1400	Oldest known eukaryotes		
1600			Proterozoic
1800	Beginning of widespread development of red beds	1% PAL O_2	
2000			Early Proterozoic
2200	Abundant B.I.F.s, stromatolites; diversification of cyanobacteria } Glaciation		
2400			
2600	Kenoran orogeny and formation of large volumes of continental lithosphere		Late Archean
2800			
3000		Anoxic Environment	
3200	Formation of greenstone belts		Middle Archean / Archean
3400			
3600	Earliest evidence of life		Early Archean
3800	Oldest dated rocks on earth		
4000	Origin of lithosphere, atmosphere, hydrosphere, biosphere; widespread volcanism and meteoric bombardment		Pregeologic history of the earth
4200			
4400			
4600	Origin of the earth		

the Grand Canyon. These basement rocks are scattered exposures of the stable nucleus of the continent—extensions of the shield to the margins of the ancient continent beneath the cover of younger, Late Proterozoic and Phanerozoic sediments. Cryptozoic rocks in the Canadian Shield and elsewhere have yielded important economic mineral deposits; e.g., iron, gold, copper, platinum, and nickel.

Late Proterozoic sedimentary rocks of the Appalachian and Cordilleran belts are the deposits of rifted continental margins and signify the breakup of a supercontinent.

The Cryptozoic sedimentary rock record provides important clues regarding the evolution of the earth's lithosphere, hydrosphere, atmosphere, and biosphere. Archean sedimentary rocks are texturally and mineralogically immature, indicating an unstable, predominantly basaltic crust. Proterozoic sedimentary rocks show progressive enrichment in quartz and potassium feldspar, indicating major increase in granitic, continental crust. Compared with Archean rocks, Proterozoic sedimentary rock suites are more texturally and mineralogically mature, contain more abundant shallow-water sedimentary structures and more carbonates, and indicate widespread deposition in more stable continental shelf settings.

The early Archean atmosphere was probably devoid of oxygen but rich in CO_2, as suggested by the presence of unoxidized unstable minerals such as uraninite and pyrite in sediments. The origin of life came about under oxygen-deficient conditions. The oldest known fossils are bacterialike structures preserved in cherts of greenstone belts in South Africa and Western Australia. Free

oxygen in the oceans presumably was furnished by water splitting and later by oxygen-releasing photosynthesis. Banded iron formations suggest oxygen was quickly tied up with iron and silica in aquatic environments that acted as oxygen "sinks." Extensive development of banded iron formations during the Early Proterozoic coincides with significant buildup of carbonate rocks and stromatolites.

The earliest true red beds formed sometime after 2000 million years ago and are signatures of filling of oxygen sinks and release of free oxygen into the atmosphere in amounts of perhaps 1 percent present atmospheric level. The onset of red-bed deposition coincides closely with the cessation of deposition of banded iron formations and represents increased production of oxygen by aerobic photosynthesis.

The origin of the nucleated cell and thus the emergence of the superkingdom Eukaryota was the most important, far-reaching evolutionary step in the history of life. All known eukaryotes require oxygen for their metabolism. Their first appearance in the geologic record, some time between 2000 and 1500 million years ago, is evidence that significant amounts of free oxygen, conducive to respiration, were present in the environment. Increasing atmospheric oxygen levels coupled with the rise of sexual reproduction provided the evolutionary triggers that promoted Late Proterozoic eukaryotic diversification. Change in life and the chemistry of the oceans and atmosphere during the Cryptozoic is largely a story of the oxygen revolution. Life as we know it could not have arisen in the presence of oxygen, but it could not have evolved beyond bacteria without it.

SUGGESTIONS FOR FURTHER READING

Cloud, P. E., 1983, The Biosphere: *Sci. American,* vol. 249, no. 3, pp. 176–189.

Dott, R. H., Jr., and Batten, T. L., 1981, *Evolution of Earth,* 3rd ed.: McGraw-Hill, New York.

Gould, Stephen Jay, 1980, An Early Start, in *The Panda's Thumb: More Reflections in Natural History:* W. W. Norton, New York, pp. 217–227.

Groves, D. I., Dunlop, J. S. R., and Buick, R., 1981, An Early Habitat of Life: *Sci. American,* vol. 245, no. 4, pp. 64–74.

Moorbath, S., 1977, The Oldest Rocks and the Growth of the Continents: *Scientific American* Offprint No. 357, W. H. Freeman, San Francisco.

Piper, J. D. A., 1976, Paleomagnetic Evidence for a Proterozoic Supercontinent: *Phil. Trans. R. Soc. London,* vol. A280, pp. 469–490.

Schopf, J. W., 1978, The Evolution of the Earliest Cells: *Scientific American* Offprint No. 1402, W. H. Freeman, San Francisco.

Schopf, J. William, ed., 1983, *Earth's Earliest Biosphere: Its Origin and Evolution:* Princeton University Press, Princeton, N. J.

Walter, M. R., 1977, Interpreting Stromatolites: *Am. Scientist,* vol. 65, pp. 563–571.

10

THE "EOPALEOZOIC": Proterozoic–Paleozoic Transition

KEY TERMS

Ediacarian
Craton
Tommotian Stage
Epeiric sea
Transcontinental arch
Miogeosyncline
Eugeosyncline
Laurentia
Baltica
Gondwana
Iapetus

THE BURGESS SHALE FAUNA

During a field excursion to the majestic Canadian Rockies in 1909, Charles D. Walcott, the great American geologist and student of the Cambrian, made a fantastic discovery high on the west face of Mt. Wapta near Field, British Columbia. His pack horses stumbled over a fallen chunk of shale and when the specimen split apart, Walcott's trained eyes caught the gleam of reflection from some glossy markings on the bedding plane surfaces. A closer examination revealed *carbonaceous* imprints and films of soft-bodied creatures preserved in amazing detail (Fig. 10-1). Further search and the eventual opening of a quarry produced thousands of specimens that represent an unusual association of species, most of which are unique to this locality. These remarkable fossils occur in the Burgess Shale member of the Stephan Formation of Medial Cambrian age. The biota includes more than 150 species representing at least eight known and several previously unknown phyla; most of the fossils represent taxa that became extinct by the end of the

Cambrian. Approximately 40 percent of the presently recognized fauna consists of arthropods, but sponges, coelenterates, echinoderms, molluscs, annelid worms, and even a primitive chordate number among the finds. Not only is the fauna unique, it is amazingly well preserved and many of the arthropod specimens show preservation of appendages, bristles, and even internal organs.

Despite the early major research effort by Walcott and some of his associates, significant gaps remained in knowledge of the Burgess Shale paleoenvironment, conditions of preservation, faunal composition, and phylogenetic relationships. In the late 1960s, under the direction of the Canadian Geological Survey, renewed interest in the Burgess Shale fauna, stimulated in large part by investigations of the depositional paleoenvironments of Cambrian strata in this region, resulted in temporary reopening of Walcott's quarries and a new period of collecting. Both the new material and major portions of Walcott's original collections have been studied by systematic paleontologists at Cambridge University, England. This work, combined

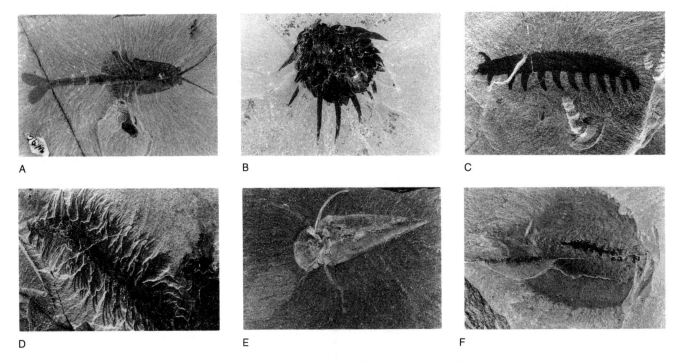

A B C

D E F

Figure 10-1. Burgess Shale fossils. A. *Waptia fieldensis,* an arthropod. B. *Wiwaxia corrugata,* affinity uncertain. C. *Aysheaia pedunculata,* an arthropod. D. *Canadia irregularia,* a polychaete worm. E. *Hyolithes carinatus,* a mollusc (?). F. *Naraoia spinifer,* an arthropod.
(Photos courtesy of National Museum of Natural History. Smithsonian Institution Photo Nos. 114259/52.FS, 65056/29.FS, 83942/18.FS, 17.FS, 137509/37.FS, 83946/20.FS)

with analysis of the depositional and postdepositional setting of the Burgess Shale by Canadian Survey geologists, has narrowed the areas of uncertainty about this remarkable fauna.

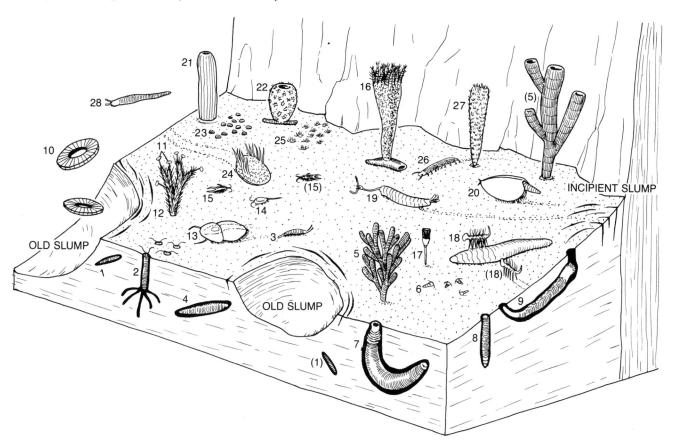

Figure 10-2. Reconstruction of the Burgess Shale fauna. The fauna is depicted here as inhabiting a muddy bottom at the base of an algal reef that stood more than 100 m high and whose top was near sea level. The scene depicts a pre-slide environment showing the scars of previous slumps as well as incipient failures, both signatures of the impending catastrophe that will transport part of the fauna to the less hospitable post-slide environment of deposition. No attempt has been made to show the animals in numbers proportional to their fossil abundance. The taxa are identified by number, starting at the bottom left; only about a fifth of the species fossilized in the shale are shown. Most of the immobile animals of the sea floor are sponges: *Pirania* (12), seen with brachiopods attached to its spicules; *Eiffelia* (22); the gregarious *Choia* (25); a gracile species of *Vauxia* (5), with a more robust species at the top right, and *Chancelloria* (27). Three other immobile animals are *Mackenzia* (21), a coelenterate; *Echmatocrinus* (16), a primitive crinoid, seen attached to an empty worm tube, and *Dinomischus* (17), one of the Burgess Shale species that represent hitherto unknown invertebrate phyla. The burrow-dwelling animals are *Peronochaeta* (1), a polychaete worm that fed on food particles in the silt; *Burgessochaeta* (2), a second polychaete that captured food with its long tentacles; *Ancalogon* (4), a priapulid worm possibly ancestral to some modern parasites; *Ottoia* (7), another priapulid, seen at the center feeding on the mollusc *Hyolithes* (6) and at the right burrowing; *Selkirkia* (8), a third priapulid, seen here in a burrow front end down, and *Louisella* (9), a fourth priapulid that inhabited a double-ended burrow and undulated its body to drive oxygenated water over its gills. *Peytoia* (10) is a free-swimming coelenterate shaped like a pineapple ring. The sea-floor-dwelling molluscs in addition to *Hyolithes* are *Scenella* (23), its soft parts hidden under "Chinese hat" shells, and *Wiwaxia* (24), with its covering scales and defensive spines, seen here plowing a trail through the silt. Among the many arthropod genera of the sea floor are *Yohoia* (3), with its distinctive grasping appendages; *Naraoia* (13), an atypical trilobite that retained some larval characteristics; *Burgessia* (14), with its long tail spine; *Marrella* (15), which may have swum just above the sea floor; *Canadaspis* (20), an early crustacean, and *Aysheaia* (26), a stubby-legged animal suggestive of the living land dweller *Peripatus*. Other representatives of new phyla seen in addition to *Dinomischus* are *Hallucigenia* (18), one preparing to feed on a dead worm and two others approaching it, and *Opabinia* (19), seen here grasping a small worm with its single bifurcated appendage. Finally, seen swimming alone at the top left, is *Pikaia* (28), the sole representative of the chordate phylum in this Middle Cambrian fauna. *Pikaia* probably used its zigzag array of muscles to propel itself above the sea floor. The phylum of chordates includes the subphylum of vertebrates, which evolved later.
(From Simon Conway Morris and H. B. Whittington, The Animals of the Burgess Shale, *Scientific American,* July 1979, p. 72. Copyright © 1979 by Scientific American. All rights reserved).

As described by British paleontologists H. B. Whittington and S. Conway Morris, the Burgess Shale soft-bodied fauna represents a rare mode of preservation (see discussion, Chapter 4). This fauna reflects a combination of circumstances in a paleoenvironment with optimum accumulation rates of organic-rich mud and lack of bottom-scavenging organisms to devour the carcasses that had settled there. More recent investigation of the unique deposit paints a paleoenvironmental picture of a moderately deep marine bottom situated at the toe of an embayed carbonate algal reef (Fig. 10-2). The bottom waters in this setting probably were restricted in circulation and poisoned by the buildup of hydrogen sulfide. Talus mounds of mud accumulated against the base of the reef and had surfaces presumably elevated above the noxious bottom waters. These perched surfaces provided habitats for a complex community of predominantly soft-bodied, but also some shell-bearing, invertebrates. However, the tranquility of this community structure periodically was interrupted by slumping of these mud mounds, with the catastrophic result that the bottom dwellers were swept into the stagnant and poisonous bottom waters where they were annihilated and rapidly buried. Later, slow compaction of the muds to shale and the eventual uplift of a mountain range made the deposit available for discovery.

The Burgess Shale fauna is one of the most important paleontological discoveries ever made. This example of rare preservation provides a hint of the nature, complexity, and full array of Cambrian life. It is only fitting that Walcott discovered the Burgess Shale fauna because his name ranks high in the annals of North American geology, particularly in studies of the Cambrian Period and its life. Adam Sedgwick named the Cambrian, but it was Walcott who really "put the Cambrian on the map" through his many published reports on fauna and correlations.

THE EDIACARIAN SYSTEM

Ediacara Fauna

The rare and unusual but important soft-bodied type of preservation of the Burgess Shale fauna presents an alluring challenge—if there is one such site, there must be others waiting to be discovered. Indeed, a significant soft-bodied animal fauna was discovered by paleontologist R. C. Sprigg in 1947 in the Ediacara Hills of South Australia (Fig. 10-3). Here were found circular impressions resembling jellyfish preserved in strata of the Pound Quartzite, a stratigraphic unit of presumed Cambrian age. Somewhat later, private collectors found more jellyfishlike imprints and fossils of wormlike forms and traces, as well as some fossils bearing no resemblance to any kind of known organism, fossil or living. These finds stirred up considerable interest and prompted the South Australian Museum and the University of Adelaide, under the direction of paleontologist Martin Glaessner, to undertake a joint-effort investigation of the area. Reexamination of the geology showed that the fossil-bearing strata are nearly 200 meters below the oldest definitive Cambrian shelly fossils in the succession and are separated from them by several unconformities. Also, preliminary analysis of the soft-bodied fauna indicated that none of these fossils occur with known Cambrian fossils. These observations conclusively demonstrated a Late Proterozoic age for the Ediacara fauna. Regional correlation with rocks bracketed by radiometric dates indicated that the sediments of the Pound Quartzite were deposited about 650 to 570 million years ago.

The Late Proterozoic age determination had important ramifications for what the pre-Cambrian–Cambrian boundary signified in terms of the history of life. Previously, the apparently abrupt termination of the fossil record below this boundary was considered to be a par-

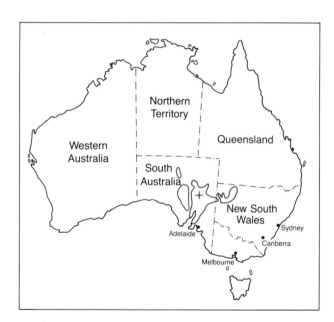

Figure 10-3. Area of exposure of Ediacarian rocks of the Adelaide Basin, Australia, and location of the Ediacara Hills, the type locality of the Ediacara fauna.
(From M. F. Glaessner, 1961, Pre-Cambrian Animals, p. 63, *in* L. F. LaPorte, ed., *The Fossil Record and Evolution: Readings from Scientific American,* W. H. Freeman, San Francisco; and M. Wade, 1968, Preservation of Soft-bodied Animals in Precambrian Sandstones at Ediacara, South Australia: *Lethaia,* vol. 1)

adox of distinct importance. Why did the first (earliest) occurrence of multicellular life appear so abruptly and include such diverse and complex forms? Through the years many different explanations for this mystery had been advanced. Darwin himself felt obliged to search for an answer because such a nongradualistic pattern was not at all compatible with his theory of evolution. Discovery of the Ediacara fauna was the beginning of laying to rest the speculations about what kind of life existed prior to the Cambrian. It also was the beginning of recognition of a distinct phase in the history of metazoan life earlier than the Cambrian.

From the more than 1500 specimens that have been collected at the type locality, about 30 species in 20 genera have been described. Two-thirds of the specimens have been classified tentatively as coelenterates, about one-quarter as annelid worms, and about one-twentieth as arthropods. According to Glaessner, the coelenterates include abundant and moderately diverse medusoid jellyfishlike forms, as well as leaflike or frondlike stalked forms similar to modern-day sea pens (Fig. 10-4A, B). The annelids are represented by small wormlike forms; some had flexible, segmented bodies, paired appendages, and head shields, as displayed by the genus *Spriggina* (Fig. 10-4C), while others, like *Dickinsonia,* were bilaterally symmetrical, elliptical forms covered with distinctly patterned ridges and grooves (Fig. 10-4D). Primitive arthropodlike animals are represented by *Parvan-*

corina (Fig. 10-4E), which has a kite-shaped body form with a midridge and hints of legs or gills. The strange-looking disc-shaped *Tribrachidium,* possibly a primitive echinoderm, has three raised hooked and tentacle-fringed arms radiating at equal angles (Fig. 10-4F); nothing like it has ever been described among the approximately 2 million species of extinct and living organisms. A variety of feeding trails possibly made by wormlike creatures is also present.

Most of the specimens are preserved as casts or molds on the bottom surfaces of sandstone beds—a seemingly more difficult way to preserve soft-bodied structures than by carbonaceous films as in the Burgess Shale. Many of the sandstone layers are ripple marked and cross-bedded, indicating fluctuating currents and shifting substrate—hardly the kind of depositional conditions conducive to preservation of soft-bodied creatures. However, between many of the sandstone layers are seams and stringers of mudstone. As interpreted by Glaessner, the animals themselves probably lived on or were stranded on mud flats or mud patches that formed between megaripples and sand ridges during quiet-water periods; the soft animal bodies were molded or their impressions cast on the bottom surfaces of layers of shifting sand that washed across the mud. The in-life setting was probably a shallow, nearshore marine environment, perhaps a tidal flat and lagoon. According to Glaessner, some of the animals, such as the frondform,

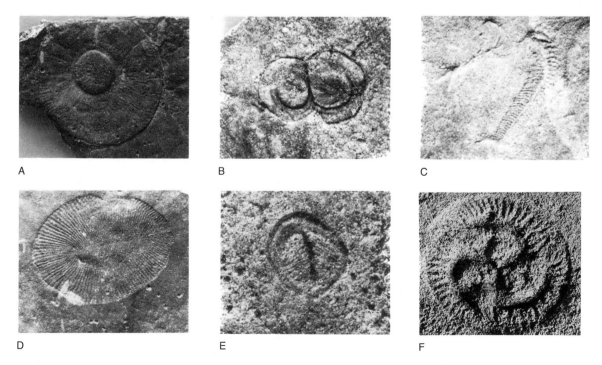

Figure 10-4. Ediacarian fossils from the Pound Quartzite, South Australia. Primitive coelenterates: A. *Charodiscus;* B. *Cyclomedusa.* Primitive annelid flatworms: C. *Spriggina floundersi;* D. *Dickinsonia costata.* Primitive arthropod(?): E. *Parvancorina minchami.* Primitive echinoderm(?): F. *Tribrachidium.* (Photos courtesy of M. F. Glaessner)

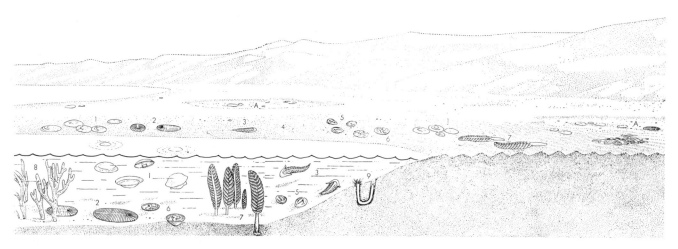

Figure 10-5. Interpreted in-life setting of Ediacara fauna. Some creatures shown stranded in small, muddy tidal ponds (A) and on sand flats. Others are shown below the water line. The scene includes jellyfish coelenterates (1); flatworms like *Dickinsonia* (2); segmented worms like *Spriggina floundersi* (3), as well as worm(?) traces (4); the primitive arthropod(?) *Parvancorina* (5); *Tribrachidium,* a possible primitive echinoderm(?) (6); *Rangea* and *Charnia* (sea-penlike coelenterates) (7); hypothetical algae and sponges (8); and a worm in a U-shaped burrow in sandy substrate (9). (From Martin F., Glaessner, 1961, Precambrian Animals, pp. 72–73: *Scientific American* offprint no. 837. All Rights Reserved. Reprinted by permission of W.H. Freeman and Co., Publishers, San Francisco)

sea penlike *Rangea,* lived attached to the sea bottom; some, such as the annelids, crawled along the bottom or burrowed into the substrate; and others, like the medusoid coelenterates, were free-floating or swimming (Fig. 10-5).

Recently, Adolf Seilacher of West Germany, one of the world's leading authorities on trace fossils and animal–sediment relationships, has argued that interpretation of Ediacara-like fossils in terms of modern soft-bodied taxa must be reconsidered. Seilacher questions the jellyfish nature of the medusoidlike fossils, preferring instead to relate them to a heterogeneous group of trace fossils and discoid benthic organisms. He attributes impressions of supposed sea pens and annelid worms to unknown benthic organisms with a quasi-autotrophic mode of nutrition. Seilacher believes the Ediacara fauna represents a distinct low-diversity episode in the evolution of multicellular organisms and not the beginning of metazoan radiation.

Even if the Ediacarafauna represents an evolutionary dead end, these soft-bodied animals formed a part of the earliest known metazoan community. Although not as diverse or spectacular as the Burgess Shale fauna, the Ediacara animals nevertheless occupy a unique position in the evolutionary history of the Metazoa.

Basis for A New Period

Ediacarian fossils have been recognized in southwestern Africa, England, Scandinavia, northern and southwestern Russia and Siberia, China, and North America. The southwestern Africa fauna was known be-

fore the type Ediacara fauna was discovered, but its pre-Cambrian age was not appreciated until later regional correlations were established. The actual number of Ediacara faunal localities is small, but in all these places the stratigraphic position beneath the lowest known Cambrian has been demonstrated.

The widespread Ediacara fauna characterizes a distinct interval of pre-Cambrian history. If Seilacher is correct, the fauna is totally unique in composition. Preston Cloud and Glaessner have proposed the **Ediacarian** Period for the geologic time interval beginning with the earliest appearance of soft-bodied metazoans and ending with the earliest appearance of skeletonized metazoans. All rocks formed within this period of geologic history constitute the Ediacarian System. These rocks lie between the uppermost Proterozoic tillites and the stratigraphically lowest occurrence of shelly fossils that mark the base of the Cambrian System (Fig. 10-6). The geochronologic data derived from the radiometric dates on glauconite from Ediacarian strata and on igneous rocks that bracket the Ediacarian succession in different parts of the world imply a range in time from about 670 million years ago at the base to about 570 million years ago at the top. The proposed stratotype of the Ediacarian System is about 380 km north of Adelaide, South Australia.

The Ediacarian partly fills the hiatus that many early stratigraphers long felt existed at the supposed worldwide unconformity (Fig. 10-7) between the pre-Cambrian and Cambrian. This hiatus, called the Lipalian Interval by Walcott, provided an explanation for the abundant shelly animal megafossils in Cambrian rocks

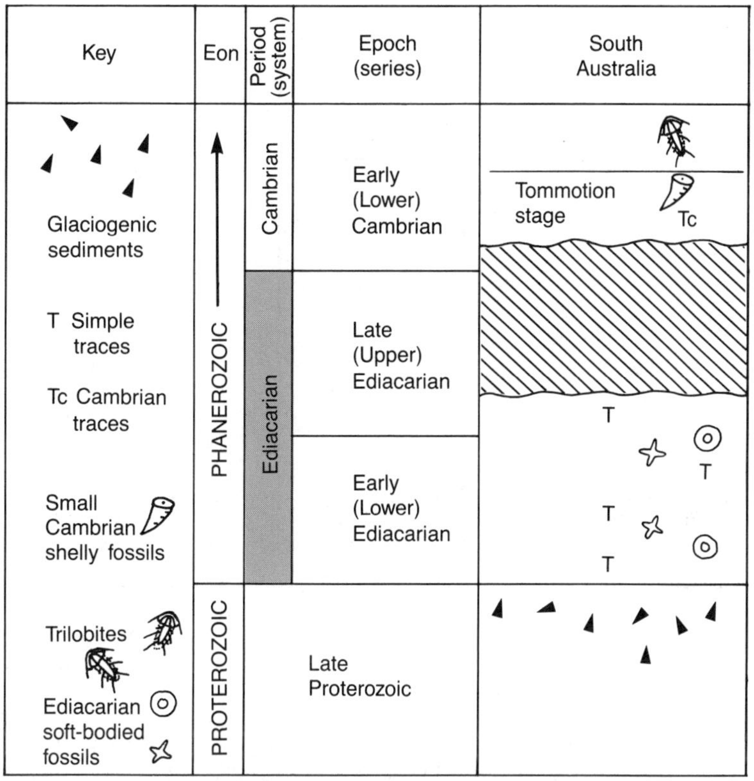

Key	Eon	Period (system)	Epoch (series)	South Australia
Glaciogenic sediments	PHANEROZOIC	Cambrian	Early (Lower) Cambrian	Tommotion stage Tc
T Simple traces		Ediacarian	Late (Upper) Ediacarian	(hiatus)
Tc Cambrian traces				T
Small Cambrian shelly fossils			Early (Lower) Ediacarian	T T T T
Trilobites	PROTEROZOIC		Late Proterozoic	
Ediacarian soft-bodied fossils				

Figure 10-6. Stratigraphic relationship between the type Ediacarian, underlying Proterozoic rocks, and overlying Cambrian rocks, South Australia. Also shown are the occurrences of tillites, soft-bodied metazoans, metazoan traces, and the lowest appearance of skeletal fossils, defining the base of the Cambrian. Diagonal lines represent hiatuses.
(Data from M.D. Brasier, 1979, The Cambrian Radiation Event, *in* M. R. House, ed., *The Origin of Major Invertebrate Groups:* Academic Press, London; P.E. Cloud, Jr., and M.F. Glaessner, 1982, The Ediacarian Period and System: Metazoa Inherit the Earth: *Science,* vol. 217, no. 4562)

Figure 10-7. Nonconformity between Upper Cambrian sandstone and pre-Cambrian crystalline rocks, Colorado River, near Glenwood Springs, Colorado
(Photo by J. Cooper)

(Fig. 10-8) immediately above pre-Cambrian igneous and metamorphic rocks. Walcott and others were impressed by the seemingly abrupt and mysterious appearance of rather advanced fossil animals in the geologic record. One explanation for this paradox was that the first metazoans originated and underwent a long period of evolution in ocean basin areas where sediments accumulated that did not become accessible for examination. Much later, according to this idea, as seas encroached onto the continental platform interiors during the Cambrian, organisms migrated into environments that produced sedimentary rocks that are accessible. During the twentieth century, and particularly during the last several decades, increased knowledge of pre-Cambrian rocks and the magnitude of time involved has stimulated a search for the biological conditions that existed prior to the explosive appearance of complex shell-bearing organisms. According to Cloud and Glaessner, this search has revealed the existence of some well preserved sedimentary successions that extend beneath the base of the Cambrian without apparent major interruption. Evidence in these sub-Cambrian rocks of marine origin fueled the expectation that if there were an-

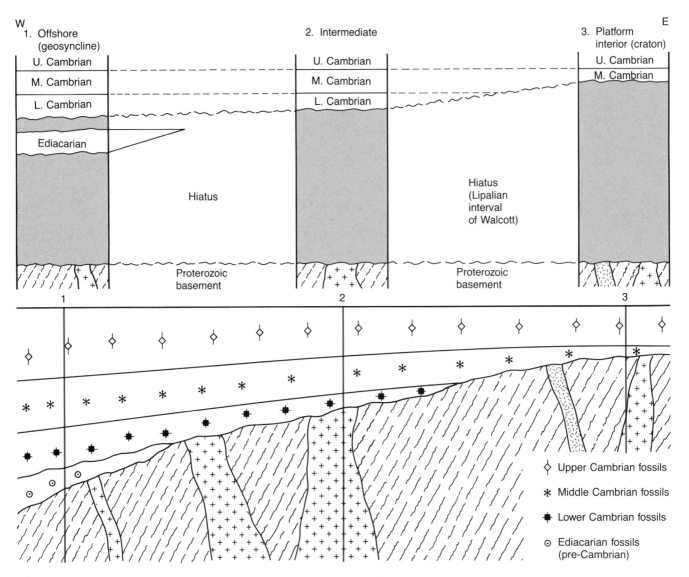

Figure 10-8. Lipalian interval of Walcott. A. Time-stratigraphic relationship (magnitude of hiatus). Hiatus at nonconformity in platform interior (craton) of continent is partly filled by fossiliferous older Cambrian as well as by strata of Ediacarian and Late Proterozoic age in stratigraphic sections representing offshore (geosynclinal) areas once believed to be inaccessible. B. Natural onlap stratigraphic relationship.

tecedents to the Cambrian biota, they would be found in these rocks. The Ediacara fauna represents the realization of that expectation and gives us a critical look at the earliest known metazoan life forms and some of the traces they made. However, it should be remembered that there is great difficulty in establishing the synchroncity of both the lower and upper boundaries of the Ediacarian. By definition, a time-stratigraphic unit must have isochronous boundaries.

The Ediacarian World

How long before 670 million years ago (age of the oldest known metazoan body fossils) did the Metazoa originate? The length of prior history of metazoan evolution is unknown. The earliest metazoans must have been the multicellular descendants of protists, but no fossil record of this evolutionary transition has been found. However, if late Proterozoic metazoans were present, we should see some evidence of their existence in the form of trace fossils. The oldest definitive animal traces occur in association with documented Ediacara faunas. Some pre-Edicarian structures described as traces have been reported, but their metazoan, not to mention organic, affinity has not been proved. Trace fossils diversity across the Proterozoic–Cambrian transition shows a trend parallel to the increase in diversity of body fossils (Fig. 10-9). Thus trace fossils provide a good clue to metazoan activity and relative diversity.

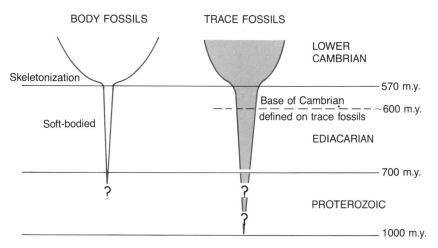

Figure 10-9. Diversity trends in trace fossils and body fossils from Proterozoic to Early Cambrian.
(Data from S. M. Stanley, 1976, Fossil Data and the Precambrian-Cambrian Evolutionary Transition: *Am. Jour. Sci.,* vol. 200; S. P. Alpert, 1977, Trace Fossils and the Basal Cambrian Boundary, *in* T. P. Crimes and J. C. Harper, eds., *Trace Fossils,* vol. 2: Seel House, Liverpool; M.D. Brasier, 1979, The Cambrian Radiation Event, *in* M. R. House, ed., *The Origin of Major Invertebrate Groups:* Academic Press, London)

The prime prerequisites for attaining a metazoan grade of evolution were probably oxygen level in the atmosphere and sexual reproduction. The time constraints on these prerequisites are not well known, but certainly sexual reproduction had developed by 1000 million years ago (see discussion, Chapter 9). Necessary oxygen levels of 5 percent present atmosphere level (PAL) would have been attained any time between about 1500 and 670 million years ago. However, even if metazoans did not evolve until sometime between about 700 to 670 million years ago, this would be consistent with the punctuated equilibrium evolutionary model; a long history of undocumented metazoan evolution is not required for the level of diversity we see in the Ediacaria fauna.

Paleontologic and paleomagnetic evidence suggests that during Ediacarian time two main clusters of continents existed: proto-Gondwana (Fig. 10-10) and proto-Laurasia. The classic benthic Ediacara fauna appears to have been centered around the margins of proto-Gondwana. Medusoid and frondlike Ediacara fossils have a more cosmopolitan distribution. As mentioned earlier, Ediacara faunas, wherever found, occur above the youngest Proterozoic tillites (Fig. 10-6). General similarity of Ediacara faunal associations all over the world probably reflects the absence of major ecological barriers on proto-Gondwana continental shelves during extensive postglacial Ediacarian transgression.

Ediacarian of North America

In North America, Cambrian strata generally rest noncomformably on pre-Cambrian crystalline basement rocks; however, in some places along the margins of the

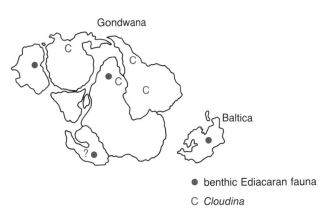

Figure 10-10. Continental elements of proto-Gondwana showing occurrences of classic benthic Ediacarian assemblages and *Cloudina,* a slightly skeletonized tubular fossil of Ediacarian age.
(From M.A.S., McMenamin, 1982, A Case for Two Late Proterozoic–Earliest Cambrian Faunal Province Loci, Fig. 1, p. 291: *Geology,* vol. 10. Reprinted by permission of Geological Society of America and author)

continent, Cambrian rocks appear to be essentially conformable with underlying pre-Cambrian strata. Documented accounts of Ediacara fossils are rare but do include specimens from stratigraphic successions in Newfoundland and North Carolina. In a number of places, strata of Ediacarian age are presumed to be present on the basis of stratigraphic position and apparent conformity with overlying definitive Cambrian rocks. Rocks filling the hiatuses of a pre-Cambrian–Cambrian unconformity are believed to be the deposits of rifted continental margins (Fig. 10-11). These margins evolved into major subsiding depositional basins or troughs called *geosynclines.* Two major geosynclines, the *Appa-*

lachian and *Cordilleran* (Fig. 10-12), rimmed the more tectonically stable interior platform region or nucleus of the continent—the **craton**—during late Proterozoic and early Paleozoic. Throughout their history the geosynclines were active loci of accumulation of sedimentary rocks and were sites of active mountain building. The geosyncline concept will be discussed more fully in Chapter 11.

Good exposures of the upper Proterozoic–Ediacarian–Cambrian transition stratigraphic interval occur in geosynclinal successions in the Appalachian Mountain belt, the Canadian Cordillera, east-central Canada and bordering Alaska, and the southwestern Great Basin of eastern California and southern Nevada. The last region, particularly in the southern Nopah and White-Inyo Mountain ranges, displays an apparently continuous section composed of sandstones, siltstones, and shales, with minor conglomerates and interbedded limestones and dolomites (Fig. 10-13). Here the base of the Cambrian is defined on the lowest occurrence of skeletonized body fossils, although several kinds of trace fossils occur lower in the section. No Ediacara fossils have been found, but the conformable stratigraphic relations suggest that strata of Ediacarian age most likely are present.

These strata also lie above upper Proterozoic rocks such as the Pocatello Formation of southeastern Idaho and the Kingston Peak Formation of southeastern California, both of which contain sediments of glacial origin. Both the Nopah and White-Inyo Mountains sections have been proposed by Cloud as good candidates for a North American stratotype for the Proterozoic–Phanerozoic boundary (Fig. 10-14).

EARLIEST CAMBRIAN HISTORY
The Cambrian Radiation Event

The Proterozoic to Lower Cambrian sedimentary succession has been studied intensively in recent years in the quest for data on which a clearer definition of the base of the Cambrian might be developed. This work, particularly in the Russian platform region, has shown clearly that the Ediacara fauna of soft-bodied animals disappeared from or became very scarce in the preserved record after Ediacarian time. The Ediacara forms were succeeded by a fauna of small shelly fossils representing organisms *not* descended from members of the Ediacara assemblage.

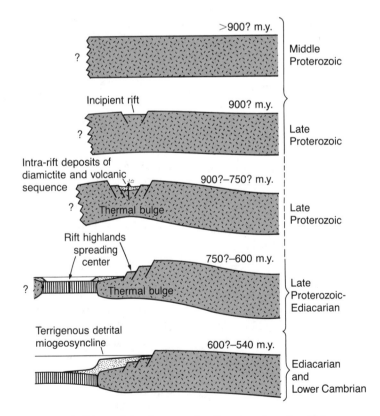

Figure 10-11. Late Proterozoic continental rifting and development of sedimentary deposits of Late Proterozoic, Ediacarian, and Early Cambrian ages along a rifted continental margin—the Cordilleran geosyncline.
(After J. H. Stewart, 1982, Regional Relations of Proterozoic Z and Lower Cambrian Rocks in the Western United States and Northern Mexico, Fig. 11, p. 183: Geol. Soc. America Cordilleran Section Guidebook)

The appearance of exoskeletons represented an evolutionary milestone in overall general metazoan diversification. According to paleontologist Steven Stanley at Johns Hopkins University, the apparent sudden appearance of skeletal faunas is not so sudden—as commonly portrayed—and is not mysterious—as once believed—when viewed in light of the stratigraphic occurrences. Major skeletonized faunas appeared sequentially, not simultaneously, over an interval of perhaps 20 to 30 million years. In this light the so-called pre-Cambrian–Cambrian boundary presents a rapid transition more than a distinct faunal discontinuity. The **Tommotian Stage** (Fig. 10-6) includes the earliest known shelly faunas: small primitive molluscs, brachiopods, archaeocyathids, arthropods, echinoderms, and sponges as well as tubular and conical fossils of uncertain systematic position (Fig. 10-15). This fauna predates the first appearance of trilobites, which historically has been used as the main criterion to draw the base of the Cambrian. The discovery of pre-trilobite skeletal faunas in a number of places around the world not only underscored the problem of recognizing a synchronous base of the Cambrian, it added a new perspective to the transition from nonskeletal to trilobite-bearing skeletal faunas.

The Tommotian shelly fauna is significantly more diverse than the Ediacarian soft-bodied fauna. The diversity of trace fossils shows a similar pattern. Known Ediacarian trace fossils, probably of detritus feeders, are mostly simple, two-dimensional shallow burrows and relatively poorly oriented horizontal search trails on bedding surfaces. In contrast, Early Cambrian trace fossils are more complex three-dimensional forms and include diverse deep, vertical burrows, complex burrows, and a variety of trackways.

The earliest representatives of the Tommotian fauna were small, cone-shaped shelly fossils confined to the proto-Laurasia continental assembly (Fig. 10-16). Before the end of the Tommotian, there was an explosive increase in the diversity of skeletonized organisms, with

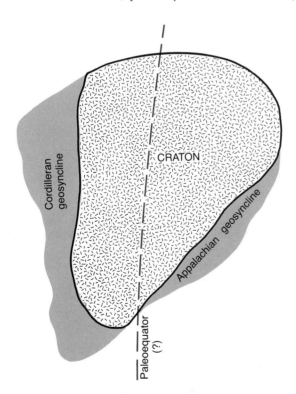

Figure 10-12. Proto-North America during Ediacarian time showing relationship between exposed craton of Cryptozoic basement flanked by newly developed geosynclines.

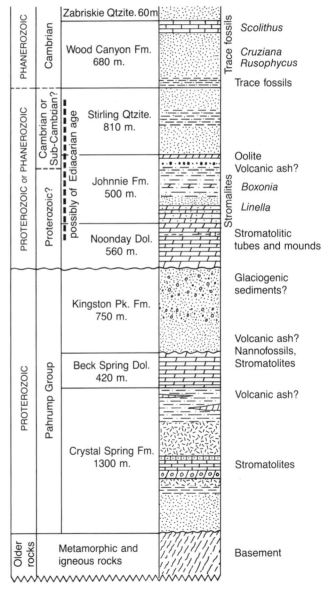

Figure 10-13. Composite stratigraphic column for Proterozoic, Ediacarian(?) and Lower Cambrian, southern Death Valley region, Nopah Range, California. (Modified slightly from P. E. Cloud, 1973, Possible Stratotype Sequences for the Basal Paleozoic of North America, Fig. 3, p. 201: *Am. Jour. Science*, vol. 273. Reprinted by permission of *American Journal of Science* and author)

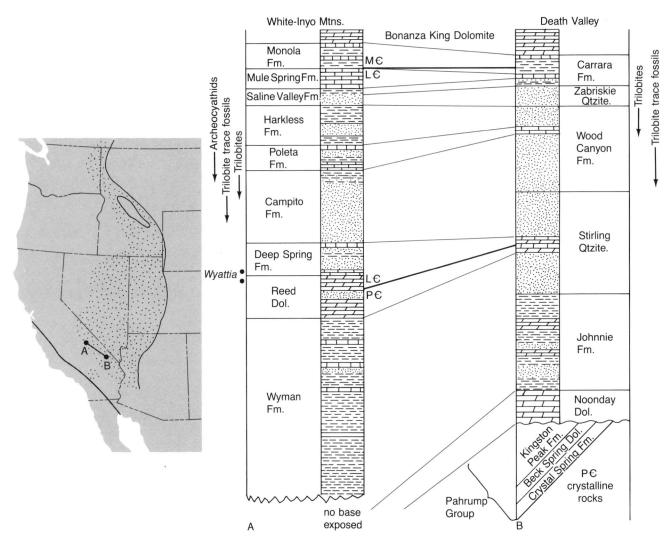

Figure 10-14. Upper pre-Cambrian and lower Cambrian strata from White-Inyo Mountains to Death Valley region. A. Thickness in kilometers of upper pre-Cambrian (Proterozoic Z of U.S. Geol. Survey scheme) and Lower Cambrian strata. Line of section for B is shown. B. Correlation between the White-Inyo Mountains and Death Valley. *Wyattia* is an enigmatic shelly fossil.
(B. From C. A. Nelson, 1976, Late Precambrian–Early Cambrian stratigraphic and Faunal Succession of Eastern California and the Precambrian-Cambrian Boundary, Fig. 4, p. 42: Pacific Section SEPM Paleogeography Symposium, Fieldguide, vol. 1. Reprinted by permission of Pacific Section, Society of Economic Paleontologists and Fieldguide, Mineralogists)

more-cosmopolitan faunas radiating from the proto-Laurasian center to invade proto-Gondwana. The oldest Tommotian fossils are believed to be about 570 million years old; rapid diversification occurred perhaps 560 million years ago. By about 550 million years ago the Tommotian ended, and diverse trilobite and archaeocyathid faunas existed in the Early Cambrian seas of nearly every continent.

Rapid skeletonization was an integral part of metazoan diversification and illustrates a pattern that is consistent with the punctuated equilibrium model of evolution discussed in Chapter 7. The metazoan radiation from Tommotian into later Early Cambrian probably was related to rising atmospheric oxygen levels, as well as to continental fragmentation and flooding of continental shelves by transgressing seas. These changes created a greater diversity of shallow marine habitats and niches. Also, the evolutionary emergence of herbivores, which resulted in extensive cropping of cyanobacteria mats and eukaryotic algae, opened up new niches for a variety of other organisms. Skeletonization became selected for as free oxygen levels continued to increase (perhaps to 10 percent PAL) and more advanced organic respiratory and circulatory systems evolved. Exoskeletons of calcareous, phosphatic, and chitinous composition would have provided an adaptive advantage for bottom

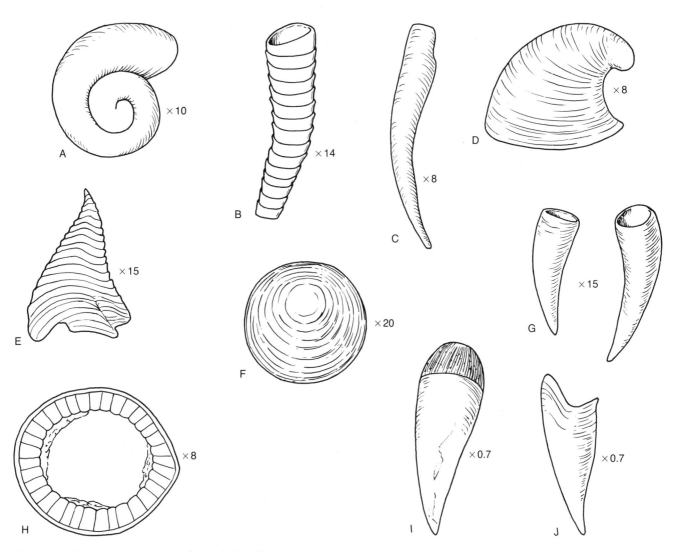

Figure 10-15. Enigmatic earliest shelly fossils from Tommotian strata, U.S.S.R. A–I. Mollusclike calcareous cone and coiled shells. J. Transverse section of archaeocyathid showing double-walled structure. (From M. E. Raaben, 1981, The Tommotian Stage and the Lower Cambrian Boundary (trans. from Russian): U.S. Dept. Interior and National Science Foundation, Washington)

living, leverage, articulation, protection, security, and accommodation for size increase. The world was changing. Attainment of critical oxygen levels, sexual reproduction, algae cropping, and breakup of the proto-Laurasia continental assembly all combined to foster early metazoan diversification.

LATER CAMBRIAN HISTORY

Sedimentation Patterns and Paleogeography

The newly formed Cordilleran and Appalachian geosynclines received predominantly terrigenous clastic sediments derived from erosion of exposed Cryptozoic crystalline rocks in the interior part of the craton (Fig. 10-17). As seas slowly spread out from the geosynclinal margins, the craton, which had been subject to a prolonged interval of weathering and erosion, was gradually inundated by shallow marine waters. This major transgression and flooding of the continental interior platform by shallow seas, called **epeiric seas,** were nearly complete by Late Cambrian time except for a few emergent island areas that are collectively referred to as the **transcontinental arch** (Fig. 10-18).

The innermost belts of the marginal geosynclines are called **miogeosynclines.** Sedimentary rock types from North American miogeosynclines (Fig. 10-19) are essentially identical in composition and texture to those deposited in the shallow epeiric seas of the cratonic interior. The main difference in these two suites of strata is the stratigraphic thickness, which is greater in the miogeosynclinal sections because of more continuous subsidence. Sedimentation kept pace with subsidence and shallow-marine environments were predominant.

Figure 10-19 shows that transgressing seas spread nearly pure quartz sand, derived from the craton, as a widespread blanket that ranges in age from Early Cambrian in the geosynclinal regions to Late Cambrian in the cratonic interior. A more seaward belt of finer detri-

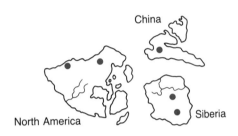

Proto-Laurasia

● *Protohertzina* and/or
Anabarites

Figure 10-16. Continental elements of proto-Laurasia showing occurrences of shelly *Tommotian* (basal Cambrian) faunas characterized by genera *Protohertzina* and/or *Anabarites*.
(From M. A. S. McMenamin, 1982, A Case for Two Late Proterozoic–Earliest Cambrian Faunal Province Loci, Fig. 1, p. 291: *Geology*, vol. 10. Reprinted by permission of Geological Society of America and author)

tal sediments including shale and siltstone accumulated in subtidal environments. During the Late Cambrian, after most of the Cryptozoic basement-rock source areas had been drowned by shallow seas in one of the highest sea-level stands of the entire Phanerozoic, carbonate sedimentation predominated in the geosynclinal and cratonal regions; and there was progressive overlap of successive Lower, Middle, and Upper Cambrian deposits. In the miogeosynclinal belt, carbonate sedimentation was established earlier because the carbonate-rich seas were further removed from the emergent parts of the craton and calcium carbonate could be deposited without being overwhelmed by terrigenous clastic detritus.

Miogeosynclinal and cratonal sedimentary rocks show abundant evidence of shallow-water deposition. Ripple marks and cross-stratification are common sedimentary structures (Fig. 10-20A, B). Limestones contain such shallow-water features as stromatolites and oncolites, oolitic beds and lenses, bioclastic layers, and intraformational limestone-pebble conglomerates (Fig. 10-20C–E). Abundant and laterally extensive intraformational conglomerates in Cambrian subtidal shelf sequences reflect frequent storms that shredded cohesive thin carbonate-mud sea-bottom layers into tabular clasts that were quickly redeposited, commonly in edgewide orientations (Fig. 10-20F, G). Limestone and dolomite beds commonly display various kinds of desiccation features (Fig. 10-20H). Trace fossils are present as feeding burrows (Fig. 10-20I). The shallow epeiric seas that invaded the craton most likely ranged in depth from shoal waters to several tens of meters, and large areas of tidal flats were commonplace.

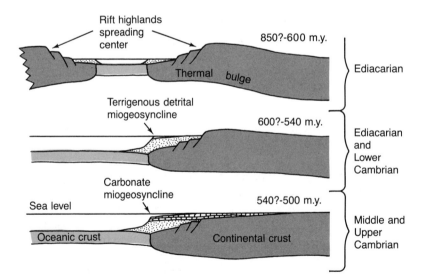

Figure 10-17. Evolution of Cordilleran margin of North America from a late pre-Cambrian–Early Cambrian terrigenous clastic miogeosyncline to a mid-Late Cambrian carbonate miogeosyncline as terrigenous source lands were denuded and progressively drowned by shallow seas advancing from the geosyncline into the craton.
(After J. H. Stewart, 1982, Regional Relations of Proterozoic Z and Lower Cambrian Rocks in the Western United States and Northern Mexico, Fig. 11, p. 183: Geol. Soc. American Cordilleran Section Guidebook)

An illustration of the rock record and depositional patterns produced as the result of gradual encroachment of Cambrian seas is shown in Figures 10-21 and 10-22. Ediacarian and Lower Cambrian rocks of the Cordilleran miogeosyncline progressively thin and pinch out toward the craton and are overlapped by Middle and Upper Cambrian deposits. Note also the drastic change in thickness between the miogeosynclinal, platform-edge, and cratonic sedimentary successions. By the time we reach the Grand Canyon, relatively flat-lying sandstone of late Early Cambrian to Medial Cambrian age nonconformably overlies Cryptozoic basement in some places and in others overlies tilted Upper Proterozoic sedimentary rocks of the Grand Canyon Supergroup.

In north-central Nevada, parts of California, and the Pacific Northwest, there are some very different-looking Cambrian rocks that include graded sandstone and impure limestone beds, argillite, chert, and greenstone, as well as deposits bearing slump features and transported blocks. These rocks are believed to represent deeper marine slope and basin deposits and are related to an outer, deeper water belt—the **eugeosyncline** (Fig. 10-23)—within the geosyncline.

A Facies Model: Cambrian of the Grand Canyon

The Cambrian rocks of the Grand Canyon provide an instructive model for appreciating sedimentation patterns in the advancing sea and for illustrating Walther's Law of the correlation of facies (Chapter 6). The section of rock includes three major lithofacies, which correspond generally to the formally named Tapeats Sandstone, Bright Angel Shale, and Muav Limestone (Fig. 10-24). The Tapeats Sandstone represents the basal transgressive shoreface deposits of beach, bar, and sand-

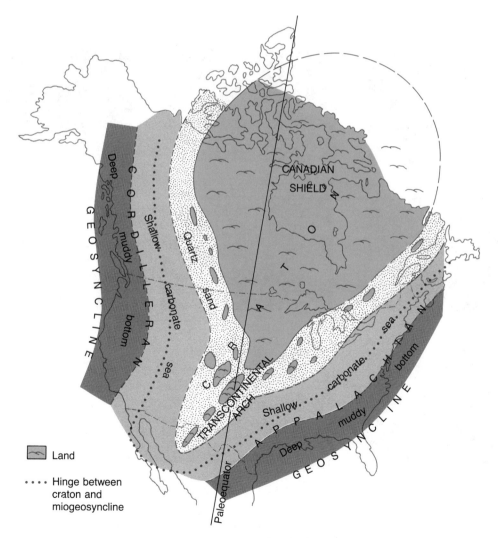

Figure 10-18. Paleogeographic map of North American continent during the Late Cambrian. (Data from C. Lochman-Balk, 1971, Cambrian System, Fig. 14, p. 74, *in Geologic Atlas of the Rocky Mountain Region:* Rocky Mountain Association of Geologists, Denver)

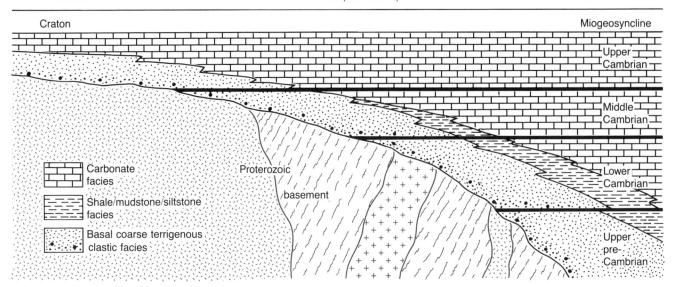

Figure 10–19. Three main Cambrian depositional facies from Appalachian geosyncline to cratonic interior. Time-transgressive nature and progressive development of carbonate facies are shown as terrigenous source areas were gradually drowned by advancing Late Cambrian seas. High vertical exaggeration and not to scale.

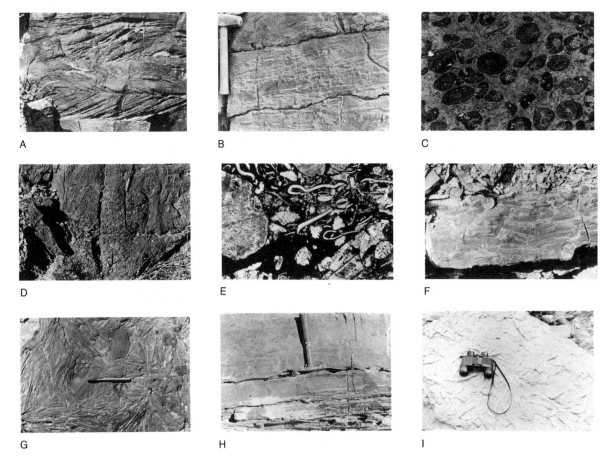

Figure 10-20. Shallow-water features from Cambrian sedimentary rocks. A. Tidal herringbone cross-bedding. B. Cross-bedded oolite, C. Oncolites (shallow, subtidal rolled stromatolites. D. Desiccated laminated stromatolites. E. Domal intertidal stromatolites. F. Loaf-shaped thrombolite mound (nonlaminated stromatolite). G. Tabular intraclasts in skeletal calcarenite matrix. H. Edgewise intraformational limestone-pebble conglomerate. I. Trace fossils (feeding burrows).
(Photos by J. Cooper)

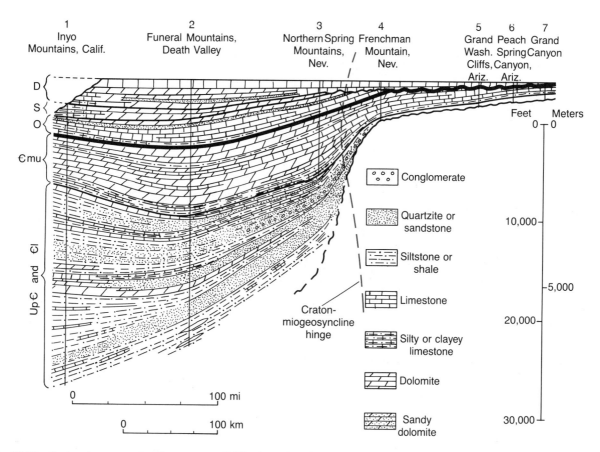

Figure 10-21. Restored cross section from eastern California to Grand Canyon showing transition from miogeosyncline to craton margin to western craton.
(After J. H. Stewart and F. G. Poole, 1974, Lower Paleozoic and Uppermost Precambrian Cordilleran Miogeocline, Great Basin, Western United States, Fig. 6, p. 37: *SEPM Spec. Pub.* No. 22. Reprinted by permission of Society of Economic Paleontologists and Mineralogists)

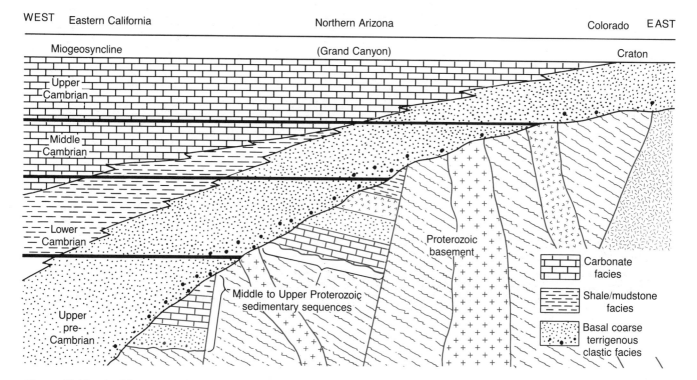

Figure 10-22. Cordilleran geosyncline to cratonic interior, highlighting time-transgressive nature of depositional facies, the product of gradually advancing marine encroachment into the craton during the Cambrian. High vertical exaggeration and not to scale.

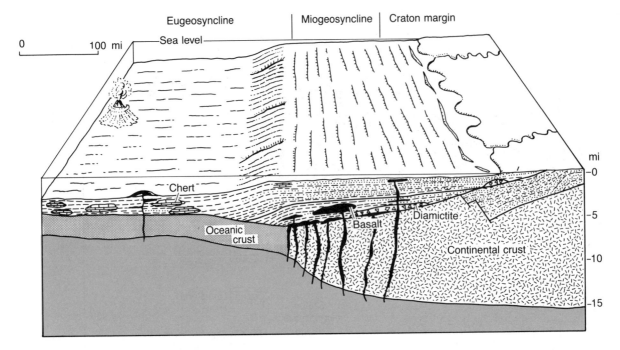

Figure 10-23. Depositional settings of the Cordilleran margin for upper Proterozoic and Lower Cambrian rocks in the northern Great Basin, Nevada and Utah.
(From J. H. Stewart, 1972, Initial Deposits in the Cordilleran Geosyncline: Evidence of a Late Precambrian (<850 m.y.) Continental Separation, Fig. 3, p. 1350: *Geol. Soc. America Bull.* vol. 83, no. 5. Reprinted by permission of Geological Society of America and author)

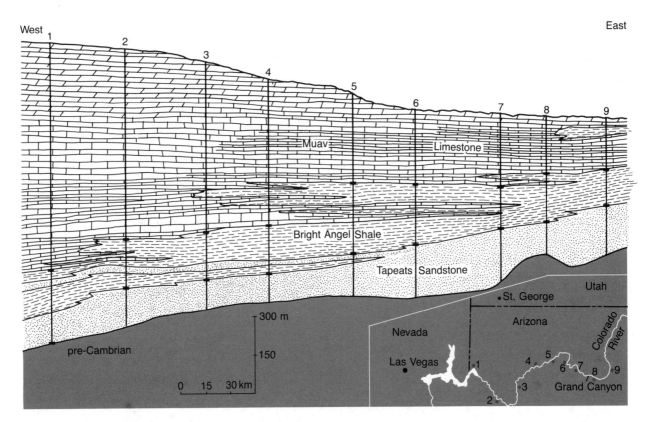

Figure 10-24. Stratigraphic cross section from Lake Mead to the eastern part of the Grand Canyon showing the classic Grand Canyon onlap cycle of Cambrian facies. Note the mappable boundaries in each section between the three formations and their relationships to the complex onlap-offlap facies boundaries. Dashed line is time line based on faunal horizon. High vertical exaggeration.
(From E. H. McKee, 1945, Cambrian History of the Grand Canyon Region, Part I. Stratigraphy and Ecology of the Grand Canyon Cambrian, Fig. 1, p. 14: *Carnigie Inst. Washington, Pub.* 563. Reprinted by permission of Carnegie Institution of Washington)

Figure 10-25. Cambrian section, Marble Mountains, eastern California. This sequence, farther geosynclineward than the Cambrian of the Grand Canyon, shows similar succession of major facies (sandstone–shale–carbonate) but is Early to Medial Cambrian in age. (Photo by J. Cooper)

flat environments—sediments that accumulated as marine waters slowly inundated the western margin of the craton during late Early and Medial Cambrian time. As transgression continued during the Medial Cambrian, terrigenous mud and silt of the Bright Angel Shale and carbonate sediments of the Muav Limestone were deposited over the Tapeats in an onlap succession. The vertical succession of sandstone–shale–limestone exposed in any one section comprises the well-known Grand Canyon cycle, which records the cratonward migration of progressively more seaward depositional environments through time. Figure 10-24 depicts how the three major lithofacies are intertongued; they represent the products of depositional environments that coexisted at any one time. Figure 10-25 shows similar facies further geosynclineward. The dynamics of cyclic sedimentation have been responsible for the intertonguing stratigraphic patterns that emerge in such a reconstruction. The formally named formations—superposed layer-cake units of convenience in geologic mapping and regionl correlation—are but local expressions of these intertonguing facies.

Paleogeography of the Cambrian World

Paleogeography, as discussed in Chapter 6, is a prime goal of historical geology. To know what the world looked like at various times in the past involves reconstruction of ancient geographies including land–sea relationships, character of land areas, location of mountain belts, and inferences about paleoclimatic patterns and oceanic circulation. The general reconstruction of Late Cambrian paleogeography of North America (Fig. 10-18) is made possible by careful geologic mapping and analysis of the rock and fossil record and interpretation of paleoenvironments of deposition and of unconformities. But what about the larger, global context of which ancient North America was only a part? What did the Cambrian *world* look like? Figure 10-26 suggests that it was a world strange and different from the present-day configuration of continents and ocean basins. What kinds of geologic evidence permit reconstructing various patterns of geographic change? Recall from the discussion in Chapter 6 that reconstructing global paleogeography involves (1) positioning the paleocontinents, (2) determining the geographic features and paleoclimates, (3) interpreting the paleoenvironmental conditions on each paleocontinent, and (4) depicting the above features within the limits of a time *age,* the corresponding time interval for a time-stratigraphic stage. The data bases for these determinations include paleomagnetic information; rocks and structures indicative of

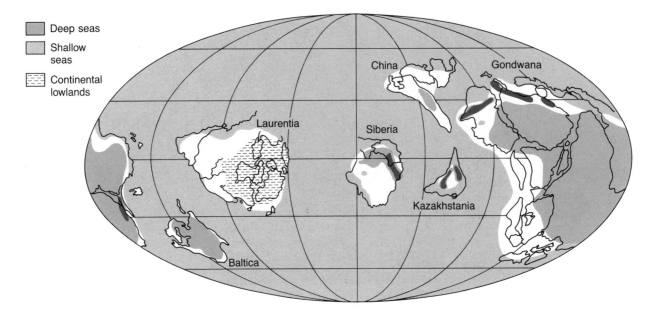

Figure 10-26. Paleogeography of the Late Cambrian world.
(From R. K. Bambach et al., 1980, Before Pangaea: The Geographies of the Paleozoic World, Fig. 5, p. 29: *Am. Scientist,* vol. 68. Reprinted by permission of *American Scientist*)

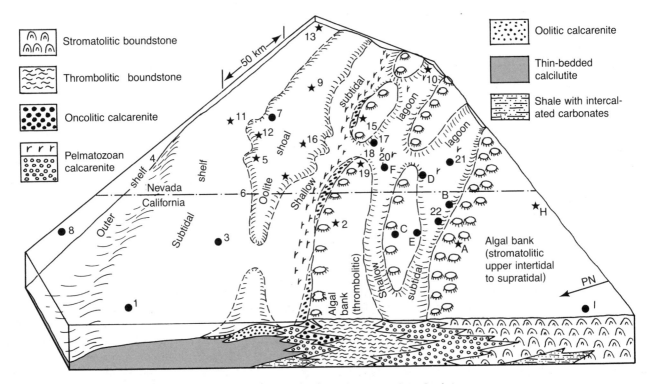

Figure 10-27. Relationship between lithofacies and interpreted paleoenvironments, Late Cambrian, southern Nevada and eastern California. Numbers and letters represent stratigraphic control sections restored to interpreted original Late Cambrian geographic positions. Paleonorth is shown.
(From J. D. Cooper et al., 1982, Late Cambrian Paleogeography of Southeastern California and Southern Nevada, Fig. 17, p. 112: Geol. Soc. America Cordilleran Section Guidebook)

tectonic belts; and climatically, environmentally, and temporally sensitive rock types and fossils. Let us see how these apply to our Cambrian story.

According to paleogeographers C. R. Scotese, A. M. Ziegler, R. K. Bambach, and J. T. Parrish at the University of Chicago, six major continental blocks existed during the Cambrian (Fig. 10-26). Much of what is now North America and Greenland, together with parts of the British Isles and western U.S.S.R., comprised the ancient continent of **Laurentia.** Figure 10-26 shows Laurentia lying astride the equator and in an orientation roughly 90 degrees counterclockwise from the present-day orientation of North America. This determination is made possible by paleomagnetic evidence derived from Upper Cambrian rocks on the present-day continent. Nearly horizontal inclinations of magnetized particles tell the ancient latitude, and orientation of the particles permits positioning of Laurentia with respect to the magnetic poles. Paleomagnetic evidence suggests that all the paleocontinents except **Baltica** and China straddled the equator during the Late Cambrian and occupied much of the space available in the equatorial belt. Their relative order of longitudinal sequence has been determined by the distribution of faunal provinces. Although North America (Laurentia) cannot be related to an absolute longitude (relative to the prime meridian), the space constraints imposed by paleobiogeography and paleomagnetism fix longitudinal position within rather narrow limits.

Through the synthesis of many diverse kinds of data derived from different fields of geology and geophysics there has emerged a picture of a Late Cambrian world in which continents were dispersed around the globe in tropical latitudes; ocean basins were extensively interconnected; all the continents except **Gondwana** were awash with shallow seas; and the polar regions lay in open ocean. These paleogeographic features made Late Cambrian time climatologically unique. If the general structure of atmospheric circulation has not changed radically, prevailing winds and ocean currents would have been almost wholly parallel to latitude, especially above 50°–60° north and south. Recent heat-budget calculations suggest that the peculiar distribution of land and sea in the Late Cambrian meant that the amount of absorbed solar radiation was probably lower than for any other time during the Phanerozoic and that the Cambrian was a relatively cool period. This is in marked contrast to the modern world, in which the continents are grouped into three extensive north–south oriented masses that partly isolate three equatorially centered oceans; continents are largely emergent, with major mountain belts representing zones of earlier plate collisions; polar regions are either covered or surrounded by land masses; and climates are strongly zoned.

The distribution of Late Cambrian paleoenvironmental conditions in a small part of the western United States (northern Laurentia during Late Cambrian) is shown in Figure 10-27. The mosaic of depositional environments represents a synthesis of stratigraphic, paleontologic, and sedimentologic data and interpretations generated by study of the sedimentary facies within this time-stratigraphic interval.

Cambrian Life

The Cambrian is the lowermost subdivision of the geologic time scale having a well-preserved, abundant fossil record that can be recognized and correlated on a worldwide scale. Remember, it is the Cambrian and younger record—containing skeletonized faunas—whose time-stratigraphic subdivision is deeply rooted in the principle of fossil succession, and subdivision boundaries are defined by time-significant evolutionary changes. Although the base of the Cambrian is defined on the first appearance of skeletonized faunas, this criterion does not represent a synchronous event, as we have shown, and this boundary is not as accurate as those at the bases of succeeding stratigraphic systems. The base of each succeeding system, although recognized on first occurrences of particular faunas, is determined also on the comparison of fossil faunas with those of the preceding system.

By the end of the Cambrian, all living phyla—except perhaps the bryozoans—that are now well skeletonized had appeared. These phyla came into existence during the Ediacarian and Cambrian and, except for the sponges, which are believed to have descended independently from a protistan ancestor, probably represent evolution from four or five major ancestral worm groups (Fig. 10-28). The fossil record sheds little light on the origin of the phyla.

The Cambrian fossil record (Table 10-1) is dominated by the trilobites (Figs. 10-29, 30) which are an extinct class of the phylum Arthropoda. Trilobites were present early in the period with amazing diversity and include about 75 percent of known Cambrian life forms that secreted exoskeletons. In their initial adaptive radiation, the trilobites explored a variety of niches in the marine environment. Morphologic diversity is expressed by differences in size (most were only a few centimeters long, but some attained lengths of 20 cm), shape, and ornamentation, and by eyes; some trilobites had very advanced compound eyes and some were completely blind. Trilobites shed their exoskeletons much like their living cousins, shrimp and crayfish, and an individual had the potential for leaving not one but several fossils through a succession of molt stages.

Trilobite assemblages provide most of the data for biostratigraphic zonation and correlation of Cambrian rocks as well as information on Cambrian paleobiogeography. Cambrian trilobites in eastern North American belong to two distinct faunal provinces, the American

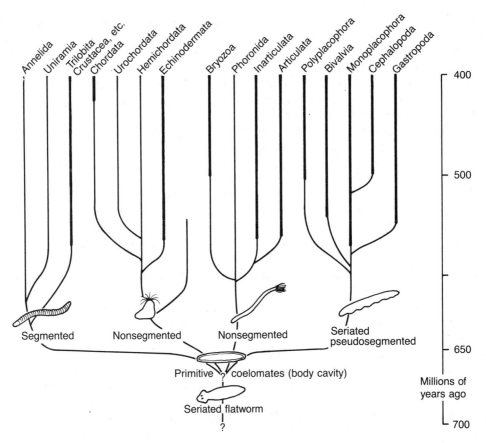

Figure 10-28. Evolution of the metazoan phyla. The wide bars indicate that the lineage of the phylum had acquired a mineralized skeleton.
(After J. W. Valentine, and Kathryn Campbell, 1975, Genetic Regulation and the Fossil Record; Fig. 3, p. 478: *Am. Scientist,* vol. 63. Reprinted by permission of *American Scientist*)

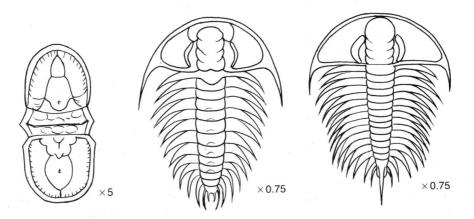

Figure 10-29. Representative Cambrian trilobites.
(From A. R. Palmer, 1974, In Search of the Cambrian World, Fig. 1, p. 217: *Am. Scientist,* vol. 62. Reprinted by permission of *American Scientist*)

and European (Fig. 10-31). Both provinces contain a diverse assemblage of trilobite genera but share few genera in common. These two distinctive faunal complexions suggest that some kind of barrier was present to prevent mixing. There is both geologic and paleontologic information that strongly suggests that Laurentia and Baltica were separated during Cambrian time by an ocean called **Iapetus.** But how do we explain the fact that this boundary separating the two faunal provinces is recognized in both present-day western Europe and

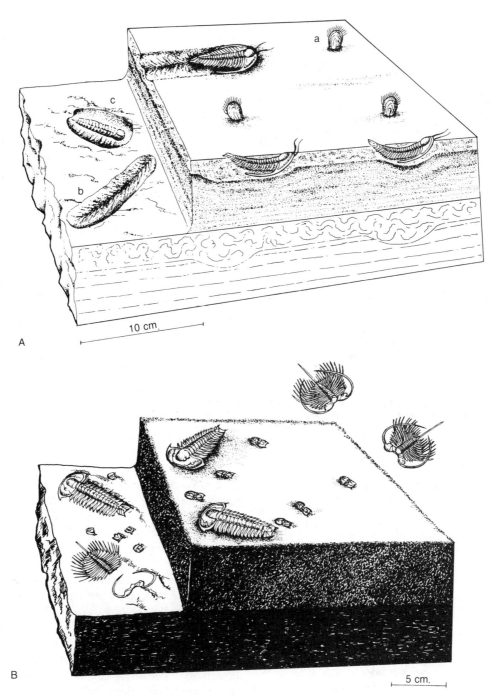

Figure 10-30. In-life Cambrian faunal associations. A. *Cruziana–Lingulella* community: a, *Lingulella*, an inarticulate brachiopod; b, *Cruziana*, a trilobite crawling trace; c, *Rusophycus*, a trilobite resting trace. B. Community of trilobites belonging to the family Olenellidae. (From W. S., McKerrow, 1978, *Ecology of Fossils*, Figs. 1, 4, pp. 53, 61: Gerald Duckworth, London. Reprinted by permission of author)

eastern North America? We shall defer discussion of this question to Chapter 11 where we can consider it in the context of post-Cambrian events.

The phylum Brachiopoda makes up 15–20 percent of the Cambrian skeletonized fossil record. Most Cambrian brachiopods (Fig. 10-32A–C) had phosphatic shells and belong to the class Inarticulata; the great evolutionary expansion of forms bearing calcium carbonate shells and belonging to the class Articulata was yet to come in the explosive Ordovician faunal radiation.

An enigmatic Cambrian group consisted of sponge-like creatures with a double-walled calcium carbonate

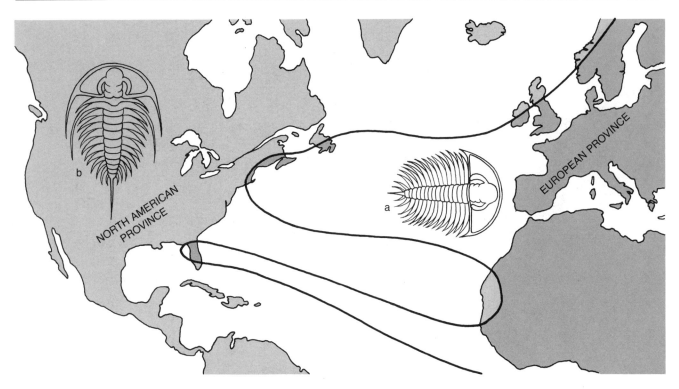

Figure 10-31. Trilobite provinces. The solid line represents the boundary between two distinctly different Cambrian faunal realms (represented by the trilobite genera *Paedeumius*, in North America, and *Holmia*, in Europe). An oceanic barrier originally separated the two biogeographic provinces which are connected today in Scandinavia, the British Isles, parts of eastern North America, and northwestern Africa. The odd pattern of contact between the two provinces is the result of post-Cambrian continental collisions and later separations.
(Data from several sources)

Table 10–1. Uniqueness of Ediacarian and Cambrian Faunas

	Time	Fauna and Environments
PALEOZOIC	Ordovician	Evolutionary expansion of the $CaCO_3$ shell; evolutionary stabilization Stability of O_2 in seas
		——— ~500 m.y. ———
EOPALEOZOIC	Cambrian	Initial metazoan radiation; origin of skeletonization; dominance of calcium phosphate shells; trilobite-dominated fauna; evolutionary "experimentation" Marine transgression; instability of O_2 levels in seas
		——— ~570 m.y. ———
	Ediacarian	Origin and early evolution of metazoans; primitive soft-bodied forms; limited range of adaptations Separation of continents and marine transgression
		——— ~700 m.y. ———
	Late Proterozoic	Abundant cyanobacteria and eukaryotic algae; protistans; abundant stromatolites Glaciation on a supercontinent

shell (Fig. 10-32J, K); these forms were assigned to a separate, extinct phylum, the Archaeocyathya. The cup-shaped archaeocyathids were moderately abundant during the Early and Medial Cambrian and formed meadowlike patches on shallow limy sea bottoms. They became extinct by the end of the Medial Cambrian—the victims of environmental foreclosure.

In addition to the Archaecyatha, a number of other Cambrian animal groups were short lived; several minor arthropod and echinoderm classes (Fig. 10-32G–I) and primitive molluscan classes (Fig. 10-32E, F), as well as at least a half-dozen primitive and poorly known phyla (Fig. 10-32D) became extinct before the period ended. Also, algal stromatolites, so prevalent during the late Proterozoic, became significantly reduced in Cambrian rocks, most likely the result of large-scale cropping of algae by grazing herbivorous metazoans. Cambrian stromatolites developed mainly in hypersaline environments where grazing herbivorous predators were not common.

The explosive Cambrian faunal radiation of diverse skeletonized taxa may be accentuated by the better preservability (versus Ediacarian soft-bodied forms); however, the trace fossils, which have a constant preservability throughout the column, show a similar pattern of abundance and increased diversity (Fig. 10–9) from Ediacarian to Cambrian (although, globally, trace fossil diversification begins stratigraphically slightly below the radiation of shelly fossils). This pattern of trace fossil diversification strongly suggests that the Cambrian fossil

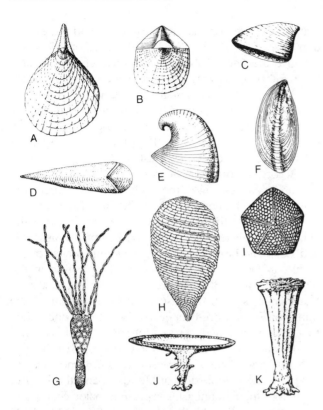

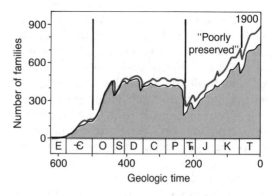

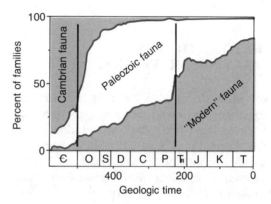

Figure 10-32. Representative nontrilobite Cambrian fossils. A–C. Brachiopods. D. Hyolithid (primitive mollusc or separate phylum?). E, F. Molluscs. G–I. Echinoderms. J, K. Archaeocyathids.
(From A. R. Palmer, 1974, In Search of the Cambrian World, Fig. 1, p. 217: *Am. Scientist,* Vol. 62. Reproduced by permission of *American Scientist*)

record reflects the initial adaptive radiation of marine animals in general and not simply the advent of skeletonization. At the other end of the spectrum, trace fossils of pre-Ediacarian (Riphean) age have been reported from several localities, but are equivocal traces of infaunal deposit-feeding organisms and may in fact be inorganic sedimentary structures. Ediacarian sediments contain a low-diversity trace assemblage of burrows of more definitive infaunal deposit feeders as well as a variety of crawling tracks and trails. Cambrian trace fossils include, in addition to the above, abundant arthropod crawling and resting traces as well as moderately diverse burrows of infaunal suspension feeders. Thus the scant and equivocal traces in sediments older than 700 to 800 million years, together with the pattern of diversification in the Ediacarian and Cambrian, strongly suggest an evolutionary trend rather than a chance occurrence or facies controls.

To what degree does the Burgess Shale reflect the norm of Cambrian paleocommunity structure and composition? Although the composition of the fauna is unique in terms of the preserved fossil record, the faunal assemblage probably was not unique in the Cambrian. The uniqueness is more a reflection of the remarkable preservation than of the composition of the fauna. According to Morris, if the extraordinary condi-

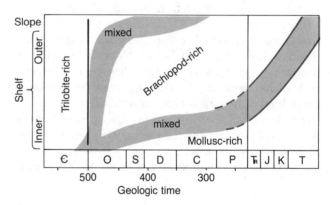

Figure 10-33. Faunal changes through time. A. Phanerozoic history of diversity of marine animal families. The total standing number of families described from the marine fossil record is indicated by the upper curve of the graph. The lower curve, bounding the stippled field, represents the family diversity of "shelly" taxa that constitute the bulk of the fossil record. The unshaded field between the two curves indicates the diversity of nonskeletal taxa. Note the abrupt increase in diversity in the early Ordovician after a pronounced leveling off during the Middle and Late Cambrian. B. Changes in the taxonomic composition of marine shelly faunas through time. The Cambrian fauna, represented mainly by trilobites, inarticulate brachiopods, archaeocyathids, hyolithids, and echinoderms, participated but little in the radiation of the "Paleozoic fauna" inaugurated in the Ordovician. C. Changes in general environmental distribution of marine invertebrate communities. The Cambrian trilobite-rich shelf communities were largely replaced during the Ordovician by the brachiopod-rich communities of the later Paleozoic.
(From J. J. Sepkoski, Jr., 1981, The Uniqueness of the Cambrian Fauna, Figs. 1–3, pp. 203, 205, 206, *in* Short Papers for the Second International Symposium on the Cambrian System: *U.S. Geol. Survey Open File Rept.* 81–743. Reprinted by permission of author)

tions of fossilization had not occurred, so that only animals with shells survived to be fossilized, the resulting assemblage would consist of components of most Cambrian faunas: trilobites, brachiopods, sponges, a few echinoderms, and hyolithids (primitive molluscs?). In this light, perhaps the Burgess Shale fauna can be rightly considered an approximate guide to the original diversity of at least some Cambrian invertebrate communities that lived in comparable moderately deep, muddy bottom environments.

In many respects the Cambrian was a harsh time that brought many different selective pressures to bear on animal groups. Only the more hardy, ecologically flexible taxa survived.

EARLY HISTORY OF THE METAZOA—A SEPARATE ERA?

Some workers share the opinion that the Ediacarian and Cambrian together comprise a sufficiently unique and long interval of geologic time to be considered as a separate era. This interval of at least 175 million years duration has a truly unique fauna in terms of diversity, taxonomic composition, and ecological organization when compared to the pre-Phanerozoic (Cryptozoic) and post-Cambrian Phanerozoic. Certainly the Ediacara fauna, displaying the world's first multicellular grade of organization, is very different from the algal-dominated Proterozoic biota. In general, the following features are characteristic of the Ediacarian metazoan biota that preceded the Cambrian biota: lack of mineralized skeletons: comparatively large size with sprawling, flimsy, flattened body plans; moderate diversity; apparent differentiation at the level of high-rank taxa; and lack of evident ancestors of the small shelly organisms that appeared explosively in the stratigraphic record (base of Cambrian) due to evolution of exoskeletons.

The explosive emergence and diversification of new

higher taxa, the dominance of trilobites, and the extinction of several primitive phyla and classes of low diversity give the Cambrian a rather unique character. After the Cambrian, no new phyla appeared, and none became extinct. After an initial burst early in the period, diversification slowed markedly through the remainder of the Cambrian. Diversification increased exponentially during the Ordovician (Fig. 10-33) as the more typical Paleozoic fauna became established.

According to paleontologist J. J. Sepkoski, Jr., of the University of Chicago, the differences between the faunas of the Cambrian and of the later Paleozoic are rivaled only by the differences observed between the faunas of the Paleozoic and of the combined Mesozoic and Cenozoic Eras (Fig. 10-33). This uniqueness of the Cambrian fauna arises from its special position in the history of life; it is the earth's first diverse metazoan fauna and, as such, represents the completion of the transition from a pre-Phanerozoic autotroph-dominated to a Phanerozoic heterotroph-dominated world ocean.

Recall from the discussion in Chapter 3 that the Phanerozoic eras (Paleozoic, Mesozoic, and Cenozoic) are based on the complexion of life forms and the level of evolution; their boundaries represent major changes in the history of life. An "Eopaleozoic" Era (Table 10-1) made up of the Ediacarian and Cambrian Periods would be consistent with what an era represents and perhaps would reflect more accurately the major chapters in the history of life. However, if the Ediacara fauna represents an evolutionary dead end and if the disappearance of this fauna is due to mass extinction—the end result of an "experiment" gone wrong, as Seilacher has proposed—the older, more conventional idea of the era boundary separating Ediacarian and Cambrian might be supported. We shall leave it to the experts on the Ediacarian–Cambrian interval to argue for or against a new era designation; here we merely suggest the intriguing possibility and by doing so underscore the uniqueness and importance of this interval of earth history.

SUMMARY

The Ediacarian and Cambrian together represent the transition from Late Proterozoic to Paleozoic. This stratigraphic interval has a sufficiently unique history and fauna to justify its being considered a separate era, herein referred to informally as the Eopaleozoic. The Ediacarian Period (and corresponding system) was formally proposed in 1982 as the time interval (about 700 to 570 million years B.P.) beginning with the earliest ap-

pearance of soft-bodied metazoans and ending with the earliest appearance of skeletonized metazoans. It is based on the distinct soft-bodied Ediacara fauna from the Ediacara Hills in South Australia. Rocks of the Ediacarian System are confined to geosynclinal continental margins and lie between the uppermost Proterozoic tillites and the stratigraphically lowest occurrence of shelly fossils, which mark the base of the Cambrian System.

Ediacarian soft-bodied fossils have been described from Africa, England, Scandinavia, northern and southwest Russia, Siberia, China, and North America, in addition to the classic localities in South Australia. In North America, documented accounts of Ediacarian fossils are rare, but strata of Ediacarian age are presumed to be present on the basis of stratigraphic position and apparent conformity with overlying Cambrian rocks. These strata lie above glaciogenic sediments and are early deposits of both the Appalachian and Cordilleran geosynclines, which developed along the newly rifted margins of the continent.

The appearance of exoskeletons represents an important evolutionary milestone and a major aspect of early metazoan diversification. Skeletonized faunas appear sequentially; thus the base of the Cambrian is not ideally synchronous. The oldest skeletal fossils belong to the Tommotian stage and include small conical calcium carbonate shells of uncertain biologic affinity as well as sponges, archaeocyathids, primitive molluscs, coelenterates, echinoderms, and brachiopods. The Tommotian shelly fauna and associated trace fossils are significantly more diverse than the Ediacara fauna. Tommotian forms do not appear to have evolved from Ediacara-type animals and probably radiated from a proto-Laurasia continental assembly. Skeletonization, which had many adaptive advantages for bottom living, probably became selected for as oxygen levels continued to increase—perhaps to about 10 percent PAL—and more advanced respiratory and circulatory systems evolved.

Through the remainder of the Cambrian, metazoan diversification continued as continental margins and interiors were flooded by marine transgression. The transgressive Cambrian epeiric sea spread a blanket of quartz sand ranging in age from Early to Late Cambrian. By Late Cambrian, most of the cratonic interior of North America was flooded by shallow epeiric seas, and widespread carbonate deposition was the rule.

North American cratonal and miogeosynclinal strata contain abundant features of shallow-water deposition, and areally extensive tidal flats were widespread. Turbidites, black shales, chert, and slump deposits are lithologic associations of eugeosynclinal slopes and deep basins.

Cambrian world paleogeography was such that six identified continents were positioned at low latitudes. Much of what is now North America and Greenland, together with parts of the British Isles and western U.S.S.R., made up the ancient continent of Laurentia, which lay astride the equator in a position roughly 90 degrees counterclockwise from the present-day orientation of North America.

Post-Tommotian Cambrian skeletal faunas are dominated by trilobites, which make up about 75 percent of the taxa and provide most of the biostratigraphic zonation. Of the remainder, inarticulate brachiopods (15–20 percent), archaeocyathids (about 10 percent), primitive molluscs, and echinoderms are the most abundant groups. Trace fossils are abundant and diverse and show a distinct parallelism with the pattern of increased body fossil diversity from Ediacarian to Cambrian. One of the most unique fossil assemblages in the geologic archives is the Middle Cambrian Burgess Shale fauna, which includes more than 150 species representing at least eight known and several previously unknown phyla. Most of the fossils are unique to the single locality and represent taxa that became extinct by the end of the Cambrian.

The Cambrian, containing the world's first skeletal faunas, was not only a time of rapid diversification and appearance of high-level taxa, but was also a time of extinction of some short-lived, low-diversity higher taxa.

SUGGESTIONS FOR FURTHER READING

Bambach, R. K., Scotese, C. R., and Ziegler, A. M., 1980, Before Pangaea: The Geographies of the Paleozoic World: *Am. Scientist,* vol. 68, no. 1, pp. 26–38.

Cloud, P. E., Jr., and Glaessner, M. F., 1982, The Ediacarian Period and System: Metazoa Inherit the Earth: *Science,* vol. 217, no. 4562, pp. 783–792.

Dott, R. H., Jr., and Batten, R. L., 1981, *Evolution of the Earth,* 3rd ed.: McGraw-Hill, New York.

Glaessner, M. F., 1961, Precambrian Animals: *Scientific American* Offprint No. 837, W. H. Freeman, San Francisco.

Hallam, A., 1972, Continental Drift and the Fossil Record: *Scientific American* Offprint No. 903, W. H. Freeman, San Francisco.

McAlester, A. L., 1977, *The History of Life,* 2nd ed. Foundations of Earth Science Series, Prentice-Hall, Englewood Cliffs, NJ.

Morris, S. C., and Whittington, H. B., 1979, The Animals of the Burgess Shale: *Sci. American,* vol. 241, no. 1, pp. 122–135.

Palmer, A. R., 1974, Search for the Cambrian World: *Am. Scientist,* vol. 62, pp. 216–225.

Raup, D. M., and Sepkoski, J. J., Jr., 1982, Mass Extinctions in the Marine Fossil Record: *Science,* vol. 215, pp. 1501–1502.

Valentine, J. W., 1978, Evolution of Multicellular Plants and Animals: *Scientific American* Offprint No. 1403, W. H. Freeman, San Francisco.

11

PALEOZOIC HISTORY

CONTENTS

KEY TERMS

Miogeosyncline
Eugeosyncline
Wilson cycle
Epeiric sea
Cratonic sequence
Taconic orogeny
Clastic wedge
Iapetus
Acadian orogeny
Laurussia
Ophiolite
Antler orogeny
Flysch
Alleghenian orogeny
Pangaea
Cyclothem
Panthalassa
Tethys Sea

AMERICA'S FIRST MAJOR GEOLOGIC CONCEPT

In his presidential address to the American Association for the Advancement of Science in 1859, James Hall, a great pioneer American geologist and state geologist for New York, submitted the observation that Paleozoic strata exposed in the Appalachian Mountain belt, in addition to being much more deformed, were comparatively much thicker than their time-equivalent counterparts in the Mississippi Valley region of the midcontinent. Hall reasoned that there must be a causative relationship between the greater thickness of sediments in the Appalachians and their deformed character. From this he drew the conclusion that all large mountain chains must represent areas of inordinately thick accumulation of sediment. He also observed that the character of the sediments in both the Appalachians and midcontinent was unmistakably similar—they bore the earmarks of shallow-water deposition.

Hall maintained that the greater thickness of shallow-water sediments in the Appalachians had to be an expression of great crustal subsidence there than in the continental interior (craton). He believed the crust gradually yielded under the weight of the sediments themselves; subsidence then kept pace with sedimentation so that the depositional surface was constantly in the shallow-water realm. Mountain building was envisioned as a cyclic process whereby sedimentation resulted in subsidence, which in turn accommodated more sediment. Eventually, along the axis of the subsiding belt, sediments were folded and faulted, occasionally intruded by plutons, and, in places, metamorphosed.

In 1873, James Dwight Dana, professor of geology at Yale University, gave Hall's theory a name—"geosynclinal" (later changed to *geosyncline*)—but not without some additional ideas of his own. The geosynclinal concept, as theorized by Hall, was coming to be appreciated more and more by geologists. Dana agreed with the facts, but he felt that another explanation was in order. He was critical of Hall's analysis as a "a theory on the origin of mountains, with the origin of mountains left out." In other words, Dana took exception to the role that sediments were assumed to play in the deformation of geosynclinal sequences into mountain belts. Dana was one of the first geologists to recognize the fundamental differences between oceanic and continental crust, and

he subscribed to the notion that the earth's crust was subjected to constant compression due to the general contraction (through cooling) of the interior. He believed that the greatest yielding to this compression should be along the interfaces between continents and ocean basins. Such yielding would manifest itself by downwarping to form "geosynclinals." In Dana's hypothesis, thick masses of sediment were able to accumulate because the crust, downwarped by forces *within* the earth, subsided to create a place for them.

Hall's and Dana's original ideas on geosynclinal theory came from studies in the Appalachian Mountain belt. The geosynclinal theory is also the first major geologic concept developed in America. European geologists were quick to adopt the notion of thick accumulations of sediment as the ancestors of mountains. From observations in the Alps, however, they disagreed with the shallow-water origin of geosynclinal sedimentary sequences. Instead they contended that geosynclines began as downwarped furrowlike depressions on the sea floor; the depressions served as deep marine basins for accumulation of sediments (Fig. 11-1). European geologists also pointed out that some mountain belts, like the Alpine–Himalayan chain and the Urals of Russia, are not situated along present-day continental margins.

As the study of mountain belts progressed through the turn of the century, the idea of thick geosynclinal

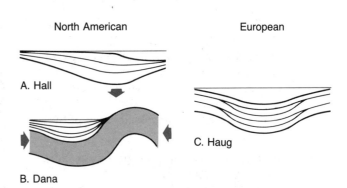

Figure 11-1. Early conceptions of the geosyncline: American shallow-water versus European deeper marine views. A. James Hall's subsidence by sedimentation. B. J. D. Dana's subsidence due to crustal buckling with complementary upwarping of a crustal ridge (geanticline). C. European Emil Haug's idealized view of a symmetrical intracratonic trough.
(After R. H. Dott, Jr., 1974, The Geosynclinal Concept, Fig. 1, p. 3, *in* Modern and Ancient Geosynclinal Sedimentation: *SEPM Spec. Pub.* No. 19. Reproduced by permission of Society of Economic Paleontologists and Mineralogists)

sediments' being the precursor to mountains became more accepted. However, the geosyncline concept itself took on a variety of meanings that eventually included many types of thick sedimentary accumulations in a variety of both intracratonal and marginal basins. The notion of a geotectonic cycle involving geosyncline depositional and postdepositional tectonic phases became popular. Despite the range of opinion on what

geosyncline meant, a consensus emerged about depositional and stratigraphic phases of geosynclines.

Most geologists agreed that ancient geosynclines, at least those defined in the classic sense as marginal to cratons, were long, essentially linear depositional basins that could be separated into an inner belt—the **miogeosyncline**—and an outer, more oceanic belt— the **eugeosyncline** (Fig. 11-2). In this context,

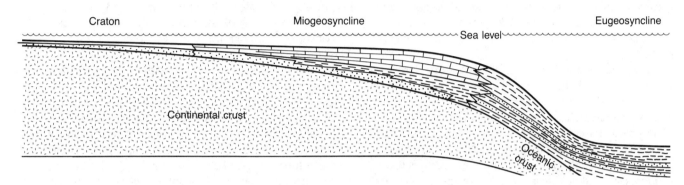

Figure 11-2. Miogeosyncline–eugeosyncline couplet. Vertical profile exaggerated.

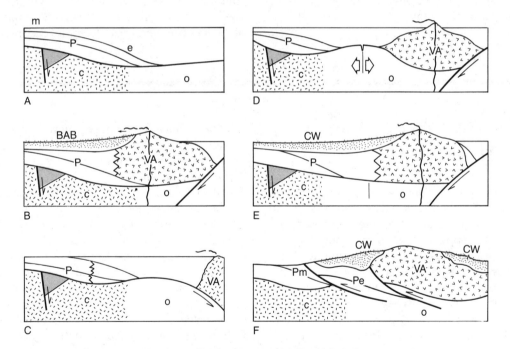

Figure 11-3. Possible genetic relations between geosynclines and orogen (not to scale): A. Continental terrace (miogeosynclinal-eugeosynclinal couplet) developed along trailing edge of rifted continental margin. B. Subsequent development of volcanic-arc (VA) system and back-arc basin (BAB) when continental margin becomes orogenically active due to continentward subduction. C. Reversal of polarity of subduction with continental terrace prism destined to collide with volcanic island-arc system. D. Continental terrace prism facing a small ocean basin having a local spreading site and an orogenic outer margin. E. Marginal ocean basin after cessation of spreading followed by filling of basin by nonmarine and shallow-marine clastic-wedge (CW) strata. F. Orogenic telescoping of rock suites and basement rock complexes such as might accompany the closing of an ocean basin (as in C, D, and E) or the termination of a continent-margin orogenic belt (as in B). Note bilateral clastic wedges on either side of suture belt. Thrust faulting may also be bilateral, especially in intercratonic orogens.
(Modified from R. H. Dott, Jr., 1974, The Geosynclinal Concept, Fig. 4, p. 9, *in* Modern and Ancient Geosynclinal Sedimentation: *SEPM Spec. Pub.* No. 19. Reproduced by permission of Society of Economic Paleontologists and Mineralogists)

miogeosynclines were regions of tectonically quiet sedimentation and slow subsidence in which carbonates, shales, and mature sandstones were deposited in shallow waters of a continental shelf setting. Eugeosynclines were envisioned as developing in a more tectonically active or mobile-belt setting that involved comparatively more rapid crustal subsidence, local warping and uplift, and periodic volcanism. The deep-water rock suite was notably lacking in carbonate rocks and clean sandstones but included an abundance of dark mudrocks and graywackes commonly associated with submarine volcanic rocks and chert beds.

Rock successions that turn-of-the-century European workers believed characterized geosynclinal deposition in general belong to the broad concept of eugeosynclinal facies, whereas James Hall's original concept was based on the miogeosynclinal rocks of the Appalachian belt. The recognition that volcanic rocks comprise a sizeable proportion of many geosynclinal sequences and that volcanism plays an important role in the evolution of mountain belts gradually led many geologists to associate ancient geosynclines with island-arcs and oceanic-trench systems.

Plate tectonics has given us new insights and a new model within which to view ancient geosynclines. The theory has also provided modern (actualistic) analogues to allow for advancing the geosynclinal concept from a descriptive one to a genetic one. As an example, genetically distinct sedimentary prisms form on tectonically passive trailing edges of diverging continents in contrast to those deposits that accumulate on tectonically active leading edges of converging lithospheric plates (Fig. 11-3 on page 253). In his introduction to a 1973 symposium on the geosyncline concept, R. H. Dott, Jr., of the University of Wisconsin, stated that plate tectonics demonstrates how these two different depositional settings with their contrasting stratigraphic successions, both considered as geosynclines, ultimately may become deformed in orogenic belts called **orogens.** Characteristics of geosynclines are understood better now in terms of the **Wilson cycle:** the opening and closing of an ocean basin through initial rifting and rafting of continents; subduction of oceanic lithosphere along continental margins, terminating in continent-to-continent collisions; and renewed rifting to begin the cycle again. The cycle was named after J. Tuzo Wilson, one of the pioneers of modern plate tectonics theory.

ORDOVICIAN HISTORY

Continued Flooding of the Continent

Sedimentation patterns continued generally uninterrupted from Late Cambrian into Early Ordovician. A large volume of limestone and dolomite is exposed today in parts of the Appalachian, cratonal, and Cordilleran regions, attesting to vigorous and continuous depositional activity in calcium carbonate-rich, shallow-marine waters during the late Cambrian and Ordovician. A major regression of **epeiric seas** from the craton occurred near the end of Early Ordovician. As a result, Lower Ordovician and Upper Cambrian sedimentary rocks were exposed to erosion. Very pure quartz sand, typified by the St. Peter Sandstone (Chapter 6) of the eastern craton, was then reworked and redeposited during a major marine transgression that inaugurated the Medial Ordovician, with the consequent covering of the erosion surface (Fig. 11-4). The epeiric sea continued to advance and flood almost the entire craton by Late Ordovician. Shelly limestones were deposited over most of the continental platform in water depths of a few tens of meters or less, covering more of the erosion surface and producing a major cratonwide unconformity.

These shallow, warm, calcium carbonate-rich epicontinental or epeiric seas, so characteristic of Paleozoic time, have no modern-day counterparts. Extensive *carbonate* sedimentation does not occur today on continental shelves. At present, the North American continent is relatively emergent, with a large percentage of its surface elevated more than 100 m above sea level. Erosion of positive areas produces great volumes of terrigenous clastic detritus that is responsible for smothering carbonate production in potential carbonate environments of continental shelf areas. One important place where present-day carbonate sedimentation is the rule is the Great Bahama Bank. This is a region of high calcium carbonate productivity and one that is geographically removed from the influx of terrigenous detritus. The Bahamas are part of a major carbonate platform that has been a locus for limestone production since the late Mesozoic. This region provides an instructive actualistic model for studying carbonate sedimentation. Figure 11-5, a Holocene facies map of an area adjacent to Andros Island, is probably a reasonably accurate present-day small-scale analogue for much of the North American craton during the Late Ordovician—a large-scale shallow-marine platform containing a mosaic of carbonate environments that produced complex intertonguing facies.

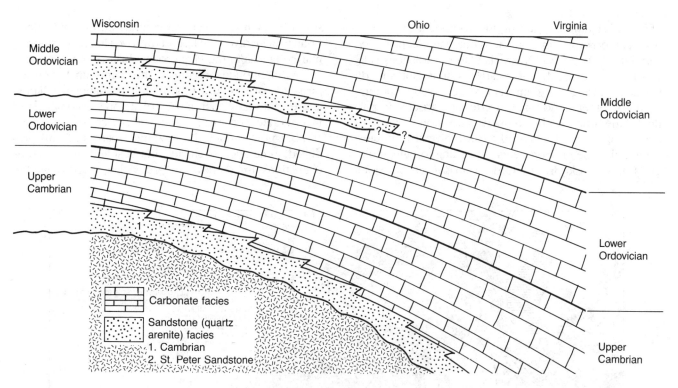

Figure 11-4. Lower and Middle Ordovician facies relationships from the Appalachian geosyncline to the central craton.

The large-scale cyclic pattern of transgression and regression, which we have already considered for the Early-Middle Ordovician transition, was repeated several more times during the Paleozoic (Fig. 11-6). American geologist L. L. Sloss has applied the term **cratonic sequence** to the large-scale (greater than supergroup) rock-stratigraphic packages that represent major onlap-offlap cycles, bounded by unconformities of cratonwide extent.

The large-scale (cratonwide) unconformity-bounded sequences have been recognized also in continental areas outside North America. Some geologists believe these major onlap-offlap cycles reflect the dynamic responses to fluctuations in holding capacity of the Paleozoic ocean basins as influenced by lithospheric plate motions. The crux of this idea is that, during times of accelerated sea-floor spreading, bulging, spreading ridges caused a decrease in ocean-basin holding capacity, with the result that continental margins were drowned by land-transgressing displaced waters. Figure 11-7 shows how these sequences relate to global cycles of relative changes of sea level during the Paleozoic.

The Calcium Carbonate Shell: A Success Story

Ordovician faunas differ considerably from Cambrian faunas, an expression of the great evolutionary de-velopment of the calcium carbonate shell as tremendous increases in diversity of shell-secreting organisms accompanied the reinvasion of epeiric seas into the cratonic interior during early Medial Ordovician. Selection pressure may have been exerted on marine invertebrates to evolve rigid, external, calcium carbonate shells as adaptations for anchoring on shallow-sea bottoms, for muscle and organ support, for possible protection from cosmic radiation in shallow water, and for protection against early predators. Whatever the reasons, many diverse groups of organisms found it to be an evolutionary advantage to be encased in a hard shell. Abundant Middle and Upper Ordovician shelly limestones attest to a sea teeming with life. Richly fossiliferous limestones in the Cincinnati, Ohio, area sparked the childhood curiosity of several prominent North American paleontologists.

The epitome of this expansion of the calcium carbonate shell was the phylum Brachiopoda. The articulate brachiopods (Fig. 11-8, 9(g),10) rare in Cambrian seas, experience a fantastic diversification in the Ordovician. They are comparatively rare in modern seas, but contributed to large volumes of shelly limestone during the Paleozoic. Medial and Late Ordovician seas were abundantly populated with diverse brachiopods, bryozoans, ostracodes, conodonts, graptolites, nautiloid cephalopods, gastropods, echinoderms, corals, and bivalve molluscs (Figs. 11-9, 10)—faunal groups that were of minor significance during the Cambrian. The dominant Cambrian groups, on the other hand, contributed very little to the great Ordovician faunal radiations (Fig. 11-8). Trilobites, although nearly as abundant and diverse as during the Cambrian, were greatly outnumbered by other Ordovician invertebrate groups, in particular the articulate brachiopods, and, along with inarticulate brachiopods, enjoyed greatest diversity in outershelf and slope environments. According to paleontologist J. J. Sepkoski, Jr., the Ordovician pattern of expan-

sion of new major taxa and contraction of typical Cambrian taxa involved the largest turnover in composition of marine faunas in the Phanerozoic history of the oceans.

As mentioned previously, the Cambrian Period appears to have been a time of initial adaptive radiation and evolutionary trial and error among major phyla of invertebrate organisms. Nature's search for workable combinations among the new metazoan life forms produced some success stories, but resulted in numerous extinctions. In contrast, the Ordovician, when viewed in

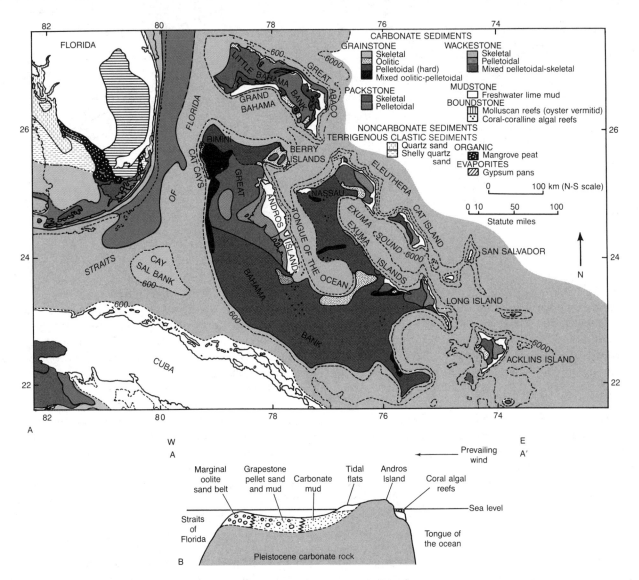

Figure 11-5. Present-day sediments in South Florida and the Bahamas. A. Distribution of sediments. Grainstones are cleanly washed particle-supported deposits lacking lime mud matrix. Packstones are particle-supported deposits with lime mud matrix. Wackestones are deposits containing particles supported by lime mud matrix. B. Western part of the Bahama platform showing distribution of modern carbonate facies.
(A after R. B. Halley, P. M. Harris, and Albert C. Hine, 1983, Bank Margin Environment, *in* Carbonate Depositional Environments, Fig. 2, p. 245: *Am. Assoc. Petroleum Geologists Mem.* 33. Reproduced by permission of American Association of Petroleum Geologists) (B from R. E. Garrison, 1975, Carbonate Sedimentation on Shelves and Platforms, Fig. 4-4, p. 4-4, *in* Current Concepts of Depositional Systems with Applications for Petroleum Geology: Short course Volume. Reproduced by permission of San Joaquin Geological Society)

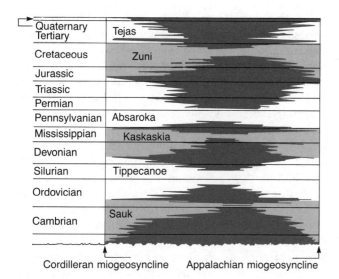

Figure 11-6. Time-stratigraphic relationships of the major onlap-offlap sequences of the North American craton. Black represents hiatuses (erosion + nondeposition time magnitude of sequence-bounding unconformities). White and stippled patterns represent deposition of tangible rock record.
(After L. L. Sloss, 1963, Sequences in the Cratonic Interior of North America, Fig. 6, p. 110; *Geol. Soc. America Bull.,* vol. 74. Reprinted by permission of author and Geological Society of America)

the history of life, was a time of *secondary adaptive radiation* and accelerated diversification, as successful evolutionary lines became established and stabilized. This increase in diversity and evolutionary success may, in part, have been an expression of stabilization of oxygen levels and environments in the marine realm. By late in the period, all of the *invertebrate phyla* and most of the *classes* that we are familiar with today were firmly established. Since that time, the changes in invertebrate life have been mostly smaller scale evolutionary radiations and extinctions within these classes at lower taxonomic levels. This contrast further accentuates the difference between the Cambrian fauna and the more typical Paleozoic fauna (Fig. 11-8) and lends credence to a separate "Eopaleozoic" Era (Fig. 10-33).

Of great importance in dating and correlating Ordovician strata are the *graptolites* (Fig. 11-8B), colonial organisms that evolved rapidly during the Ordovician and Silurian and became extinct during the Carboniferous. Graptolites are preserved most abundantly and characteristically as carbonaceous films on the surface of dark shale layers that were deposited in eugeosynclinal areas. This association of graptolites with a particular lithology has inspired the expression "graptolite facies."

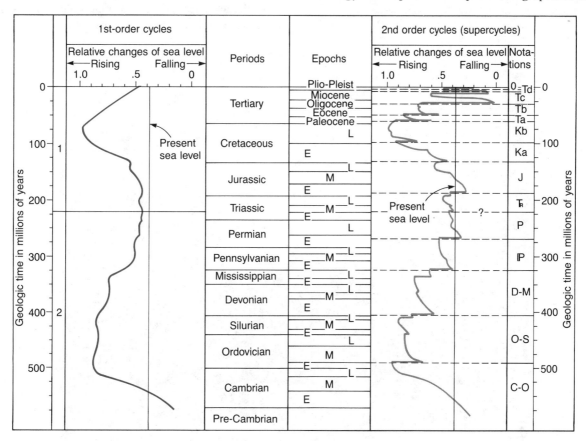

Figure 11-7. First- and second-order global cycles of relative change of sea level during Phanerozoic.
(From R. R. Vail, R. M. Mitchum, and S. Thompson III, 1977, Seismic Stratigraphy and Global Changes of Sea Level, Fig. 1, p. 84, *in* Global Cycles of Relative Changes of Sea Level: *Am. Assoc. Petroleum Geologists Mem.* 26. Reproduced by permission of American Association of Petroleum Geologists)

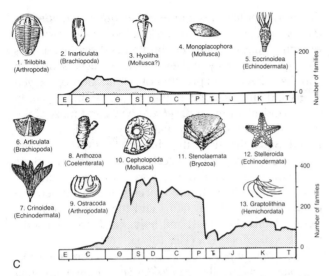

C

Figure 11-8. Phanerozoic history of taxonomic diversity of marine animal families comprising the two principal evolutionary faunas of the Paleozoic. A. The Cambrian fauna, characterized by the classes (phyla shown in parentheses) Trilobita, Inarticulata, Hyolitha, Monoplacophora, Eocrinoidea. B. The Paleozoic fauna, characterized by the classes (phyla shown in parentheses) Articulata, Anthozoa, cephalopoda, Stenolaemata, Stelleroida, Crinoidea, Ostracoda, Graptolithina.
(From J. J. Sepkoski, Jr., 1984, A Kinctic Model of Phanerozoic Taxonomic Diversity. III. Post-Paleozoic Families and Mass Extinctions, Fig. 2: p. 250. *Paleobiology*, vol. 10, no. 2. Reproduced by permission.)

True *vertebrate* organisms, represented by primitive jawless fish of the class Agnatha, first appear in the geologic record in rocks of Late Cambrian age in North America. Abundant fragments of bone and external plates of an agnathan group called *ostracoderms* occur in Middle Ordovician sandstone in Colorado and represent the oldest known diverse vertebrate fauna in North America. Were agnathans the first vertebrates, and if so, what were their evolutionary roots? The fossil record sheds little light on this important evolutionary step.

Conodonts—those enigmatic, microscopic "toothlike" structures (see Chapter 4) composed of calcium phosphate—are locally abundant in Ordovician rocks. Although we are not certain of the biologic affinity of conodonts, the suggestion has been made that they are structural parts of primitive vertebrates; if this is true, yet another line of evidence suggests that the vertebrate record extends back at least as far as Late Cambrian. Conodonts are useful for biostratigraphic correlation. Because they commonly survive recrystallization and dolomitization of limestones (unlike calcareous body fossils) they are particularly good biostratigraphic tools in carbonate sequences. Much of the Ordovician-through-Devonian time-stratigraphic subdivision of the Cordilleran and cratonal stratigraphic successions is being reexamined by use of conodonts.

Figure 11-9. Diorama of Ordovician shallow-marine bottom scene: a, algae; b, straight-shelled nautiloid cephalopods; c, solitary coral; d, colonial coral; e, crinoids; f, bryozoans; g, brachiopods; and h, trilobites.
(Photo courtesy National Museum of Natural History. Smithsonian Institution Photo No. 653.)

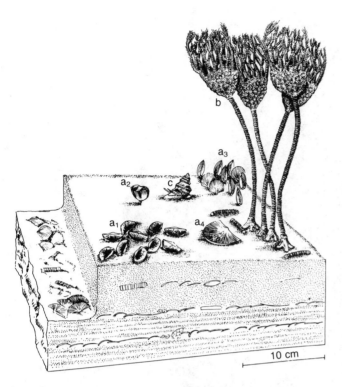

Figure 11-10. Shallow-marine Ordovician bottom community: a, articulate brachiopods; b, crinoid echinoderms; c, gastropod mollusc. (From W. S. McKerrow, 1970, *The Ecology of Fossils*, Fig. 11, p. 81: Gerald Duckworth, London. Reproduced by permission of author)

Mass extinctions of marine invertebrates at the family level occurred near the end of the Ordovician (Fig. 10-33). Raup and Sepkoski have computed an extinction rate of 19.3 families per million years for the Late Ordovician. This figure represents a diversity drop of 12 percent and a rate significantly greater than the "normal" background rate of about 8.0 families per million years. Possible causes may have included increased predation and competition and cooling of marine waters due to glaciation.

Tectonic Unrest in the Appalachians

During the last half of the Ordovician and extending into Early Silurian, a mountainous land mass emerged in the northern part of the Appalachian geosyncline (Fig. 11-11). The growth of this structurally complex mountain range was a result of the first in a series of major regional Paleozoic orogenic disturbances that affected the Appalachian belt and deformed rocks deposited in the geosyncline. This Ordovician event, called the **Taconic orogeny,** is named after the Taconic Mountains in eastern New York, Vermont, and central Massachusetts. The mountains that were produced during a complex series of Taconic disturbances have long since eroded, but their existence, location, and growth can be deduced from the geologic evidence both within and peripheral to the disturbed belt.

The Taconic orogeny was a significant event in Paleozoic history and provides a model for understanding mountain building and the unraveling of geologic history through interpretation of the rock record. One critical aspect of the orogeny involved the tectonic emplacement (Fig. 11-12) of a large mass of previously formed Cambrian and Ordovician eugeosynclinal rocks on top of a miogeosynclinal section of about the same age. The eugeosynclinal rocks are characterized by rhythmically interbedded graded sandstones and shales, called **flysch**, and associated volcanics; the miogeosynclinal rocks are mostly shallow-marine, shelf carbonates with shelly fossils. The stratigraphic and structural relationships have been worked out through painstaking field investigations, aided significantly by graptolite biostratigraphic studies. Here we see the advantage of paleontology: it provides a critical tool that is not available for working out details of pre-Paleozoic orogenies.

Additional structural, stratigraphic, petrologic, and sedimentologic evidence for orogeny has provided much information on timing and magnitude as well as origins of this Early Paleozoic disturbance. For example, in a number of places within the Taconic belt, steeply tilted Lower Ordovician and older rocks are overlain by less deformed Silurian and younger rocks presenting a pronounced angular unconformity. Yet another line of evidence includes volcanic activity, which was more prevalent in the geosynclinal belt in New England during the Ordovician period than during the Cambrian. Rocks produced by this volcanism include eugeosynclinal sea-floor lava flows as well as successive levels of widespread volcanic ash beds that punctuate Ordovician successions of the miogeosynclinal belt. Igneous activity also included emplacement of *plutonic* bodies from Georgia to Newfoundland but particularly in New England, with radiometric dates clustering around 480 to 440 million years. Pervasive regional metamorphism, much of which coincided with the Taconic orogeny, has overprinted New England eugeosynclinal rocks and represents yet another manifestation of orogenic-belt activity.

A final body of evidence for the Taconic orogeny is the great volume of sediment that was shed from the western front of the Taconic Mountains, from North Carolina to southern maritime Canada. A voluminous **clastic wedge** of gravel, sand, and mud spread westward from the belt of deformed eugeosynclinal rocks and was thickest and coarsest along the eastern margin of the miogeosynclinal belt (Fig. 11-13). Red shales and sandstones of the Queeston Formation in New York and southern Ontario, Canada, have given rise to the name "Queeston clastic wedge." Significant amounts of terrigenous clastic material were derived from the east for the first time in the history of the Appalachian geosyncline; this pattern was to be repeated throughout the remainder of the Paleozoic.

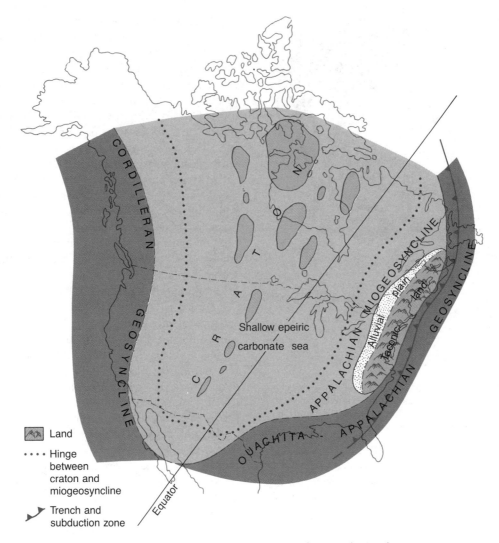

Figure 11-11. Paleogeography of Late Ordovician superimposed on outline map of present-day North America.
(Data from several sources)

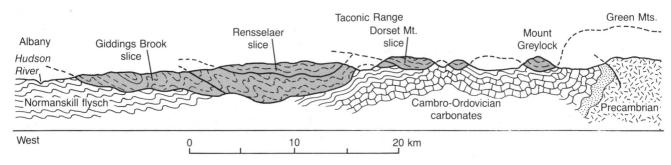

Figure 11-12. Taconic eugeosynclinal slices overlying miogeosynclinal rocks of about the same age.
(From Philip B. King, *The Evolution of North America*. Copyright © 1959, rev. ed. 1977 by Princeton University Press. Fig. 36, p. 60, reprinted by permission of Princeton University Press)

For more than 100 years, geologists have wrestled with various aspects of the Taconic orogeny, in particular with the Taconic eugeosynclinal terrane and the mechanism of its emplacement. What was responsible for the tremendous crustal shortening and the westward displacement, for tens of kilometers, of one colossal

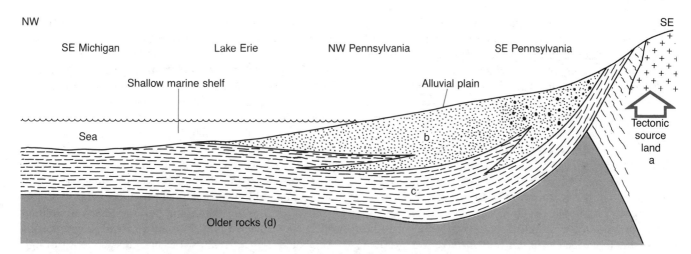

Figure 11-13. Relationship between tectonic source land of Taconic orogen (a); clastic-wedge deposits consisting of coarse terrigenous clastics of alluvial fan, stream, and marine shore-face environments (b); shallow-marine shelf shale facies (c); and older rocks (d).
(Modified from G. Marshall Kay, 1948, North American Geosynclines: *Geol. Soc. America Mem.* 48)

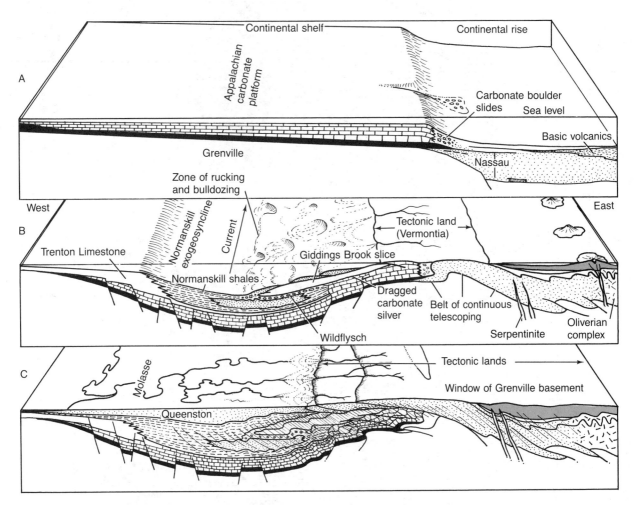

Figure 11-14. Pre-Taconic and Taconic orogeny evolution of the continental margin of North America in western New England. A. Pre-Taconic. B. Early phases of Taconic orogeny. C. Late phases of Taconic orogeny.
(From J. F. Bird and J. M. Dewey, 1970, Lithosphere Plate-Continental Margin Tectonics and the Evolution of the Appalachian Orogen; Fig. 7, p. 1043: *Geol. Soc. America Bull.,* vol. 81, no. 4. Reproduced by permission of authors)

rock suite over another? Plate tectonics has provided a new model within which to view the various lines of geological evidence. With respect to present-day orientation, the Taconic orogeny was caused by westward subduction of oceanic lithosphere beneath the northeastern margin of the North American (Laurentian) continental plate and possible collisions of a volcanic island arc and one or more microcontinents. Plate convergence caused westward overthrusting, surficial volcanism, perhaps in part related to island-arc activity, and deep-seated igneous intrusion. Plate convergence resulted also in the constriction of **Iapetus,** the proto-Atlantic ocean basin (in Greek mythology Iapetus was the father of Atlas, for whom the Atlantic was named), as the continental blocks *Laurentia* and *Baltica* were drawn closer together (Fig. 11-14 on page 261).

The northern part of the Appalachian geosyncline, from its beginning as a rifted continental margin in the late Proterozoic to its crustal mobility in Late Ordovician and Early Silurian, evolved from depositional phase to orogenic phase. What had been a passive, trailing-edge setting during the Cambrian and Early Ordovician, when Baltica and Laurentia were moving apart, became an active convergent plate margin during the later part of the Ordovician as a subduction zone developed and underthrusting of oceanic lithosphere produced compression that caused deformation of the geosynclinal rocks. The American and European faunal provinces (Fig. 10-31), so evident during the Cambrian, became less well defined in late Ordovician and Silurian as once geographically separated faunal realms became telescoped through plate convergence.

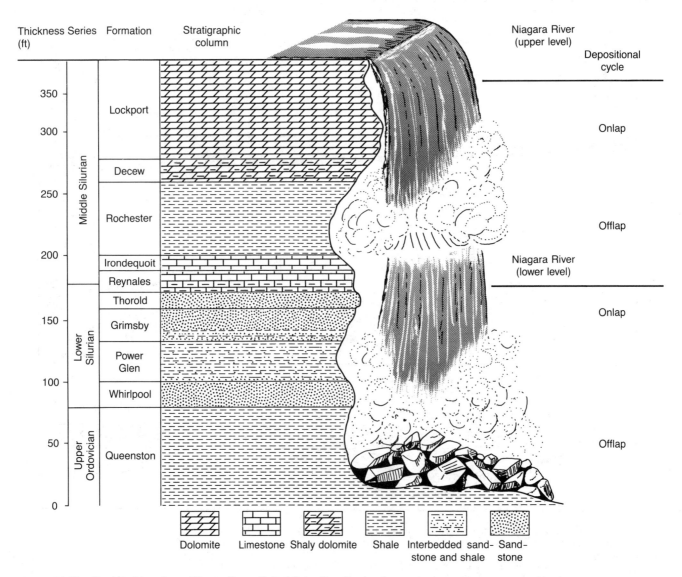

Figure 11-15. Stratigraphic column, Niagara Gorge, United States-Canadian border. (Redrawn from several sources)

SILURIAN AND DEVONIAN HISTORY

Salt Deposits, Reefs, and Cratonic Basins

The Taconic Mountains continued to be worn down during the Silurian Period, and the post-Taconic clastic wedge containing abundant red beds built out from east to west. Hematite-rich sandstones and mudstones of this sequence serve as the source of iron for the Birmingham, Alabama, steel mills. By the end of the Silurian, the tectonic highlands had been reduced to a low relief, and carbonate sedimentation succeeded deposition of terrigenous clastics in the Appalachian miogeosynclinal area.

Except for the waning stages of Taconic disturbance in the Appalachian geosynclinal belt, Silurian history in North America was comparatively quiet tectonically. Major parts of the continental interior were covered by shallow epeiric seas that produced widespread limestone and dolomite deposits. A well-known section exposed along the walls of the gorge below Niagara Falls contains a basal unit of beach sandstone overlain by shallow-marine shale and dolomite (Fig. 11-15). Silurian shallow-marine faunas include abundant articulate brachiopods, bryozoans, cephalopods, crinoids, and corals; planktonic graptolites populated the near-surface waters of geosynclinal seas (Figs. 11-16, 17). Unusual arthropods, the eurypterids (Fig. 11-18) and ostracodes, also are locally abundant in some Silurian deposits.

Thick deposits of salt and gypsum accumulated in restricted basins that developed within the eastern part of the craton in a large area encompassing most of Michigan and parts of Ohio, Pennsylvania, New York, and southern Ontario (Fig. 11-19). Evaporite basins, such as the Michigan Basin, were cut off from the open seas by a complex of organic reefs that were built by corals, stromatoporoid sponges, and calcareous algae. The reefs were rigid, wave-resistant, moundlike frameworks of calcium carbonate skeletons that created barriers to circulation. The physical setting that promoted reef growth involved the development of several basins and adjacent arches within what previously had been a stable craton. This warping of the craton was a byproduct of the Taconic orogeny. Abnormally thick cratonic Silurian sections in the Michigan and Williston Basins attest to subsidence and sediment accumulation in these areas.

Reef communities flourished and the wave-resistant, moundlike structures that formed profoundly affected the development of adjacent sedimentary facies. The reefs are important paleoecologic and stratigraphic features. During the Silurian tabulate and rugose corals, stromatoporoid sponges, and bryozoans contributed to the reef ecosystem. This important marine ecosystem represents a major evolutionary advance at the community level and was tied strongly to physical changes in the architecture of the craton.

Figure 11-16. Silurian shallow-marine bottom community: a, Articulate brachiopod; b, tabulate coral; c, rugose coral; d, bryozoan. (From W. S. McKerrow, 1970, *The Ecology of Fossils,* Fig. 22, 109: Gerald Duckworth, London. Reproduced by permission of author)

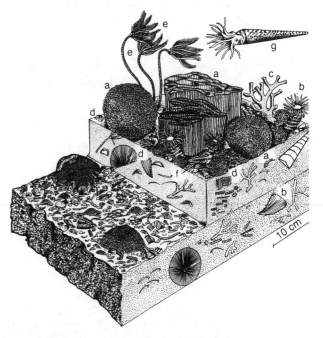

Figure 11-17. Silurian reef assemblage: a, tabulate corals; b, rugose coral; c, bryozoan; d, articulate brachiopod; e, crinoid echinoderm; f, trilobite; g, nautiloid cephalopod mollusc. (From W. S. McKerrow, 1970, *The Ecology of Fossils;* Fig. 24, p. 111: Gerald Duckworth, London. Reproduced by permission of author)

Figure 11-18. Diorama of Silurian brackish-marine bottom scene near Buffalo, New York. Shown are algae, eurypterids, worms, and shrimp.
(Photo courtesy National Museum of Natural History, Smithsonian Institution Photo No. 659A)

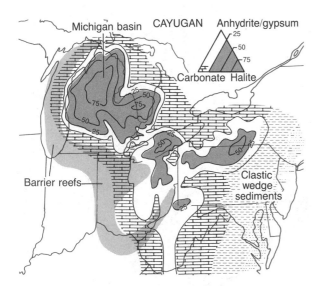

Figure 11-19. Late Silurian facies map of northeastern United States showing distribution of three principal chemical sediment types: carbonate rock (limestone and dolomite), anhydrite/gypsum, and halite, the products of an evaporite setting.
(After H. L. Alling, and L. I. Briggs, 1961, Stratigraphy of Upper Silurian (Cayugan) Evaporites; Fig. 8, p. 541: *Am. Assoc. Petroleum Geologists Bull.*, vol. 45. Reproduced by permission of American Association of Petroleum Geologists)

Organic reefs continued to flourish well into Devonian time. Devonian reef complexes buried in the subsurface of western Canada are important petroleum reservoirs. The reef cores represent carbonate facies that commonly had favorable *porosity* and *permeability* and were situated updip from basinal deposits rich in organic matter, which subsequently became converted to hydrocarbons. The Golden Spike reef complex in Alberta is a representative example (Fig. 11-20). Roughly one-half of the petroleum reservoirs of the world are in carbonate rocks (mainly dolomite) and a goodly proportion of these are associated with reef facies.

The widespread, carbonate-rich epeiric seas that characterized the Silurian receded and much of the continental interior underwent erosion during Early and Medial Devonian. Early Devonian was a time of continued warping of the craton as basins and arches continued to grow and influence stratigraphy and sedimentation. The Michigan Basin and the Williston Basin of North Dakota and southern Canada continued to subside and received deposits of dolomite, anhydrite, and some gypsum and salt. In these areas downwarping attendant with deposition preserved sedimentary sequences so that Upper Silurian

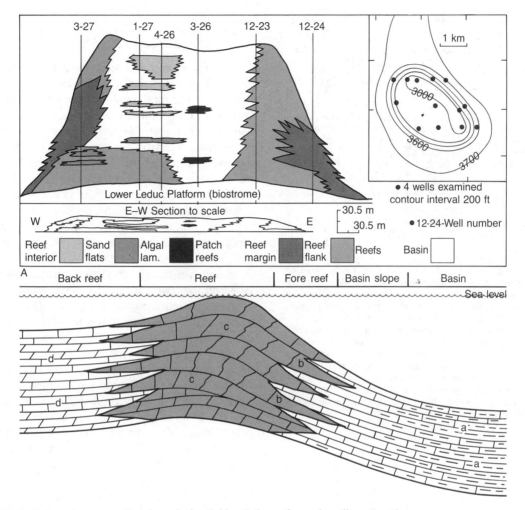

Figure 11-20. A. East–west cross section through the Golden Spike reef complex, illustrating the major facies. B. Schematic cross section across reef tract showing back-reef lagoonal facies (d), reef core (c), reef flank (b), and basinal facies (a). Hydrocarbons (oil and gas) are generated from raw organic matter that accumulates under low-oxygen bottom conditions in (a). Maturation of organic material to petroleum involves burial under sediment load and subsequent heating, with explusion of hydrocarbons into basin-slope and reef-flank carrier beds (b). Hydrocarbons become trapped mainly in highly porous and permeable carbonates of the reef core, which is situated along edge of platform in attractive position to receive migrating fluids. Favorable reservoir porosity and permeability characteristics of reef core are related to initial organic framework fabric of reef as well as post depositional solution and dolomitization.

(A from R. A. Walls, 1983, Golden Spike Reef Complex, Alberta, Fig. 3, p. 446: *in* P. A. Scholle, D. G. Bebout, and C. H. Moore, eds., Carbonate Depositional Environments: *Am. Assoc. Petroleum Geologists Mem.* 33, Reproduced by permission of American Association of Petroleum Geologists)

and Lower Devonian rocks are conformable. Through much of the rest of the cratonic region, Lower and Middle Devonian rocks are absent. The Middle Devonian transgressive sea deposited fossiliferous sandstones, shales, and limestones. This transgression inaugurated another major onlap-offlap stratigraphic *sequence* cycle (Fig. 11-6) in the cratonic interior.

Much of the Cambrian–Devonian carbonate section in North America consists of dolomite [$CaMg(CO_3)_2$], which formed mainly by secondary replacement of original limestone. Dolomitization of much of the lower Paleozoic carbonate section in the Great Basin province of the Cordillera is believed to have resulted from shallow

subsurface mixing of marine and fresh water, the latter having filtered into the subsurface from exposed tracts of tidal flats (Fig. 11-21). Significant amounts of dolomite may also have formed as early postdepositional replacement of limestone in supratidal environments where magnesium-rich brines were concentrated (Fig. 11-22).

Devonian limestones have furnished some of the finest fossil collecting in North America. It was primarily a comparison of fossiliferous Devonian strata in New York and Iowa that led James Hall to his ideas on geosynclines. Abundant articulate brachiopods, corals, bryozoans, sponges, crinoids, and trilobites comprise an invertebrate fauna that reached its greatest diversity in

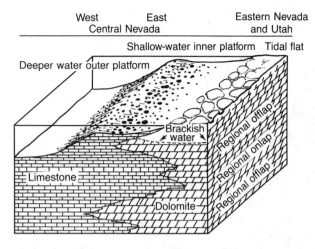

Figure 11-21. Generalized paleogeography of early Paleozoic Cordilleran miogeosyncline and its influence on regional dolomitization of subtidal limestone deposits. Open tracts of supratidal environments and areas exposed during regressive phases acted as areas of freshwater recharge. Subsurface mixing of fresh and marine waters provided optimum Mg/Ca ratios and slow crystallization rates, promoting extensive dolomitization. Lateral migration of recharge areas with regional onlap and offlap resulted in migration of limestone–dolomite boundary with time. Limestones in west-central Nevada escaped dolomitization because of distance from areas of freshwater recharge. Horizontal dimension represents tens to hundreds of kilometers; vertical dimension represents tens to hundreds of meters. (From J. B. Dunham, E. R. Olsen, 1978, Diagenetic Dolomite Formation Related to Paleozoic Paleogeography of the Cordilleran Miogeocline in Nevada; Fig. 1, p. 557: *Geology*, vol. 6. Reproduced by permission of the authors and Geological Society of America.

Middle Devonian seas (Figs. 11-23, 24). Ammonoid cephalopods underwent a major adaptive radiation, and the earliest forms, those with *goniatite sutures,* are important taxa for biostratigraphic subdivision and correlation of Devonian and Carboniferous rocks. Devonian carbonate rocks also are commonly rich in conodonts, which have been important to biostratigraphic studies.

The annual and daily growth increments (analogous to growth rings of tree trunks) of some Middle Paleozoic rugose corals can be compared with those of modern scleractinian corals. The number of daily layers between annual layers generally ranges from 410 to 420 in Devonian species, a significant difference from the 360 to 370 in modern specimens. This comparison suggests that there were more days in a Devonian year than in our present year. The inference here is that if the path of the earth's orbit around the sun has not changed significantly, the speed of the earth's rotation about its own axis has. If such is the case, the cause was probably the braking action of the gravitational attraction of the moon on the earth's rotation. If we safely extrapolate from the Devonian coral data, then the slowdown in the earth's rotation has been on the order of one second every 50,000 years since the origin of the earth–moon system.

The Age of Fish and the Vertebrate Transition to Land

Vertebrate life also diversified during the Devonian. Five classes of fish had evolved by early in the period. Because of the rapid evolutionary adaptive radiation the Devonian has commonly been referred to as "the age of fish." The five fish classes include the jawless Agnatha, represented today by the parasitic lampreys and hagfish; the jawed armored fish—Placodermi and Acanthodii—both extinct; the Chrondrichthyes, including the true sharks, skates, and rays; and the Osteichthyes, which are the bony fishes (Fig. 11-25). Because of their great efficiency in the aquatic environment, the last two groups rapidly replaced the other classes in Late Devonian and have been dominant ever since. Geologic evidence suggests that fish initially evolved in marine environments. However, as part of the Devonian radiation, several groups invaded freshwater habitats. One of the milestones in the evolutionary history of the vertebrates was development of the jaw. The jaw of primitive placoderms is believed to have evolved from the anterior gill arches of agnathan ancestors. By early in the Devonian, fish had evolved jaws lined with cutting teeth and represented a new level of predation.

The bony fishes are divided into two main groups, one of which, the ray-finned fish, includes the forms we are most familiar with today. The other group of bony fishes, the so-called *lobe-finned fish,* were the ancestors of nonfish vertebrate classes. A group of Devonian lobe-finned fish called the *crossopterygians* possessed paired fins that consisted of bones around a strong central axis and attached to the body by a shoulder girdle, structurally different from those of ray-finned fishes (Fig. 11-26). This bone arrangement allowed the lobe fins more freedom of movement at the point of attachment to the body, and their muscles extended into the fin to permit greater control of fin movement. In addition to the specialized fin, crossopterygians had lungs for breathing air, presumably an adaptation for existence in oxygen-depleted terrestrial ponds.

Lobe fins, ideally suited to develop into flexible supportive limbs, and lungs for breathing air proved to be beautifully *preadapted* for making vertebrate life on land a possibility. Before the end of the Devonian, primitive amphibians evolved from a group of crossoptarygians called the *rhipidistians.* Having adapted to freshwater environments that occasionally became dried up or depleted in oxygen, the rhipidistians were constantly in a position of vulnerability. Their lobed fin and lung evolved to cope with such harsh conditions. These structures were adaptations for survival, not for the luxury of walking on land. But the lobed fin itself possessed the basic tetrapod (four-legged) limb plan (Fig. 11-26) and this paved the way for the evolutionary transforma-

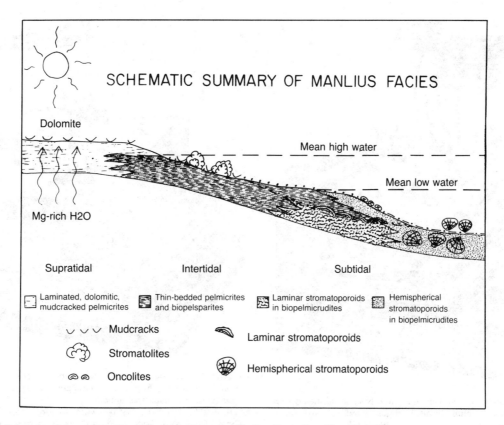

Figure 11-22. Relations of three depositional facies in Devonian Manlius Formation, New York. Lateral migration of facies within Manlius depositional regime resulted in complex facies mosaic in Manlius today. Supratidal deposits periodically prograded across intertidal and subtidal facies; at other times, the supratidal facies were onlapped by intertidal and subtidal deposits, thus forming complex three-dimensional facies relationships.
(From L. F. Laporte, 1967, Carbonate Deposition near Mean Sea-level and Resultant Facies Mosaic: Manlius Formation (Lower Devonian) of New York State; Fig. 29, p. 91: *Am. Assoc. Petroleum Geologists Bull.*, Vol. 51. Reproduced by permission of American Association of Petroleum Geologists)

tion of a water-dwelling creature to a land creature and for the emergence of a new class of vertebrates, the Amphibia.

Living crossopterygians are represented by the *coelacanths,* an evolutionary branch that diverged during the Devonian (Fig. 11-27). Coelacanth fossils have been reported from rocks as young as Cretaceous, but were thought to be extinct. However, in 1939, off the coast of South Africa, a fisherman landed a live coelacanth in his nets; several dozen have been reported since. These "living fossils" have helped clarify the structures of the Paleozoic lobe-finned fish.

The amphibians were not the first inhabitants of the terrestrial environment, however. Land plants, which probably evolved from green algae (see Chapter 14 for thorough discussion of plant evolution), first appear in the Upper Silurian rock record and are represented by forms no longer than matchsticks. During the Devonian, several divisions of *seedless vascular plants* evolved rapidly; by late in the period, diverse forests, including large trees, covered extensive lowland areas. Many elements of this diverse flora provided food for the amphibian invaders. Among the animals, early insects, spiders, and some gastropods preceded the first amphibians in the invasion of the land.

Terrestrial Vertebrates: The Earliest Amphibians

The oldest known amphibians come from Old Red Sandstone deposits in eastern Greenland and are well represented by the genus *Ichthyostega.* Similar in form to their crossopterygian ancestor, the ichthyostegids had streamlined bodies, long tails, and fins, all well suited for efficient swimming. But, in addition, they had four legs and strong hip girdles, shoulders, and rib cages, all structural adaptations for walking on land. Adaptations must have taken place slowly and involved changes in a variety of physiological and structural characters. Problems of breathing, hearing, and feeding, in addition to

Figure 11-23. In-life Middle Devonian muddy-shelf marine-bottom community: a, crinoid echinoderms; b, rugose coral; c, bryozoan; d, trilobite; e, tabulate coral; f, articulate brachiopods.
(From W. S. McKerrow, 1970, *The Ecology of Fossils;* Fig. 34, p. 136: Gerald Duckworth, London. Reproduced by permission of author)

Figure 11-24. In-life Middle Devonian off-reef community: a, nautiloid cephalopod mollusc; b, tabulate corals; c, stromatoporoid sponge; d, articulate brachiopods.
(From W. S. McKerrow, 1970, *The Ecology of Fossils;* Fig. 35, p. 137: Gerald Duckworth, London. Reproduced by permission of author)

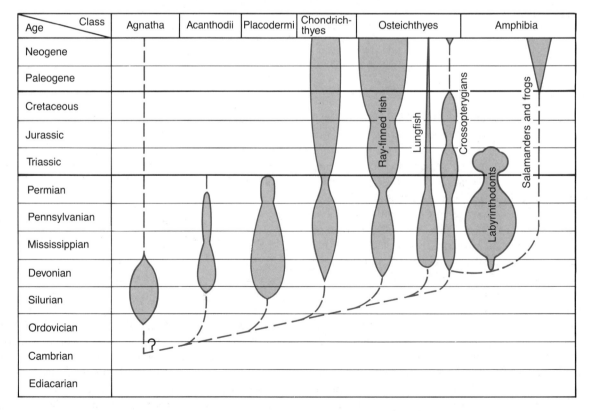

Figure 11-25. Large-scale evolutionary history of fishes and amphibians. Dashed lines show the most probable evolutionary relations. The width of the patterned areas indicates the approximate diversity and abundance of each group.
(Information from several sources)

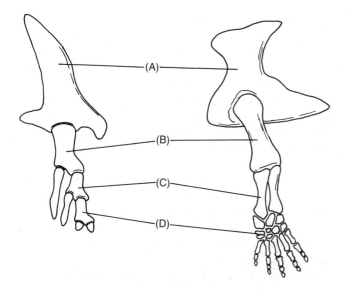

locomotion, had to be overcome. Such changes occurred through a succession of populations before the amphibian condition became a reality. The transition is well documented in the fossil record. During the Carboniferous, the amphibians that followed the ichthyostegids were the *labyrinthodonts,* so named for the intricately grooved, labyrinthine structure of their teeth. A group of labyrinthodonts called *rachitomes* (Fig. 11-27) were squat, bulky creatures with stout limbs and looked somewhat like stubby alligators. They were the dominant land vertebrates on the scene when the first reptiles evolved from them in the Carboniferous.

Today's amphibians, including toads, frogs, newts, and salamanders, play a subordinate role among land vertebrates. Like their ancestors they are still linked to the aquatic environment. They were able to solve the crucial difficulties of air-breathing and locomoting on land, but one final requirement of land life eluded them: Amphibians must return to the water to reproduce; their delicate, naked eggs, like those of fish, would quickly desiccate if deposited in a subaerial environment.

Figure 11-26. Comparison of pelvic region and the hind limb in Devonian crossopterygian fish *Eusthenopteron* (left) and Permian labyrinthodont amphibian *Trematops* (right): (A) pelvis, (B) femur, (C) tibia-fibula, (D) pes.
(From E. H. Colbert, *Evolution of the Vertebrates—A History of the Backboned Animals through Time,* 3rd ed; Fig. 26, p. 71. Copyright © 1980 by John Wiley & Sons, Inc. Reprinted by permission of John Wiley & Sons, Inc.)

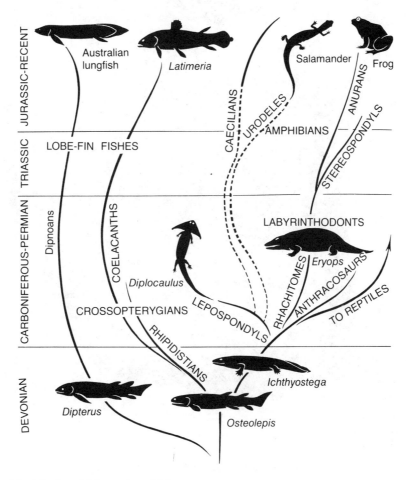

Figure 11-27. Evolution of the lobe-finned fishes and amphibians.
(From E. H. Colbert, 1980, *Evolution of the Vertebrates—A History of the Backboned Animals through Time,* 3rd ed; Fig. 29, p. 87. Copyright © 1980 by John Wiley & Sons, Inc., New York. Reprinted by permission of John Wiley & Sons, Inc., and Lois M. Darling)

More Tectonic Unrest

Middle Paleozoic orogenic disturbances resulted in yet another phase of mountain building in eastern Laurentia (northeastern North America and Greenland) and northwestern Baltica (Europe). This major tectonic event, the **Acadian orogeny,** was superimposed across the deeply eroded roots of the Taconic orogenic belt; from South Carolina to Newfoundland, the Appalachian eugeosyncline, which had been compressed during the Taconic disturbance, was deformed again. The resulting Acadian tectonic land covered an even larger area and shed a tremendous volume of terrigenous clastic debris. Evidence of this major orogenic event is read from the rock record in the same fashion as described for the Taconic orogeny.

This disturbance involved events typical of orogenic belts: folding, faulting, metamorphism, and igneous in-

trusion. Profound angular unconformities separating Lower Devonian and older rocks from Upper and post-Devonian rocks tell part of the story. Granite intrusives and regional metamorphism also were more widespread than during the Taconic orogeny. Isotopic dates from throughout the crystalline Appalachians indicate extensive plutonism and crustal disturbance between about 360 and 330 million years ago, during the later part of Devonian and the early part of Mississippian time.

Lithospheric plate interactions, involving Laurentia, Iapetus oceanic lithosphere, and possibly an island-arc system and ultimately Baltica, deformed the Appalachian geosyncline. By Medial Devonian, suturing of the northeastern part of Laurentia and the northwestern part of Baltica was nearly complete, and the nothern part of the Iapetus ocean basin had closed, thus completing a Wilson cycle. The convergent plate margins that existed as

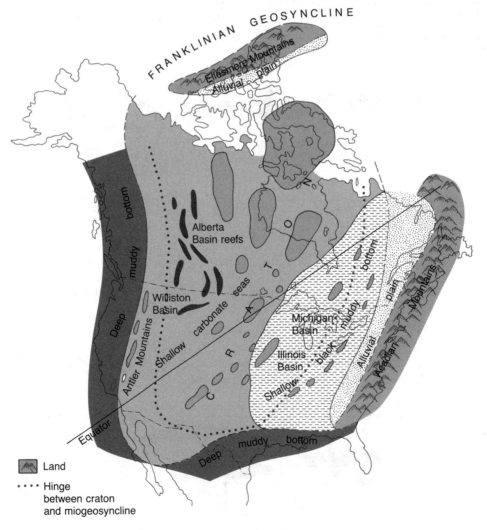

Figure 11-28. Paleography of Late Devonian superimposed on outline map of present-day North America.
(Data from several sources)

a *continent-to-ocean basin* relationship in the Late Ordovician changed to a *continent-to-continent* relationship in the Late Devonian (Fig. 11-28). The present-day Appalachian Piedmont Province contains a complex assemblage of plutonic and metamorphic rocks that are generally Devonian and older. The main part of this province is made up of terranes representing Paleozoic Appalachian eugeosynclinal rocks that were subjected to repeated deformation and intrusion, remnants of volcanic island arcs, and perhaps one or more microcontinents accreted onto the continental margin during the Acadian orogeny (Fig. 11-29). The Caledonian orogenic belt of the British Isles and Scandinavia represents a now-disjunct segment of the Acadian, and perhaps even the Taconic, orogenic belt. The Acadian and Caledonian belts match up well when North America and Europe are restored to their Devonian positions. The contact between Cambrian trilobite provinces (Fig. 10-31), seen

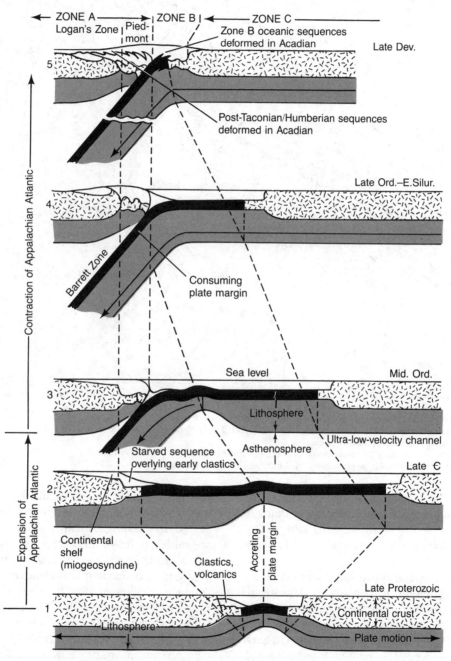

Figure 11-29. Progressive closing of Iapetus during Early Paleozoic, culminating in continent–continent collision and formation of the Acadian orogen in Late Devonian.
(From J. M. Bird and J. F. Dewey, 1970, Lithosphere Plate-continental Margin Tectonics and the Evolution of the Appalachian Orogen; Map B, p. 1046: *Geol. Soc. America Bull.,* vol. 81, No. 4. Reproduced by permission of authors and Geological Society of America)

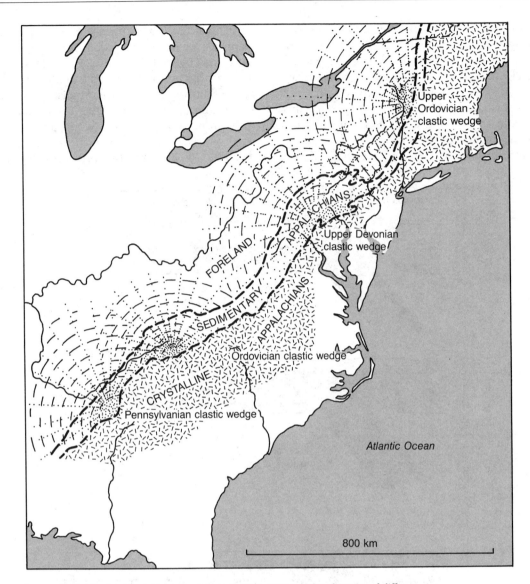

Figure 11-30. Eastern United States showing extent of the principal clastic-wedge deposits of different ages in the sedimentary Appalachians (miogeosynclinal area) and the foreland (Cumberland and Allegheny Plateau Provinces) of the Appalachian belt.
(From Philip B. King, *The Evolution of North America.* Copyright © 1959, rev. ed. © 1977 by Princeton University Press. Fig. 38, p. 63, reprinted by permission of Princeton University Press)

on both sides of the Atlantic, was formed during the Middle Paleozoic suturing of the continental margins of northeastern Laurentia and northwestern Baltica. This collision formed a larger paleocontinent called **Laurussia.**

The Acadian orogeny also produced its own clastic wedge. In fact, the volume of detritus shed westward from growing Acadian highlands was roughly twice that of the clastic wedge complex produced from the Taconic mountains (Fig. 11-30). The Late Devonian to Early Mississippian clastic wedge is named for the Catskill Mountains in upstate New York, where good exposures of these postorogenic continental deposits and shallow, nearshore-marine deposits are exposed. The Catskill clastic wedge provides a most instructive model within

which to appreciate the fundamental distinctions between litho-stratigraphic and time-stratigraphic units and to understand how formally named mappable rock units are related to major lithofacies. From east to west, major facies grade laterally from coarse sandstones and conglomerates, containing abundant red beds of terrestrial origin, to shallow-marine sandstones and siltstones, and finally to marine shales and limestones (Fig. 11-31). The Catskill clastic wedge has received considerable attention through the years and some 75 years after the concept was formulated in Europe, was the subject of one of the first serious studies in this country incorporating the notion of sedimentary facies.

The famous Old Red Sandstone of the British Isles is a Devonian clastic wedge that was shed eastward from

NW

Ohio   NW Pennsylvania   SE Pennsylvania   SE

Shallow marine shelf   Deltas   Alluvial plain   Alluvial fans

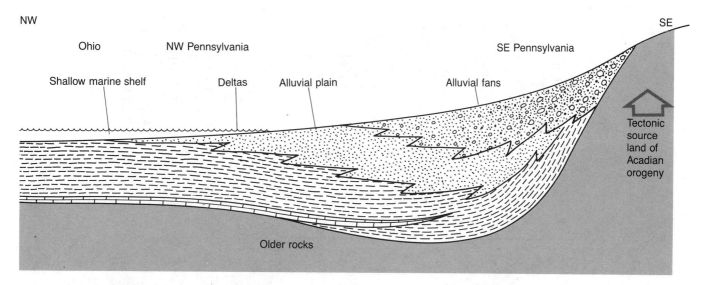

Tectonic source land of Acadian orogeny

Older rocks

Figure 11-31. Catskill clastic wedge and its relationship to Acadian orogen (not to scale). Note onlap-offlap cycles, overall offlap nature of the facies, and progressive fining of sediment with distance from Acadian highlands source area.
(Data from several sources)

Laurussia

Laurentia   Acadian–Caledonian Mountains   Baltica

Catskill clastic wedge   Old Red sandstone

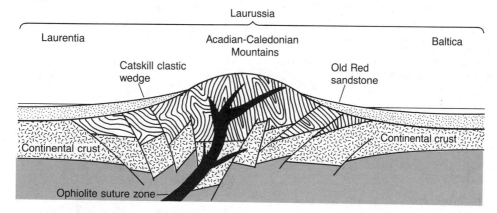

Continental crust   Continental crust

Ophiolite suture zone

Figure 11-32. Bilateral symmetry (mirror imagery) of Catskill clastic wedge and Old Red Sandstone and their relationship to the Acadian–Caledonian mobile belt and suture zone.

the Caledonian highlands. Like its North American Catskill counterpart, the Old Red contains fossil land plants, freshwater fish, and early amphibian remains and interfingers eastward with marine shelly limestones. The apparent mirror imagery of the Old Red and Catskill clastic wedges (Fig. 11-32) is no accident; it resulted from the bilateral symmetry of the Acadian–Caledonian orogenic belt, suggesting that the northern part of the Appalachian geosyncline was bordered by two cratons, the North American (Laurentian) and the European (Baltican). This bilaterally symmetrical orogenic belt is further suggested by such geologic features as opposite direction of overthrusting and zones of intense metamorphism bounded on either side by less deformed rocks, and by the presence of **ophiolite** suites that represent slivers of oceanic lithosphere lying between continent-margin rocks. It is interesting to note that the zone of suturing does not coincide with the seam of desuturing and fragmentation that formed in the Mesozoic when Pangaea broke up. Pieces of Baltica and Gondwana remained as part of the North American continental block. This is why the contact between Cambrian trilobite provinces, formed during the Acadian orogeny, now is found on both sides of the Atlantic (Fig. 10-23).

There is an unmistakable correspondence between the increase in land area as a result of orogeny, the explosive evolution of Devonian land plants, and the origin of the first terrestrial vertebrates, the amphibians. Mountain building, together with the resulting voluminous incursion of clastic wedge sediments to form alluvial plains and deltas, resulted in emergent expenses of continental areas, particularly lowlands. This physical setting provided a host of new, previously unexplored ecological niches. Rapid niche filling and niche parti-

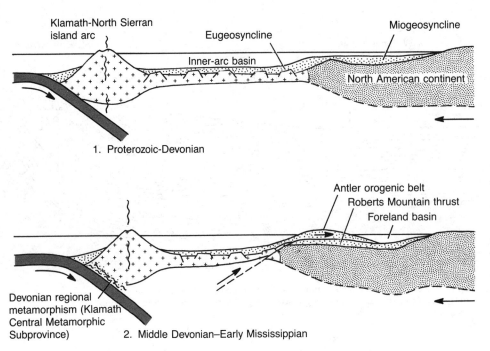

Figure 11-33.   1. A. Relationship between early Paleozoic island arc and the North American continent based on the idea of an east-dipping subduction zone. 2. Overthrust of the Roberts Mountain allochthon from inner-arc basin.
(From F. G. Poole, 1974, *SEPM Spec. Pub.* 22, Fig. 26, p. 80. Reproduced by permission of Society of Economic Paleontologists and Mineralogists)

tioning promoted major adaptive radiation among the new invaders. Enterprising floras and faunas were quick to seize such evolutionary opportunities.

The Cordillera

During the Cambrian Period, North America was similar to present-day Africa in that both eastern and western margins originated from late Precambrian rifting and were either static or trailing edges; the developing geosynclines were similar to the present-day Atlantic margin of North America. We have already discussed how this comparatively passive setting was interrupted in the Appalachian belt as eugeosynclinal rocks were deformed repeatedly from Ordovician through Devonian time. The Cordilleran geosyncline was relatively quiet until the Late Devonian (Fig. 11-33), when subduction of oceanic lithosphere along the western continental margin resulted in pre-Carboniferous eugeosynclinal deposits' being thrust eastward over miogeosynclinal carbonate rocks, in a manner somewhat reminiscent of the Taconic orogeny. This event, called the **Antler orogeny,** occurred along a belt known as the Roberts Mountains thrust and is best displayed in northern Nevada and Idaho (Fig. 11-34). Here the thrust relationships can be seen, and a pronounced angular unconformity separates the structurally deformed rocks from overlying, less disturbed Carboniferous strata. The overthrusting presumably resulted

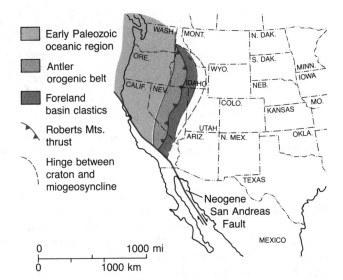

Figure 11-34.   Western United States showing principal tectonic features of the Devonian–Mississippian Antler orogeny.
(Data from several sources)

from the constriction of a small marginal ocean basin as a volcanic island-arc complex converged on the continental block. The Antler event was on a smaller scale than the Taconic and Acadian disturbances, and deposits shed from Antler highlands accumulated mainly in marine environments, with little or no development of redbed alluvial plains.

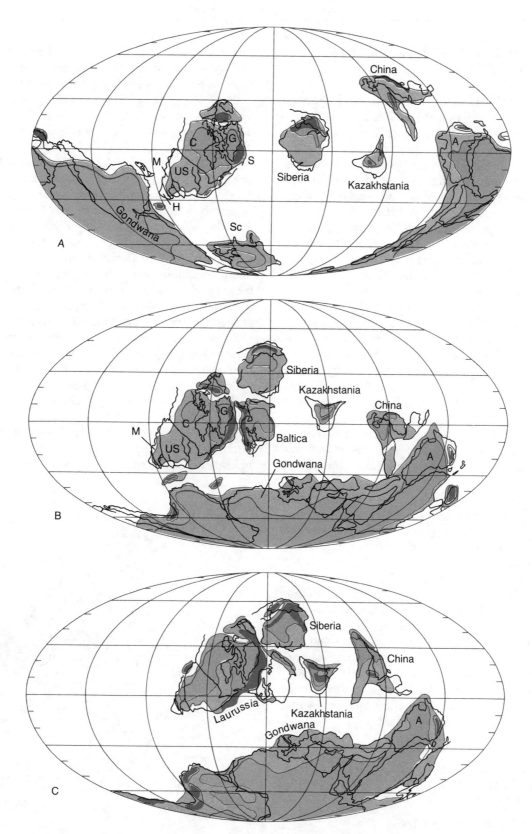

Figure 11-35. Paleogeography of early Paleozoic: deep oceans unshaded; shallow seas light shaded; lowland, intermediate shading; mountains, dense shading; A, Australia; C, Canada; G, Greenland; M, Mexico; S, Scotland; Sc, Scandinavia; US, United States; A. Middle Ordovician world. B. Middle Silurian world. C. Early Devonian world.

(After C. Scotese, R. K. Bambach, C. Barton, R. VanderVoo, and A. Ziegler, 1979, Paleozoic Base Maps; Figs. 9, 15, 21, pp. 243, 249, 255: *Jour. Geology,* vol. 87, no. 3. Reproduced by permission of University of Chicago Press)

PALEOGEOGRAPHY OF THE EARLY PALEOZOIC WORLD

Let us consider North America's position with respect to the other continents during the first half of the Paleozoic. The maps in Figure 11-35 (page 275) show three successive paleogeographic configurations for short intervals of time during the Medial Ordovician, Medial Silurian, and late Early Devonian. The positions of the six continental masses (see Chapter 10) are markedly different from positions in the Late Cambrian.

Gondwana moved southward from its Late Cambrian equatorial position (Fig. 10-26), and its north African portion was the first land area to drift into polar latitudes. This region straddled the South Pole during the Late Ordovician as evidenced by *tillites* of this age in what is now the Sahara Desert! These tillites represent the oldest Paleozoic record of glaciation and, in accordance with our present-day climatic models, suggest that a polar region became landlocked for the first time during the Paleozoic. This resulted in a zonation of climates and refrigeration at the South Pole. The geographic poles of the earlier Ordovician and Cambrian presumably were located in open ocean areas; the shift from equable Cambrian and Early Ordovician climates to more zoned later Ordovician and Silurian climates was a consequence of the more widespread dispersal of continents.

The distribution of evaporite deposits and organic reefs during the Silurian and Early Devonian also provides a basis for latitudinal positioning of the continents; these kinds of deposits in the modern world generally are concentrated between latitudes 30° north and 30° south. As previously discussed, geologic evidence indicates that Baltica and Laurentia had collided to form Laurussia by Late Devonian (Fig. 11-35C), thus closing a portion of Iapetus. Large volumes of deeply oxidized detrital sediments (Catskill clastic wedge and Old Red Sandstone), containing plant fossils lacking seasonal *growth rings,* were shed from the Caledonian and Acadian mountain chains that developed along the suture zone. This suggests that joining of the continental masses occurred in the equatorial belt.

CARBONIFEROUS HISTORY: THE MISSISSIPPIAN AND PENNSYLVANIAN

Mississippian Deposits

The Mississippian System is named for exposures in the Mississippi Valley, particularly in Iowa and Missouri, where limestones rich in oolites and crinoids are characteristic (Figs. 11-36, 37). Some refer to the Mississippian as the age of crinoids because of the widespread abundance of columnals and stem fragments (Fig. 11-38). We should remember that, unlike the other geologic system names (with the exception of the Pennsylvanian), Mississippian is a provincial name, enjoying usage only in North America; it represents the lower part of the worldwide Carboniferous System, whose stratotype is in England.

The Mississippian includes some unusual sedimentary deposits. Throughout much of the eastern craton, west of the Acadian Mountains, accumulation of black, organic-rich mud—the Chattanooga Shale and related deposits—occurred under somewhat restricted anaerobic conditions in a thermally stratified inland equatorial sea during the Late Devonian and Early Mississippian. These conditions were probably related to interactions between climatic and tectonic factors; namely, during times of tectonic activity, the rising Acadian Mountains created a barrier to moisture-laden easterly trade winds, reducing clastic input and favoring deposition of organic-rich muds. These black shales interfinger eastward with coarser clastics of the Catskill clastic wedge, which prograded cratonward during times of tectonic quiescence. Another carryover from Late Devonian was extensive formation of chert. Some of the chert occurs as nodules in limestones and probably represents localized pockets of secondary replacement of host carbonate mudstone. Other chert occurrences include thin- to

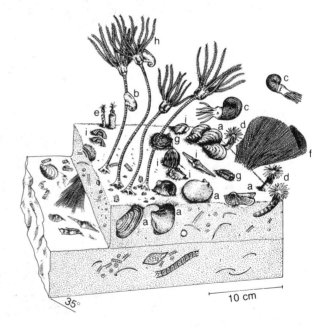

Figure 11-36. In-life restoration of Mississippian shallow-marine bottom community: a, bivalve molluscs; b, gastropod mollusc; c, ammonoid cephalopod mollusc; d, rugose coral; e, serpulid annelid worm; f, bryozoan; g, trilobite; h, crinoid echinoderm; i, articulate brachiopods. ·
(From W. S. McKerrow, 1970, *The Ecology of Fossils;* Fig. 46, p. 163: Gerald Duckworth, London. Reproduced by permission of author)

Figure 11-37. Diorama of Mississippian echinoderm community on sandy-mud bottom in northwestern Indiana. Shown is a sea lily "garden" with starfish (on bottom) and several kinds of crinoids and blastoids.
(Photo courtesy of National Museum of Natural History. Smithsonian Institution Photo no. 658)

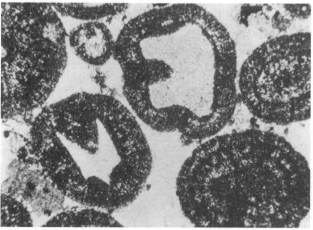

A   B

Figure 11-38. A. Crinoid stem fragments in Mississippian limestone. B. Thin-section photomicrograph of oolitic limestone. Most of the nuclei of ooids are crinoid columnals.
(A, photo courtesy of Ward's Natural Science Establishment; B, photo courtesy of American Associaton of Petroleum Geologists)

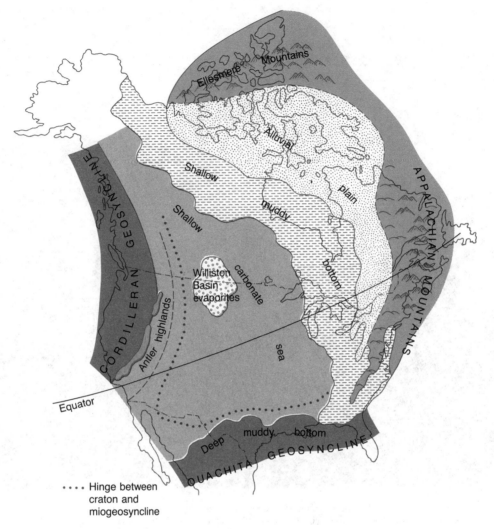

Ellesmere Mountains
Alluvial plain
Shallow muddy bottom
Shallow carbonate sea
Williston Basin evaporites
CORDILLERAN GEOSYNCLINE
Antler highlands
APPALACHIAN MOUNTAINS
Equator
Deep muddy bottom
OUACHITA GEOSYNCLINE
Hinge between craton and miogeosyncline

Figure 11-39. Paleogeography of mid-Mississippian superimposed on outline map of present-day North America.
(Data from several sources)

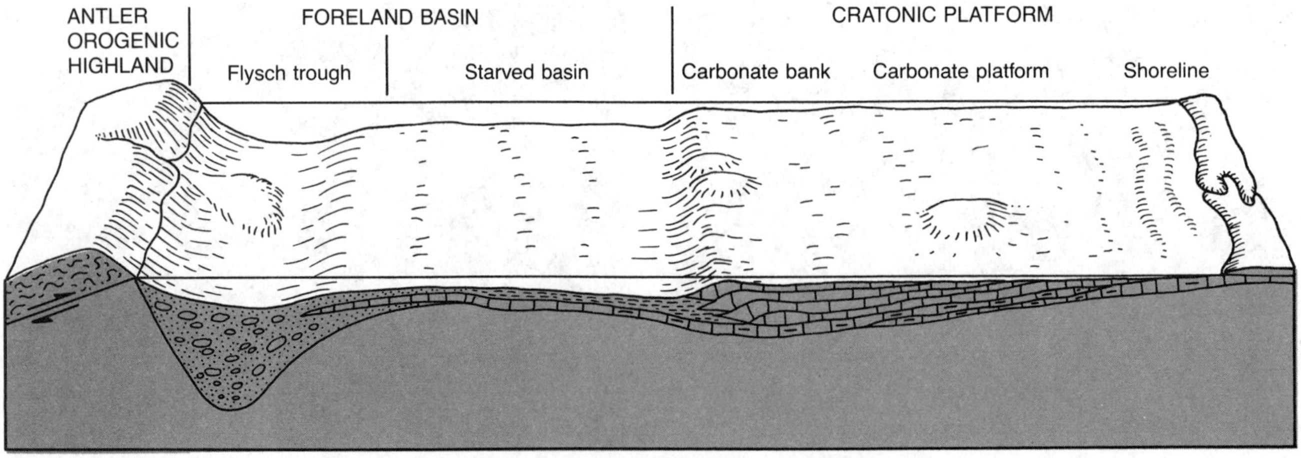

ANTLER OROGENIC HIGHLAND
FORELAND BASIN
Flysch trough
Starved basin
CRATONIC PLATFORM
Carbonate bank
Carbonate platform
Shoreline

Figure 11-40. Depositional setting of Cordilleran foreland basin and cratonic platform during Early Mississippian.
(From F. G. Poole, C. A. Sandberg, 1977, Mississippian Paleogeography and Tectonics of the Western United States; Fig. 3, p. 74: Pacific Section SEPM Paleozoic Paleogeography Symposium Volume. Reproduced by permission of Pacific Section, Society of Economic Paleontologists and Mineralogists)

thick-bedded units, presumably of primary origin. The source of silica was probably mostly biogenic, having been derived from the siliceous skeletal material of sponges and radiolarians. The origin of the thick Caballos novaculite in the Marathon Basin segments of the Ouachita belt (Fig. 11-49) has been the subject of friendly but lively debate between two sedimentologists, R. L. Folk and E. F. McBride, of the University of Texas, Austin. McBride favors a deep, starved geosynclinal basin primary origin; Folk argues for a postdepositional silicification of original peritidal carbonate rocks.

The Mississippian Period witnessed the last widespread Paleozoic carbonate-rich epeiric sea in the cratonic interior of North America (Fig. 11-39). Creamy white oolitic and foraminiferal limestone near Salem, Indiana, has been quarried extensively for choice building stone. Thick ridge- and cliff-forming limestone units such as the Redwall in the Grand Canyon and correlative formations in the northern Rockies, the southern Great Basin, and the Canadian Rockies contain rugose corals, brachiopods, crinoids, bryozoans, amminoids, and sponges, and attest to richly populated carbonate seas in the western part of North America.

Thick deposits of Upper Mississippian terrigenous clastic sediments, mainly of flysch style, accumulated in the *Ouachita geosyncline* (Fig. 11-39) and are visible in the Ouachita Mountains of Arkansas and Oklahoma. Voluminous incursions of terrigenous clastics also accumulated in a foreland basin and shelf area east of the Antler orogenic belt (Fig. 11-40). These deposits include shales to the east and thick flysch accumulations to the west. Late Mississippian deposition throughout the Appalachian miogeosyncline changed from predominantly carbonates to terrigenous muds and sands, a harbinger

of extensive uplift and mountain building that was to climax deformation of the Appalachian geosyncline throughout the remainder of the Paleozoic.

Late Paleozoic Orogenic Disturbance

During the Pennsylvanian Period large areas of the craton were uplifted, and great volumes of clastic sediments were shed off the highlands. This uplift in the ancestral Rocky Mountains elevated large masses of Cryptozoic crystalline basement that served as source areas for thick deposits of conglomerate, arkosic sandstone, and red shale (Fig. 11-41). Several marine basins within the ancestral Rockies received thick deposits of Pennsylvanian clastics and evaporites (Figs. 11-42, 43).

In eastern North America, the **Alleghenian orogeny** climaxed the final phase of a succession of orogenic events that had been occurring in the Appalachian mobile belt since the Ordovician. We have already examined how post-Lower Ordovician miogeosynclinal successions were developed as alternating carbonate and terrigenous clastic packages, signifying times of relative quiescence or mobility in the eugeosynclinal belt. During the Pennsylvanian and Permian, the miogeosynclinal rocks themselves, particularly in the region from New York southward to Alabama, were folded and thrust faulted. The folds in general are asymmetrical and in places even overturned toward the craton (Fig. 11-44A). The thrust faults are east-dipping and occur mainly in Virginia and Tennessee (Fig. 45B). This Appalachian fold-and-thrust belt is the present-day geologic and physiographic Valley and Ridge Province.

What was responsible for this late Paleozoic wave of deformation? The intensity of both folding and thrusting

A   B

Figure 11-41. Erosional remnants of upturned Pennsylvanian arkoses, Front Range of Colorado Rockies. A. Garden of the Gods near Colorado Springs. B. Red Rocks Theatre near Morrison, Colorado. (Photos by J. Cooper)

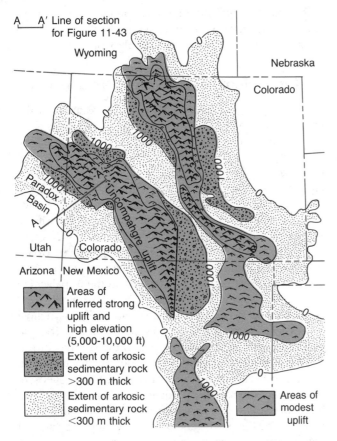

Figure 11-42. Restoration of ancestral Rockies during Pennsylvanian and distribution and thickness of associated arkosic sedimentary rocks.
(Redrawn from W. W. Mallory, 1972, Pennsylvanian Arkoses and the Ancestral Rocky Mountains, Fig. 1, p. 132, *in* Geologic Atlas of the Rocky Mountain region: Rocky Mountain Association of Geologists. Used by permission of Association)

diminishes toward the craton; the Appalachian Plateau Province includes relatively undisturbed upper Paleozoic strata that thicken to the east. The cratonic basement acted as a buttress against which the geosynclinal rocks were pushed from the southeast. If we are consistent in looking to major lithospheric plate interactions for an answer, then perhaps it is reasonable to invoke a model of continental collision whereby Gondwana impinged against the southern margin of Laurussia, thus completing the closing up of the northern part of Iapetus. This plate interaction was part of the late Paleozoic global suturing of continental masses into a supercontinent assembly called **Pangaea.** Just as the Taconic and Acadian orogenic structures have their counterparts in the Caledonian mobile belt of western Europe, so does the Alleghenian structural belt of eastern North America match the Late Pennsylvanian Hercynian orogenic trend of western Europe, northwestern Africa, and southern Asia.

In recent years, seismic reflection profiling data from the Consortium for Continental Profiling Project (COCORP) and detailed geologic studies across the southern Appalachian belt in the Carolinas and Georgia have revealed some startling relationships. The Cryptozoic and lower Paleozoic crystalline rocks that make up the Blue Ridge and Piedmont Provinces appear to overlie an eastward-thickening wedge of younger sedimentary rocks at depth. The implications are that a colossal slab of crystalline rocks has been thrust westward for several hundred kilometers. One preliminary model interprets the thrust sheet as having been emplaced initially during the Taconic orogeny when part of the floor of a marginal basin together with part of the crust of a

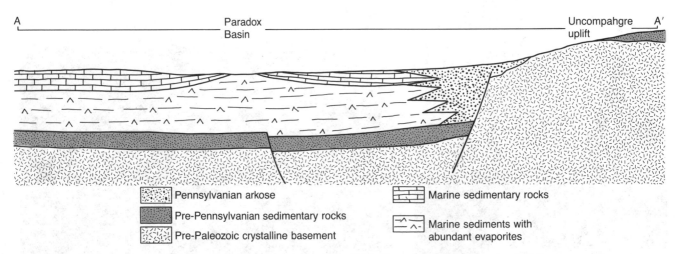

Figure 11-43. Ancestral Rocky Mountains uplift of pre-Paleozoic basement blocks and development of adjacent basin and depositional facies.
(Redrawn from R. J. Hite and F. W. Carter, 1972, Pennsylvanian Rocks and Salt Anticlines, Paradox Basin, Utah and Colorado, Fig. 48, p. 138, *in* Geologic Atlas of the Rocky Mountain Region: Rocky Mountain Association of Geologists. Used by permission of Association)

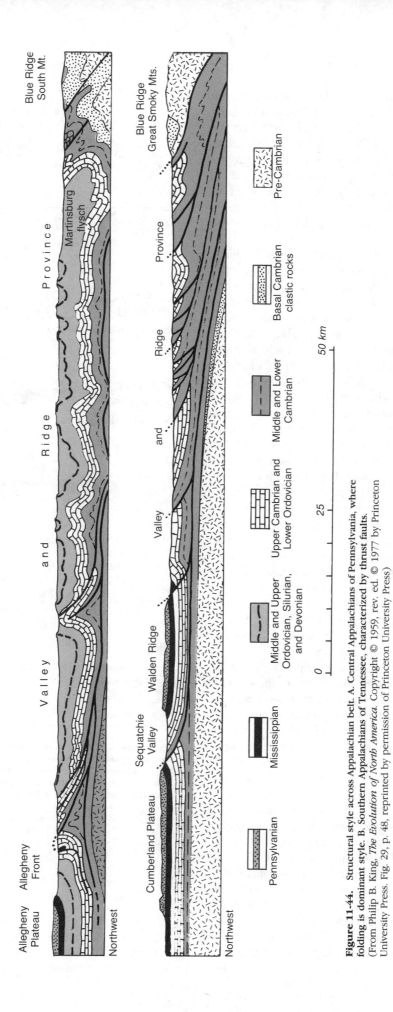

Figure 11-44. Structural style across Appalachian belt. A. Central Appalachians of Pennsylvania, where folding is dominant style. B. Southern Appalachians of Tennessee, characterized by thrust faults. (From Philip B. King, *The Evolution of North America.* Copyright © 1959, rev. ed. © 1977 by Princeton University Press. Fig. 29, p. 48, reprinted by permission of Princeton University Press)

281

microcontinent were thrust onto the continental margin of Laurentia (Fig. 11-45). Later, during the Acadian orogeny, an island-arc system, which had formed to the east of the microcontinent, collided with, became accreted to, and moved with the thrust sheet. Finally, during the Alleghenian orogeny, the African part of Gondwana collided with Laurentia, moved the thrust sheet even farther westward, and folded and faulted the miogeosynclinal strata into their present configuration in the Valley and Ridge Province.

The Appalachian orogenic belt plunges beneath younger deposits of the continental shelf on its southeastern side and under the deposits of the Gulf coastal plain in Georgia and Alabama. It would appear at first glance that the southern terminus of the outcrop of the Appalachian belt is the last we see of it, but is it? A separately named structural trend, the Ouachita, begins near the southwestern terminus of the Appalachian belt and extends for at least 1800 km across the southern United States and probably into Mexico as well (Fig. 11-46). Unlike the almost continuously exposed Appalachian belt, the Ouachita belt is exposed for only 450 km of this distance—in the Ouachita Mountains of Arkansas and Oklahoma and in the Marathon region in western Texas. The remainder is covered by Mesozoic and Cenozoic deposits of the Gulf Coastal Plain; much is known, however, about the covered part from deep drilling. Could it be that the Paleozoic orogenic belts that border the eastern and southern margins of North America are part of a continuous geosynclinal and mobile-belt trend?

The exposed rocks and geologic structures of the Ouachita and Marathon regions suggest a history that began with deposition of a modest thickness of pre-Carboniferous sandstone, shale, and finally chert. This entire pre-Carboniferous section is noticeably thinner than the contemporaneous shelf and platform carbonate sections cratonward! P. B. King, a geologist with the U.S. Geological Survey and longtime student of the Ouachita trend, attributes this odd relationship to slow accumulation of sediments in a subsiding geosynclinal trough. In both the Marathon and Ouachita Mountains segments there is a dramatic change from chert to Mississippian and Pennsylvanian flysch deposits (Fig. 11-47, 48). The thick Carboniferous flysch is composed of sand and mud and occasional boulders derived mainly from tectonic source areas to the south and southeast. The tectonic source areas resulted from subduction of oceanic lithosphere beneath the northern margin of an approaching continental block and deformation of earlier deposits (Fig. 11-49). During Late Pennsylvanian the deformation spread northward and westward and folded and faulted the flysch deposits of the Ouachita trough. This deformation culminated during Late Pennsylvanian and Early Permian time by overthrusting of deformed geosynclinal rocks many kilometers toward the continent.

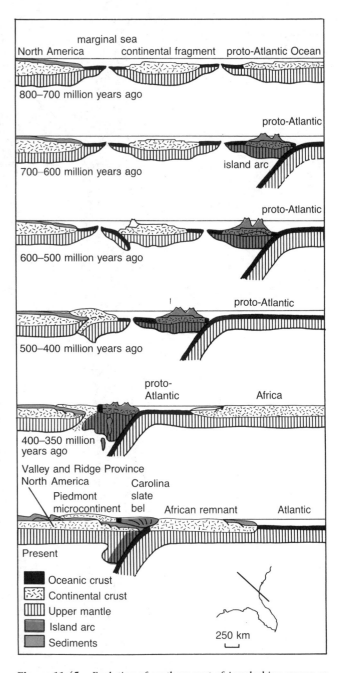

Figure 11-45. Evolution of southern part of Appalachian orogen as a result of collisions between ancient North America, the Piedmont microcontinent, an island arc, and a piece of either Africa or South America. A series of collisions resulted in the overthrusting of parts of the continental margin of North America. This tectonic model is based on geological data and COCORP seismic profiling. (From Jack Oliver, 1980, Exploring the Basement of the North American Continent; Fig. 5, p. 682: *Am. Scientist,* vol. 68, and Zvi Ben-Avraham, 1981, The Movement of Continents; Fig. 8, p. 297. *Am. Scientist,* vol. 69. Reproduced by permission of *American Scientist*)

Coal-Bearing Cycles

By the close of the Paleozoic, eastern North America had become sutured to western Europe and northwestern Africa. The plate collisions that brought about the assembly of continents to form Pangaea had definite

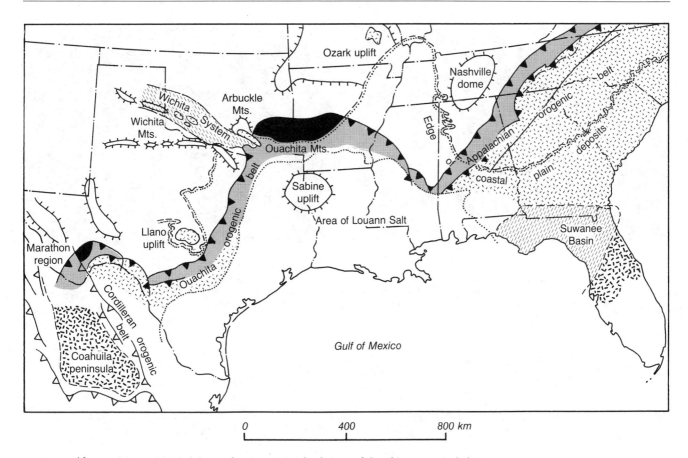

Figure 11-46. South-central United States showing regional relations of Ouachita orogenic belt in surface exposures (black) of Marathon region, West Texas, and Ouachita Mountains, Oklahoma and Arkansas, and in subsurface beneath younger sediments of Gulf Coastal Plain Province.
(From Philip B. King, *The Evolution of North America:* Copyright © 1959, rev. ed. © 1977 by Princeton University Press. Fig. 44, p. 71, reprinted by permission of Princeton University Press and author)

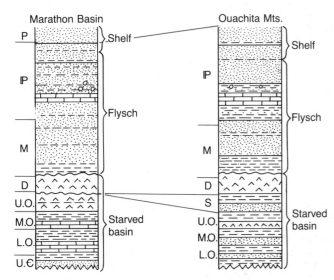

Figure 11-47. Marathon Basin and Ouachita Mountains, showing similarity in Paleozoic sections. Depositional history shows change from deep starved basin sediments in early Paleozoic to thick turbidite flysch sedimentation in late Paleozoic.
(Data from several sources)

mountain-building effects, which in turn exerted much influence on the patterns of sedimentation. The Pennsylvanian records a marked change in the dynamics of sedimentation in the eastern part of the North American craton. Enormous volumes of sand and mud, derived from emerging highlands in the Appalachian–Ouachita mobile belt, were spread westward and northward (Fig. 11-50). Nonmarine, deltaic, and marginal marine deposits of brackish-water bays and lagoons became progressively more widespread as the continent was tilted up along its collision edges. But the retreat of epeiric seas was not a simple one; numerous oscillations of sea level and changes in position of shoreline occurred that produced rhythmically repetitious sedimentary successions called **cyclothems.** Larger scale cycles containing bundles of cyclothems represent tectonic-climatic cycles—related to episodic uplift in the Appalachian–Ouachita belt and to Gondwana glaciation—and changes in sediment budget. The individual cyclothems are more closely related to the delicate interplay between nonmarine deltaic and shallow-marine interdeltaic and shelf environments. Switching of deltaic distributaries resulted in numerous small-scale onlap-offlap cycles. The cyclothem model (Fig. 11-51), produced by the migra-

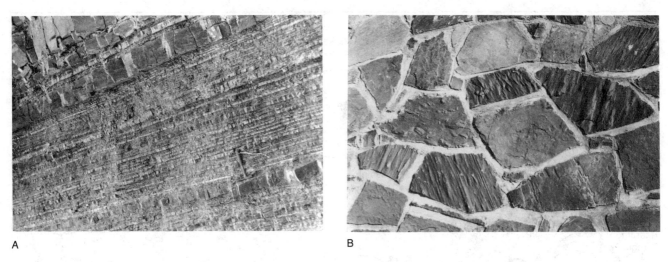

A   B

Figure 11-48. A. Rhythmically bedded turbidite flysch, Haymond Formation, Marathon Basin, Texas. B. Ouachita Mountains flysch, southern Oklahoma, used in wall of building in Talahini, Oklahoma and artistically masoned to display soles of Pennsylvanian sandstone beds exposing flute and groove casts. (Photos by J. Cooper)

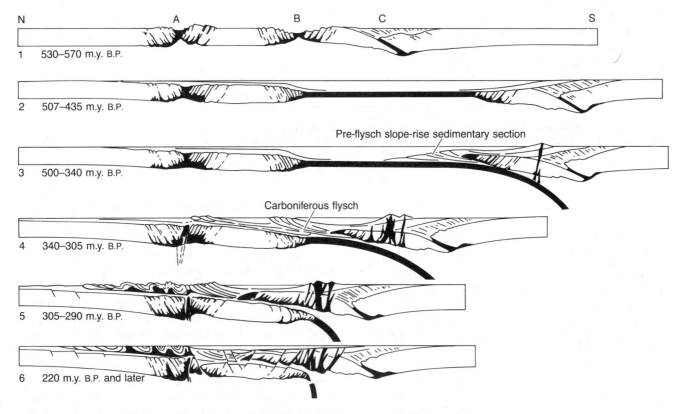

Figure 11-49. Plate tectonic model of the Ouachita orogenic belt. Oceanic crust shown in black. 1. Middle to Late Cambrian; rifting to form Anadarko aulacogen (A) and Mississippi aulacogen (B). 2. Ordovician; open spreading across the Mississippi aulacogen to form a southern extension of Iapetus ocean, with subsidence and deposition along continental margins. 3. Ordovician to Mississippian; continued open spreading of rift ocean, with southward subduction along northern margin of South America. 4. Mississippian to Pennsylvanian; incipient continental collision. The orogenic front along the southern continental margin overrides continental remnants of the northern margin, supplying increasing volumes of clastics to form thick flysch deposits. 5. Pennsylvanian; continued continental collision. Dislocated masses of flysch and preflysch are thrust cratonward. 6. Permian; rifting, normal faulting and fault-block tilting incipient to sea-floor spreading in the Gulf of Mexico Basin.
(After G. Briggs, and D. Roeder, 1975, Sedimentation and Plate Tectonics, Ouachita Mountains and Arkoma Basin, Fig. 6, *in* A Guidebook to the Sedimentology of Paleozoic Flysch and Associated Deposits, Ouachita Mountains and Arkoma Basin, Oklahoma: Dallas Geological Society. Reprinted by permission of Dallas Geological Society)

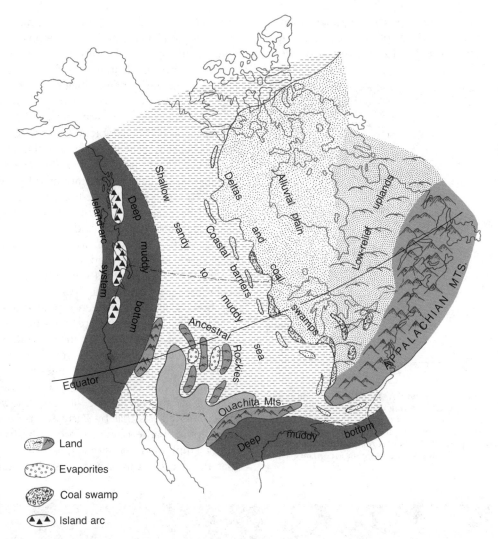

Figure 11-50. Paleogeography of mid-Pennsylvanian superimposed on outline map of present-day North America.
(Data from several sources)

tion of closely related environments and expressed by a characteristic superposition of facies, is a good illustration of Walther's Law. These Pennsylvanian cyclic successions of the Appalachian Plateau Province and craton display rapid vertical and lateral facies changes involving sediments that were deposited at or near sea level in alternating nearshore marine and terrestrial coastal environments. They also characteristically contain coal beds.

The coal beds of the Pennsylvanian cyclothems are the compacted accumulations of stems and leafy vegetable matter in coastal swamps, delta plains, and lagoons—settings similar to the present-day dismal swamps of Virginia and North Carolina and the bayou country of the Mississippi delta in southern Louisiana. These coal deposits have given rise to the abundant coal fields of Pennsylvania, West Virginia, and Illinois, as well as other parts of the Applachian plateaus and midcontinent. Land plants were abundant and varied during the

Pennsylvanian, and coal-forming swamps included lush stands of scale trees, scouring rushes, and ferns (Fig. 11-52) (see Chapter 14). Insects were uncommonly large and abundant in the coal swamps; winged forms similar to dragonflies as well as giant cockroaches were numbered among a diverse fauna. Deltaic-plain deposits at Mazon Creek, Illinois, contain small concretions that entomb exquisite fossils of a variety of plant compressions as well as spiders, centipedes, and insects (Fig. 11-53).

Terrestrial Vertebrates: Emergence of the Reptiles

The oldest reptiles discovered thus far were found in sedimentary rocks of Early Pennsylvanian age near the Bay of Fundy in Nova Scotia. Today the Bay of Fundy is rimmed by cliffs that are battered daily by 15-meter tides, but 280 million years ago, the district lay near the equator and was a tropical lowland covered by dense swamps. Hollow trunks of scale trees and giant club mosses (see Chapter 12), relics of the lush flora that grew in the swamps, were fossilized in the sediments

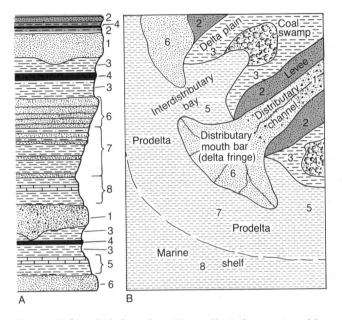

Figure 11-51. Cyclothem deposition. A. Vertical succession of facies (not to scale) of cyclothem showing characteristic cyclic repetition. B. Delta to shallow-marine environments that result in cyclothem deposition. The numbers relate the type of environment to the type of sediments deposited. Facies in A represent the migration of environment of B through time at one place, in accordance with Walther's Law.

exposed in the cliffs; entombed within the hollows of some of the trunks were the skeletons of a small, primitive reptile called *Hylonomus*. These earliest known reptiles were small animals that lived in a terrestrial realm dominated by amphibians. However, amphibians were limited in their colonization of the land because of their dependence on water to lay their delicate eggs. Reptiles solved this problem with an egg that could resist drying out. Thus began one of the greatest adaptive success stories in the history of life. The problem of reproduction on land was surmounted by the evolution of a shelled egg within which the embryo could float in a liquid-filled sac, the *amnion* (Fig. 11-54). The protective layers provided strength and resistance to desiccation. The amnion provided the liquid environment and acted as a shock absorber. Reptilian offspring were able to emerge from the eggs as miniature adults, amphibians had to progress through a larval stage.

Key differences between reptile and amphibian are mostly physiological and are not well preserved in the fossil record, especially for the evolutionary transition from amphibian ancestor to reptile descendant. However, fossil evidence showing differences in skull structure, limb bone construction, jaw and dental patterns, position of the ear, shape of ribs, and vertebral construction suggests that by Early Pennsylvanian time the first reptiles had evolved from closely similar labyrinthodont amphibian ancestors. It is only in later reptiles that the more easily detected skeletal differences be-

Figure 11-52. Restoration of Pennsylvanian coal-forest scene in Illinois showing scale trees, club mosses, and ferns.
(Photo courtesy of Field Museum of Natural History, Chicago Photo No. 75400)

tween reptiles and amphibians developed. These are differences that befit an animal required to move about on land more efficiently than an amphibian can.

Hylonomus and its close relatives from the Bay of Fundy tree stumps belong to the group involved in the initial reptile radiation, the order Cotylosauria (Fig. 11-55). The early cotylosaurs were small forms that most likely fed on insects and grubs. The evolutionary innovations that accompanied the initial radiation of the cotylosaurs were so successful that by early in the Permian several other major reptile groups had evolved from this stock (Fig. 11-56).

PERMIAN HISTORY: THE PALEOZOIC DRAMA ENDS

An Arid Continent

During the Permian Period, the final chapter of Paleozoic history, major regression of seas and gradual emergence of the continent continued. Lower Permian rocks in the Allegheny Plateau and in the New Brunswick Basin in maritime Canada are only gently folded, suggesting that compressional forces that folded and faulted the rocks of the Appalachian miogeosyncline had diminished by Early Permian time. No Middle or Upper Permian rocks have been recognized east of the Mississippi River, but the Lower Permian strata show essentially a continuation of Pennsylvanian depositional pat-

terns. These rocks consist of cyclically bedded sandstone, siltstones, shale, and some coal, and are the products of coastal stream, lake, and swamp continental environments. In the Ouachita and Marathon regions, later Permian deposits, such as fossiliferous limestone in the Glass Mountains of West Texas, unconformably overlap the older Permo-Carboniferous deformed rocks and provide an upper time limit to major deformation.

The Permian inaugurates a long interval of extensive *red bed* deposition in the western part of the craton (Fig. 11-57). The occurrence of red beds signifies depositional environments periodically exposed to the atmosphere—environments such as coastal tidal flats, river floodplains, alluvial fans, and lakes. During the Permian, interior North America was the scene of widespread distribution of such environments, fed by northward- and westward-draining streams that carried iron-rich sediment loads from marginal highlands. The various hues of the red beds—red, maroon, orange, vermilion, purple, and lavender—have been brought about by oxidation through direct exposure to the atmosphere.

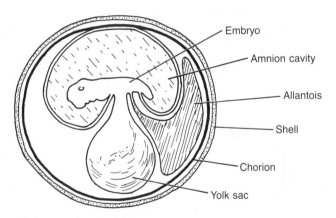

Figure 11-54. Amniotic egg.
(From various sources)

Figure 11-53. Fossil plants in concretions from Pennsylvanian deposits, Mazon Creek locality, Will County, Illinois. A. *Neuropteris.* B. *Annularia.*
(Photo courtesy of Field Museum of Natural History, Chicago. Photos No. 81443-A 81011)

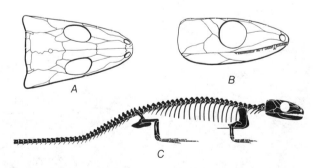

Figure 11-55. The oldest known reptile, *Hylonomus,* a small, slender, and agile cotylosaur. A. Dorsal view of skull. B. Lateral view of skull. C. Skeleton.
(From E. H. Colbert, *Evolution of the Vertebrates—A History of the Backboned Animals through Time;* 3rd ed., Fig. 42, p. 113. Copyright © 1980 by John Wiley & Sons, Inc., New York. Reproduced by permission of John Wiley & Sons, Inc.)

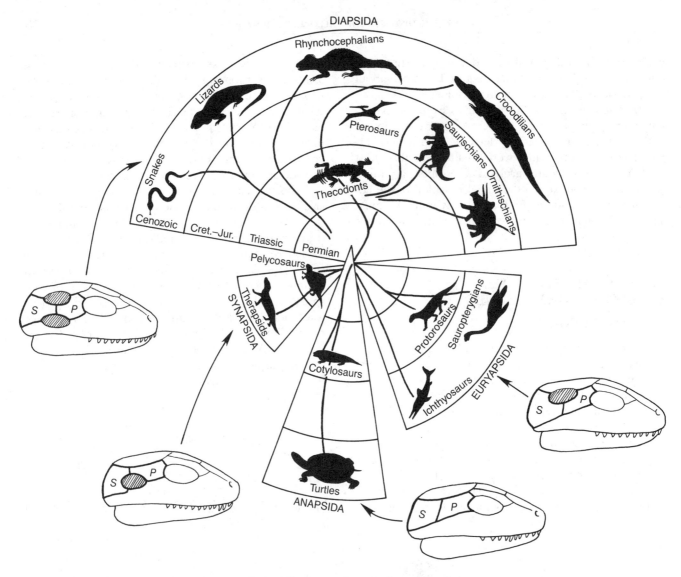

Figure 11-56. Evolution of major reptile groups. Basic skull structure shown for four major subclasses of skulls are differentiated mainly by number and position of temporal openings (shaded); S, squamosal bone; P, postorbital bone.

(Evolution diagram from E. H. Colbert, *Evolution of the Vertebrates—A History of the Backboned Animals through Time;* 3rd ed., Fig. 45, p. 119. Copyright © 1980 by John Wiley & Sons, Inc., New York. Reproduced by permission of John Wiley & Sons, Inc., and Lois M. Darling)

A   B   C

Figure 11-57. A. Cross-bedded Permian Coconino Sandstone, Colorado Plateau, Arizona. B. Monument Valley Arizona–Utah. Mesas and buttes composed of Organ Rock Shale and DeChelly Sandstone, Permian fluvial and eolian deposits. C. Fisher Towers, Utah, composed of sandstones and siltstones Cutler Formation (Permian) and Moenkopi Formation (Triassic).

(A, C, photos by J. Cooper; B, photo by John S. Shelton; used by permission)

Red beds can form under both humid and dry conditions; however, during the Permian, widespread conditions of aridity are suggested by evaporite deposits of anhydrite and gypsum. Extensively cross-bedded sandstone units (Fig. 11-57A) with well-rounded, sorted, and frosted quartz grains suggest ancient sand-dune deposits and evoke a picture of coastal deserts. Perhaps the aridity was the result of continentality that came about through the continental collisions and suturing of fragments into the super landmass of Pangaea. The mountain belts that developed along the sutures no doubt acted as barriers to moisture, thus creating rain shadows that contributed to this change in world climatic patterns.

Last Vestiges of Paleozoic Seas

In North America open-marine deposition during the Permian was confined principally to the western part of the United States and Canada (Fig. 11-58). The Middle Permian Kaibab Limestone forms the spectacular rimrock of the Grand Canyon and is widely exposed over the surrounding plateau and as far west as southern Nevada. The Kaibab is the deposit of a marine invasion from the Cordilleran miogeosyncline onto the cratonic shelf. Sandwiched between the Kaibab and the Mississippian Redwall Limestone, in the walls of the Grand Canyon, are Pennsylvanian and Permian red beds that signify deposition on coastal mudflats and river floodplains. A prominent cross-bedded quartz sandstone, the Coconino, represents ancient sand dunes of a Permian desert (Fig. 11-57A). Northward from the Kaibab sea, a moderately deep and restricted basin (Fig. 11-58) was the site of accumulation of the phosphate-rich sediments of the Phosphoria Formation. This unit has been extensively mined for fertilizer in the northern Rocky Mountain states of Wyoming, Idaho, and Montana. A tremendous thickness of limestone and sandstone accumulated

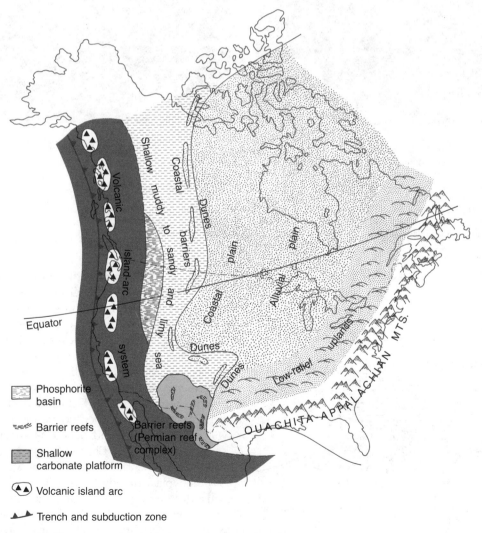

Figure 11-58. Paleogeography of Late Permian superimposed on outline map of present-day North America.
(Data from several sources)

Figure 11-59. Permian reef complex, southern end of Guadalupe Mountains, West Texas and southeastern New Mexico. A. Looking northwest at El Capitan Peak, Guadalupe Mountains, exposing forereef and reef facies of Capitan Limestone. B. Looking east at western escarpment of Guadalupe Mountains, showing offlap succession of basinal (Bell Canyon Formation) and fore-reef and reef (Capitan Formation) facies. Top of Guadalupe Peak is composed of limestones of back-reef facies (Carlsbad Group). C. Looking northeast at back-reef carbonates of Carlsbad Group, Guadalupe Mountains. D. Anhydrite (light) and calcite (dark) showing varve lamination, Castile Formation, southeastern New Mexico.
(Photos by J. Cooper)

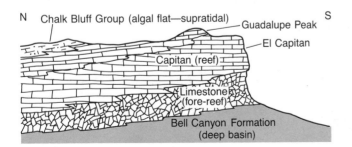

Figure 11-60. Relationship between stratigraphy and environments, upper Permian of Guadalupe Mountains, Texas and New Mexico.

in the Oquirrh Basin in northwestern Utah and southern Idaho. Thick eugeosynclinal deposits in northern California, Oregon, British Columbia, and Alaska are composed largely of volcanic material. Permian volcanism in the eugeosyncline was an expression of oceanic plate subduction in the western borderland (Fig. 11-58).

To the south in the Midland and Delaware basins (together called the Permian Basin) of West Texas and southeastern New Mexico, thick deposits of the famous Permian reef and associated lithofacies were formed (Fig. 11-59). A splendid platform-to-deep basin succession (see discussion, Chapter 6) is exposed in the east-facing escarpment of the Guadalupe Mountains (Fig. 11-60), and platform-edge carbonate rocks buried in the subsurface have produced abundant hydrocarbons. The stratigraphic succession in this region serves as the North American standard Permian reference section.

Figure 11-61. Diorama of Permian sandy-mud sea-bottom scene, north-central Texas. Shown are algae, sponges, corals, crinoids, brachiopods, scallops, mussels, snails, and nautiloid and ammonoid cephalopods.
(Photo courtesy of National Museum of Natural History. Smithsonian Institution Photo No. 657)

Figure 11-62. Diorama of Permian patch-reef community, Glass Mountains, West Texas. Shown are algae, sponges, rugose corals, spiny brachiopods, and nautiloid and ammonoid cephalopods.
(Photo courtesy of National Museum of Natural History. Smithsonian Institution Photo No. 652)

Life of the Permian

Invertebrate life in the seas during Permian time (Figs. 11-61, 62) was highlighted by the great abundance of fusulinid Foraminifera. After their first appearance in the Late Mississippian, fusulinids flourished during the Pennsylvanian and Permian and provide excellent index taxa for the upper Paleozoic. For example, as illustrated in Chapter 6, their biostratigraphic usefulness has contributed importantly to time-stratigraphic subdivision and understanding of the depositional history of the petroleum-rich, reef-rimmed Permian Basin. Paleobiogeographic differences in Permian fusulinid faunas within the Cordilleran geosynclinal belt suggest post-Permian convergence of what were presumably separate Asiatic and American faunal provinces—a consequence of lithospheric plate interactions (see later discussions, Chapter 12). Spinose productid brachiopods, such as the beautiful silicified specimens from the Glass Mountains of West Texas (Fig. 11-62), are characteristic of the Permian; calcareous sponges, molluscs, and corals were also abundant.

The dominant tetrapods during the Early Permian were the pelycosaurs. They evolved into a diverse assemblage of plant-eating and predatory forms, and many were up to 3 m long. The genera *Edaphosaurus* and *Dimetrodon,* from the Permian red beds of north-central Texas, exemplify the main characteristics of the pelycosaur line (Fig. 11-63). It is interesting that the teeth of the powerful carnivore *Dimetrodon* show a fairly marked degree of adaptation for different functions, a hint of the mammalian condition of well-differentiated incisors, molars, and canines that was yet to come. The real hallmark of the pelycosaurs, however, was the erect sail, formed by vertebral spines covered with skin, along the back of the animal (Fig. 11-64). This elaborate structural feature probably functioned as a heat receptor and radiator that evolved as a special adaptation to meet requirements determined by selection pressures of the environment.

By Late Permian, the pelycosaurs had been replaced largely by their more successful descendants, the therapsids, better known as the mammallike reptiles (Fig. 11-63). The therapsids evolved from the carnivorous pelycosaur line, but following the pattern repeated over and over again by groups of tetrapods, they rapidly diversified into separate carnivore and herbivore lines. The therapsids made up at least 90 percent of the known reptile genera during the late Permian and occupied diverse ecological niches.

End-of-Paleozoic Faunal Crisis

In spite of this seemingly rich array of life, the Permian was a time of calamity and crisis for many groups of organisms, both invertebrate and vertebrate,

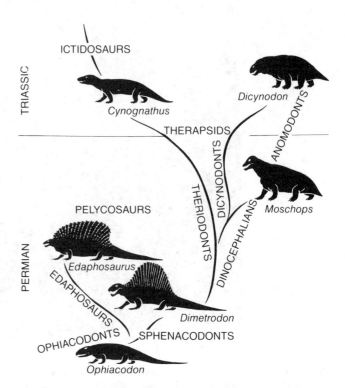

Figure 11-63. Evolution of the synapsid reptiles. (From E. H. Colbert, *Evolution of the Vertebrates—A History of the Backboned Animals through Time;* 3rd ed., Fig. 48, p. 127. Copyright © 1980 by John Wiley & Sons, Inc., New York. Reproduced by permission of John Wiley & Sons, Inc., and Lois M. Darling)

marine and terrestrial. Before the period ended roughly one-half of the invertebrate families, 75 percent of amphibian families, and 80 percent of reptile families became extinct (Fig. 11-65). Raup and Sepkoski have calculated a Late Permian extinction rate of about 14.0 families per million years, significantly greater than the "normal," background, rate of about 8.0 families per million years. Plants (Chapter 14) did not suffer quite the calamity at the end of the Paleozoic as did the animals, although Permian plant forms do decline in comparison to the lush, diverse Pennsylvanian coal-forest flora.

Trilobites, which had been on the wane since the end of Devonian, did not survive the end of the Paleozoic. Fusulinid Forminifera, rugose and tabulate corals, many kinds of brachiopods, two orders of bryozoans, and several groups of attached echinoderms became extinct. All of these groups had been successful and some even dominant during the Paleozoic.

Physical and biological causes for mass extinctions were discussed in Chapter 7. Regarding the end of Paleozoic extinctions, the unifying concept of plate tectonics may well hold the key. We have already mentioned the possible causal relationship between plate tectonics and the onset of climatic aridity as the Paleozoic drew to a close. This may have been responsible for some of

Figure 11-64. Skeleton of pelycosaur *Edaphosaurus* from Permian red beds, north-central Texas. (Photo courtesy of National Museum of Natural History. Smithsonian Institution Photo No. 36248-A)

the terrestrial floral and faunal extinctions, but what about the large number of marine invertebrates? Continental convergence toward the end of the Paleozoic and the formation of Pangaea caused a draining away of epeiric seas and the consequent loss of marine habitats and ecospace. Previously separated, even isolated, faunas were brought into closer proximity and had to compete for the same dwindling resources. Highly specialized organisms like the ornately spinose productid brachiopods, with a limited range of environmental tolerance, were not able to cope with the pressures of the changing world.

From Late Devonian to the end of Permian there is a marked gradual decrease in total diversity among marine invertebrates, presumably a biological consequence of continental convergence, the closing of seaways, and total reduction of provinciality. The Paleozoic faunal patterns suggest a general correlation between comparatively higher diversity and continental separation (Fig. 11-65) and lowered diversity (involving extinctions) during times of continental assembly. Perhaps changes of large magnitude in the relative configuration of the continents may have had far-reaching effects on environments and, in turn, environmental changes exerted fatal stress on numerous groups of organisms. For example, continental clustering, resulting in a more arid climate, was the underlying cause of the formation of extensive

marginal-marine evaporite deposits. The tying up of salts in these environments may have caused the oceans to become less saline, a condition that would have impacted populations of organisms with narrow salinity tolerance ranges.

PALEOGEOGRAPHY OF THE LATE PALEOZOIC WORLD

Figure 11-66 shows a succession of paleogeographic reconstructions—leading to the formation of Pangaea in the Permian—of the late Paleozoic world. The Carboniferous reconstructions, like those of the earlier Paleozoic, are based on a variety of paleomagnetic and geologic data compiled and synthesized by Bambach, Scotese, Ziegler, and Parrish of the University of Chicago. The pageant of changes in these reconstructions shows that Gondwana continued to move across the South Pole from its earlier Paleozoic positions and gradually converged on Laurussia, closing the ocean basin between them. The Hercynian and Alleghenian orogenies produced mountain belts in what are now eastern North America, North Africa, and central Europe. During collision between Gondwana and Laurussia, the Baltica part of Laurussia was displaced northward along a series of megashears that developed on the original suture

A

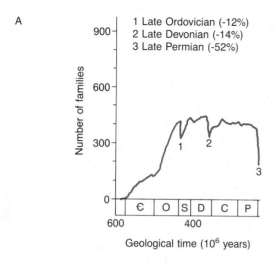

1 Late Ordovician (-12%)
2 Late Devonian (-14%)
3 Late Permian (-52%)

Number of families

900

600

300

0

€ | O | S | D | C | P

600   400

Geological time (10^6 years)

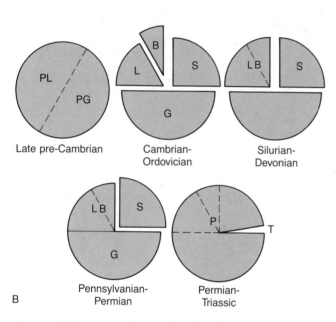

Late pre-Cambrian

Cambrian-
Ordovician

Silurian-
Devonian

Pennsylvanian-
Permian

Permian-
Triassic

B

Figure 11-65. Faunal diversity. B, Baltica; G, Gondwana; L, Laurentia; LB, Laurussia; P, Pangaea; PG, proto-Gondwana; PL, proto-Laurasia; S, Siberia; T, Tethys. A. Standing diversity through Paleozoic for
families of marine invertebrates and vertebrates, highlighting abrupt
drop in diversity (mass-extinction event) at the end of the Permian.
Late Ordovician mass-extinction event may have been influenced by
rapid cooling of marine waters brought on by glaciation. B. Schematic continental configuration during Paleozoic showing relationship between diversity and degree of continental separation.
(A from D. M. Raup and J. J. Sepkoski, Jr., 1982, Mass Extinctions in the
Marine Fossil Record; Fig. 2, p. 1502: *Science*, vol. 215. Copyright 1982
by the American Association for the Advancement of Science. B. from
J. W. Valentine and E. M. Moores, 1972, Global Tectonics and the Fossil
Record; Fig. 2, p. 170: *Jour. Geology*, vol. 80, no. 2. Reproduced by
permission of University of Chicago Press)

zone between Baltica and Laurentia. These large-scale
strike-slip faults presently can be mapped from coastal
New England across Newfoundland and through Scotland. The Great Glen Fault of Scotland and the Cabot
Fault of Newfoundland match up when western Europe

is restored to its post-Devonian pre-Late Carboniferous
position with respect to northeastern North America.

The Late Carboniferous orogenic belt straddled the
equator as suggested by distribution of climatically sensitive Upper Carboniferous deposits. As with the preceding Caledonian-Acadian disturbances, great volumes of
terrigenous clastic sediments were shed from the mountainous highlands. This time the deltaic plains and
coastal-lagoon and marsh paleoenvironments of the
clastic-wedge successions were the sites of extensive
coal swamps, where vegetation accumulated under conditions of heavy rainfall. These environments gave rise
to the great coal deposits of the eastern United States,
western Europe, and Donetz Basin of the U.S.S.R. The
plant fossils in the coal-bearing cyclothems of these regions do not show strong seasonal growth rings; this
implies that the vegetation grew in a wet but constantly
warm tropical belt. Coal deposits developed also in Siberia and China, but here the plant fossils have growth
rings, indicating seasonal climates, which, in turn, imply
that these continental blocks occupied a north temperate belt during the Late Carboniferous (Fig. 11-66).

Tillites in southern Gondwana attest to major continental glaciation during the Permo-Carboniferous time
interval, supporting the reconstruction of the south polar position of this region at that time. Plant fossils with
seasonal growth rings also are found in the south temperate latitudes of Gondwana, marginal to the ancient
ice sheets. Recall that it was this evidence of late Paleozoic glaciation in modern-day South America, South Africa, India, and Australia that inspired Edward Suess, in
the late nineteenth century, to support the notion that
these continents were once part of a larger assembly
(Chapter 1). Eustatic sea-level changes, brought about by
the waxing and waning of these Gondwana ice sheets,
are recorded in the Carboniferous and Permian cyclothems on the continents.

Paleomagnetic information suggests a Late Permian
geographic configuration that is supported by distribution of evaporites and other climatically sensitive deposits. The formation of Pangaea resulted in an enormous
ocean, **Panthalassa,** which spanned the globe from
pole to pole and encompassed nearly 300° of longitude!
Such a configuration would have exerted a major control on oceanic circulation, which, in turn, would have
profoundly influenced Permian climates. These climates
were characterized by aridity. In accordance with present-day climatic models, the equatorial currents, driven
by the trade winds, would have flowed uninterrupted
around five-sixths of the circumference of the earth, im-
pinging against the east-facing coast of Pangaea, making
it extremely warm. The warm waters of the **Tethys Sea,**
the indented eastern margin of Pangaea, probably were
circulated by gulf streams that extended warm conditions into higher latitudes.

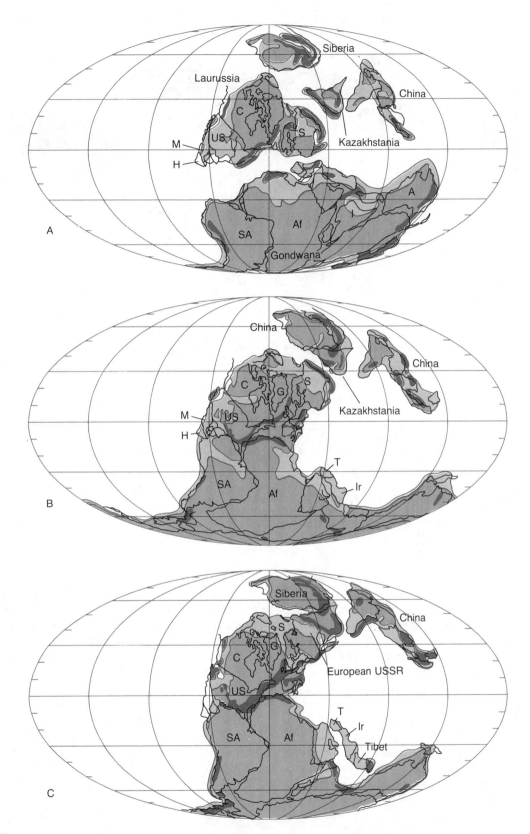

Figure 11-66. Paleogeography of late Paleozoic. Deep oceans unshaded; shallow seas, light shaded; lowland, intermediate shading; mountains, dense shading. A, Australia; Af, Africa; C, Canada; H, Honduras; I, India; Ir, Iran; M, Mexico; S, Spain; SA, South America; T, Turkey; US, United States. A. Mississippian world. B. Pennsylvanian world. C. Late Permian World.

(After C. Scotese, R. K. Bambach, C. Barton, R. VanderVoo, and A. Ziegler, 1979, Paleozoic Base Maps; Figs. 33, 38, 39, pp. 267, 272, 273: *Jour. Geology,* vol. 87, no. 3. Reproduced by permission of University of Chicago Press)

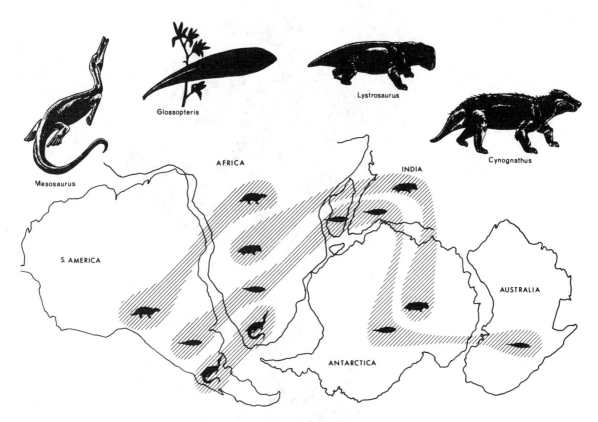

Figure 11-67. Gondwanaland reassembled and some of the paleontological links that bind it together. *Mesosaurus*, a Permian reptile, occurs in southern Brazil and South Africa; *Glossopteris*, a Permian plant, occurs across all of the Gondwana components; *Lystrosaurus*, a Lower Triassic therapsid reptile, occurs in South Africa, peninsular India, southeast Asia and Antarctica; *Cynognathus*, a Lower Triassic reptile, occurs in Argentina and South Africa. Late Paleozoic continental glacial deposits are also present on the five continents shown.

(From E. H. Colbert, 1973, *Wandering Lands and Animals;* Figs. 30, 31, pp. 68, 72: E. P. Dutton, New York. Reproduced by permission of author)

The paleobiogeography of some Permian cotylosaurs and pelycosaurs has contributed importantly to the interpretation of continental positions. The Permian cotylosaur *Mesosaurus* has been found in Brazil and South Africa. The plant-eating pelycosaur *Edaphosaurus* and an associated fauna of other reptiles, crossopterygian fish, and labyrinthodont amphibians have been found in Permian deposits of both northern Texas and Czechoslovakia. These now disjunct occurrences, which reinforce the paleomagnetic data and other lines of geological evidence, support the reconstruction of continents for Permian time (Fig. 11-67).

The mountain belts between Gondwana and Laurussia formed barriers that created giant rain shadows, even in tropical latitudes. These desert conditions are recorded by extensive dune deposits and evaporites. Major withdrawals of shallow seas—brought about by plate collisions—together with changes of volume of sea-floor spreading ridges, left large areas of exposed land, which contributed to the climatic extremes of severe aridity. Abundant evaporites accumulated along the margins of these retreating seas, and red beds blanketed wide areas of the exposed supercontinent.

We have already seen how this inhospitable world presented a real survival test for a major part of the biosphere. Only the hardy, adaptable groups of organisms survived.

SUMMARY

Carbonate deposition in the miogeosynclines and craton continued generally uninterrupted from Late Cambrian into Early Ordovician, but near the end of Early Ordovician major regression of epeiric seas produced a craton-wide erosion surface. By Late Ordovician nearly the entire craton was inundated once again by shallow epeiric seas that contained a complex mosaic of carbonate environments.

This renewed Ordovician transgression was accompanied by an evolutionary explosion of the calcium carbonate shell, reflecting major adaptive radiation of such shell-bearing taxa as articulate brachiopods and bryozoans. Other important faunal groups that participated in the Ordovician radiations included molluscs, echinoderms, and corals, and noncalcareous-secreting taxa such as graptolites, conodonts, and ostracodes. Primitive jawless fish, the ostracoderms, were present in modest diversity, and represent an early radiation of vertebrates after their origin in the Cambrian.

From Late Ordovician to Early Silurian, the Taconic orogeny deformed the northern part of the Appalachian geosyncline. The Taconic, the first of several phases of deformation in the Appalachian belt, resulted from subduction of oceanic lithosphere beneath the eastern margin of Laurentia and telescoping of the continental margin as Iapetus began to close. Erosion of the Taconic mountain land mass produced a voluminous clastic wedge.

During the Silurian, Taconic highlands continued to be worn down, an alluvial plain blanketed a strip of eastern Laurentia, and carbonate epeiric seas covered much of the cratonic interior. Evaporite deposits accumulated in several intracratonic basins that were rimmed with organic reefs.

Withdrawal of epeiric seas during Early and Medial Devonian resulted in extensive erosion in the craton, with consequent removal of much of the Silurian section. Renewed invasion of cratonic seas in Medial Devonian produced fossiliferous carbonate deposits. Coral and stromatoporoid reefs continued to flourish around the margins of intracratonic basins. The Paleozoic record of North America consists of four large-scale onlap-offlap cycles, called sequences, that are bounded by unconformities of cratonwide extent. The sequences are the first-order rock-response products of interactions between global tectonics, climate, and eustatic sea-level changes.

Vertebrate organisms diversified greatly during the Devonian and by early in the period five fish classes were well represented. A group of freshwater bony fish—the lobe-finned crossopterygians—gave rise to the class Amphibia late in the period, thus beginning the vertebrate transition to land. Primitive vascular plants, which had appeared during the Silurian, also experienced dramatic diversification and by Late Devonian much of the land surface was cloaked with vegetation. Devonian was a time of maximum Paleozoic diversity. Marine invertebrates, however, experienced a wave of mass extinctions late in the period; from Late Devonian onward there was a gradual decline in diversity, culminating in the end-of-Paleozoic faunal crisis.

In the Appalachian region, Late Devonian mountain building—the Acadian phase of the Appalachian orogeny—resulted from final closing of the northern part of Iapetus as Baltica and Laurentia collided to form the continental mass Laurussia. The Catskill sediments of northeastern North America and the Old Red Sandstone of Greenland and western Europe are parts of clastic wedge facies associations that developed symmetrically on opposite sides of the suture orogen.

In Late Devonian and Early Mississippian, subduction along the Cordilleran margin of the continent caused telescoping of a marginal basin and produced the Antler orogeny. The Devonian–Mississippian transition also was a time of abundant chert formation and deposition of black, organic-rich shales. Clean oolitic and crinoidal deposits characterize the central craton and much of the Cordilleran belt and attest to widespread invasion of the last great carbonate-producing epeiric sea during the Mississippian. Deposits of flysch accumulated in the Ouachita geosyncline signaling the beginnings of orogeny.

Pennsylvanian was a time of tectonic activity in the ancestral Rocky Mountains and the Appalachian–Ouachita belt and of cyclic sedimentation in tectonic foreland basins and the craton. Thick piles of Upper Mississippian and Pennsylvanian flysch in the Ouachita geosyncline were derived from erosion of tectonic sourcelands raised along the southern margin of Laurussia as a result of plate collisions. Compressional folding and thrust faulting in the Appalachian miogeosynclinal belt resulted from collision between southeastern Laurussia and Gondwana and the final closing of Iapetus. The Alleghenian orogeny represents the climax of a long Appalachian orogeny involving interactions between a number of oceanic and continental plates and the final forging of the Appalachian orogen. Folding and thrusting of Carboniferous flysch and older deposits in the Ouachita belt during Late Pennsylvanian to Permian signify the completion of suturing of Gondwana and Laurussia.

Coal-bearing cyclothems of the midcontinent and Appalachian foreland basins reflect global effects of cyclicity in Gondwana glaciation, regional cyclicity in tectonic activity, and local to regional cyclicity of sediment input. Coal-forming swamps represent deltaic-plain and coastal-lagoon environments that developed along the margins of prograding clastic wedges. The swamps contained lush stands of scale trees, club mosses, rushes, and ferns, which provided the raw organic matter for the coal. The oldest known reptiles have been found in Lower Pennsylvanian cyclothem deposits and signify a new adaptive breakthrough in the vertebrate conquest of land.

Permian was a time of final suturing of continental masses to form the supercontinent Pangaea. Elevated mountain chains and withdrawal of epeiric seas brought about widespread aridity on Pangaea; this favored extensive deposition of red beds and evaporites on alluvial plains and coastal tidal flats. In present-day North America, post-Lower Permian deposits are confined to the western craton, where continental and coastal red beds, dune sands, and shallow-marine mudstones and carbonates formed, and to the Cordilleran belt, where shallow- to deep-marine deposits accumulated. The Permian Basin of West Texas and southeastern New Mexico was rimmed by a major barrier reef; extensive marine phosphorite deposits formed in the northern Rockies; and thick volcaniclastic eugeosynclinal deposits formed in the western Cordillera.

Permian was a period of crises in the history of life. Near the end of the period, roughly one-half of marine invertebrate families and more than 75 percent of terrestrial vertebrate families became extinct in the most devastating wave of mass extinctions of the Phanerozoic. A number of higher taxa became extinct, including fusulinid Foraminifera, the echinoderm subphylum Blastozoa, several subclasses of crinoids, trilobites, several orders of bryozoans and articulate brachiopods, rugose and tabulate corals, and the reptilian order Pelycosauria. Drastically increased continentality, increased aridity, withdrawal of epeiric seas with consequent loss of ecospace, salinity changes, and increased competition combined to place exhorbitant demands on many highly specialized, vulnerable taxonomic groups. Most, if not all of the environmental changes that contributed to the end-of-Paleozoic faunal crises were byproducts of global tectonics—the suturing of continents and closing of ocean basins.

SUGGESTIONS FOR FURTHER READING

Bambach, R. K., Scotese, C. R., and Ziegler, A. M., 1980, Before Pangaea: The Geographies of the Paleozoic World: Am. Scientist, vol. 68, no. 1, pp. 26–38.

Dietz, R. S., 1972, Geosynclines, Mountains, and Continent-Building: Scientific American Offprint No. 899, W. H. Freeman, San Francisco.

Dott, R. H., Jr., 1974, The Geosynclinal Concept, in Modern and Ancient Geosynclinal Sedimentation: Soc. Econ. Paleontologists and Mineralogists, Spec. Pub. No. 19, Tulsa, OK, pp. 1–13.

Hallam, A., 1972, Continental Drift and the Fossil Record: Scientific American Offprint No. 903, W. H. Freeman, San Francisco.

King, P. B., 1977, The Evolution of North America, rev. ed.: Princeton University Press, Princeton NJ.

Laporte, Leo, 1979, Ancient Environments, 2nd ed.: Foundations of Earth Science Series, Prentice-Hall, Englewood Cliffs NJ.

McAlester, A. L., 1977, The History of Life, 2nd ed.: Foundations of Earth Science Series, Prentice-Hall, Englewood Cliffs, NJ.

Newell, N. D., 1963, Crises in the History of Life: Scientific American Offprint No. 901, W. H. Freeman, San Francisco.

Newell, N. D., 1972, The Evolution of Reefs: Scientific American Offprint No. 901, W. H. Freeman, San Francisco.

Oliver, Jack, 1980, Exploring the Basement of the North American Continent: Am. Scientist, vol. 68, no. 6, pp. 676–683.

Raup, D. M., and Sepkoski, J. J., Jr., 1982, Mass Extinctions in the Marine Fossil Record: Science, vol. 215, pp. 1501–1504.

Runcorn, R. K., 1966, Corals as Paleontologic Clocks: Scientific American Offprint No. 871, W. H. Freeman, San Francisco.

Valentine, J. W., 1978, Evolution of Multicellular Plants and Animals: Scientific American Offprint No. 1403, W. H. Freeman, San Francisco.

Valentine, J. W., and Moores, E. M., 1974, Plate Tectonics and the History of Life in the Oceans: Scientific American Offprint No. 912, W. H. Freeman, San Francisco.

12

MESOZOIC HISTORY

CONTENTS

KEY TERMS

Blueschist
Seamount
Sonoma orogeny
Mélange
Cordilleran orogeny
Farallon plate
Nevadan orogeny
Granitization
Bentonite
Sevier orogeny
Endothermy
Ectothermy
Homiothermy
Microplate
Exotic terrane
Suspect terrane

THE BONE HUNTERS

Mesozoic, translated literally, means "middle life," an apt description for the intermediate stage of evolution of life present in that era as compared to the more primitive life of the Paleozoic and the more modern life of the Cenozoic. The Mesozoic is also the age of reptiles—not the kinds of reptiles we see today, but forms that included the largest beasts ever to walk the earth. The Mesozoic was a fantastic time of dinosaurs, giant sea "monsters," and pterodactyls—creatures that would sooner suggest creation by the minds of science fiction writers than by nature. Nonetheless the bizarre reptiles of the Mesozoic were real. Their bones and skeletons adorn the halls of the great natural history museums in this country and abroad, captivating the imaginations of millions. But who were the people responsible for collecting these tremendous specimens and where were the fossils found? These questions take us back to the golden age of vertebrate paleontology and to those heroes of the American west—the bone hunters.

During the pre-Civil war period of the nineteenth century there was no organized program for collecting fossils in North America. But after the Civil War, when the great expanses of the west were being opened, the really serious business of collecting fossils on a grand scale mushroomed. The U.S. government, in an effort to assess the natural resource of this new land, instituted a series of territorial surveys. Because much of the emphasis of these surveys was on geology, many fossils were discovered, and word soon spread of the fantastic fossil fields of the west. Thus began a new and highly exciting period in American paleontology, centered around two intriguing individuals.

Othniel Charles Marsh (1831–1899) (Fig. 12-1A), nephew of the wealthy banker and philanthropist George Peabody, acquired an early interest in paleontology; he was inspired by the richly fossiliferous rocks exposed by the diggings of the Erie Canal near his home in Lockport, New York. Having broken away from what he considered the drudgery of farm life, Marsh, with financial assistance from rich Uncle George, got a first-rate education at Phillips Academy and later at Yale, where he studied geology under J. D. Dana of geosynclinal fame. While at Yale, Marsh began amassing a collection of fossils that would later become one of the finest of its kind in the world. During the summer of 1861, he explored the gold fields of Nova Scotia, and his writing style in his published report on

the new discovery clearly demonstrated his penchant for capitalizing on the dramatic. His report aroused attention even in Europe. In 1863 George Peabody gave Yale a museum of natural history, the famous Peabody Museum, with the proviso that his nephew be given a professorship. Marsh gratefully accepted the unpaid professorship because such a post left him without teaching obligations and gave him free rein to conduct research. Allowances and a bequest from his uncle gave him enough money to live comfortably and to amass his rich collection of fossils.

Edward Drinker Cope (1840–1897) (Fig. 12-1B), nine years younger than Marsh, was a child prodigy. At the age of 6 he began recording his own journal, and by 10 was making scientific observations and sketches. His formal education at Quaker schools ended at 16 when his wealthy father decided that the frail, undersized youngster should prepare for the life of a practical farmer. Like Marsh, however, Cope had other ideas; he was impatient with the slow process of crop cultivation and longed for the excitement of science. In the winter of 1860, Cope enrolled in a series of lectures given by the eminent Joseph Leidy at the University of Pennsylvania. The following year Cope became a member of the Philadelphia Academy of Sciences, and four years later accepted an unpaid curatorial post at the Academy. From 1864 to 1868 he was Professor of Natural History at Haverford College. Cope supported himself and his family with the rent from a farm his father had given him a few years earlier, and he plunged headlong into his exciting career of science. Although he had been publishing for several years on modern vertebrates, he published his first scientific paper on fossil vertebrates in 1870. He roamed the East in search of fossils and made at least one collecting trip with a friend from Yale University—O. C. Marsh. In 1871, Cope made his first trip to the western fossil fields.

When the railroad opened up the west and made its fossils accessible, Cope and Marsh, who began their careers as friends, were employed by competing scientific surveys. Cope was collecting fossils with the Hayden Survey, led by Ferdinand Hayden, while Marsh was affiliated with the King Survey, led by Clarence King.

The competitive spirit of both men exploded into full-fledged rivalry, and by 1873 they were using that ultimate battleground—the scientific literature—for airing their differences. Both men became obsessed with being the first to discover and publish. From

Figure 12-1. The great bone hunters. A. Othniel Charles Marsh (1831–1899). B. Edward Drinker Cope (1840–1897).
(Redrawn from E. H. Colbert, 1945, *The Dinosaur Book,* Figs. b, c, p. 19: Am. Mus. Natural History, New York)

1870 to 1875 staggering amounts of material were obtained by the collecting parties directed by the rivals. However, much of the systematic work on the specimens was hastily done, and the rapid-fire publications that followed were occasionally laced with inaccuracies.

After 1874, Marsh spent little time in the field, being content to delegate responsibility to trusted subordinates. He devoted most of his time to research and lab work, and he revolutionized existing procedures for fossil collection and preparation. Cope, on the other hand, realized the importance of seeing the vertebrate fossil in its relation to the outcrop; he continued to make forays, often alone, into the fossil fields of the west. The bone-hunting teams of Cope and Marsh raked across the Dakota badlands, the Smoky Hill River country of Kansas, the Morrison beds of Como Bluff, Wyoming, and Canon City, Colorado, and numerous other virgin fossil localities of the Great Plains and Rocky Mountain basins. The number of species of Mesozoic reptiles and Tertiary mammals described by Cope and Marsh is staggering. Perhaps the rivalry between the two men stimulated each to herculean efforts much greater than they would have made if all had been pleasant and serene.

In their last years, the feud abated somewhat as both men fell into poor health and financial straits. Marsh had to give up his home and had to ask for a salary for his professorship at Yale. When he died in 1899, he left to Yale and to the U.S. National Museum in Washington, D.C., perhaps the greatest collection of vertebrate fossils ever assembled. He described more

than 450 new species, and his restorations of dinosaurs will always be a monument to his genius. Marsh was a sure and methodical thinker, of lesser intellect than Cope, but nonetheless a scholar of ability; he possessed a rare genius for organization, for recognizing the key elements of situations, and for delegating responsibility and capitalizing on the hard work of others.

Cope died at his home in Philadelphia, which, along with the Marsh laboratories at Yale, served as one of the important world centers for research into vertebrate paleontology during the last half of the nineteenth century. His deathbed was a cot that lay amid a veritable plethora of fossil bones. Cope is remembered as a quick, witty, incisive individual who had boundless energy and enthusiasm. He was a man of extraordinary brilliance and ability, perhaps one of the greatest true scholars this country has ever produced.

To Cope and Marsh we owe much about the knowledge of life of the past, in particular of the great reptiles (Fig. 12-2) of the Mesozoic and the mammals of the early Cenozoic. Because of the new techniques of collecting, preparing, and studying fossils established by these men, their work formed the basis for modern vertebrate paleontology in this country and, for that matter, the entire world. They also transformed vertebrate paleontology from a passive, chance-collecting activity into a vigorous, dynamic science. Discovery of the great dinosaur fossil-hunting grounds in the west and collection of great numbers of skeletons completely revolutionized both the concept and construction of natural history museums.

Figure 12-2. Mounted skeleton and restoration of Jurassic sauropod dinosaur *Apatasaurus (Brontosaurus)*.

TRIASSIC HISTORY

More Red Beds

The Triassic, the initial period of the Mesozoic, was named for a three-part stratigraphic subdivision in Germany. This sequence of red sandstone and shales separated by fossiliferous limestones does not closely resemble the Triassic succession in other parts of the world, but the presence of red beds is characteristic of many sections and attests to much continental deposition. In North America, Mesozoic history began rather undramatically with a continuation of Permian depositional patterns, characterized by continental red beds in the western interior part of the craton. Variegated Triassic sediments, in particular the red beds, underlie some of the most spectacular and colorful scenery in the American west (Fig. 12-3).

One of the most varied and interesting rock units in western America is the Moenkopi Formation (Fig. 12-4), a succession of red and chocolate mudstones of Early Triassic age, laid down in a variety of coastal nonmarine and shallow-marine environments. Mudcracked and ripple-marked siltstone beds are common, and local beds of gypsum and casts of halite crystals indicate a relatively arid climate, a carryover from the dry Permian Period.

Several tongues of limestone that thicken westward punctuate the Moenkopi mudstone succession (Fig. 12-

5) and attest to periodic marine encroachments across coastal mudflats and floodplains. Marine fossils in these carbonate tongues provide good biostratigraphic tie-ins for the sparse amphibian and reptile fossils in the continental facies.

Unconformably overlying the Moenkopi is a widespread unit consisting predominantly of mudstone and siltstone in a dazzling array of colors ranging from red, pink, and purple to chocolate, blue, ash grey, and even white. This variegated unit, the Chinle Formation of Late Triassic age, is widely exposed over the Colorado Plateau and is famous for its petrified wood. The abundant and beautifully preserved conifer logs (Fig. 12-6) in Petrified Forest National Park, Arizona (discussed in Chapter 14) are products of silicification of the wood, commonly in exquisite detail, by red, yellow, orange, and purple agate. Much of the silica was probably from volcanic ash beds within the fluvial and delta-plain Chinle sediments. The logs were not preserved as a fossilized forest but rather were transported during flood stages, eventually being deposited in sand bars and on floodplains and later petrified by silica that precipiated from ground water.

Fossil cycads and ferns also have been recovered from the Chinle and together with the abundant conifer logs provide a fair representation of what Triassic land floras were like. The Chinle is not known for its fossil

Figure 12-3. Triassic sedimentary rocks A. Triassic succession at Capitol Reef National Park, Utah. B. Fluvial deposits near St. George, Utah. Note lens geometry of channel deposits. (Photos by J. Cooper)

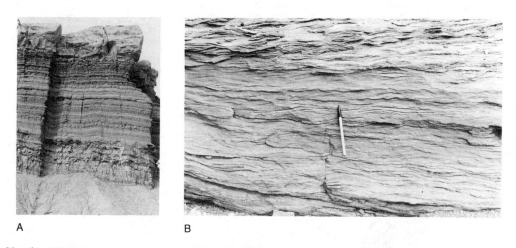

Figure 12-4. Moenkopi Formation. A. Outcrop at Capitol Reef National Park, Utah. B. Climbing ripple lamination in fluvial deposits, Shafer Trail, Canyonlands area, near Moab, Utah. (Photos by J. Cooper)

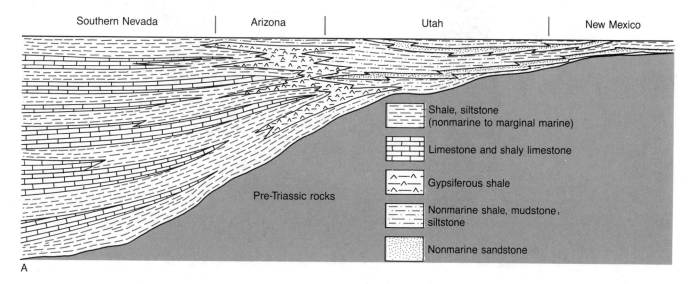

Figure 12-5. Restored east–west cross section of Moenkopi Formation showing nonmarine to marine transition from western craton to geosynclinal belt.
(Data from M. E. MacLachlan, 1972, *Geologic Atlas of Rocky Mountain Region:* Rocky Mountain Assoc. Geologists)

A

B

Figure 12-6. Chinle Formation at Petrified Forest National Park, Arizona. A. Exposure showing lens-shaped fluvial channel deposits with petrified log protruding from outcrop face. B. Petrified log-strewn landscape.
(Photos by J. Cooper)

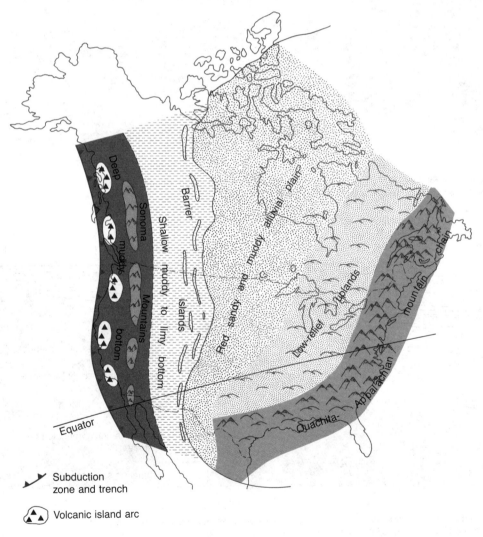

Figure 12-7. Paleogeography of Early Triassic superimposed on outline map of present-day North America.
(Data from several sources)

vertebrates but in a few places it has produced remains of labyrinthodont amphibians, peculiar crocodilelike phytosaurs, and some small bipedal dinosaurs.

Tectonic Unrest in the Western Cordillera

The miogeosynclinal and eugeosynclinal belts that had characterized the Cordilleran geosyncline during much of the Paleozoic underwent major changes during the late Paleozoic and Triassic. Intense volcanism within an island-arc system that extended from California to Alaska and orogeny throughout much of the western Cordillera affected the geosyncline. The Permo-Triassic volcanic belt was probably similar to the present-day Japanese archipelago and was related to a major zone of subduction. This volcanic island-arc system (Fig. 12-7) was separated from the mainland by a small ocean basin similar to the modern Sea of Japan.

The tectonic belt is well expressed in the Pacific Northwest, particularly in British Columbia and in the Klamath Mountains, where deformed Permian rocks intruded by acidic and mafic plutons together with remnants of ophiolites, volcanic-arc andesites, and **blueschist** attest to major orogenesis. Throughout much of the western Cordillera, angular unconformities separate Permian and older rocks from Upper Triassic strata. The Upper Triassic sediments are varied and complex. Some

show the influence of the last vestiges of volcanism; some are thick sequences of shales, graywacke turbidite sandstones, and conglomerates; and some are developed as local carbonate banks and reefs that grew around foundering volcanic islands and topped **seamounts** much as some modern-day small Pacific Islands form.

The Permo-Triassic orogenic event involved subduction of oceanic lithosphere, the closing of the Late Paleozoic back-arc basin, and the convergence of the volcanic arc against the western edge of the continental plate (Fig. 12-8). In Nevada, Upper Paleozoic rocks of the marginal back-arc basin were overthrust eastward in a style reminiscent of the Late Devonian–Early Mississippian Antler Orogeny (Fig. 11-34). Overthrusting occurred along the major westward-dipping Golconda thrust, which overrode the Roberts Mountains thrust (Fig. 12-9). This episode of deformation in Nevada is commonly referred to as the **Sonoma orogeny.**

After early Mesozoic destruction of the volcanic island-arc system in the Cordilleran eugeosynclinal belt, the western margin of North America became part of a plate tectonics regime similar to that of the western margin of present-day South America, where the Andes mountain range has formed (Fig. 12-48). Throughout the remainder of the Mesozoic, a deep trench and associated magmatic arc, related to subduction of oceanic lithosphere, characterized the western edge of the con-

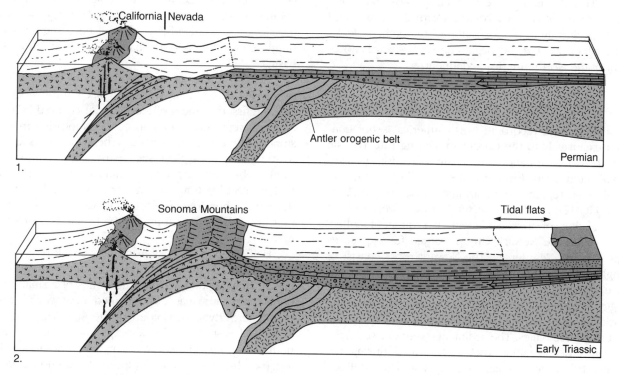

Figure 12-8. Inferred paleogeography and tectonic activity culminating in Permian-to-Triassic Sonoma orogeny in western North America.
(Data from several sources)

LAST PHYSICAL EVENT CYCLE ④

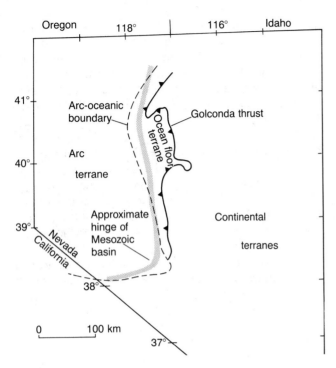

Figure 12-9. Tectonic elements involved in Sonoma orogeny, Nevada.
(From R. C. Speed, 1977, Island Arc and Other Paleographic Terrones of Late Paleozoic Age in the Western Great Basin; in Paleozoic Paleogeography of Western United States: Pacific Section SEPM Paleography Symposium. Reprinted by permission of Pacific Section, Society of Economic Paleontologists and Mineralogists)

tinent. This change in plate interactions most likely was related to the early Mesozoic breakup of Pangaea and more specifically to the formation of the embryonic Atlantic rift-ocean and the beginning of westward movement of the North American continent.

Triassic Events in the Appalachians

The Triassic record in North America is not confined exclusively to the Cordillera. During Late Triassic (Fig. 12-10) a discontinuous chain of variously sized downfaulted basins formed in the crystalline part of the Appalachian region, from maritime Canada to the Carolinas (Fig. 12-11). These structural troughs were sites of accumulation of thick sedimentary sequences collectively called the Newark Group, named for exposures near Newark, New Jersey. The Newark rocks consist of conglomerate and sandstone of arkose composition and shale and mudstone, commonly developed as red beds. Flaggy brown sandstones have furnished the building blocks of the well-known brownstone houses of the eastern United States. The sediments were derived from highlands of crystalline rock that formed during the middle Paleozoic Acadian orogeny and later uplifted during the Late Triassic along high-angle normal faults. The rocks are coarsest adjacent to the upfaulted basin

margins, where sands and gravels were deposited in alluvial fans. Farther out in the continental basins, finer sand and muds were deposited in stream channels, floodplains, and lakes. Associated with the sedimentary rocks are basaltic lava flows, diabase dikes and sills (Fig. 12-12).

The Late Triassic block faulting and near-surface mafic igneous activity have important implications for the early Mesozoic tectonic history of eastern North America. The high-angle normal faults indicate tensional stress or extension of continental lithosphere—a structural style superimposed on the older Appalachian compressional mountain structures. This Late Triassic disturbance, also evidenced in northwestern Africa and western Europe, was a prelude to later rifting of North American, European, and African continental fragments from the Pangaea assembly.

Terrestrial Life and the Beginning of a Dinosaur Dynasty

No evidence of marine life has been found in the Newark beds, but fossils of land plants and freshwater fish are locally abundant, particularly in gray, organic-rich beds. The plant remains are mainly foliage (as opposed to the petrified logs in the Chinle Formation) and are most common in Virginia and the Carolinas, where gray mudstones are more abundant. The plant fossils represent a swamp flora of ferns and scouring rushes (see Chapter 14). Leaves and needles from cycads and conifers were washed into the swamps from forests on the slopes and uplands. A few coal beds suggest more humid climatic conditions than existed in the northern basins.

The Newark strata have yielded more dinosaur footprints (Fig. 12-13) than any other place in the world. In fact, the first evidences of dinosaurs discovered in North America were odd-looking three-toed footprints in sandstone in the Connecticut Valley. These footprints are the theme of a Connecticut state park where some sandstone slabs contain prints in patterns that suggest gregarious reptiles traveling in herds. Curiously enough, despite the hundreds of footprints, very few early dinosaur bones have been found in the Upper Triassic of the East.

Half a continent away from the Newark Basins, brilliantly colored Chinle beds at a locality named Ghost Ranch, New Mexico, yielded the bones of a small light-boned dinosaur that Cope named *Coelophysis.* Many years later, Edwin H. Colbert of the American Museum of Natural History headed an expedition that produced numerous bones and several complete skeletons of *Coelophysis,* thus increasing significantly the knowledge of one of the earliest dinosaurs. *Coelophysis* ran on almost birdlike hind limbs and hunted prey with clawed fore-

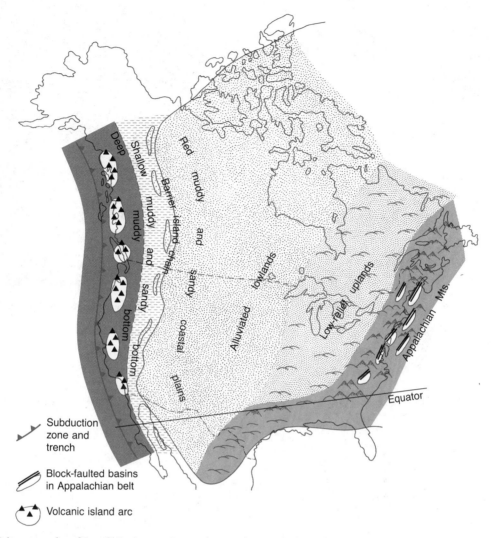

Subduction zone and trench

Block-faulted basins in Appalachian belt

Volcanic island arc

Figure 12-10. Paleogeography of Late Triassic superimposed on outline map of present-day North America.
(Data from several sources)

limbs. Unlike the giants of the Jurassic and Cretaceous (Fig. 12-14), most Triassic dinosaurs were similar to their thecodont ancestors, being nimble, light-boned, rather small carnivores (Figs. 12-15, 16). They represent the beginning of a fantastic dynasty that lasted for more than 100 million years.

Marine Life

The original Triassic type section, with its preponderance of red beds and sparse fossil record, was an unfortunate choice because of its inadequacy in providing a good standard of reference. Subsequently, a sequence of predominantly fossiliferous marine carbonate rocks in the Alps was designated as the principal standard for Triassic reference and correlation. This underscores the difficulties posed by a number of the original

type sections of geologic systems and also the importance of fossiliferous sections of marine strata that give the concept of system its integrity. Some of the original type sections turned out to be poor choices because of their lack of fossils. Later, after the importance of paleontologic control was realized and it was appreciated that systems are interpretive units based on aggregates of fossils, additional sections were selected as standards of reference and correlation.

By far the best Triassic fossiliferous sequences in North America are in the western Cordilleran belt, particularly in the foothills of northern British Columbia and the Sverdrup Basin of the Arctic islands. Middle and Upper Triassic sequences in these and other areas contain ammonoid zones that provide excellent correlation and a composite standard reference section for the marine Triassic of North America. The Triassic marine rec-

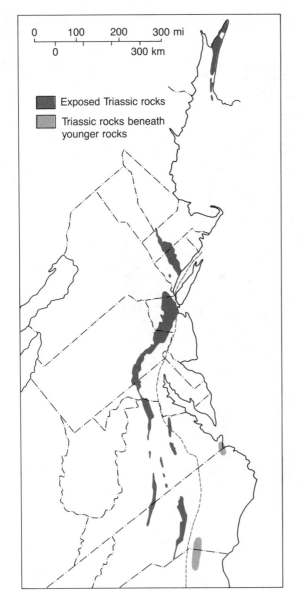

Figure 12-11. Northeastern United States and maritime Canada showing distribution of Triassic basins.
(From Morris S. Petersen, J. Keith Rigby, and Lehi F. Hintze, *Historical Geology of North America,* 2nd ed., Fig. 10.4, p. 143. © 1973, 1980 Wm. C. Brown Publishers, Dubuque, Iowa. All Rights Reserved. Reprinted by permission)

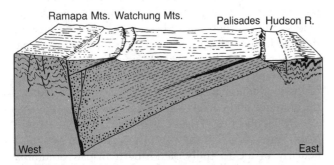

Figure 12-12. Present-day Triassic basin showing nonmarine sedimenary facies and diabase sills.
(From Morris S. Petersen, J. Keith Rigby, and Lehi F. Hintze, *Historical Geology of North America,* 2nd ed., Fig. 10.5, p. 144. © 1973, 1980 Wm. C. Brown Publishers, Dubuque, Iowa. All Rights Reserved. Reprinted by permission)

Figure 12-13. Dinosaur footprint on sandstone bed of Newark Group, Connecticut Valley.

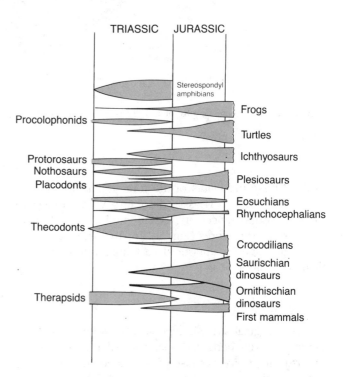

Figure 12-14. Range and relative abundance of tetrapods (four-legged terrestrial vertebrates) within and beyond the Triassic.
(From E. H. Colbert, *Evolution of the Vertebrates—A History of the Backboned Animals through Time;* 3rd ed., Fig. 64, p. 162. Copyright © 1980 by John Wiley & Sons, Inc., New York. Reprinted by permission of John Wiley & Sons, Inc.)

Figure 12-15. Restoration of scene in eastern United States during the Late Triassic Period (200 million years ago). Dominant plants are conifers and cycadeoids. Animals, left to right: small, slim dinosaurs, *Coelophysis;* larger dinosaurs (two) in background, *Trilophosaurus;* crocodilelike forms to right, phytosaurs.
(Photo courtesy of National Museum of Natural History. Smithsonian Institution Photo No. 2526C)

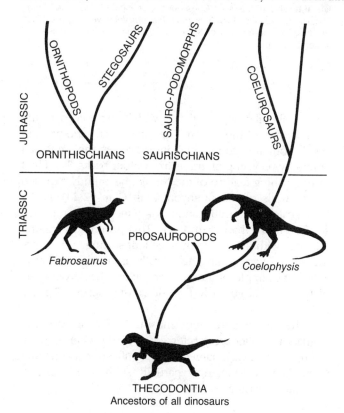

Figure 12-16. Early evolution of the dinosaurs.
(From E. H. Colbert, 1980, *Evolution of the Vertebrates—A History of the Backboned Animals through Time;* 3rd ed., Fig. 61, p. 160. Copyright © 1980 by John Wiley & Sons, Inc., New York. Reprinted by permission of John Wiley & Sons, Inc.)

ord consists of isolated remnants. Most of the rocks were uplifted and eroded or buried during later Mesozoic and Cenozoic events. Fortunately, abundant ammonoid faunas have allowed good correlation among these separated sequences as well as general correlation with the Alpine section of the European Triassic.

The main differences between Triassic and Permian rocks is expressed in the fossil assemblages. The major faunal extinctions that marked the close of the Paleozoic were followed by repopulation of Triassic seas by very different faunas. This change is indelibly impressed in the stratigraphic record and, of course, inspired the drawing of a boundary that separates the Paleozoic and Mesozoic Eras of earth history. The rich populations of ammonoid cephalopods (Fig. 12-17) were characteristic of Triassic seas worldwide and contributed importantly to the composition of marine early Mesozoic invertebrate faunas. However, the ammonoids, whose swimming mode of life made them so well adapted and successful during the Triassic, nearly died out near the end of that period. The entire subclass Ammonoidea apparently survived as one lone family that gave rise to the great adaptive radiation of ammonites in the Jurassic.

The Triassic also marks the beginning of a major adaptive radiation of bivalve molluscs (Fig. 12-18). New groups evolved and achieved marked ecological success in the aftermath of the late Paleozoic extinctions. Part of this success stemmed from ecological replacement of extinct epifaunal brachiopods, but most of the diversifi-

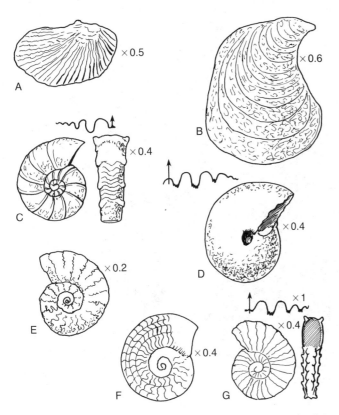

Figure 12-17. Representative Triassic marine invertebrate fossils. A,B. Bivalve molluscs. C–G. Ceratite ammonoid cephalopods. (From R. C. Moore, editor, 1957, 1969. Lawrence, Kansas: Geological Society of America and University of Kansas Press. *Treatise on Invertebrate Paleontology)*

SUPERFAMILIES

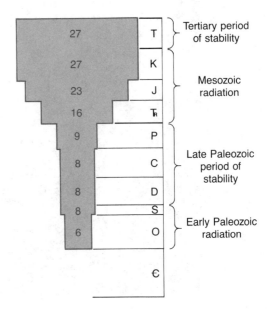

Figure 12-18. Diversity of marine bivalve molluscs through Phanerozoic, measured in number of superfamilies existing per geologic period. Most of the Mesozoic radiation involved new groups of infaunal siphon-feeding forms. (From S. M. Stanley, 1968, Post-Paleozoic Adaptive Radiation of Infaunal Bivalve Molluscs—A Consequence of Mantle Fusion and Siphon Formation; Text-Fig. 1, p. 215: *Jour. Paleontology,* vol. 42, no. 1. Reproduced by permission of Society of Economic Paleontologists and Mineralogists)

cation can be linked to evolution of the siphon, an anatomical feature that developed from fusion of folds in the fleshy mantle (Fig. 12-19). This evolutionary breakthrough allowed for both filter feeding from the water column and an infaunal existence.

Modern types of reef-building corals belonging to the order Scleractinia replaced the tabulate and rugose orders of the Paleozoic. Mobile echinoderms such as sea urchins and starfish became more abundant and largely replaced the predominantly attached forms—crinoids, blastoids, and cystoids—of the Paleozoic.

JURASSIC HISTORY
Sand, Sand Everywhere

In western North America, Jurassic strata, like the underlying Triassic, form parts of many familiar landmarks. Magnificent arches, alcoves, spires, and pinnacles have been sculptured in cross-bedded sandstone and red beds and are featured in national parks and monuments throughout the Rocky Mountain area. Among the best known of these rocks is the Navajo Sandstone, a vast blanketlike deposit that is widely exposed over the western part of the craton. The quartz-rich sand was derived from the continental interior and the Canadian Shield by recycling of grains from older rocks. The chief distinguishing feature of the Navajo, beautifully exposed in the walls of Zion Canyon, Utah (Fig. 12-20), is the large-scale cross-stratification; some individual cross-bed sets are more than 25 m thick. This prominent feature, together with the generally excellent sorting and rounding of the quartz grains, historically has made the Navajo a classic example of a dune deposit and has evoked the picture of a large Early Jurassic coastal desert (Fig. 12-21).

The Navajo is the uppermost unit of a succession of formations collectively referred to as the Glen Canyon Group (Fig. 12-22), named for magnificent exposures in Glen Canyon, Utah, now partially submerged beneath the waters of Lake Powell.

Jurassic Marine Sediments

Except for some continental deposits in Michigan and the upper part of the Newark Group, Jurassic sedimentary rocks are not exposed in eastern North Amer-

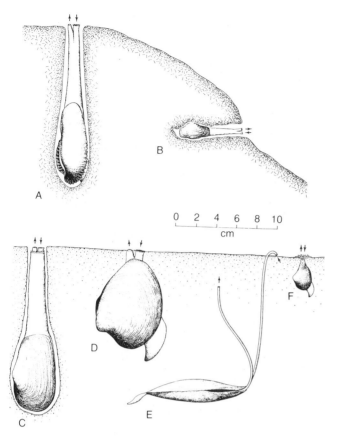

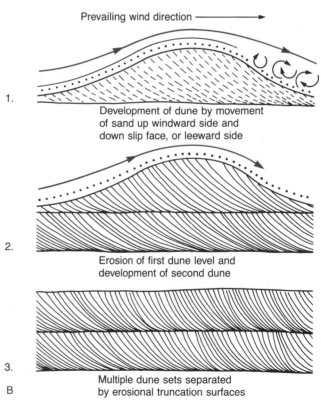

A

Figure 12-19. Representative kinds of infaunal siphon-feeding bivalve molluscs. A,B. Inhabiting rock; C–F. In soft sediment. Arrows indicate direction of water currents.
(From S. M. Stanley, 1968, Post-Paleozoic Adaptive Radiation of Infaunal Bivalve Molluscs—A Consequence of Siphon Formation; Text-Fig. 6, p. 219: *Jour. Paleontology,* vol. 42, no. 1. Reproduced by permission of Society of Economic Paleontologists and Mineralogists)

ica. However, their presence beneath the cover of younger rocks in the Atlantic and Gulf Coast area has been demonstrated by deep drilling. In the Atlantic shelf margin, Jurassic terrigenous clastics interfinger seaward with carbonates (Fig. 12-23). This relationship has its actualistic analogue in the facies patterns in the Gulf of Elat (Aqaba) (Fig. 12-24), an early-stage rift-zone setting that is the northern continuation of the Red Sea rift. In the United States Gulf coast, Jurassic rocks occupy a crescent-shaped subsurface belt extending from Alabama to northeastern Mexico (Fig. 12-25) and comprise a seaward-thickening wedge that grades from a shoreward facies of red beds to shallow-marine carbonates. The Smackover Limestone in the Louisiana subsurface has yielded significant amounts of petroleum. Evaporites are locally common in the section, and a thick evaporite unit, the Louann Salt, was deposited in the poorly circulated shallow waters of the embryonic Gulf of Mexico Basin during the early stages of separation of southeastern North America from northwestern Africa and north-

Figure 12-20. Large-scale, high-angle cross-stratification. A. Beds of eolian (dune) origin, Navajo Sandstone, Zion National Park, Utah. B. Succession of truncation surfaces in eolian sandstones.
(A, photo by J. Cooper; B, data from several sources)

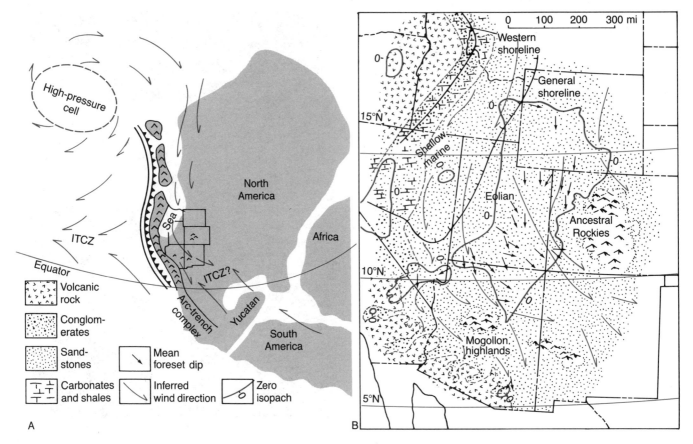

A

B

Figure 12-21.   Early Jurassic paleogeography. A. Western interior of United States. ICTZ is intertropical convergence zone. B. Detail for southwestern United States.
(From G. Kocurek, and R. H. Dott, Jr., 1983, Jurassic Paleogeography and Paleoclimate of the Central and Southern Rocky Mountain Region, Fig. 3, p. 106, *in* Mesozoic Paleogeography of the West-central United States, M. W. Reynolds, and E. D. Dolly, eds.: Rocky Mountain Section SEPM. Reproduced by permission of Rocky Mountain Section, Society of Economic Paleontologists and Mineralogists)

ern South America. The Jurassic rocks of the Atlantic and Gulf Coast regions are overlapped by Cretaceous deposits, but their presence in the subsurface indicates the earliest marine deposition on the trailing margin of the North American continent as Pangaea began to split apart.

Beginning in the Early Jurassic, marine encroachments advanced widely into the western craton on at least three different occasions and deposited a complex association of sandstones, shales, and limestones. The sedimentary succession is comparatively thin, generally not exceeding several hundred meters, and indicates tectonic stability of the craton. The Jurassic interior seaway (Fig. 12-25), commonly referred to as the Sundance Sea, extended from the Arctic almost to the newly formed Gulf of Mexico. This cratonic shelf sea spread eastward from a miogeosynclinal trough that is well represented by thick sections of fossiliferous marine shale in western Canada. Life flourished in the Sundance sea-

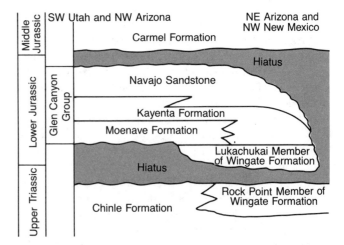

Figure 12-22.   Relationship between Navajo Sandstone and other units of Glen Canyon Group. Note the unconformities expressed by hiatuses.
(Data from F. Peterson, and G. N. Pipiringos, 1979, Stratigraphic Relations of the Navajo Sandstone to Middle Jurassic Formations, Southern Utah and Northern Arizona: *U.S. Geol. Survey Prof. Paper* 1035-B)

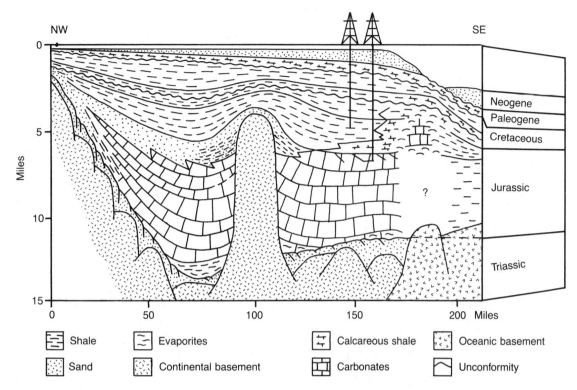

Figure 12-23. Baltimore Canyon trough, Atlantic coast of United States, showing Jurassic terrigenous clastics interfingering with carbonates.
(From E. W. Poag, 1979, Stratigraphy and Depositional Environments of Baltimore Canyon Trough, Fig. 2, p. 1454: *Am. Assoc. Petroleum Geologists Bull.,* vol. 63, no. 1. Reproduced by permission of American Association of Petroleum Geologists)

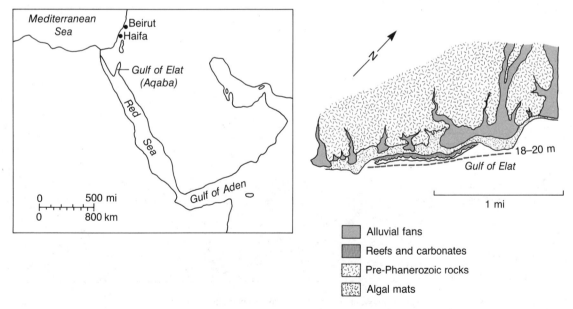

Figure 12-24. Modern Gulf of Elat (Aqaba), Red Sea, and local area along shore of Gulf of Elat showing distribution of terrigenous alluvial fans and carbonate reefs. Dashed line is 18- to 20-m bathymetric contour, marking shelf break.
(From S. A. Epstein, and G. M. Friedman, 1983, Depositional and Diagenetic Relationship between Gulf of Elat (Aqaba) and Mesozoic of United States East Coast Offshore; Figs. 1,2, p. 954: *Am. Assoc. Petroleum Geologists Bull.,* vol. 67, no. 6. Reproduced by permission of American Association of Petroleum Geologists)

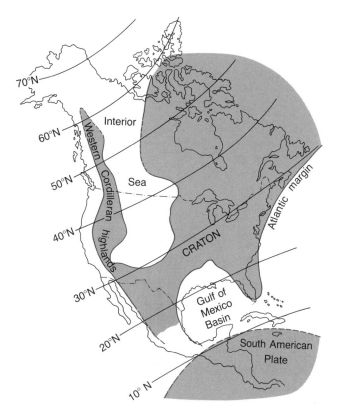

Figure 12-25. Paleogeography of North America showing paleolatitudes and distribution of Late Jurassic land and sea.
(After R. Brenner, 1983, Late Jurassic Tectonic Setting and Paleogeography of Western Interior, North America, Fig. 1, p. 120, *in* Mesozoic Paleogeography of the West-central United States, M. W. Reynolds and E. D. Dolly, eds.: Rocky Mountain Section SEPM. Reproduced by permission of Rocky Mountain Section, Society of Economic Paleontologists and Mineralogists)

way, and abundant fossils, characterized especially by belemnite and ammonite cephalopods (Fig. 12-38D–I), are found in its deposits.

Along the very western margin of the continent from California to Alaska thick sequences accumulated that were typical of a eugeosynclinal setting. Graywackes, dark shales, bedded cherts, and conglomerates, together with volcaniclastic sediments and submarine basalt flows, make up a great thickness of deposits. But as in the Triassic eugeosynclinal section, later Mesozoic and Cenozoic orogenies have overprinted much of the western Cordillera and have disrupted the original continuity in rock assemblages.

Sedimentation and Plate Convergence

As the westward-moving North American continent continued to override the Pacific oceanic lithosphere during the Late Jurassic, an unusual rock assemblage, the Franciscan Formation of the California Coast Ranges, began to accumulate in a submarine trench (Fig. 12-26). The Franciscan is a complex unit of unknown thickness (in places at least 7000 m) that ranges in age from Late Jurassic to Late Cretaceous. It consists predominantly of graywacke but also includes siltstones, conglomerates, black shales, chert, pillow basalts, and greenstones. Some sections contain rhythmically bedded flysch and peculiar pebbly mudstones and boulder beds. Franciscan lithologies indicate a mixing of rocks originally formed in abyssal-plain, continental-slope, and continen-

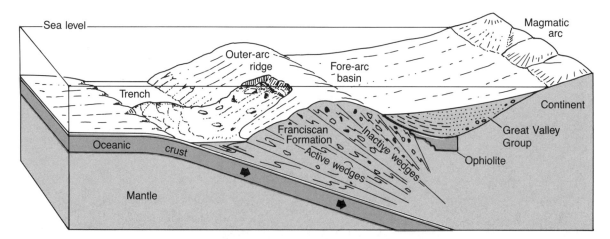

Figure 12-26. Inferred depositional setting of Franciscan Formation in active trench along Cordilleran margin. Olistostromes—deposits of submarine slumps—incorporate variously sized fragments called olistoliths (various kinds of sedimentary rocks, blueschist, and oceanic basalt), derived from previously formed rock. These fragments are reinvolved in subduction and reincorporated into successively younger accretionary wedges during continued underthrusting. Mélanges differ from olistostromes in that they involve chaotic fragments in pervasively sheared matrix—fragments that are stirred, mixed and sheared in the subduction zone. Sediments, mostly turbidites, of the Great Valley Group were deposited in a fore-arc basin, behind the outer-arc ridge.
(After B. M. Page, 1977, Effects of Late Jurassic-Early Tertiary Subduction in California, Fig. 5-9, p. 66, *in* Late Mesozoic and Cenozoic Sedimentation and Tectonics in California: San Joaquin Geol. Soc. Short Course. Reproduced by permission of San Joaquin Geological Society)

tal-shelf environments and brought together in the submarine trench. Chaotic mixtures of various-sized blocks of Franciscan material contained within a pervasively sheared matrix are called **mélange.** Parts of the Franciscan terrane have been metamorphosed to the blueschist facies, and ophiolite complexes also are present. The mélange and blueschist indicate subduction along the western edge of the continent. Fossils are rare in the Franciscan, but radiolarians in the thin chert beds indicate a Late Jurassic age for the lower part of the assemblage.

To the east of the Franciscan complex, a thick succession of sandstone and mudstone comprising the Great Valley Group (Fig. 12-26) was deposited on the continental slope and shelf in a fore-arc basin setting from late Jurassic through the Cretaceous. No volcanics are associated with the Great Valley Group. In comparison to the Franciscan, the stratigraphy is more regularly organized and benthic fossils are moderately abundant. Much of the Great Valley Group was deposited as turbidite facies in subsea fan environments.

This Pacific margin setting generally was separated from the interior seaway by land masses that were uplifted by lithospheric deformation. The deformation heralded the beginning of a wave of orogenic disturbance that would continue through the Mesozoic into the Cenozoic. This major phase of deformation is called the **Cordilleran orogeny,** and, like the Appalachian orogeny of the Paleozoic, it was a protracted event, consisting of separate pulses that manifested themselves in particular regions. The history of the Cordilleran orogeny is related primarily to the underthrusting of Pacific Ocean lithosphere beneath the western margin of the North American Plate along an eastward-dipping subduction zone. Subduction, as you recall, also was largely responsible for late Paleozoic orogeny in the western Cordillera; however, the rate of plate convergence, intensity of deformation, orientation of structures, and plate boundary architecture were quite different during the Cordilleran orogeny.

The initial phase of the Cordilleran orogeny, the **Nevadan orogeny,** was related to an increase in the relative rate of convergence of Pacific oceanic lithosphere (the **Farallon Plate**) and the North American continent, produced mainly by the westward movement of the latter as the mid-Atlantic Ridge spreading center developed (Fig. 12-27). This increase in plate convergence was responsible for a change from a Japanese-type to an Andean-type margin of western North America (Fig. 12-28). The Japanese margin analogue, which was expressed as a volcanic island-arc system and inter-arc basin, characterized the western edge of North America from late Paleozoic through early Mesozoic. The Andean-margin setting, manifested by a continental margin magmatic-arc and fore-arc basin, characterized the western edge of North America from mid-Mesozoic through early Cenozoic.

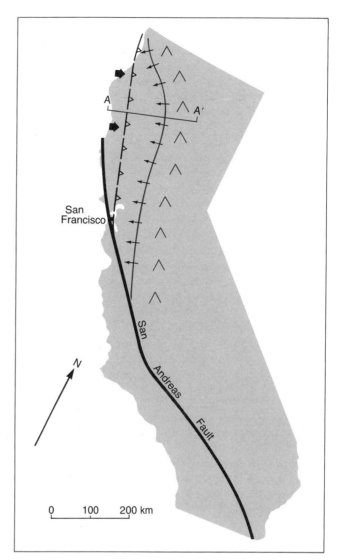

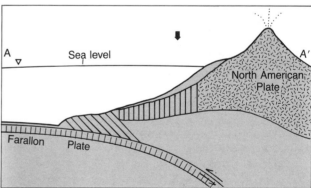

Figure 12-27. Late Jurassic, northern California. Double arrows indicate direction of subduction of Farallon Plate. Single arrows indicate sediment-dispersal directions. Colored line indicates location of paleoshoreline. Sawtooth pattern indicates position of trench and direction of dip of subduction zone. San Andreas fault zone (Neogene) truncates Late Mesozoic paleotectonic trends. Mountain symbols indicate location of the magmatic front as expressed by the westward limit of plutonism.

(After R. V. Ingersoll, 1978, Paleogeography and Paleotectonics of the Late Mesozoic Forearc Basin of Northern and Central California; Figs. 3,4, p. 475: *in* Mesozoic Paleogeography of the Western United States: Pacific Section SEPM Paleogeography Symposium, vol. 2. Reproduced by permission of Pacific Section, Society of Economic Paleotologists and Mineralogists)

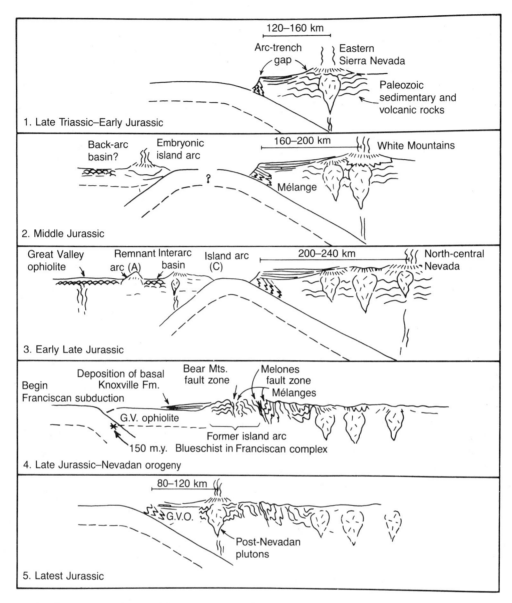

Figure 12-28. Postulated tectonic evolution of the Sierra Nevada during early Mesozoic. Note the change from Japanese- to Andean-type margin with initiation of Nevadan orogeny in Late Jurassic. (After R. A. Schweikert, and D. S. Cowan, 1975, Early Mesozoic Tectonic Evolution of the Western Sierra Nevada, California; Fig. 3, p. 1334: *Geol. Soc. America Bull.,* vol. 86. Reproduced by permission of authors)

The Morrison Formation: Graveyard of the Dinosaurs

The Nevadan orogeny was characterized by intrusive igneous activity and associated volcanism and produced tectonic highlands that contributed great volumes of sediment to the east. Following a pattern we have already seen in the Paleozoic record of the eastern part of the continent, orogeny in the western Cordillera during Late Jurassic is shown by the sedimentary record of the western interior part of the continent. Colorful gray, green, red, and maroon sediments of the Morrison For-

mation (Fig. 12-29) were deposited over an immense region (Fig. 12-30). A great regression of the Sundance Sea occurred northeastward into Canada as the Morrison clastic wedge built eastward. Most of the Morrison consists of mudstone and siltstone, with local beds and lenses of sandstone and conglomerate, punctuated by a few volcanic ash horizons, all deposited on an expansive floodplain that was built up by streams that carried detritus eastward from rising mountains. Stream-channel facies within the Morrison, as well as similar sediments in the Triassic Chinle Formation, have been the sites of significant uranium mineralization in the Colorado Plateau Province.

Figure 12-29. Morrison Formation, Colorado National Monument, Colorado. Note lenticular fluvial channel sandstone enveloped by floodplain shales. (Photo by J. Cooper)

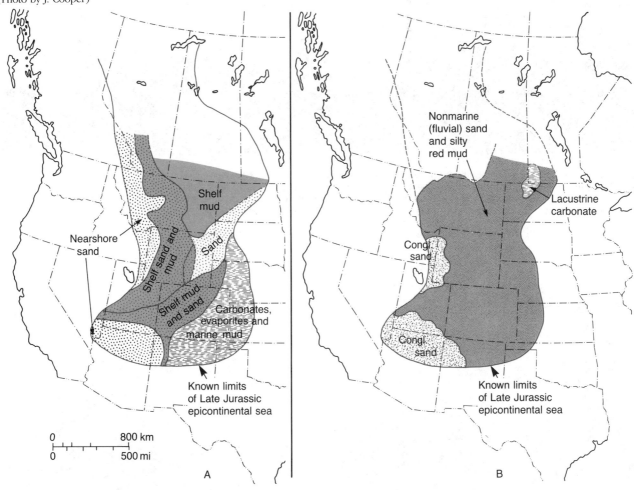

Figure 12-30. Major lithofacies and environments during late part of Late Jurassic in western interior of North America. A. Approximately 150 m.y. B.P. B. Approximately 155 m.y. B.P.
(From R. L. Brenner, 1983, Late Jurassic Tectonic Setting and Paleogeography of Western Interior, North America, Fig. 9, p. 129, *in* Mesozoic Paleogeography of West-central United States, M. W. Reynolds and E. D. Dolly, eds.: Rocky Mountain Section SEPM. Reproduced by permission of Rocky Mountain Section, Society of Economic Paleontologists and Mineralogists)

The Morrison Formation has achieved its fame from the great wealth of dinosaur remains. Dinosaurs in great numbers walked the surfaces of Morrison floodplains and ate vegetation along lakeshores and in swamps (Fig. 12-31). After death, their bones were entombed and preserved in stream-channel sand bars. Museums of the world feature Morrison dinosaur fossils (Fig. 12-32). Di-

nosaur National Monument near Vernal, Utah, has a splendid display of a diverse assemblage of bones chiseled into relief on the surface of a thick sandstone bed (Fig. 12-32C). Another famous Morrison locality is the Bone Cabin quarry in southern Wyoming. Here, near the turn of the century, a collecting party from the American Museum of National History happened across

Figure 12-31. Restoration of scene in western interior of United States on Morrison alluvial plain during Late Jurassic Period (140 million years ago). Plants include cycadeoids, ferns, and horsetails. Dinosaurs, left to right: *Antrodemus* (rear), flesh eater; *Stegosaurus* (left front, with plates along back), plant eater; *Diplodocus* (far right, large), plant eater; *Camptosaurus* (small form, right front and middle rear), plant eater.
(Photo courtesy of National Museum of Natural History, Smithsonian Institution Photo No. 79-11134)

A   B

Figure 12-32. Morrison Formation fauna. A. *Allosaurus* (left), a theropod, and *Camptosaurus* (right), a sauropod. B. Restored scene from Morrison alluvial plain; combatants are *Camptosaurus* and *Allosaurus*.
(A, B courtesy of Utah Museum of Natural History)

an isolated sheepherder's cabin only a few miles from the famous Como Bluff site where Marsh's diggers had discovered great quantities of dinosaur bones a quarter century before. But this was no ordinary cabin! It was constructed entirely of agatized dinosaur bones from a nearby site that eventually yielded freight cars full of dinosaur fossils.

Terrestrial Life

The dinosaurs diversified rapidly during the Jurassic into an amazing array of forms (Fig. 12-33). Perhaps the most popularized of the Jurassic dinosaurs are the great herbivorous sauropods *Apatasaurus* (formerly *Brontosaurus*), *Brachiosaurus,* and *Diplodocus* and the voracious flesh-eating theropod *Allosaurus.* Translated literally dinosaur means "terrible lizard." Not only were they not lizards, but many may have been quite peaceable giants. To dispel another myth, all dinosaurs were not large. Many were of only modest size and some were no bigger than a chicken. Sir Richard Owen (1804–1892), who established the science of vertebrate paleontology in England, was the first to recognize that these extinct reptiles needed a name, and it was he who

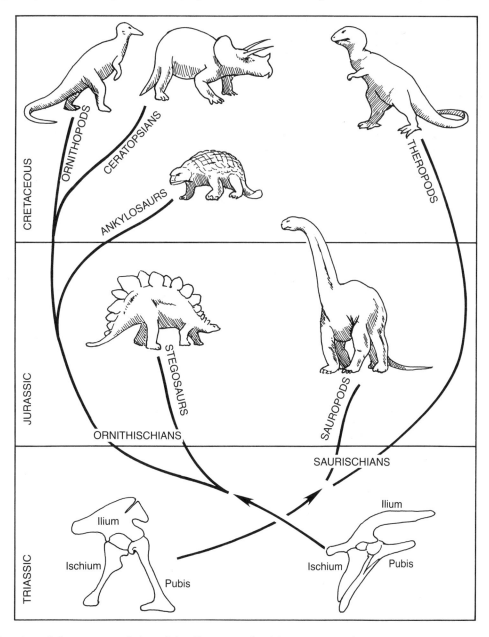

Figure 12-33. Jurassic and Cretaceous evolution of the dinosaurs and pelvis structure in the two orders of dinosaurs: A. Saurischian pelvis, with a forwardly directed pubis. B. Ornithischian pelvis, with the pubis parallel to the ischium.
(From E. H. Colbert, 1969, *Evolution of the Vertebrates—A History of the Backboned Animals through Time;* 2nd ed., Figs. 68,70, pp. 197, 200. Copyright © 1980 by John Wiley & Sons, Inc. Reprinted by permission of John Wiley & Sons, Inc. and Lois M. Darling)

Figure 12-34. *Archaeopteryx* preserved in Jurassic Solenhofen Limestone. Note impressions of feathers. Skeleton is like that of a small, running insectivorous dinosaur.
(Photo courtesy of National Museum of Natural History. Smithsonian Institution Photo No. 15771)

coined the word Dinosauria, later anglicized to dinosaur. Dinosauria is not a formal taxonomic name in modern classification of vertebrates. Dinosaurs include two extinct orders of the reptilian subclass Diapsida: Saurischia and Ornithischia (Fig. 12-33). In the process of filling the available ecological niches, some reptiles took to the air, as exemplified by the gliding pterosaurs.

Jurassic land floras were dominated by seed-bearing plants, including cycads, gingkos, and conifers (see Chapter 14). Jurassic mammals include four principal orders, known best from small bits of skeleton, jaws, and teeth obtained mostly from the Morrison Formation. These creatures left a very meager Jurassic fossil record but enough has been learned to determine that they underwent an initial radiation of small, primitive forms after their origin in Late Triassic.

All the known specimens of the earliest bird, *Archaeopteryx* (Fig. 12-34), are from the Upper Jurassic Solenhofen Limestone in Bavaria, West Germany. The soft lime mud that provided the burial grounds for these fossils preserved feathers in exquisite detail. However, as some vertebrate paleontologists have pointed out, without the feathers *Archaeopteryx* looks very reptilian. It has been suggested that *Archaeopteryx* was actually a

Figure 12-35. Origin of powered flight from the ground up (cursorial theory) as envisioned by Yale University paleontologist John Ostrom. Proto-Archaeopteryx (top) illustrates an early stage in the enlargement of feathers on the hands and arms as an aid in catching insects (insect-net theory). The enlarged tail feathers are hypothesized as aerodynamic stabilizers enhancing agility and quick maneuvering during chase after prey. *Archaeopteryx* (bottom) presumably was at or just past the threshold of powered flight, but is shown here in a similar predaceous pose. Ostrom's version of the cursorial theory has the "insect nets" providing the lift for flight. Three scientists at the University of Northern Arizona have conducted experiments that suggest jumping may have supplied the necessary lift.
(From John H. Ostrom, 1979, Bird Flight, How Did It Begin? Fig. 10, p. 55: *Am. Scientist,* vol. 67, Reproduced by permission of *American Scientist*)

theropod dinosaur whose feathers were more an adaptation for thermal insulation than for powered flight. Perhaps *Archaeopteryx* represents an evolutionary transition from theropod saurischian to bird (Figs. 35, 36).

Marine Life

The Jurassic is the least widespread system in North America. Although some sections are abundantly fossiliferous (see Fig. 12-37 for representative community), strata on this continent are not nearly as rich in fossils or as well represented as in Europe. Jurassic strata in Great Britain provided the substance for William Smith's monumental principle of fossil succession. The effect of this principle on the science of stratigraphy can hardly be exaggerated. Somewhat later, from studies of Jurassic strata, refinement of Smith's ideas led to the concepts of stage and zone (Chapter 3), and the observation that rock types change laterally within certain Jurassic ammonite zones led to the formulation of the facies concept. Because Jurassic strata and their contained fossils fostered these fundamental stratigraphic principles, the Jurassic System commonly is referred to as the cornerstone of stratigraphy.

The organisms largely responsible for recognition of stratigraphic units based on fossil content were the ammonites. These swimming cephalopods were the most abundant invertebrates in Jurassic seas and displayed a remarkable degree of diversity in shell form and in intricacy of shell sutures. Jurassic ammonite successions also illustrate several detailed evolutionary lineages, and the rapid vertical stratigraphic change in ammonite species composition has allowed for a refined zonation scheme that has resolved Jurassic time into increments of approximately one million years duration.

The belemnites were squidlike cephalopods whose bullet-shaped internal shells are common Jurassic fossils. Paleotemperature determinations on the calcite shells of belemnites point to sea-water temperatures at midlatitudes about 15°C warmer during the Jurassic than at present. The quantitative paleotemperature data are derived from oxygen-18 : oxygen-16 ratios that vary according to the temperature of the seawater in which the organisms lived. The amount of oxygen-18 in shells decreases as temperature increases.

Jurassic invertebrate faunas, exclusive of the ammonites, definitely had a strong semblance of modernization. Bivalve and gastropod molluscs continued to thrive and diversify, as did the sea urchins and corals. Also numbered among the marine fauna were abundant vertebrates. Several groups of reptiles that had invaded the sea during the Triassic were well represented by the ichthyosaurs (Fig. 12-38), plesiosaurs, and marine turtles, all of which shared the seas with a great variety of bony fishes.

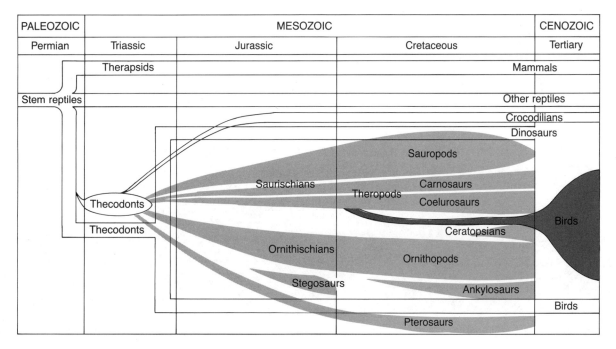

Figure 12-36. Evolution of birds from theropod dinosaurs.
(After John H. Ostrom, 1978, New Ideas about Dinosaurs; p. 166: *Nat. Geographic,* vol. 154, no. 2. Used by permission of National Geographic Society)

Figure 12-37. Jurassic calcareous sand-bottom community: a₁, ammonoid cephalopod mollusc; a₂. Nautiloid cephalopod mollusc; b, bivalve molluscs; c, articulate brachiopods; d, echinoid echinoderm; e, trace fossil (burrow) containing crustacean.
(From W. S. McKerrow, 1970, *The Ecology of Fossils*; Fig. 78, p. 245: Gerald Duckworth, London. Used by permission of author)

CRETACEOUS HISTORY
An Abundant Record

The Cretaceous System is well represented in North America. Anyone who has cast even a cursory glance at the geologic map of the United States cannot help but be impressed by the amount of area colored green, representing Cretaceous sedimentary rocks. Significant areas of Cretaceous plutonic rocks also are present. Cretaceous sediments were deposited over a wide region by marginal and interior seas that spread beyond the limits of Jurassic deposition. With the gradual encroachment of Cretaceous seas onto the continent, the Pacific coast and Atlantic and Gulf coasts were submerged and thick stratigraphic sections accumulated. When considering Cretaceous depositional patterns, one must keep in mind the plate tectonics of the period. The Pacific margin was the leading edge and the Atlantic and Gulf of Mexico margins represented the passive trailing edge of the North American continent. We shall discuss the effects of plate tectonics on sedimentary patterns by first examining the trailing margin.

Sedimentation on a Trailing Margin

On the Atlantic coastal plain, Cretaceous sediments are exposed in a narrowing band from Georgia to Long Island (Fig. 12-39) and have been traced in the subsur-

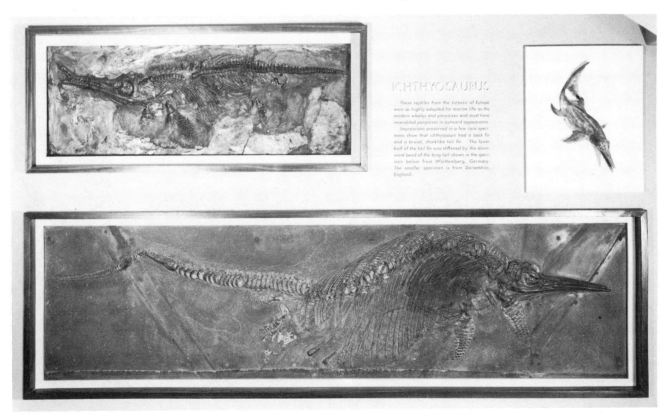

Figure 12-38. Fossilized skeletons of ichthyosaurs, Jurassic of western Europe.
(Photo courtesy of National Museum of National History. Smithsonian Institution Photo No. 1142)

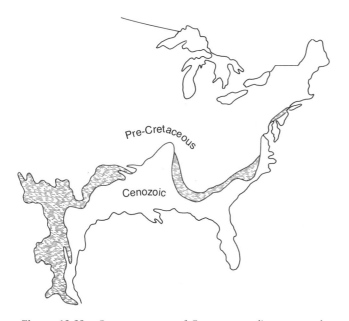

Figure 12-39. Outcrop pattern of Cretaceous sedimentary rocks, eastern United States and northeastern Mexico.
(Data from P. B. King and H. Beikman, comps., 1974, Geological Map of the United States: U.S. Geol. Survey; P. B. King, comp., 1969, Tectonic Map of North America: U.S. Geol. Survey)

face as far north as Newfoundland. The outcrop belt represents the exposed, gently upturned edges of a wedge-shaped succession that thickens seaward and comprises a sizeable bulk of the present Atlantic continental shelf. The sediments (Fig. 12-40) include sandstones, with local development of *glauconitic* sand (greensand), siltstones, and mudstones, and minor amounts of conglomerate. Units of marl (clayey limestone) are present, but few bona fide limestones occur in the succession. The terrigenous clastic material was derived from erosion of the Appalachian Mountains.

The morphology of the Atlantic coastline during the Cretaceous was probably somewhat similar to that of the present and included beach, bay, lagoon, barrier-island, delta, and marine-shelf environments. A present-day prominent erosional scarp, called the Fall Line, marks the boundary between the Coastal Plain and Piedmont Provinces; the scarp is an expression of the contact between the harder and older crystalline rocks of the Piedmont Province and the edge of the softer Cretaceous deposits along the western margin of the Coastal Plain Province.

In the Gulf Coast Province, Cretaceous sediments are exposed in a large crescent-shaped belt from the southern margin of the Atlantic coastal plain in Georgia

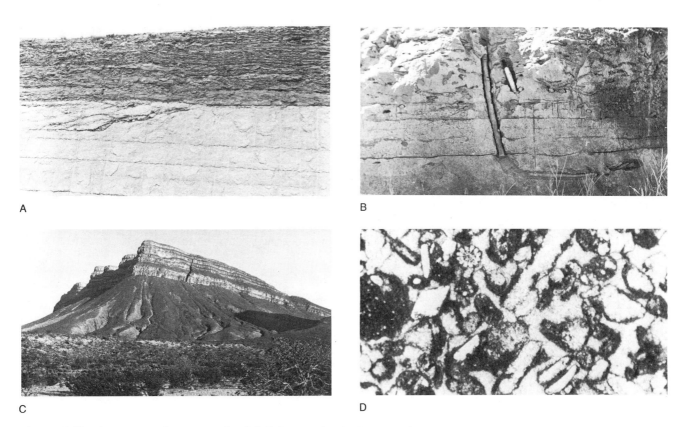

Figure 12-40. Cretaceous sedimentary rocks of Gulf Coast region. A. Cretaceous limestone, Pecos River valley, Texas. B. Upper Cretaceous shallow-marine sandstone with vertical burrow structure. C. Lower Cretaceous reef limestone, northeastern Mexico. D. Thin-section photomicrograph of Lower Cretaceous beach-rock limestone composed of rounded skeletal fragments.
(Photos by J. Cooper)

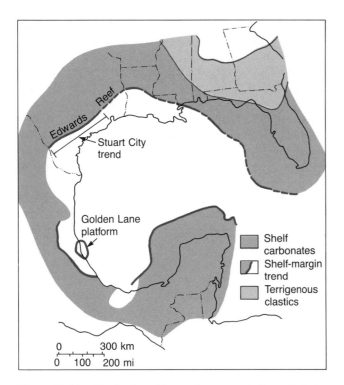

Figure 12-41. Distribution of Lower Cretaceous shelf margin facies around Gulf of Mexico Basin. Reef trend shown as dark band.
(After A. A. Meyerhoff, 1967, Future Hydrocarbon Provinces of Gulf of Mexico–Caribbean Region, Fig. 8; p. 231: *Gulf Coast Assoc. Geol. Societies Trans.,* vol. 17; D. G. Bebout and C. H. Moore, Jr., eds., Carbonate Depositional Environments, Fig. 1, p. 441: *Am. Assoc. Petroleum Geologists Mem.* 33. Used by permission of Gulf Coast Association of Geological Societies and American Association of Petroleum Geologists)

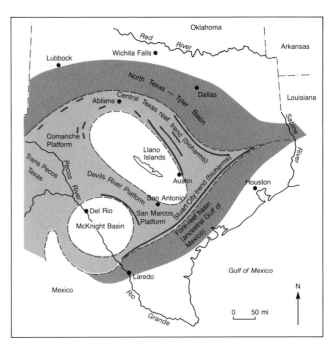

Figure 12-42. Early Cretaceous paleogeography of Texas, showing barrier-reef trend, back-reef platform, and shelf basins.
(After D. G. Bebout, and R. G. Loucks, 1974, Stuart City Trend, Lower Cretaceous, South Texas—A Carbonate Shelf Margin Model for Hydrocarbon Exploration; Fig. 2, p. 442; *in* Carbonate Depositional Environments: *Am. Assoc. Petroleum Geologists Mem.* 33; W. L. Fisher and P. U. Rodda, 1969, Edwards Formation (Lower Cretaceous), Texas: Dolomitization in a Carbonate Platform System, Fig. 1, p. 56: *Am. Assoc. Petroleum Geologists Bull.,* permission of American Association of Petroleum Geologists)

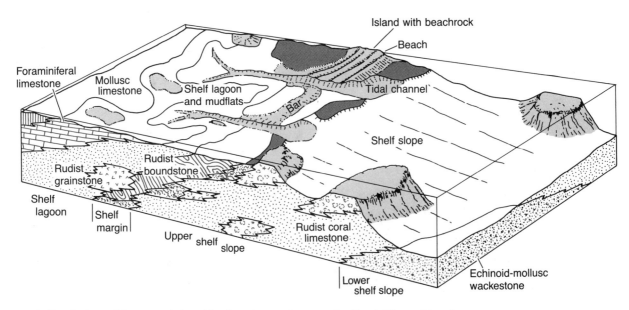

Figure 12-43. Facies and interpreted depositional environments across the Stuart City reef trend, South Texas. Wackestones are limestones that have a lime mud-supported fabric.
(After D. G. Bebout, and R. G. Loucks, 1974, Stuart City Trend, Lower Cretaceous, South Texas—A Carbonate Shelf Margin Model for Hydrocarbon Exploration, Fig. 3, p. 443, *in* Carbonate Depositional Environments: *Am. Assoc. Petroleum Geologists Mem.* 33, Used by permission of American Association of Petroleum Geologists)

to the Mexico–Guatamala border. The outcrop belt is interrupted around the Mississippi Embayment (Fig. 12-39), where a prominent reentrant of coastal Cenozoic deposits overlaps the Cretaceous. As in the Atlantic coastal region, Gulf Coast Cretaceous sediments form a seaward-thickening wedge, but unlike the Atlantic section much of the Gulf section is developed as carbonates. In the eastern Gulf Coast areas of Alabama and Mis-

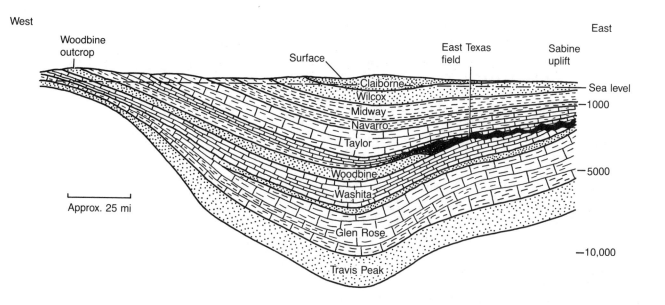

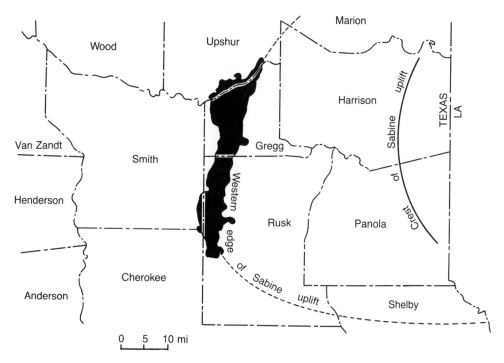

Figure 12-44. East Texas petroleum accumulation. A. Wedging of eroded Woodbine Sandstone between Austin Chalk and Washita Limestone, forming traps for East Texas field on western flank of Sabine uplift. B. Surface projection of petroleum accumulation in wedgeout of Woodbine Sandstone. (After M. T. Halbouty, and J. J. Halbouty, 1982, Relationships between East Texas Field and Sabine Uplift in Texas; Figs. 2,3, p. 1045: *Am. Assoc. Petroleum Geologists Bull.,* vol. 66, no. 8. Used by permission of American Association of Petroleum Geologists)

sissippi only Upper Cretaceous rocks crop out at the surface, but to the west an expansive terrane of Lower Cretaceous rocks is exposed in the Edwards Plateau of central Texas and the Sierra Madre Oriental of northeastern Mexico.

From Alabama to southeastern Mexico, Early Cretaceous carbonate deposition was influenced by a major barrier-reef complex called the Edwards Reef (Fig. 10-41). The reef trend had a particularly strong influence on carbonate platform deposition in the Texas region. The barrier-reef trend itself lies in the subsurface of the Coastal Plain Province, but the varied flat-lying deposits of the back-reef area are widely exposed throughout the dissected Edwards Plateau (Fig. 12-42). Here the carbonate rocks are every bit as complex in facies patterns as Paleozoic carbonates of the great platform-interior seas. Most of the rocks were deposited in shallow water in a mosaic of carbonate tidal flats, oolite and skeletal banks and bars, lagoons, bays, and small patch reefs (Fig. 10-43). One formation, the Glen Rose Limestone, contains some individual beds that can be traced for hundreds of square kilometers, suggesting deposition on broad tidal flats. Other shallow-water features include ripple marks that commonly show interference patterns, cross-bedding, flat-pebble intraformational conglomerates, mud cracks, evaporites, and even dinosaur footprints, some of which contain raindrop impressions. During Glen Rose deposition the 600-km trip from Dallas to Del Rio could have been made in a rowboat and a good pair of wading boots.

In contrast to the Lower Cretaceous, the Upper Cretaceous in the western part of the Gulf Coast region contains large amounts of terrigenous clastic material, probably derived from Late Cretaceous uplift in the Rocky Mountain region. The Upper Cretaceous of the eastern Gulf Coast contains more carbonates, indicating that the region was largely isolated from major sources of terrigenous influx. Today major carbonate deposition is confined principally to the Florida Bay area of the west Florida shelf and to the Yucatan shelf, reflecting, again, lack of proximity to terrigenous influence. Both the eastern and western Gulf Coast Upper Cretaceous sections contain prominent units of fine-grained limestone called chalk. It was the chalk (creta is the Celtic word for chalk) deposits on both sides of the English Channel that served as the original type section for the Cretaceous System.

Cretaceous rocks in the Gulf Coast Province have yielded abundant hydrocarbons. The productive oil fields of eastern and northeastern Mexico have tapped carbonate reservoir rocks of the Edwards Reef, Golden Lane, and other reef trends and associated platform facies. The truncation of a prominent sandstone unit, the Woodbine Sandstone, beneath an unconformity has given rise to the East Texas Field (Fig. 12-45), a multibillion barrel producer that has been one of the largest

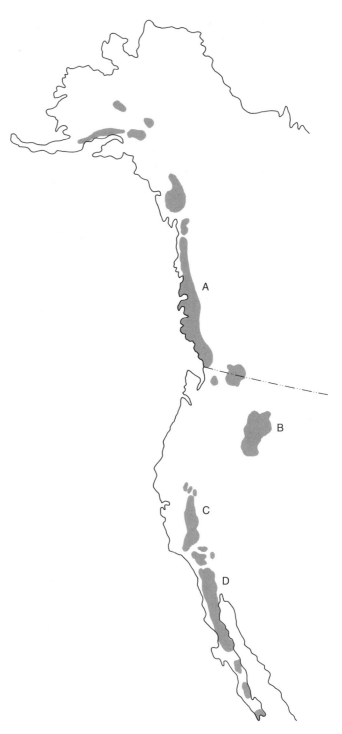

Figure 12-45. Western North America showing trend of Cretaceous batholiths. A. Coast Ranges batholith. B. Idaho and Boulder batholiths. C. Sierra Nevada batholith. D. Southern California batholith.
(Data from P. B. King, comp., 1969, Tectonic Map of North America: U.S. Geol. Survey)

oil fields on the North American continent. The Woodbine is a shallow-marine shoreline facies in which are trapped hydrocarbons that migrated from more seaward and basinal organic-rich mudstones.

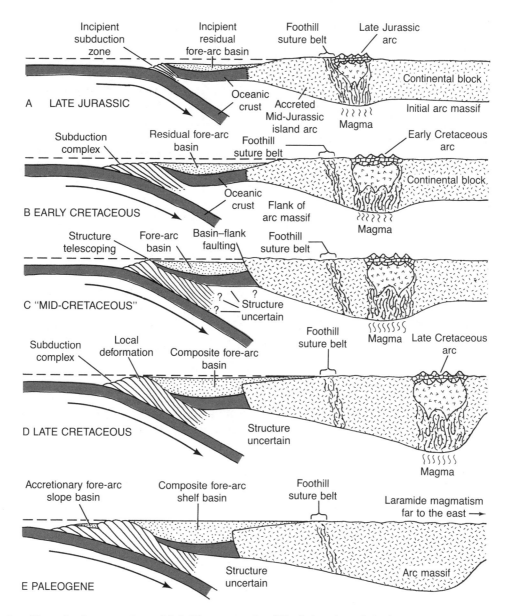

Figure 12-46. Late Mesozoic plate tectonic model. A. Western margin of North America at latitude of northern California. Note subduction zone and associated trench (site of deposition of Franciscan Formation), formation of accretionary ridge and fore-arc basin (Great Valley Group) in arc-trench gap, and magmatic arc (which coincides with batholithic trend of Fig. 12-41). Foothill suture belt is zone of collision between mid-Jurassic island-arc system and continental-margin magmatic arc to initiate Nevadan phase of Cordilleran orogeny and the change from a Japanese to an Andean (South American) type of margin.
(From W. R. Dickinson and D. R. Seely, 1979, Structure and Stratigraphy of Fore-arc Basins, Fig. 11, p. 25: *Am. Assoc. Petroleum Geologists Bull.,* vol. 63. Used by permission of American Association of Petroleum Geologists)

The Continent's Leading Edge

Let us now examine some Cretaceous events in the western Cordillera along the dynamic, leading edge of the continent. The beginnings of the Cordilleran orogeny in Late Jurassic continued into Cretaceous time as immense volumes of granite were formed from Alaska to Baja California (Fig. 12-45). These granitic plutons include the Southern California, Sierra Nevada, Idaho, and

Coast Range batholiths. This chain of batholiths is interpreted as representing a Late Mesozoic magmatic arc related to subduction of Pacific oceanic lithosphere along the western margin of the North American continental plate (Fig. 12-46). Gold nuggets that lured prospectors to California in the Gold Rush of 1849 came from gold-bearing quartz veins in the plutonic rocks, the Mother Lode. Metallic ores of the Coeur d'Alene district in Idaho; Butte, Montana; and British Columbia were deposited by hydrothermal solutions that accompanied the magmatic activity and formation of the batholithic rocks. The formation of granite, originally thought to be a product of slow cooling of parent magma at depth, also has been ascribed to an essentially metamorphic process, **granitization,** whereby preexisting rocks have been transformed into granite through replacement. The granites of orogenic belts may be related to subcrustal melting and replacement involving a heat source generated from subduction below the edge of the con-

tinental lithosphere. The Andes belt of South America contains abundant granite and andesite as part of the magmatic arc related to the subduction of Pacific lithosphere beneath the western edge of the continent (Fig. 12-47). As mentioned previously, the western margin of North America was probably an Andean-type margin during most of Late Mesozoic time. Parts of the batholith complexes are as old as mid-Jurassic, but most of the dates indicate an Early to Medial Cretaceous age.

Deposition of the chaotic Franciscan Formation in

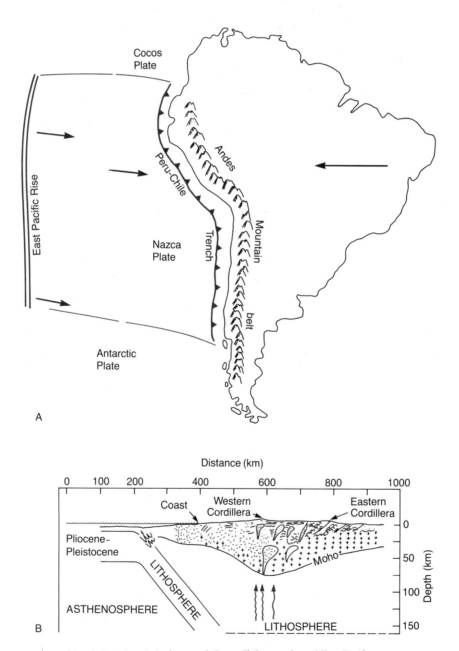

Figure 12-47. Western margin of South America. A. Andes trend, Peru–Chile trench, and East Pacific rise spreading center—a modern plate tectonic analogue for Late Mesozoic–Early Cenozoic western margin of North America. B. Cross section showing similarity to western margin of North America during Late Cretaceous Sevier orogeny.
(A courtesy Critter Creations; B after D. E. James, 1971, Plate Tectonic Model for the Evolution of the Andes, Fig. 10, p. 3341: *Geol. Soc. America Bull.,* vol. 82. Reproduced by permission of author)

the trench and Great Valley turbidite-fan assemblage in the fore-arc basin of an active plate margin continued generally uninterrupted through the Cretaceous (Fig. 12-48). The western part of the thick Great Valley Group was probably deposited on oceanic basement. However, the original physical relationship between the Great Val-

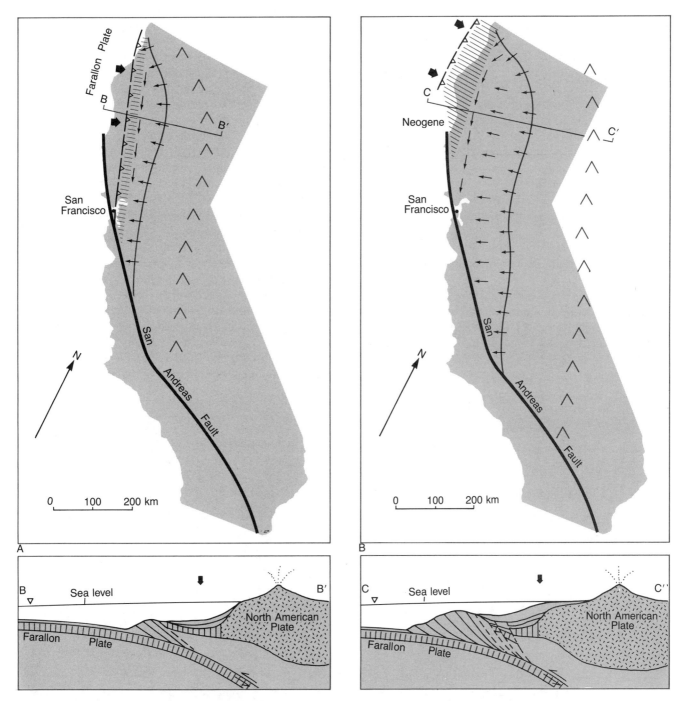

Figure 12-48. Cretaceous, northern California. A. Early Cretaceous. B. Late Cretaceous. Symbols the same as for Figure 12-27. By Cretaceous time the subduction complex (Franciscan Formation) had enlarged and the fore-arc basin had subsided sufficiently so that arc-derived detritus was trapped behind a bathymetric barrier. Most sediments (Great Valley Group) accumulated in deep-marine basin-plain environments within the arc-trench gap during Early Cretaceous, but by Late Cretaceous, sedimentary progradation of turbidite-fan, slope, and shelf deposits was beginning to fill the fore-arc basin.
(After R. V. Ingersoll, 1978, Paleogeography and Paleotectonics of the Late Mesozoic Forearc Basin of Northern and Central California; Figs. 5–8, pp. 476–477, *in* Mesozoic Paleogeography of the Western United States: Pacific Section SEPM Paleogeography Symposium, vol. 2. Reproduced by permission of Pacific Section, Society of Economic Paleontologists and Mineralogists)

ley and Franciscan rock assemblages has been masked by Cenozoic faulting. The late Mesozoic rocks of central California consist of three side-by-side assemblages that evolved during the same time span: the Franciscan—a poorly organized trench and subduction zone complex; the Great Valley—a more rhythically bedded fore-arc basin accumulation; and the Sierra Nevada batholith—a magmatic arc complex. In southern California, these belts of Jurassic and Cretaceous igneous, metamorphic, and sedimentary rocks cut across the trend of the major Paleozoic paleotectonic features and facies belts (Fig. 12-49). This relationship probably resulted from the different angle of convergence of the North American and Pacific Plates once the mid-Atlantic rift opened and the North American continent began its westward movement in mid-Mesozoic time.

As with the preceding Triassic and Jurassic Periods, reconstructions of Cretaceous paleogeography of the Pacific Coast are sketchy because of overprinting by Cenozoic tectonics. The evidence, however, points to continued subduction with development of shallow-water coastal deposits along a narrow shelf and rapid subsidence and thick sediment accumulations in deeper water offshore. Turbidite-fan complexes are well represented in northern California and Oregon, and abundant coal-bearing deltaic deposits are found in the north slope of Alaska.

Cretaceous Interior Seaway

In the western part of the craton, flooding of marine waters from both the Arctic and Gulf of Mexico formed a broad interior seaway (Fig. 12-50)—somewhat reminiscent of the Sundance Sea of the Jurassic—by mid-Cretaceous time. This mid-Cretaceous transgression, also documented on the other continents, probably resulted from accelerated sea-floor spreading as Pangaea underwent major fragmentation; marine waters, displaced by inflated spreading centers, flooded the margins and interiors of drifting continents. The mid-Cretaceous flooding represents the last major marine transgression and the highest sea-level stand of the entire Phanerozoic, slightly exceeding that estimated for the Late Cambrian (Fig. 12-51). The North American continent has been in a condition of emergence during the last 80 million years. At its maximum extent, during the early Late Cretaceous, the seaway was more than 1500 km wide, stretching from Utah to Iowa. The eastern margin of the seaway sloped gently and subsided slowly. Comparatively little sediment was introduced from the emergent parts of the craton to the east, resulting in thickness of generally less than 100 m for the Cretaceous deposits in the eastern part of the interior seaway. The Dakota Sandstone, a famous aquifer in the western interior of the United States, is a nearshore deposit of the initial transgression of the sea.

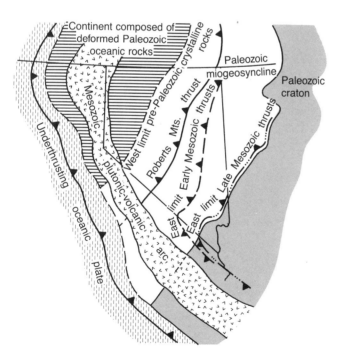

Figure 12-49. Late Mesozoic, showing truncation of Paleozoic geosynclinal and deformational trends by a Mesozoic plutonic-volcanic arc of Andean type. An inferred southward convergence of early and late Mesozoic thrust zones is also shown as is the change in trend of such faults and their departure from the geosynclinal terrane in southeastern California.
(From B. C. Burchfiel and G. A. Davis, 1972, Structural Framework and Evolution of the Southern Part of the Cordilleran Orogen, Western United States; Fig. 7, p. 111: *Am. Jour. Science,* vol. 272)

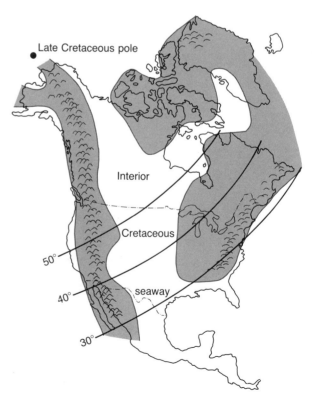

Figure 12-50. Generalized Late Cretaceous paleogeography of North America showing extent of interior Cretaceous seaway.
(Data from several sources)

The depositional and stratigraphic patterns of this interior seaway bear evidence of the continuously growing Cordilleran mountain belt to the west (Fig. 12-52). Abundant thin volcanic ash beds called **bentonites** are present in many Cretaceous sections in Montana, Wyoming, and Colorado and attest to continued volcanism in the western and central Cordilleran belt. The bulk of the sediment was supplied from Cordilleran highlands raised during the **Sevier orogeny,** along the western margin of the seaway. Here pronounced subsidence and sediment influx resulted in sequences more than 6000 m thick in places. The facies relationships show a progressive change from coarse sandstone and conglomerates adjacent to the source areas on the west to finer sandstones, siltstones, and shales with a few tongues of limestone to the east. Figure 12-53 depicts the dynamic transgressive-regressive patterns as recorded in onlap-offlap stratigraphic cycles and the progressive offlap pattern of the sedimentary succession. These facies reflect

alluvial-fan, coastal-plain, fluvial, beach, lagoon, barrier bar, delta, and marine-shelf paleoenvironments (Fig. 12-54). The stratigraphic patterns faithfully record the migration of these various environments over one another through time in response to the relationship between tectonic activity (sevier orogeny) in the source area, subsidence, and the varying amounts of detritus delivered to the depositional systems.

Rocky Mountain Clastic Wedge

This thick sequence of Cretaceous terrigenous clastics that prograded from west to east is the Rocky Mountain clastic wedge. The relationship of this clastic wedge to the Cordilleran orogeny is reminiscent of the Ordovician, Devonian Catskill, and Pennsylvanian clastic wedges and their relationships to the Taconic, Acadian and Alleghenian phases, respectively, of the Appalachian orogeny. With regard to the last comparison, Cretaceous

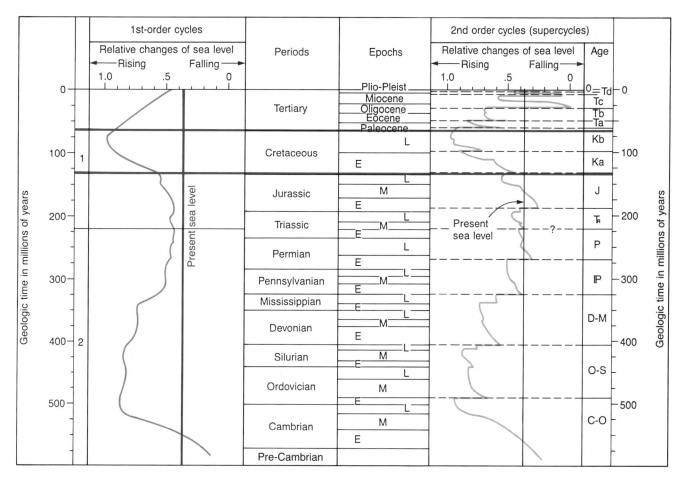

Figure 12-51. First- and second-order global cycles of relative sea-level change during Phanerozoic, highlighting early Late Cretaceous transgressive maximum. In recent years, detailed knowledge of ancient Phanerozoic sea-level changes has been enhanced by seismic stratigraphic techniques developed primarily within the petroleum industry. In such a context, subsurface discontinuities have been revealed that identify transgressive and regressive events. Cycles identified in the subsurface of continental margins commonly can be correlated with outcrop sections on land.
(After P. R. Vail, R. M. Mitchum, and S. Thompson II, 1977, Seismic Stratigraphy and Global Changes of Sea Level; Figs. 1,2, pp. 84, 85: *Am. Assoc. Petroleum Geologists Mem.* 26. Reproduced by permission of American Association of Petroleum Geologists)

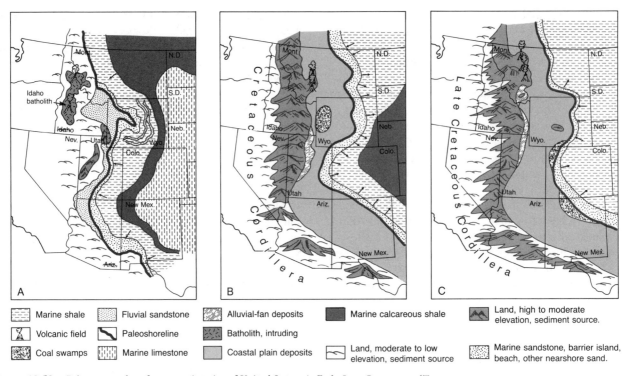

Marine shale	Fluvial sandstone	Alluvial-fan deposits
Volcanic field	Paleoshoreline	Batholith, intruding
Coal swamps	Marine limestone	Coastal plain deposits

Marine calcareous shale	Land, high to moderate elevation, sediment source.
Land, moderate to low elevation, sediment source	Marine sandstone, barrier island, beach, other nearshore sand.

Figure 12-52. Paleogeography of western interior of United States. A. Early Late Cretaceous (Turonian). B. Mid-Cretaceous (Campanian). C. Late Late Cretaceous (Maestrichtian).
(After D. P. McGookey, coord. 1972, Cretaceous System, Figs. 27,40,49, pp. 209, 218, 224, *in* Geologic Atlas of the Rocky Mountain Region: Rocky Mountain Assoc. Geologists, Denver. Reproduced by permission of Rocky Mountain Association of Geologists)

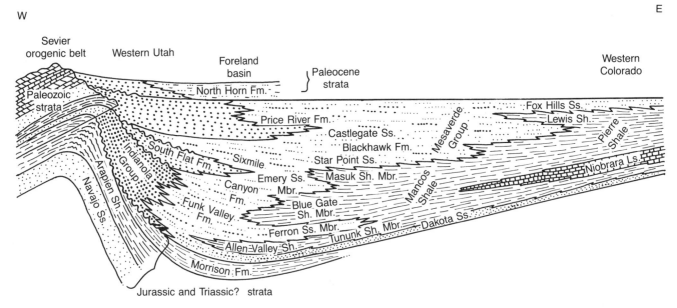

Figure 12-53. Restored west–east cross section of Cretaceous facies of western interior seaway and relationship to Sevier orogenic belt.
(From R. L. Armstrong, 1968, Sevier Orogenic Belt in Nevada and Utah; Fig. 5, p. 446: *Geol. Soc. America Bull.*, vol. 79. Reproduced by permission of Geological Society of America and author)

outcrops in the Four Corners region, such as in Mesa Verde National Park, Colorado; in the San Juan Basin of northwestern New Mexico; and in the Book Cliffs in north-central Utah are offlap sequences that record numerous fluctuations of strandlines and cyclic deposition. Much of the coal in the Rocky Mountains and Colorado

Plateau is from these Cretaceous cyclic deposits that record deltaic, coastal-barrier, and lagoonal environments. Cretaceous sandstones in the subsurface of Wyoming, Colorado, and Montana also have produced significant hydrocarbons.

As the Cretaceous drew to a close, heightened tectonic activity and massive incursions of terrigenous clastics from the west together with a slowing of global sea-floor spreading forced the withdrawal of this last great interior sea from the continent. The marine waters retreated to the north and south, and marginal marine and continental deposition took place. Widespread coastal-plain deposits formed as the seaway retreated. These deposits are locally coal bearing and include diverse fossil plants and the last of the dinosaurs.

The style of deformation during the Cretaceous in much of the Cordilleran belt was characterized largely by back-arc thrusting, which displaced gigantic slabs of Paleozoic miogeosynclinal rocks and shoved them eastward, along low-angle westward-dipping thrust planes, over Mesozoic rocks (Fig. 12-55). Imbricate thrust slices stacked like huge shingles can be traced from southern California through Nevada (Figs. 56, 57), northward into the Canadian Rockies, and on into the Brooks Range of Alaska. This crustal shortening, related to the Sevier phase of the Cordilleran orogeny, was as much as 50 to 100 km in places and was a principal manifestation of the subduction that affected the Cordilleran margin from Late Jurassic to early Cenozoic time. In recent years the Cordilleran overthrust belt has been the focus of active exploration for hydrocarbons, with targets being attractive reservoir facies and structures beneath the soles of thrust faults.

Marine Invertebrates

The nearshore and offshore marine-shelf environments of the interior sea and coastal regions teemed with life (Figs. 58, 59). Exceptionally fossiliferous deposits in the Gulf Coast and in the Great Plains areas of Alberta, eastern Montana, the Dakotas, Nebraska, and Kansas attest to seas populated with diverse cephalopod, bivalve, and gastropod molluscs (Fig. 12-60A–D). Many of the ammonite cephalopods from the Pierre Shale in the western interior are beautifully preserved and still retain the pearly luster of their aragonitic shells. Rapidly evolving Cretaceous ammonite faunas have permitted refined biostratigraphic zonations that have greatly facilitated both intracontinental and intercontinental correlation. Cretaceous ammonites show an amazing diversity (Figs. 12-60G–M, 61) related to many different adaptations and life habits in the marine realm. Unfortunately, the lack of modern-day close relatives of this fascinating group of marine invertebrates has severely limited our understanding of their paleoecology.

Cretaceous rocks also contain the first evidence of planktonic Foraminifera as well as the microscopic

A

B

C

Figure 12-54. Cretaceous marine shelf and shoreface deposits. A. Book Cliffs, eastern Utah. B. Mesa Verde National Park, Colorado. C. Mesa de los Cartujanos, Sabinas Coal Basin, northeast Mexico. (Photos by J. Cooper)

plates of calcareous floating marine algae, the coccolithophorids (Fig. 12-62). The fine-grained calcareous skeletal material of these organisms contributed to the deposits of chalk on marine shelves. Diatoms also became abundant for the first time during the Cretaceous, and

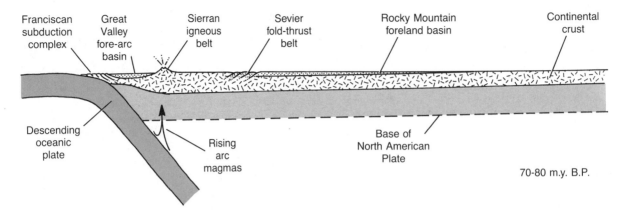

Figure 12-55. Inferred configuration of subducted slab and associated tectonic elements during early Late Cretaceous Sevier orogeny phase of evolution of the Cordillera. Note Cretaceous interior seaway occupying a foreland basin (retroarc basin) cratonward of Sevier orogenic belt, itself a zone of back-arc thrusting related to subduction along continental margin.
(From W. R. Dickinson, 1979, Cenozoic Plate Tectonic Setting of the Cordilleran Region in the United States, Fig. 2a, p. 4. *in* Cenozoic Paleogeography of the Western United States: SEPM Pacific Section, Pacific Coast Paleogeography Symposium 3, Used by permission of Pacific Section, Society of Economic Paleontologists and Mineralogists)

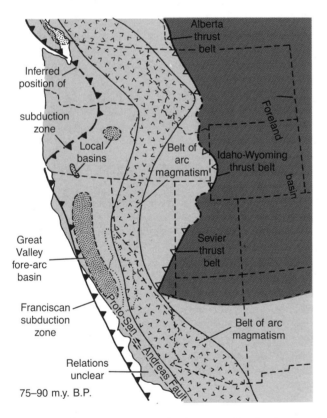

Figure 12-56. Paleotectonic and paleogeographic sketch map of Cordilleran region. Late Cretaceous, 75 m.y. B.P.
(From W. R. Dickinson, 1979, Cenozoic Plate Tectonic Setting of the Cordilleran Region in the United States, Fig. 10, p. 11, *in* Cenozoic Paleogeography of the Western United States: SEPM Pacific Section, Pacific Coast Paleogeography Symposium 3, Used by permission of Pacific Section, Society of Economic Paleontologists and Mineralogists)

their photosynthesis coupled with that of the coccolith-oporids may have resulted in significant increases of atmospheric oxygen.

Figure 12-57. East face of Spring Mountains (looking northwest), west of Las Vegas, Nevada, showing Keystone thrust, one of the major faults in the Sevier overthrust belt. Dark-colored rocks of upper plate are lower Paleozoic miogeosynclinal and craton margin carbonates; light-colored rocks of lower plate belong to eolian Lower Jurassic Aztec Sandstone (equivalent to Navajo).
(Photo courtesy of John S. Shelton)

Peculiar oysters assigned to the genera *Exogyra* (Fig. 12-60A) and *Gryphaea* are locally abundant in Cretaceous invertebrate faunas. Another interesting group of Cretaceous bivalve molluscs were asymmetrical, sessile clams called rudists (12-60D). Most bivalve molluscs, modern and ancient, have shell forms that are sensitive indicators of their life habits. The rudists were no exception; the shape, size, and robustness of rudist shells are good paleoecologic indicators of shallow water depth and high energy. Because of their superficial resemblance to certain rugose corals, rudists provide a good example of convergent evolution. Rudists, along

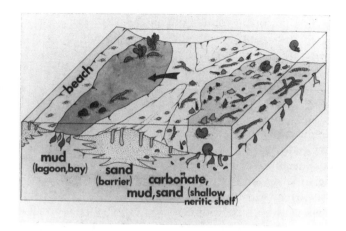

Figure 12-58. Restoration of scene from interior seaway of northern Rocky Mountain region during the Late Cretaceous (approximately 80 m.y. B.P.). Fauna includes several kinds of planispirally coiled ammonites, straight-shelled baculitid ammonites, belemnites, gastropods, and bivalve molluscs.
(Photo courtesy of National Museum of Natural History, Smithsonian Institution Photo No. 656)

Figure 12-59. Late Cretaceous shallow-marine invertebrate benthic communities. Three environmental faunal assemblages are shown: 1, low-diversity lagoonal assemblage; 2, low-diversity, deep-burrowing, shifting-substrate community of a barrier bar, reflecting stressed high-energy conditions; 3, high-diversity, open-marine assemblage consisting of ammonoid cephalopods, bivalve and gastropod molluscs, and horizontal feeding traces.

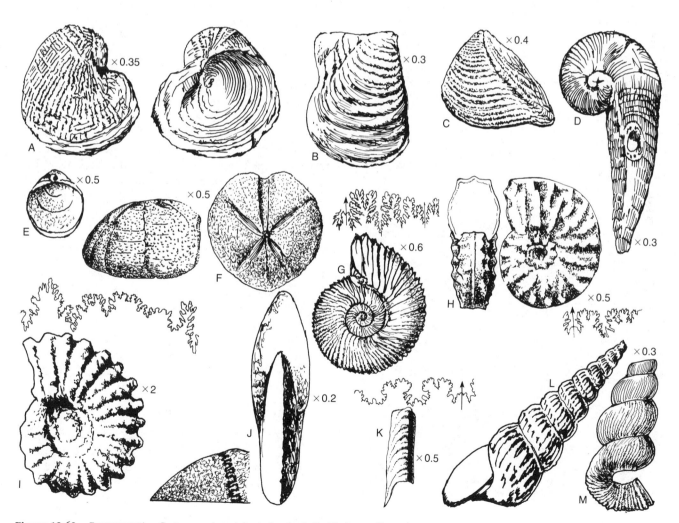

Figure 12-60. Representative Cretaceous invertebrate fossils. A–D. Bivalve molluscs (D, rudist clam); E. Articulate brachiopod. F. Echinoid echinoderm. G–M. Ammonite cephalopod molluscs (K–M are heteromorph forms).
(From *Treatise on Invertebrate Paleontology*).

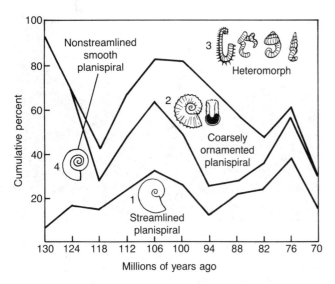

Figure 12-61. Three evolutionary trends in Cretaceous ammonite shell types—1, streamlined, 2, ornamented, 3, heteromorphic (nonplanispiral, asymmetric)—developed at the expense of 4), nonstreamlined, smooth planispiral forms that characterized most Jurassic ammonoids. Peter Ward of the University of California at Davis interprets this trend as an expression of adaptation to the selective pressures exerted by progressively more sophisticated marine predators, such as teleost fish and marine reptiles, that evolved during the Jurassic and Cretaceous. The smooth exterior and laterally compressed cross section in 1 reduced drag, and tight coiling increased hydrodynamic stability, both factors contributing to agility and speed in swimming, an advantage in outmaneuvering and escaping predators. Broad cross section and comparatively loose coiling in 2 made shell less well adapted for swimming, but coarse ornamentation may have evolved to discourage shell-crunching predators. As shown by the graph, however, this trend was the least successful. Most of the heteromorphs (3) were probably well-poised and balanced floaters. (Adapted from Peter Ward, 1983, The Extinction of the Ammonites, p. 139: *Sci. American,* vol. 249, no. 4)

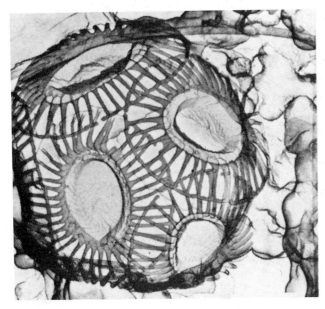

Figure 12-62. Scanning electron photomicrograph of coccolithoporid.
(Photo courtesy of Union Oil Company of California)

with scleractinian corals, coralline algae, and bryozoans, were important contributors to Cretaceous reefs and were the most important frame builders for the Edwards Reef and smaller patch reefs of back-reef platform areas.

Cretaceous reefs of the Gulf Coast region are yet another example of the reef ecosystem. The geologic history of reef communities is testimony to the resiliency of a most important association of organisms. Prior to the Cretaceous, the reef ecosystem had suffered two major calamities: near the end of the Devonian and again at the end of the Permian as part of the end-of-Paleozoic faunal crisis. After each collapse, reef communities resurged with renewed vigor as new populations and taxa of reef builders and reef dwellers filled niches in the ecosystem. Rudists possessed a shell form well adapted for existence in high-energy reefs. One species found in Cretaceous reef limestones in Jamaica has a shell more than 1 m long and more than 10 cm thick.

The presence of rudists in Cretaceous deposits of the West Indies and Gulf Coast regions reflects the spread of Tethyan-realm faunas and tropical paleoecologic conditions. When Pangaea formed during the late Paleozoic, a large oceanic indentation developed on its eastern side in tropical latitudes. This Tethys Sea was a most hospitable marine realm during the Permian and spawned some of the richest reef faunas that have ever existed. The Tethyan realm extended westward during the Mesozoic as Africa separated from Europe and North America (Fig. 12-63). The Tethys Sea, generally synonymous with the tropical marine belt during the Mesozoic, provided a region of stable conditions for invertebrate life. Extensive Triassic reefs in the Alps were nurtured in the resource-rich western end of Tethys. The break up of Gondwana, the opening of the Atlantic, the establishment of a major zone of east-to-west tropical currents, and the flooding of parts of the continents in the mid-Cretaceous paved the way for extension of the stable Tethyan-type conditions to other regions in the tropical belt. The rudists exemplify the spread of Tethys paleobiogeography from the Mediterranean into the Caribbean and Gulf Coast regions during the late Mesozoic as continental breakups reached the point where the tropical marine belt nearly circumscribed the globe.

Cretaceous Vertebrates

Marine vertebrates found the Cretaceous seas much to their liking. Ray-finned bony fishes diversified dramatically during the period, and the western interior seaway in particular included among its inhabitants a horde of aquatic reptiles. Fossils of large sea turtles (Fig. 12-64), marine lizards called mosasaurs (Fig. 65) and the ichthyosaurs and plesiosaurs grace many museum halls. Paddle-shaped limbs and streamlined bodies are the

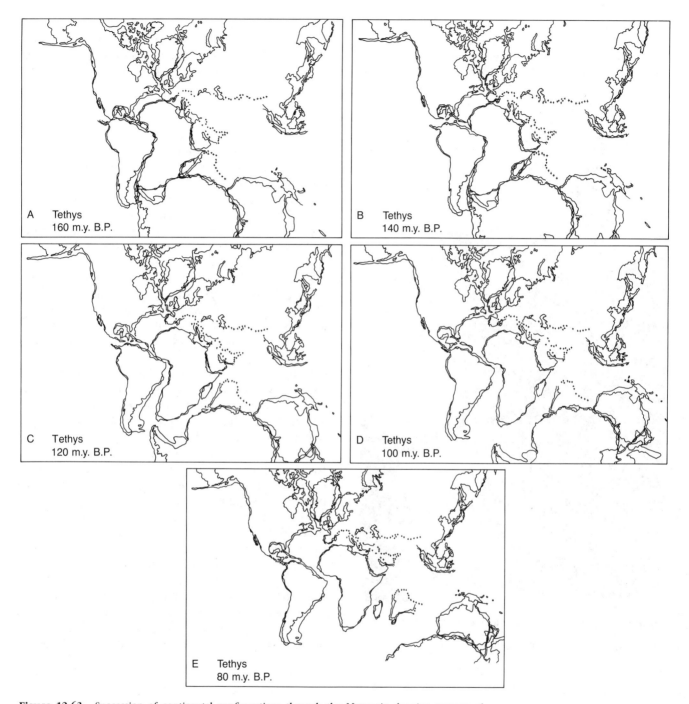

Figure 12-63. Succession of continental configurations through the Mesozoic showing pageant of changes in the Tethyan belt and the opening of the Atlantic Ocean basin. A. Latest Triassic. B. Mid-Jurassic. C. Late Jurassic. D. Early Cretaceous. E. Mid-Cretaceous. F. Late Cretaceous.
(From A. G. Smith, and J. C. Briden, 1977, *Mesozoic and Cenozoic Paleocontinental Maps;* nos. 7, 8, 9, 10, 12, pp. 18, 19, 20, 21, 23: Cambridge University Press, New York. Reproduced by permission of Cambridge University Press)

most characteristic fossilized structural adaptations that accommodated reptilian life in the sea (Fig. 12-66).

After the delicate preservation of *Archaeopteryx* in the Late Jurassic Solenhofen Limestone there is a 50-million-year gap in the fossil record of bird evolution. Cre-

taceous bird fossils, although rare, include specimens representing several of the modern orders: ducks, grebes, and pelicans. Perhaps best known is the large flightless bird *Hesperornis,* well adapted to life in the sea, where it could swim and dive. Pterosaurs (Fig. 12-

67) were also present, and were probably outcompeted by birds before the end of the Cretaceous.

Cretaceous mammals, although somewhat more advanced than their Jurassic predecessors, were nonetheless small, primitive forms that played a distinctly subordinate role to the dinosaurs. Skeletal material beyond the customary jaws and teeth has been collected at Bug Creek and Hell Creek, Montana. Marsupials, some probably ancestral to the common opossum, and early placental mammals number among the more than 25,000 specimens that have been recovered from these significant localities.

Cretaceous dinosaurs reached a range of adaptive types not previously attained. Like other groups of tetrapods the dinosaurs evolved many specialized herbivores and relatively few carnivores. The meat-eating theropods, although in the minority, served an important role in maintaining a balanced ecosystem by functioning as a form of population control. The most popularized of the Cretaceous theropods is *Tyrannosaurus rex,* a voracious flesh eater who reached lengths of more than 12 m, stood about 6 m high, and weighed 5 to 10 metric tons (Fig. 12-68). Although varied in size, the theropods typically possessed large, strong hind limbs with large hooked claws protruding from the three toes. The forelegs were much smaller, but also bore sharp claws, and the oversized powerful jaws were lined with doubly serrated stabbing teeth.

The herbivorous ornithischians diversified into a fantastic array of adaptive types and had a range of dental patterns well adapted for browsing, cutting, cropping, and chewing. The stegosaurs (Fig. 12-31) possessed a double row of about twenty alternating, erect, immovable plates along their arched backs. These dorsal plates, laced with blood vessels, probably functioned mainly for thermoregulation. They also may have afforded some protection, as did two tail spikes. Stegosaurs had an enlarged opening at the spinal base that may have housed nerves from the legs and body mass where they entered the spinal cord or may have been air space to aid in cooling during periods of heightened activity.

During the Cretaceous, the stegosaurs were succeeded by a group of bulky, squat ornithischians, the ankylosaurs, which had tough bony plates that covered their 6-m long bodies and short but sharp sturdy spikes along their flanks. The tail terminated in a bludgeonlike

Figure 12-64. Giant marine turtle from deposits of Cretaceous interior seaway.
(Photo courtesy of Peabody Museum of Natural History)

Figure 12-65. Fossilized skeleton of *Tylosaurus,* a mosasaur from the Cretaceous interior seaway.
(Photo courtesy of National Museum of National History. Smithsonian Institution Photo No. 1145)

knot of bone. With their compact construction and surface armor, the ankylosaurs show definite evolutionary advances toward protection against predators.

A third ornithischian suborder was the Ornithopoda (Fig. 12-68, 69), a bipedal group that diversified dramatically during the Cretaceous. Commonly called duck-billed dinosaurs because of the flattened jaw, the ornithopods had cheek teeth well adapted for grinding vegetation. Some dental patterns were particularly well-suited for grinding and macerating coarse, abrasive plants such as the newly evolved angiosperms. Near the end of the Cretaceous, a group of ornithopods called

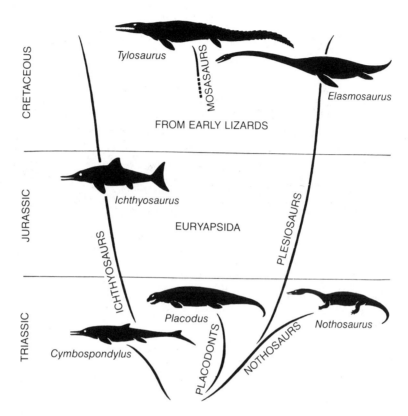

Figure 12-66. Evolutionary relationship of Mesozoic marine reptiles.
(From E. H. Colbert, *Evolution of the Vertebrates—A History of the Backboned Animals through Time;* 3rd ed., Fig. 68, p. 175. Copyright © 1980 by John Wiley & Sons, Inc., New York. Reproduced by permission of John Wiley & Sons, Inc., and Lois M. Darling)

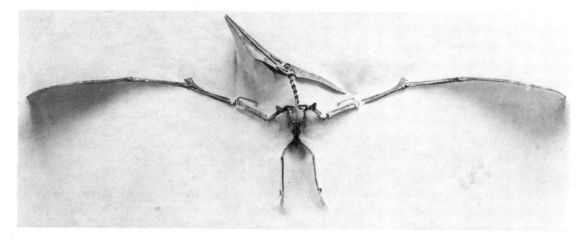

Figure 12-67. Fossilized skeleton of *Pteranodon,* a pterosaur from the Cretaceous of the western interior of the United States.
(Photo courtesy of National Museum of Natural History. Smithsonian Institution Photo No. 28138)

Figure 12-68. Restoration of scene in northern Rockies and Great Plains region during the Late Cretaceous (80 m.y. B.P.). Plants include angiosperms, some of which resemble living types. Dinosaurs, left to right: duck-billed ornithopod *Edmontosaurus,* aquatic plant eater; *Triceratops,* horned plant eater; and two voracious theropod predators—*Gorgosaurus* (left) and *Tryannosaurus* (right)—confronting each other.
(Photo courtesy of National Museum of Natural History, Smithsonian Institution Photo No. 81-16573)

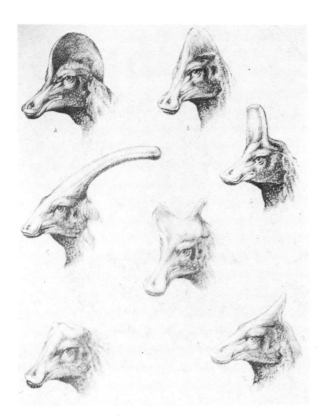

Figure 12-69. 1. Cretaceous ornithopods (left and right) and ankylosaur (center). 2. Cretaceous ceratopsian (*Triceratops,* left) and theropods (*Tryannosaurus,* right).
(C. R. Knight mural, Field Museum of Natural History, Chicago)

hadrosaurs (Fig. 12-70) became large and abundant, and many evolved bony crests, domes, and tubular outgrowths of the nasal passages on the top of the skull. The first recorded dinosaur skeleton discovered in North America was a hadrosaur from Cretaceous beds near Haddonfield, New Jersey. Different crest forms may have been adapted for a variety of functions, but were not, as commonly hypothesized, snorkels for underwater activity. Some domal outgrowths may have cushioned the impact from butting during courtship ritual. The more-tubular outgrowths may have been structural adaptations for improved sense of smell and perhaps for making hooting and honking sounds as a means of verbal communication. Remarkable preservation of nests and egg clutches in upper Cretaceous alluvial-plain sediments in Montana suggest that hadrosaurs nested in large, densely packed colonies, guarded their eggs, and also guarded and fed the young in the nest. This is behavior more characteristic of mammals than reptiles.

The fourth group of ornithischians included the quadripedal, horned forms of the suborder Ceratopsia. The earliest form, *Protoceratops,* was small and lacked horns, but had the characteristic beaklike jaw front. Fossilized *Protoceratops* eggs were found in Lower Cretaceous rocks in Mongolia during the highly publicized Central Asiatic Expedition of the American Museum of Natural History in the 1920s. *Triceratops* had a pair of horns that projected from the skull over the eyes and a

Figure 12-70. Examples of hadrosaur ornithopods.
(C. R. Knight mural, courtesy Field Museum of Natural History)

median horn just above the nostrils (Fig. 12-68). The head was large and had a shieldlike posterior extension for protection and for attachment of jaw and neck muscles. The uppermost stratigraphic occurrence of *Triceratops* fossils helps define the Cretaceous–Paleogene boundary in continental deposits of the western interior.

The Hot-Blooded Dinosaur Controversy

Historically dinosaurs have been regarded as the epitome of extinction. They have been viewed as evolutionary failures and oversized, inefficient hulks that plodded through life toward ultimate extinction. The 1970s witnessed a rebirth of dinosaur research highlighted by pronouncements that dinosaurs, like their vertebrate cousins the mammals and birds, were warm blooded. Cast into the role of unreptilelike reptiles, the dinosaurs have become the subjects of renewed research aimed particularly at obtaining a better understanding of their bioenergetics and behavior. This research has shed new light on the dinosaurs' role in past tetrapod ecosystems. This new perspective regards them as an amazing group of animals with a wide range of adaptations and as much more than just an evolutionary novelty. Regardless of the final outcome of the hot-blooded dinosaur controversy, this research has dispelled many myths concerning these fascinating beasts.

Dr. Robert Bakker, a vertebrate paleontologist at Johns Hopkins University and one of the leaders in the renaissance of dinosaur research, cites predator–prey ratios as being one of the most reliable lines of evidence to support the notion that dinosaurs were warm blooded. The idea here is that a given population of prey animals can support far fewer warm-blooded pre-

dators than cold-blooded ones becuase of the much larger energy requirements for the more active warm-blooded predators. Bakker compared the numbers of flesh-eating dinosaurs versus herbivores from several assumed predator–prey paleoecologic assemblages representing different stratigraphic levels and geographic localities. He found the ratio of predators to prey very low, comparable to ratios of modern lions and zebras. This line of evidence alone would support only the predators' being warm blooded, but even then only if the critical assumption is correct that the specimens collected from each locality accurately portay the abundance of the kinds of dinosaurs that coexisted and, more importantly, truly represent in-place assemblages.

Bakker has suggested that the erect limbs of dinosaurs were associated with high body temperatues and activity levels as they are in modern mammals. The great size actually may have been a help rather than a hindrance in keeping active. Since large animals have much less surface area in relation to body mass than smaller animals, there is relatively less surface exposed for heat gain or loss. This could have been the key to sauropod success and perhaps to dinosaur success in general— the maintenance of a rather constant body temperature. In addition to the predator–prey ratio data mentioned earlier, Bakker cites *bone histology*, in particular, as strong evidence of **endothermy** in dinosaurs. He noted that the density of capillaries and Haversian canals (loci for complex calcium–phosphate exchanges) in dinosaur bone cross sections shows striking similarity to the patterns in birds and mammals.

An opposing line of evidence regarding dinosaur endothermy concerns the internal structure of dinosaur teeth. Paul A. Johnston of the University of Alberta has reported on the presence of conspicuous dentinal growth zones in the teeth of both saurischian and ornithischian specimens from Alberta, Canada. These zones, which appear as rings in cross section and as concentric cones in longitudinal section, are interpreted as representing periodicity of growth in dinosaurs. Johnston compared these teeth to those of both recent and fossil crocodiles (known **ectotherms**) and observed little difference in the pattern and clarity of layers. The growth rates in crocodile teeth are speeded and slowed by seasonal variation in temperature and precipitation.

By contrast, mammals, which are endothermic, do not normally develop dentinal rings comparable to those of dinosaurs and crocodiles. The teeth of mammals of Arctic regions or temperate regions with marked seasonal contrasts show annual layers. Paleontological evidence, however, strongly suggests that Alberta, as well as most of the rest of the western interior of North America, did not have a frigid climate or severe seasonal contrasts during the Late Cretaceous. From these relationships, Johnston has concluded that the dinosaurs were cold blooded, with a ring forming in their teeth during each wet and dry season. Like the bone histology evidence, the dental rings do not conclusively prove ectothermy, and the controversy continues.

Perhaps the most difficult problem facing the endothermy theory concerns the gargantuan size of many dinosaurs, particularly the ponderous sauropods. If the large sauropods were endothermic, they would have required an enormous daily food consumption. True endotherms, which regulate their body heat internally, require much more fuel to generate that internal heat than ectotherms, which depend on external heat sources to warm up to optimal temperatures. The real stumbling block then becomes one of size; if an adult African elephant, weighing 7 or 8 metric tons, consumes 125 to 250 kilograms of fodder every 24 hours, should we believe that *Brachiosaurus* ate the same amount as a dozen elephants? To cloud the question even more, James Jensen of Brigham Young University has uncovered bones of what appear to be the largest dinosaur ever found. From Morrison beds in western Colorado, Jensen has removed shoulder blades and neck vertebrae belonging to a sauropod 20 percent larger than any previous find. This beast, weighing perhaps 100 metric tons, would have had a daily food requirement equalling that of fifteen bull elephants! How could such giants, with their small mouths and tiny peglike teeth, have eaten enough to sustain the high metabolic requirements of an endotherm?

In recent years, the hot-blooded dinosaur has become a popular subject of research and discussion at scientific meetings. A number of critics of endothermy are willing to accept the possibility that many dinosaurs, although not truly endothermic like mammals and birds, were nonetheless capable of achieving a nearly stable body temperature through various styles of thermoregulation, such as basking. The elephantine bodies of such sauropods as *Apatasaurus* and *Brachiosaurus* might have acted as huge heat reservoirs; once heated up they would have cooled off slowly in an equable subtropical climate and thus maintained operating temperature without a critical need for internal heat regulation. This kind of semiconstant body temperature maintenance is called **homiothermy** and may have allowed some of the larger dinosaurs to have frequent short-term high-exercise metabolism permitting bursts of high activity.

Undeniable, though, is the erect posture of dinosaurs, whereby walking and running were accomplished with the legs held in near-vertical position. This erect posture is found today in mammals and birds, which are all endothermic and capable of prolonged activity. By contrast, most living ectothermic land vertebrates have a more sprawling posture, and their walking and running gaits take the form of a more cumbersome sidewise, waddling motion. Perhaps the real answer to the endothermy versus ectothermy riddle is couched in compromise: Some dinosaurs probably were ectotherms like modern reptiles; most may well have been *homeothermic,* capable of sustaining frequent bursts of energy and activity; and perhaps the most active, smaller bipedal hunting carnivores were true endotherms.

The End of an Era and Another Faunal Crisis

The Cretaceous Period was a rich time in the history of life, and the fossil record presents a dazzling array of forms. The end of the period, however, was a time of crisis as numerous groups of organisms became extinct. Cretaceous marine invertebrate faunas had a distinctly modern look despite the abundant forms such as the ammonites that did not survive the end of the period. Separation of continents and the onset of latitudinally zoned climates promoted increased provincialism and faunal diversity (Fig. 12-71). During the Triassic and Jurassic, marine faunal provinces were few, and the number of cosmopolitan species was great. By the Late Cretaceous, however, faunas were less cosmopolitan as geographic isolation resulted in a greater number of indigenous species. The same pattern holds true in general for terrestrial faunas; Cretaceous dinosaurs reached a range of adaptive types not previously attained, and mammals diversified and evolved into forms more advanced than Jurassic predecessors.

On land the close of the Cretaceous is marked by the extinction of the dinosaurs. But as one dynasty ended another began. During the Cretaceous the flowering plants—the angiosperms—underwent explosive evolution. Colored petals and pollen represent major breakthroughs in plant reproduction. By Late Cretaceous, angiosperms dominated land floras. Not only was the emergence of angiosperms a milestone in plant evolution, their presence has had a tremendous effect on the evolution of birds and mammals.

By the end of the period, many taxa that had played important roles in Cretaceous biotas died out. Ammonites, rudist clams, *Exogyra* and *Gryphaea* oysters, inoceramid clams, dinosaurs, pterosaurs, pleisosaurs, ichthyosaurs, and mosasaurs all disappeared. Primary producers—the coccolithophorids—and primary consumers—the planktonic Foraminifera—suffered heavy losses. Why did all of these previously successful terrestrial and marine organisms and forms as different as ammonites and dinosaurs become extinct?

The end-of-Mesozoic faunal crisis is more difficult to explain than the mass extinctions near the close of the Paleozoic. We invoked plate tectonics and the formation of a supercontinent as the cause of late Paleozoic mass extinctions, and perhaps global tectonics is partly

the explanation for the end-of-Mesozoic extinctions (See Chapter 7). Certainly major retreats of the sea occurred, perhaps reducing shallow-marine ecospace, but the separation of continents was one that would seemingly favor net increase in diversity.

The close of Cretaceous was a time of major environmental change. The continued fragmentation and separation of continents and the north–south distribution of continents (Fig. 12-71B) may have brought about the beginning of modern latitudinally zoned climates and the onset of a cooling trend. Possibly cooler mean annual temperature led to the downfall of some organisms. As discussed in Chapter 7, breakdowns in primary links in the food chain may explain many late Mesozoic extinctions. There is abundant evidence in ocean-floor Deep Sea Drilling Project cores, as well as in outcrop samples, for massive reduction in the numbers of coccolithoporids and planktonic Foraminifera near the end of the Cretaceous. The loss of primary producers such as coccolithophorids might have disrupted the food-chain of marine organisms, both invertebrate and vertebrate. In addition, the significant loss of these photosynthesizing marine plankton together with reduction of productivity in the oceans might have caused excessive buildups of atmospheric carbon dioxide, triggering a greenhouse effect that resulted in a temporary warming trend. The suggestion has been made that such a sudden rise of the global temperature might have had a devastating effect on the reproductive capabilities of large land dwellers such as dinosaurs. Yet another direct casualty of massive extinctions among the marine plankton might have been the ammonites. Peter Ward of the University of California at Davis has suggested that ammonite larval stages may have been planktonic, thus rendering the ammonites vulnerable to environmental changes that wiped out other planktonic organisms. Ward further suggests that nautiloid cephalopods, who survived the end-of-Cretaceous extinction event, may have had benthic larval stages that gave them higher survival value.

Ocean-bottom sampling during the Deep Sea Drilling Project as well as sampling of outcrop sections in Italy, Spain, and Denmark have provided data that shed important light on the terminal Cretaceous extinction event. The detection of anomalous amounts of the element *iridium* in a thin interval of clay spanning the Cretaceous–Tertiary boundary is, to many workers, evidence for the impact of a large extraterrestrial body from the solar system. Initially discovered and reported by the father-son team of Luis and Walter Alvarez and their colleagues at the University of California at Berkeley, the iridium anomaly in the Cretaceous–Tertiary boundary clay layer has been interpreted by some as a signature of a comet or meteorite impact. Iridium is exceedingly rare on our planet, yet is comparatively abundant in meteorites. The possible implications of such an

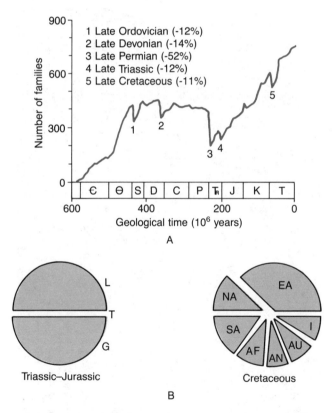

Figure 12-71. A. Standing diversity through Phanerozoic for families of marine invertebrates and vertebrates, showing abrupt drop in diversity at end of Cretaceous, a reflection of mass extinction event. B. Configuration of continental masses. AF, Africa; AN, Antarctica; AU, Australia; EA, Eurasia; G, Gondwana; I, India; L, Laurasia; NA, North America; SA, South America; T, Tethys. Note that the relationship to diversity patterns is different than for late Paleozoic (Fig. 11-65), suggesting that factors other than plate tectonics are required to explain the terminal Cretaceous extinction event.
(A from D. H. Raup, and J. J. Sepkoski, Jr., 1982, Mass Extinctions in the Fossil Record, Fig. 2, p. 1502: *Science,* vol. 215. B after J. W. Valentine, and E. M. Moores, 1972, Global Tectonics and the Fossil Record, Fig. 2, p. 170: *Jour. Geology,* vol. 80. Reproduced by permission of American Association for the Advancement of Science and the University of Chicago Press)

impact for the terminal Cretaceous extinction event are indeed intriguing. According to this scenario, impact of a meteor or comet might have resulted in mass mortality in any one of several ways: (1) by producing dust that may have retarded photosynthesis of plants, with a domino effect all the way up the food chain, (2) by producing cyanide poisoning in the oceans and killing off marine plankton, or (3) by flash heating of the atmosphere during entry of the extraterrestrial body, resulting in catastrophic extinctions. This catastrophic theory has met with considerable resistance from those who invoke a combination of more gradual causes: (1) terminal Cretaceous worldwide regression of epicontinental seas, (2) decrease in habitat space and breakdown of genetic isolating mechanisms, (3) lowering of speciation rates, (4) upward shift toward the ocean's surface of the level at which carbonate dissolves, (5) competitive re-

placement of dinosaurs by mammals, and (6) increase in seasonality at higher latitudes. The end of the Cretaceous also was a time of intense volcanic activity as arc magmatism spread eastward through much of the Cordillera. Interestingly, volcanic dust from the 1872 explosion of Krakatoa in the East Indies contained anomalously high concentrations of iridium and other platinum group metals.

The subject of mass extinctions, as brought out in Chapter 7, is a fascinating topic for research as well as for speculation. Probably nearly as many hypotheses have been advanced to explain mass mortality as there are major taxonomic groups that have become extinct. Certainly any meaningful theory or combination of theories should be in agreement with the fossil record as currently understood. Extinctions resulting from catastrophic impact should be expected to show tight syn-

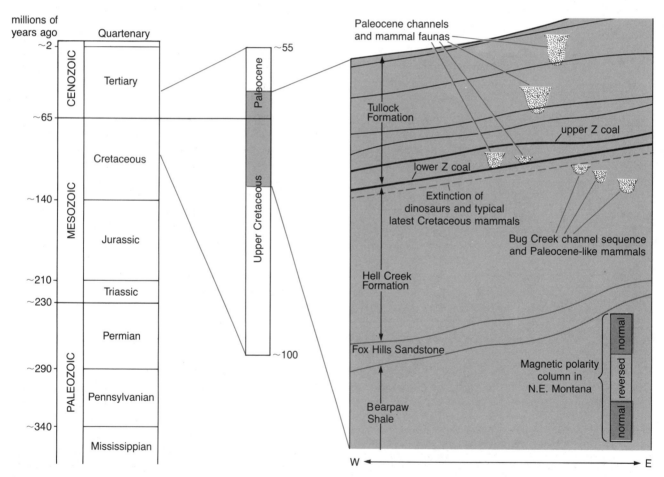

Figure 12-72. Time-stratigraphic diagram of sedimentary sequence in northeastern Montana showing some discrepancies in Cretaceous–Tertiary boundary definition and regional correlation. The boundary between the Hell Creek and Tullock Formations is the base of the lower Z coal. The Cretaceous–Tertiary boundary, as defined by the disappearance of dinosaurs, is, on the average, about 3 m below this level. The major floral changes occur well above this boundary, in places as high as the upper Z coal bed. Although it appears that the faunal and floral changes occurred in Montana and Alberta, and possibly in New Mexico, during the same epoch of reversed polarity of the earth's magnetic field, correlation with similar magnetic episodes in Europe is uncertain.
(From J. D. Archibald, and W. A. Clemens, 1982, Late Cretaceous extinctions; Fig. 4, p. 381: *Am. Scientist,* vol. 70, no. 4. Reproduced by permission of *American Scientist*)

chroneity. Those resulting from more gradually changing earthbound phenomena would, perhaps, be less synchronous. One of the problems that continue to hamper solution of the causes and effects of mass mortality at the end of the Cretaceous involves the definition of the Cretaceous–Tertiary boundary and the imprecision of correlation (Fig. 12-72). In some areas, particularly marine sections on the continents, megainvertebrates have been used to define the boundary. Within the terrestrial realm changes in fossil floras and dinosaur extinction define the boundary. In the ocean basins microfossils such as Foraminifera and calcareous nannoplankton provide the data. Most commonly these biotic changes are not concordant. For example, in a continuous stratigraphic section of marine pelagic sediments that spans the Cretaceous–Paleogene boundary near Zumaya, Spain, coccolithophorids and planktonic Foraminifera show abrupt extinctions coincident with the iridium-rich boundary clay layer. Ward, on the other hand, has documented a more gradual decline of ammonites in the same section, with the youngest specimens occurring some 12 m below the boundary clay, implying the last species may have died out some 100,000 years before the event that produced the iridium fallout. The application of radiometric dating and magnetostratigraphy (Fig. 12-72) should allow for more refined correlations to be made among disparate faunas, geographic areas, and paleoenvironments. Also, the fossil record is being examined more closely in areas where Cretaceous–Tertiary boundary stratotypes have been defined. This more detailed resolution will, no doubt, give us more information on the magnitude and selectivity, and ultimately the causes, of the extinctions and on the time interval over which they occurred.

Whatever the causes, be they catastrophic-extraterrestrial or gradual-terrestrial, or combinations of both, the affected groups of organisms were not able to adapt to the environmental changes; they succumbed to the stresses of a changing world. The mystery of the great dying out that marks the close of the Mesozoic Era still awaits an answer. The Cretaceous Period lasted some 70 million years, beginning with the continents still largely clustered, but ending with most of the continental land masses fragmented. The age of reptiles was over, but mammals would inherit the earth.

MICROPLATES AND THE ACCRETION OF WESTERN NORTH AMERICA

Exotic and Suspect Terranes

In the foregoing discussion of the Cordilleran orogeny, we related major regional tectonic events in western North America during the Mesozoic to a rather simplified model of subduction of oceanic lithosphere along the active margin of the continent. Although subduction was the main controlling influence on tectonic history and style, it does not represent the complete story of mountain building and continental accretion. Plate tectonocists David L. Jones, Allan Cox, Peter Coney, and Myrl Beck contend that "Dozens of geologic terranes in the North American Western Cordillera, lying side by side, are not genetically related as would be expected from a 'simple' ocean basin-to-continent plate interaction. Also, it appears that much more lithosphere was added to western North America during the Mesozoic than can reasonably be accounted for by volcanism along island arcs, accretion of sediments from the seafloor, or magmatic arc activity."*

Several lines of geologic evidence brought into focus during the late 1970s and early 1980s suggest that much of Western North America, from Alaska to Baja California (in particular the Alaska and Canadian Cordillera) consists of accreted lithospheric blocks—**microplates**—of various sizes (Fig. 12-73). Some of the microplates clearly have a foreign origin in that they are of unknown paleogeography with respect to the ancient Cordilleran geosyncline and western craton. Such microplates are referred to as **exotic terranes.** Others, of probable foreign origin, but not conclusively proved as exotic, are called **suspect terranes,** emphasizing paleogeographic uncertainty. Some workers are convinced that many of these accreted lithospheric blocks are prefabricated elsewhere and carried thousands of kilometers east and north on spreading oceanic lithosphere from their sites of origin in various parts of the Pacific Basin.

According to Jones, Cox, Coney, and Beck, who are among the leaders of a growing number of microplate advocates, most of the exotic terranes that have been identified are of ocean origin and consist of oceanic lithosphere that once included volcanic islands, plateaus, seamount chains, ridges, or island arcs. A few blocks represent fragments of continents. The accreted microplates are viewed as having been rafted along on subducting oceanic lithosphere into collision with the margin of the continent. As the oceanic plate plunged beneath the continent, the island arc, volcanic plateau, or microcontinent riding on the oceanic plate resisted subduction and was scraped off and accreted to the continent, after which the subduction zone migrated westward. In such a context, much of western North America (for a distance inland of an average 500 km) is viewed as a collage of accreted terranes—a western North America that has been shaped to its present configuration over the last 200 million years by the impact of oceanic plates (Fig. 12-73). For an analogy, think of the Mesozoic continental margin of western North America

---

*1982, The Growth of Western North America, p. 70: *Sci. American*, vol. 247, no. 5.

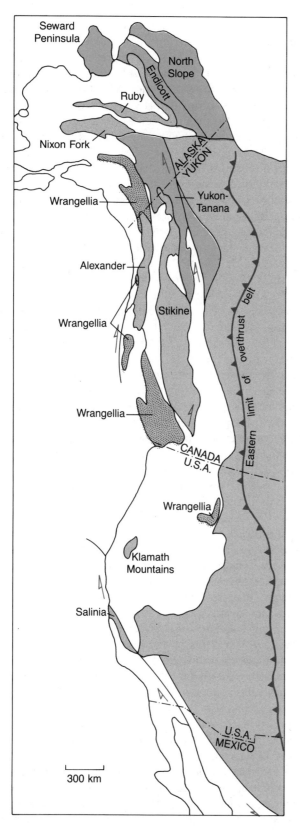

Figure 12-73. Western North America showing some of the suspect and exotic accreted terranes (dark gray). Cratonic North America is shown in light gray; Cordilleran geosynclinal deposits are shown in white. Also shown is the eastern limit of the Mesozoic–Early Cenozoic Cordilleran orogeny.
(After Zvi Ben-Avraham, 1981, The Movement of Continents; Fig. 9, p. 298: *Am. Scientist,* vol. 69, no. 3. Reproduced by permission of *American Scientist*)

as a colossal harbor receiving variously sized (horizontal dimensions ranging from hundreds to thousands of kilometers) merchant ships (oceanic microplates) each carrying a cargo of exotic rocks and each docking against the continent at different times to eventually clog the harbor and create a junkyard (collage of microplates).

As described by Jones, Cox, Coney, and Beck, terranes can be grouped into four general categories: stratified, disrupted, metamorphic, and composite. The characteristics of a microplate that give it uniqueness or individuality are rock types and ages, internal structural style and tectonic boundaries, metamorphic and plutonic histories, mineral deposits, fossil faunas and floras and their biogeographic implications, and paleomagnetic signature. In short, the terranes are internally homogeneous *geologic provinces* with features that sharply contrast with those of nearby provinces. Each terrane is bounded by major faults and records a geologic history so different from that of neighboring terranes that it is unlikely that they formed in close proximity.

The original recognition of suspect terranes came about through observations of discontinuities in stratigraphy, paleontology, and structure. Later, the explanation that these anomalies were the results of enormous displacements and dislocation of large crustal blocks was advanced on the basis of dramatic differences in paleomagnetism when compared to the stable craton of North America (Fig. 12-74). At present, many workers share the opinion that docking of microplates coincides with and was instrumental in producing deformation associated with the Sonoma, Nevadan, and Sevier orogenies.

The study of microplates is still in its infancy and is the subject of continuing and ever-increasing analysis and interpretation. At present, more than fifty separate, highly diverse exotic and suspect terranes have been identified in western North America, all lying west of the ancient continental margin (Fig. 12-74). Most of the accretion of these terranes is judged to have taken place in a relatively brief interval from about 200 million years to about 40 or 50 million years ago, mainly during the Mesozoic to early Cenozoic Cordilleran orogeny. The time of docking of terranes, either at the ancient continental margin or against each other, can be dated within reasonable limits by certain common denominators: (1) the age of similar cover rocks, called overlap assemblages, (2) the depositional age of detritus shed from one terrane upon another or from the continent onto a terrane (sedimentologic linkage), or (3) similar postdocking intrusive events (Fig. 12-75).

A New Model and New Questions

The concepts of *exotic terranes* and *collage tectonics* add a complicated wild card to our understanding of the geologic history of western North America. But they

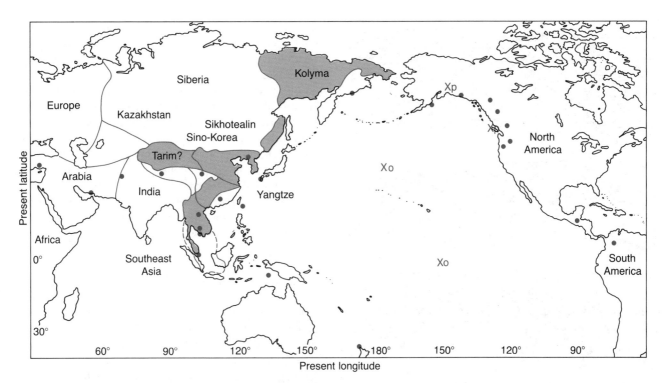

Figure 12-74. Present distribution of Tethyan fusulinid Foraminifera (shown by colored dots). Such a distribution strongly suggests large-scale tectonic dislocations of crustal blocks formerly in the Tethyan region. On the basis of reconstructions by M. W. McElhinny of The Australian National University, Tethyan crustal blocks (accreted terranes) are believed to have joined Eurasia. Pieces of the Wrangellia terrane in southern Alaska and Vancouver Island are shown by Xp, their separation being attributed to Cenozoic strike-slip faulting. Canadian workers R. W. Cole and Edward Irving, on the basis of paleomagnetic studies, have narrowed down Triassic origin of the Wrangellia terrane to two possible sites; either 16 degrees north or 16 degrees south of the Triassic equator. Thus the two present areas, which contain similar paleomagnetic signatures, are believed to have formed as part of an island in the proto-Pacific.

(After D. L. Jones, Allan Cox, Peter Coney, and Myrl Beck, 1982, The Growth of Western North America; pp. 73, 78: *Sci. American,* vol. 247, no. 5. Reproduced by permission of W. H. Freeman Co., San Francisco)

also offer an explanation of why active continental margins grow much faster than passive trailing margins; accreted terranes of oceanic origin indeed are new additions to the continent rather than the recycling of older continental material. The adherents of the microplate model believe that western North America has grown by more than 25 percent through piecemeal accretion since the Early Jurassic.

A number of terranes in the older, Appalachian orogen also are suspect. Some workers think the eastern margin of the ancient Iapetus Ocean (Chapter 11) is poorly defined and that a significant number of suspect terranes are present along the eastern flank of the orogen. Nearly all of these terranes are defined by their distinctive early Paleozoic and older rocks. These suspect terranes are not easily incorporated in a simple model of a single symmetrical Iapetus Ocean. There is no assurance that some terranes, now east of the ancient miogeosyncline, were ever previously connected to or evolved near the North American continent.

The microplate concept provides a fresh approach for gaining a better understanding of the true history

and anatomy of both the Appalachian and Cordilleran orogens, as well as others on other continents. Problems of relating episodic and more local orogenic events to continuous orogenic processes are more easily accommodated. Orogeny occurred in a particular region whenever an appropriate sea-floor feature, such as island arcs, seamounts, or submarine plateaus, impinged on that region. Finally, such a model is more compatible with the actualistic view of modern oceans cluttered with submerged plateaus and seamount chains; e.g., the tangled complexities of the Melanesian region in the western Pacific.

The more simplified, large-scale plate interactions, described in this and the previous chapters, remain basically unchanged in the microplate model; however, many of the stratigraphic, structural, paleontologic, and paleomagnetic details are radically changed. Although they represent secondary effects of global plate motion, microplate accretions may have been the primary factor in the forging of continental geology. The essence of the concept of collage tectonics is the individuality of the separate terranes; each exotic terrane records a geologic

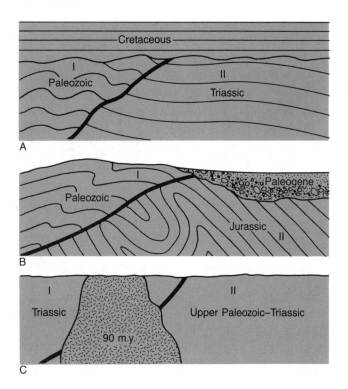

Figure 12-75. Pinning the time of docking of accreted terranes. A. Age of cover rocks. B. Age of detritus shed from one upon the other. C. Welding by intrusion. Docking cannot be earlier than youngest exotic rocks involved or later than pinning rocks.

history significantly different from that of neighboring terranes. But the recognition of terranes is only the beginning—the new concept raises many questions. From whence and by what pathways have the terranes traveled? What are the structural, stratigraphic, paleontologic, and paleomagnetic relations among accreted terranes and between terranes and the ancient continental margin of North America? How and when did the

terranes become accreted to the growing edge of the continent? What has been the postdocking history of individual terranes? Other questions involve the processes of collision, accretion, and continental growth. For example, why is there such a scarcity of subduction-zone sutures along present terrane boundaries? Why have so many terranes survived the accretion process with comparatively little internal deformation? And what roles have accreted terranes played in the formation of mountain chains along the convergent continental margin? Answers to all these questions and others call for close collaboration among geophysicists, structural geologists, stratigraphers, and paleontologists—cross-fertilization among geologic subdisciplines to tackle complex problems.

The microplate concept is not without its doubters, even among workers who firmly believe in plate tectonics. Some claim collage tectonics involves too much "arm waving" and forcing of facts into ideas rather than generation of ideas that fit the data. Some critics argue that some described differences in adjacent terranes can be explained by facies changes; others question the paleomagnetic results. The application of the microplate concept will be judged valid or invalid depending on the objectivity of studies and whether or not the interpretations are based on the preserved geologic data—the record of the rocks. In its early, simplest versions, however, plate tectonics theory suffered from the deficiency of failing to account in detail for most of the geology of the continents. Consequently, during the decade of the seventies, considerable effort went into developing and refining the theory to better explain the observations of continental geology. The microplate concept has played a major role in this refinement. It is hoped that the decade of the eighties will witness a new revolution in our understanding of continental lithosphere evolution.

SUMMARY

Triassic depositional patterns in the western craton of North America were similar to those of the preceding Permian Period and included extensive continental red beds and evaporites. Marine deposition was confined to the western part of the continental margin.

Major orogeny occurred during the Permo-Triassic in the western Cordillera. This event, the Sonoma orogeny, involved subduction of oceanic lithosphere. Convergence of a volcanic island-arc system on the western margin of the continent caused telescoping of marginal ocean basin and major overthrusting of geosynclinal

sediments along the continental margin. The Sonoma event also initiated the changeover from a Japanese-type to an Andean-type continental margin.

Although most of the Triassic record of North America is in the western craton and Cordilleran belt, a significant upper Triassic and lower Jurassic section—the Newark Group and related rocks—occurs in surface and subsurface block-fault basins within the crystalline part of the Appalachians. Basalt sills, together with the crustal extension that produced the basin-opening faults, provide a signature of early Mesozoic rifting that heralded the breakup of Pangaea.

Land life during the Triassic was highlighted by dramatic adaptive radiations of gymnosperm plants and reptiles, but also by some major extinctions of tetrapods. Dinosaurs and the first mammals made their appearance late in the period, but several other groups, including the laybrinthodont amphibians and therapsid, cotylosaurian, and thecodont reptiles, became extinct. Triassic seas swarmed with ammonoids, bivalve molluscs diversified rapidly, and scleractinian corals built extensive reefs. The richest marine province was the tropical, equatorial Tethys seaway, which indented the eastern margin of Pangaea.

During early Jurassic, the western craton of North America continued to receive nonmarine deposits. The thickly cross-stratified Navajo Sandstone of the Colorado Plateau Province represents a vast sandy desert like the Sahara. Later in the Jurassic an interior seaway, the Sundance, stretched from the Arctic almost to the newly formed Gulf of Mexico Basin. Jurassic marine sediments are present in the subsurface of the Atlantic and Gulf Coast regions and represent the earliest marine deposition along the rifted continental margin as Pangaea began to split apart. A thick evaporite unit, the Louann Salt, was deposited in the embryonic Gulf of Mexico Basin. In the western part of the Cordillera deep marine sediments were deposited from California northward to Alaska. The enigmatic Franciscan Formation of the California coast ranges began to accumulate in a deep-sea trench.

Magmatic-arc igneous activity along the Andean-type continental margin and back-arc thrusting were expressions of the early, Nevadan phase of the protracted Cordilleran orogeny. This orogeny was the response to subduction of the Farallon oceanic plate beneath the western margin of the North American continent. Growing tectonic highlands provided source areas for the fore-arc basin Great Valley assemblage of California and for the massive incursion of sediments that formed the Morrison alluvial plain of the western craton. The Morrison Formation is famous for its wealth of dinosaur fossils. Natural history museums of the world feature reconstructions of the giant Jurassic sauropod genera *Apatosaurus, Brachiosaurus,* and *Diplodocus* and the voracious meat-eater *Allosaurus. Archaeopteryx,* the first "bird," represented by about a half-dozen specimens from the Solenhofen Limestone in Bavaria, may actually be the remains of a small feathered theropod dinosaur not capable of powered flight.

Unlike North American counterparts, Jurassic marine strata in Europe are widely exposed and very fossiliferous and served as the cornerstone for development of such important stratigraphic concepts as fossil succession, stage, zone, and facies. Abundant, rapidly evolving ammonite cephalopod faunas, squid-like belemnites, and bivalve and gastropod molluscs indicate a major diversifiction of molluscs. Fossils of marine reptiles, including icthyosaurs, pleisosaurs, and turtles, bear evidence that some reptiles had invaded the sea.

Cretaceous sediments were deposited over a wide region by marginal and interior seas that spread beyond the limits of Jurassic deposition. Sedimentation on the trailing margin of the continent produced thick stratigraphic sections in the Atlantic and Gulf Coast areas. Terrigenous clastic sediments typify the Atlantic Coast region. In the Gulf Coast, abundant carbonate sediments were deposited. During Early Cretaceous a major barrier reef system rimmed much of the western gulf.

Continued subduction of the Farallon Plate beneath the western margin of the North American continent was responsible for extensive magmatic-arc activity manifested by emplacement of granitic batholiths from Alaska to Baja California during Late Jurassic to mid-Cretaceous. Deposition of the Franciscan Formation in the trench and of the Great Valley Group in the fore-arc basin in California continued through the Cretaceous. To the east, extensive back-arc thrusting produced the Sevier orogenic belt. Erosion of Sevier tectonic highlands created the Rocky Mountain clastic wedge, which prograded into an interior seaway that stretched from the Arctic to the Gulf of Mexico. The interior seaway, at times some 1500 km wide, was the result of mid-Cretaceous major marine transgression—perhaps the highest sea-level stand of the Phanerozoic.

The Cretaceous was a time of exceptionally rich faunal and floral diversity. Among the marine invertebrates, ammonite cephalopods, bivalve and gastropod molluscs, and sea urchins flourished. The peculiar sessile rudist bivalves were reef builders that lived in the tropical waters of the expanding Tethyan realm. Diatoms, planktonic foraminifera, and calcareous planktonic algae—the coccolithophorids—made a dramatic appearance; the last two groups contributed to great volumes of Cretaceous chalk. A variety of marine vertebrates—bony fish as well as large sea turtles, mososaurs, ichthyosaurs, and plesiosaurs—rounded out the gallery of marine life.

On land, flowering plants—the angiosperms—first appeared in Early Cretaceous and by late in the period had ecologically displaced most of the gymnosperms, particularly in lowland areas. Mammals continued to increase in diversity, but remained small, mainly nocturnal, forms that played a secondary role in land ecosystems. By late in the period, early marsupials and placentals had evolved. Birds, represented mainly by sea-going varieties, and pterosaurs graced the skies. It was truly a bountiful time in the history of life. Increased diversity levels reflected continued rifting and rafting of continental fragments as Pangaea broke up. What had been very cosmopolitan faunas during the Jurassic became progressively more provincial with continental breakup and separation.

The dinosaur dynasty reached its zenith during the Cretaceous with six suborders and an impressive array of adaptive types, including many specialized herbivores and relatively few carnivores. Renewed interest in research on dinosaurs during the decade of the seventies has shed light on the bioenergetics and social patterns of these amazing animals as well as on their role in tetrapod ecosystems. This new look at dinosaurs has dispelled many of the traditional myths about their being evolutionary failures and has portrayed them as some of the most successful organisms that ever lived. This renewed research has also precipitated a controversy over whether the dinosaurs were cold blooded or warm blooded.

Despite the richness and diversity of Cretaceous life, the end of the period was a time of mass extinction as many marine plankton, some marine invertebrates, most marine reptiles, and the dinosaurs died out. This wave of mass extinctions is difficult to explain in light of the plate tectonics situation of separated continents and increased provinciality. Lowered sea-level stands accompanying end-of-Cretaceous regressions probably caused stress for many marine organisms. However, the story of this mass extinction defies simple, singular explanation. One popular recent extinction theory, based in part on the anomalously high iridium content in the Cretaceous–Tertiary boundary clay in oceanic and some land sections, invokes the impact of a large meteor or comet. Such an event might have triggered a domino effect by upsetting the marine food chain, changing $O_2:CO_2$ ratios in the atmosphere and causing a short-term global warming trend that might have had disastrous consequences for large terrestrial vertebrates like the dinosaurs.

The Cordilleran orogeny shaped the western part of North America during the Mesozoic and early Cenozoic. This orogeny involved subduction of the Farallon oceanic plate beneath the edge of the continent. The overriding continental block acted like a bulldozer and scraped off parts of the ocean floor at different places and different times along much of the Cordilleran margin. Island arcs, seamounts, submarine plateaus, split-off continental fragments, and a few slivers of oceanic lithosphere itself became accreted to the edge of the continent as exotic terranes. Some traveled thousands of kilometers before their final docking. Anomalous stratigraphy, paleontology, and paleomagnetic signatures have allowed their recognition and have provided information on origins, migration routes, and times of docking. The recognition of more than fifty separate, highly diverse exotic terranes in the Cordillera as well as others in other orogens of the world has carried plate tectonics beyond definition of large-scale megaplate interactions to consideration of microplates and collage tectonics.

SUGGESTIONS FOR FURTHER READING

Alvarez, L. W., Alvarez, W., Asaro, F., and Michel, H. V., 1980, Extraterrestrial Cause for the Cretaceous-Tertiary Extinction: *Science,* vol. 208, no. 4448, pp. 1095–1108.

Archibald, J. D., and Clemens, W. A., 1982, Late Cretaceous Extinctions: *Am. Scientist,* vol. 70, no. 4, pp. 377–386.

Bakker, R. T., 1975, Dinosaur Renaissance: *Scientific American* Offprint no. 916, W. H. Freeman, San Francisco.

Ben-Avriham, Zvi, 1981, The Movement of Continents: *Am. Scientist,* vol. 69, no. 3, pp. 291–300.

Dietz, R. S., and Holden, J. C., 1970, The Breakup of Pangaea: *Scientific American* Offprint No. 892, W.H. Freeman, San Francisco.

Ganapathy, R., 1980, A Major Meteorite Impact on the Earth 65 Million Years Ago: Evidence from the Cretaceous-Tertiary Boundary Clay: *Science,* vol. 209, pp. 921–923.

Jones, D. L., Cox, Allan, Coney, Peter, and Beck, Myrl, 1982, The Growth of Western North America: *Sci. American,* vol. 247, no. 5, pp. 70–128.

King, P. B., 1977, *The Evolution of North America,* rev. ed.: Princeton University Press, Princeton, NJ.

Lanham, Url, 1973, *The Bone Hunters:* Columbia University Press, New York.

McPhee, John, 1980, *Basin and Range:* Farrar, Straus, Giroux, New York.

McPhee, John, 1982, *In Suspect Terrain:* Farrar, Straus, Giroux, New York.

Newell, N. D., 1963, Crises in the History of Life: *Scientific American* Offprint No. 867, W.H. Freeman, San Francisco.

Oliver, Jack, 1980, Exploring the Basement of the North American Continent: *Am. Scientist,* vol. 68, no. 6, pp. 676–683.

Ostrom, J. H., 1978, A New Look at Dinosaurs: *Nat. Geographic,* vol. 154, no. 2, pp. 152–185.

Ostrom, J. H., 1979, Bird Flight: How Did It Begin? *Am. Scientist,* vol. 67, pp. 46–56.

Stanley, S. M., 1984, Mass Extinctions in the Oceans: *Sci. American,* vol. 250, no. 6, pp. 64–84.

Ward, Peter, 1983, The Extinction of the Ammonites: *Sci. American,* vol. 249, no. 4, p. 136–147.

13

CENOZOIC HISTORY

CONTENTS

KEY TERMS

Base level
Laramide orogeny
Kerogen
Epeirogeny
Horst
Graben
Salt dome
Diapir
East Pacific Rise
Diatomite

CONQUERING THE COLORADO

The magnificent mountains, plateaus, and deep canyons of the Cordilleran region were formed mainly in Cenozoic time, during the last 65 million years. But every bit as colorful as the geology and scenery of this vast and grandiose region is the story of geological exploration of western North America during the last half of the nineteenth century. It is a story of adventure, hardship, and privation, but mostly of discovery.

After the Civil War, the United States government authorized a geological exploration of the 40th parallel and geological and geographical surveys of the territories of the West. These expeditions were ambitious undertakings, and even though rivalry developed among several of the surveys, their undaunted probing into the strange and ofttimes forbidding new land of the West resulted in important geologic discoveries being made in rapid succession. Of particular significance was the discovery of untold numbers of fossil localities that yielded the remains of previously undescribed species of mammals and plants (Chapter 12). Discovery of valuable mineral resources not only stimulated a period of mining activity, but also led to revolutionary theories on the formation of metallic ores. The continuity of rock exposures presented a panorama of stratigraphy and structure not previously seen in other regions of North America, and new concepts on the evolution of landscapes were developed. But perhaps the most significant byproduct of early geologic exploration of the West was the recognition that portions of this magnificent land should be preserved. The accomplishments of the Hayden survey (1870–1878), which was directed by Ferdinand Hayden and included the landscape artist

Figure 13-1. John Wesley Powell, 1898.
(Photo from National Anthropological Archives. Smithsonian Institution Photo No. 64-b-1)

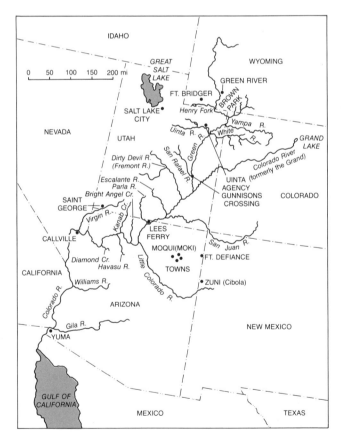

Figure 13-2. Colorado River region and route of John Wesley Powell's first expedition (indicated by dots).
(From William Culp Darrah, *Powell of the Colorado.* Copyright 1951, © renewed 1979 by Princeton University Press. Map, p. 121, reprinted by permission of Princeton University Press)

Thomas Moran and the pioneer photographer W. H. Jackson, were in large measure responsible for creation of the first national park, Yellowstone, in 1872.

One of the most remarkable figures to emerge from the geologic discovery of the West was Major John Wesley Powell (Fig. 13-1), who led two expeditions in the Colorado River and its canyons and who also was influential in the shaping of the geological sciences in America. Powell was born in western New York in 1834, but later moved with his family to Ohio, where he had the good fortune of being influenced in his early education by several naturalists. He quickly developed a strong interest in science, and when the Powell family moved to Illinois, young John nurtured this interest by attending Illinois College and Wheaton College.

The period of his early twenties was spent teaching school, exploring rivers by boat, and collecting specimens. When the Civil War broke out, he enlisted in the Union Army, serving as a military engineer and artillery officer; he lost his right forearm in the Battle of Shiloh. After the war, Major Powell became professor of geology at Illinois Wesleyan University and curator of the Illinois State Natural History Society. He also offered courses at Illinois Normal University and built a reputation as a first-rate teacher who combined learning from books with experience in the field and lab.

In 1867, while on a field trip with students to Colorado, he climbed Pikes Peak and the next summer he was one of a party of seven that made the first climb of Longs Peak. It was during these trips to Colorado that Powell became infused with the insatiable desire to explore the Colorado River by boat. And sure enough, on May 24, 1869, a Powell-led expedition, including a party of ten men, left Green River, Wyoming, in four boats. Money for this expedition came from a variety of sources, mostly as a result of Powell's urging, and from his own pockets. The trip was sponsored by the Illinois Natural History Society. Transportation of men, boats, and supplies to Green River was provided for by the Burlington and Union Pacific Railroads, and the Smithsonian Institution loaned surveying instruments. The War Department furnished the rations.

The expedition reached the mouth of the Virgin River (Fig. 13-2) on August 30, some three months after leaving Green River. Although the expedition of 1869 was a hasty reconnaissance, it nevertheless was played up as a plunge into the great unknown and it created much attention and public interest. Major Powell was a hero and he encountered no difficulty in obtaining

Figure 13-3. John Wesley Powell converses with Paiute Indian during northern Arizona Survey more than 100 years ago.
(Photo from National Anthropological Archives. Smithsonian Institution Photo No. 1591)

from Congress an appropriation for further exploration of the Colorado River country. The entire year of 1870 was spent in detailed preparation.

The Colorado River expedition of 1871 was intended to gather more scientific data than its predecessor, and more elaborate preparations to this end were made. An important addition was a photographer, E. D. Beamon of New York City. Photography was no easy task in 1871, especially on a river expedition. Imagine lugging heavy and cumbersome equipment in and out of a boat and up and down canyon walls. But Beamon had the stamina that made him equal to the task and he took more than 350 photos, many of them truly excellent. In addition to Beamon, the exploration party included a surveyor and three topographers. On May 22, 1871, almost two years to the day that the first expedition was launched, the party of twelve men in three boats left Green River, Wyoming, with Major Powell himself the only member of the 1869 crew to make the trip. This second effort spent considerably more time charting the detailed courses of the Colorado and its tributaries and in making topographic surveys and geologic maps of the canyons and plateaus. The actual boating part of the expedition ended on September 7, 1872, when the party reached Kanab Creek (Fig. 13-2), but mapping continued through 1873. Powell's effort had become recognized and was sponsored as a full-fledged survey, one of four operating in the West. In 1879, these surveys, after much cajoling and coercion of government officials for support, became merged into one—the U.S. Geological Survey. Its first director was Clarence King, and Powell served as its very capable second director for 14 years.

The Powell Survey filled in the last remaining large blank area on the geographic map of North America. This was a gigantic region that extended from the Uinta Mountains on the north to the Virgin River on the south and from the Green River westward to the high plateaus (Fig. 13-2). From Powell's work in the canyons and plateaus of the Colorado River country there grew a new understanding of the processes of erosion and the evolutionary development of land forms. Powell introduced the concept of **base level** and the classification of drainage as antecedent, consequent, and superposed. Geomorphology became a recognized discipline and the interplay between stratigraphy, structure, erosional processes, and the morphology of land forms was put into a perspective not previously appreciated.

Powell's goal was to understand the land of the West and the native Americans who inhabited the land. Known as "Kapurats," meaning "one-arm-off," Powell made friends and gained the trust of the Indians of the plateau country (Fig. 13-3 on page 353). He learned

Figure 13-4. The Grand Canyon country.
(From J. W. Powell, *The Exploration of the Colorado River and Its Canyons*, p. 16: Dover Publications, New York. The Dover edition, first published in 1961, is an unabridged and unaltered republication of the work first published in 1895 by Flood and Vincent under the title, *Canyons of the Colorado*. Used by permission of Dover Publications, Inc.)

their customs and language and became one of this country's most vigorous proponents for their protection. His zeal both for science and humanity led to the establishment of the U. S. Geological Survey and the Bureau of Ethnology, and he was one of this country's loudest spokesmen for the wise use of the public domain and natural resources.

John Wesley Powell's gifts to science and to his country are perhaps rivaled only by the beauty of the Grand Canyon itself (Fig 13-4). The last few sentences in his *The Exploration of the Colorado River and its Canyons** provide some insight into the man and his keen vision:

> You cannot see the Grand Canyon in one view, as if it were a changeless spectacle from which a curtain might be lifted, but to see it you have to toil from month to month through its labyrinths. It is a region more difficult to traverse than the Alps or the Himalayas, but if strength and courage are sufficient for the task, by a year's toil a concept of sublimity can be obtained never again to be equaled on the hither side of Paradise.

*Dover ed. 1961; first published in 1895 by Flook and Vincent under the title, *Canyons of the Colorado.*

THE CENOZOIC TIME SCALE

The Cenozoic Era is generally subdivided into two periods, but unfortunately there is not universal agreement on the period names. Although the terms *Tertiary* and *Quaternary* are more ingrained in the literature, **Paleogene** and **Neogene** currently are being used by many geologists. The use of Paleogene and Neogene allows for a more equal time subdivision of the Cenozoic (Table 3-2) because the boundary corresponds with that between the Oligocene and Miocene Epochs.

The epochs of the Cenozoic subdivisions, however, do have worldwide application. Recall from Chapter 3 that Sir Charles Lyell in 1833 defined the Eocene, Miocene, and Older and Newer Pliocene Series not only on paleontologic composition, but also on the relative proportions of living and extinct species each contained. Defined on the basis of fossiliferous strata in the Paris Basin, Eocene was judged to contain 3 percent living species; Miocene, 18 percent; Older Pliocene, 33–50 percent; and Newer Pliocene, 90 percent. Recall also that Lyell indicated that this scheme of subdivision was open to modification and that additional subdivisions might and should be recognized. Lyell himself in 1839 proposed the name Pleistocene for Newer Pliocene; later, other workers defined Oligocene (1854) and Paleocene (1874).

Many local and regional Cenozoic time-stratigraphic subdivision schemes have been erected on the basis of terrestrial mammal faunas, marine megainvertebrates, and marine and nonmarine microfossils (including foraminifera, radiolarians, and diatoms, as well as spores and pollen). In addition to and tied in with the biostratigraphic data is an ever-growing network of radiometric dates derived from interbedded volcanic ash units and continental and marine volcanic flows and shallow intrusives. *Paleomagnetic* events recorded as *geomagnetic reversals* also have furnished a powerful tool for time-stratigraphic correlations and subdivisions in some sections of Cenozoic sediments and volcanic rocks, especially when tied to biostratigraphic and radiometric dates.

The Cenozoic is amenable to refined chronologic subdivision primarily because its deposits are so widely distributed and accessible on the continents and in the ocean basins and represent such a broad spectrum of environments. The Deep Sea Drilling Project has retrieved a tremendous amount of data on sediment types, ages, and paleomagnetic signatures. Analysis of the data reveals the history of motion of the oceanic lithosphere. As mentioned in Chapter 3, the stratigraphic section at Gubbio, Italy, represents a complete Upper Cretaceous–Paleogene transition with excellent litho-stratigraphic, bio-stratigraphic, chronostratigraphic, and magnetostratigraphic control and serves as an important stratotype for correlation of coeval stratigraphic sequences in the ocean basins and on the continents.

THE WESTERN INTERIOR
The Rocky Mountains

We shall begin our examination of the Cenozoic of North America in the Rocky Mountain region of the western interior. The Cordilleran Orogeny continued from Cretaceous into early Cenozoic. The late Cretaceous to early Cenozoic phase is commonly referred to as the **Laramide orogeny;** its effects were most evident in the eastern part of the Cordilleran belt in what now are referred to as the northern, central, and southern Rocky Mountain Provinces. Remember that the Cordilleran orogeny, which began during the Jurassic in the western part of the Cordilleran belt, progressed eastward during late Mesozoic and early Cenozoic like a rippling wave, finally culminating in the Eocene (Table 13-1). The Laramide phase tilted, folded, and faulted-among other rocks—Cretaceous beds that had been deposited in the interior seaway. These Cretaceous strata, in turn, had been formed from erosional debris shed from tectonic highlands farther west, during earlier phases (e.g., the Sevier) of the Cordilleran orogeny.

In the northern Rocky Mountains—from Wyoming to the Northwest Territories in Canada and northward into Alaska—thrust faulting and folding, which began in Cretaceous time, continued to deform Proterozoic, Paleozoic, and Mesozoic strata of the Cordilleran miogeosyncline as well as Mesozoic rocks of the Rocky Mountain interior seaway. This orogenic phase, somewhat reminiscent of the late Paleozoic disturbance in the southern Appalachians (Fig. 11-45), exhibited a characteristic compressional tectonic style involving asymmetrical folding and eastward thrusting. In some parts of the overthrust belt, the thrust plates are imbricated, some of the thrusts have displacements measured in tens of kilometers (Fig. 13-5; 13-6).

In northeastern Mexico, the Sierra Madre Oriental (Fig. 13-7), a southern extension of the Rocky Mountain system, is a chain of mountains that records the Laramide deformation of Jurassic and Cretaceous marine carbonate rocks. It is primarily a fold province, characterized by tight, strongly compressed doubly plunging anticlines and synclines with a frontal (eastern) belt of broader, open folds. The individual mountain ranges are typically anticlinal in structure; in the cores of some of the anticlines, Jurassic evaporites have been deeply eroded to form picturesque hidden valleys.

In the central Rocky Mountains of Colorado, Wyoming, and New Mexico, large areas of the craton were uplifted during the Laramide event. Such was the situation in the Colorado Front Range (Figs. 13-8, 9) where

Table 13-1. Timing of phases of Cordilleran orogeny during Mesozoic and Cenozoic

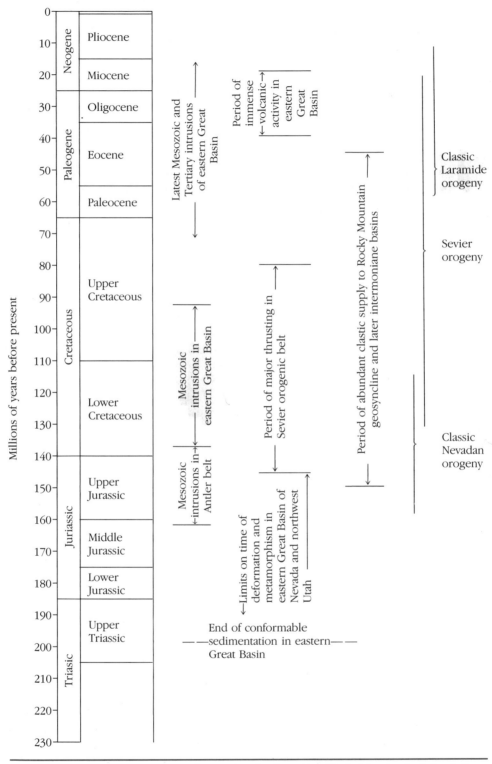

After R. L. Armstrong, 1968, Sevier Orogenic Belt in Nevada and Utah, Fig. 7, p. 452: *Geol. Soc. America Bull.,* vol. 79, no. 4.

broad-backed, asymmetrical uplifts, many with pre-Cambrian crystalline basement exposed in the cores, illustrate the principal tectonic style. Most of the majestic

Colorado peaks, such as Pikes Peak, Mt Elbert, and Longs Peak, have been sculptured from pre-Cambrian crystalline basement. These unusually high uplifts, formed by

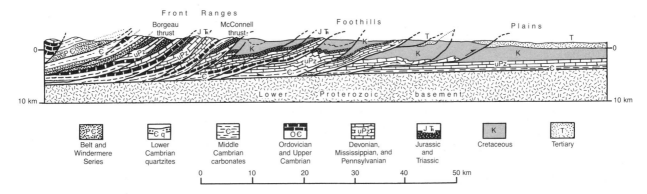

Figure 13-5. Laramide deformational style of imbricate thrusts in Alberta, Canada.
(From Philip B. King, *The Evolution of North America,* Fig. 86 (A1): Copyright © 1959, rev. ed. © 1977 by Princeton University Press, Princeton NJ. Reprinted by permission of Princeton University Press)

A

B

Figure 13-6. A. Chief Mountain, Montana, a klippe (erosional remnant of thrust sheet) of Lewis overthrust, composed of pre-Paleozoic strata of Belt Supergroup, tectonically underlain by Upper Cretaceous shale. B. Sun River Canyon, Montana, showing Mississippian Madison Limestone (cliff exposure) in upper plate of major overthrust and Cretaceous strata (in canyon) in lower plate.
(Photos by J. Cooper)

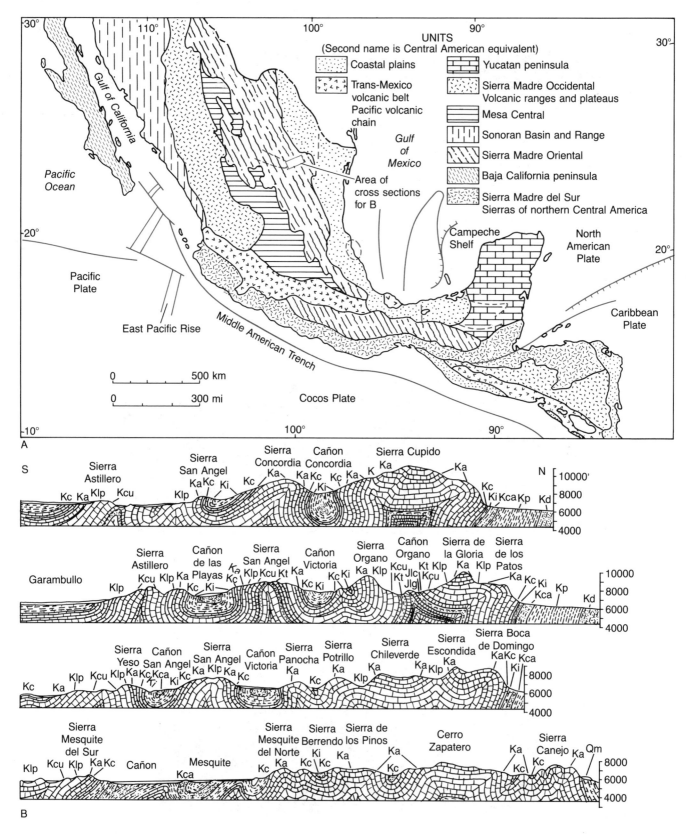

Figure 13-7. A. Tectonic regions of Mexico. B. Laramide folding style in Jurassic (J) and Cretaceous (K) rocks of Sierra Madre Oriental, northeastern Mexico.
(A from Paul Enos, 1983, Late Mesozoic Paleogeography of Mexico, Fig. 1, p. 135, *in* Mesozoic Paleogeography of West-central United States, M. W. Reynolds and E. D. Dolley, eds.: Rocky Mountain Section SEPM Paleogeography Symposium, vol. 2. Reproduced by permission of the Rocky Mountain Section, Society of Economic Paleontologists and Mineralogists. B from R. W. Imlay, 1937, Geology of the Middle Part of the Sierra de Parras, Mexico, pl. 13, p. 628: *Geol. Soc. America Bull.,* vol. 48)

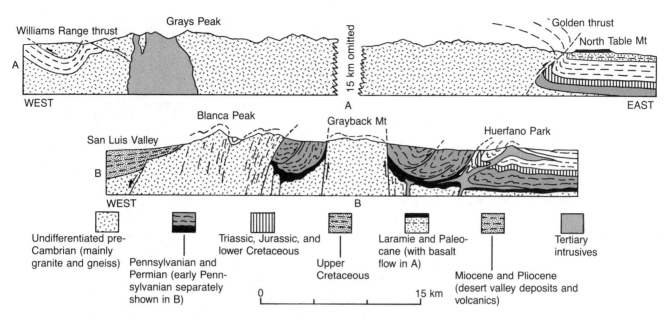

Figure 13-8. Vertical uplift of pre-Paleozoic basement style of Laramide deformation in southern Rocky Mountains. A. Colorado Front Range. B. Sangre de Cristo Mountains, Colorado. (Redrawn from P. B. King, 1977; Lovering, 1935; Van Tuyl and McLaren, 1933; Burbank and Goddard, 1937).

Figure 13-9. Looking south over Boulder, Colorado, at Rocky Mountain front. Pre-Paleozoic basement (dark, vegetated) to right (west). Pennsylvanian arkoses form steeply dipping "flat-irons" in right center along foot of escarpment; Permo-Triassic in central valley; Jurassic Morrison Formation capped by Cretaceous Dakota Sandstone form ridge to left (east). (Photo courtesy of John S. Shelton)

vertical fault movements of pre-Cambrian basement blocks, represent, in part, reactivation of ancestral Rocky Mountain fault blocks originally uplifted during the Pennsylvanian Period (Chapter 11). This uplift of cratonic basement blocks is a characteristic feature of the Laramide structural belt and illustrates that some mountain-building events have included regions not within geosynclinal settings.

The plate tectonics model that so far seems to best explain the Laramide phase of the Cordilleran Orogeny involves a change in plate convergence rates and angles that began at the boundary between North American continental lithosphere and Pacific oceanic lithosphere about 80 million years ago (Fig. 13-10A). These motions of the lithospheric plates, whose interactions controlled late Mesozoic and Cenozoic Cordilleran tectonics, have been determined by use of evidence from offshore data, particularly sea-floor magnetic anomaly patterns. In a scenario described by William R. Dickinson of the University of Arizona, Pacific oceanic lithosphere belonging to the Farallon Plate was, prior to 80 million years ago, subducted beneath the North American continental plate approximately at right angles to the Cordilleran margin and at a rate of somewhat less than 10 cm/year. The motion of the North American Plate was northwestward at rates in excess of 5 cm/year. This Late Cretaceous convergence rate is believed to be similar to standard values for modern arc–trench systems. The angle of dip of the subducted slab (Fig. 13-10B) is believed to have been about 50°, a rather normal value for modern arc–trench systems. This angle of Late Cretaceous subduction is inferred from the potassium gradient in granitic rocks of the Cretaceous batholiths (Fig. 13-10C). The leading edge of the westward-drifting North American Plate was continually pressed against the zone of flexure in the Farallon Plate. This produced contraction inland behind the magmatic arc, resulting in thrusting in the Sevier belt.

Associated with this Late Cretaceous subduction mode was a belt of *arc magmatism* that was positioned some 150 to 200 km distant from a trench that apparently extended continuously from Mexico to Alaska. The Cretaceous batholith trend, subparallel to the present-day coastline from Mexico to Alaska, represents the eroded roots of the once-continuous magmatic arc that was positioned cratonward of and parallel to the subduction zone along the continental margin.

Magnetic anomaly patterns in the Pacific suggest that from 80 million years ago (Late Cretaceous) to 40 million years ago (Late Eocene), unusually high convergent rates prevailed at the Farallon-North American trench. Subduction of the Farallon Plate (Fig. 13-11A) was abnormally rapid, perhaps about 15 cm/year, and was oblique to the Cordilleran margin. The motion of the continental block had changed to generally due west at about 5 cm/year.

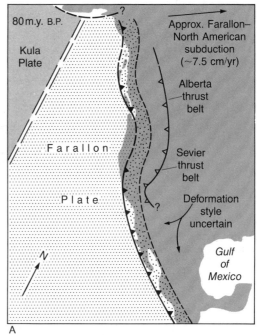

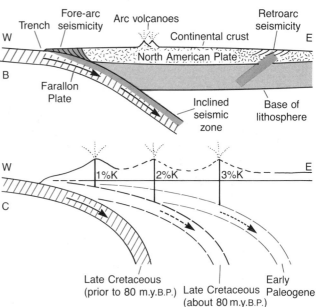

Figure 13-10. A. Plate configurations and plate motions, western North America and eastern Pacific Basic, approximately 80 m.y. B.P. B. Style of subduction (steep mode) and related arc magmatism and back-arc (Sevier) thrusting, approximately 80 m.y. B.P. C. Percentage of potassium (K) varies with distance from trench; volcanic activity farthest from trench has highest percentage of potassium. This relationship provides a guide to depth at which melting occurred and thus to determining the dip angle of the subduction zone.
(A from W. R. Dickinson, 1979, Cenozoic Plate Tectonic Setting of the Cordilleran Region in the United States, Fig. 1A, p. 2, *in* Cenozoic Paleogeography of the Western United States: SEPM Pacific Section Paleogeography Symposium, vol. 3. Reproduced by permission of Pacific Section, Society of Economic Paleontologists and Mineralogists. B from W. R. Dickinson and W. S. Snyder, 1978, Plate Tectonics of the Laramide Orogeny, Fig. 2, p. 359, *in* Laramide Folding Associated with Basement Block Faulting in the Western United States: *Geol. Soc. America Mem.* 151. Reproduced by permission of authors. C, data from S. B. Keith, 1978, Paleosubduction Geometries Inferred from Cretaceous and Tertiary Magmatic Patterns in Western North America: *Geology,* vol. 6, pp. 516–521)

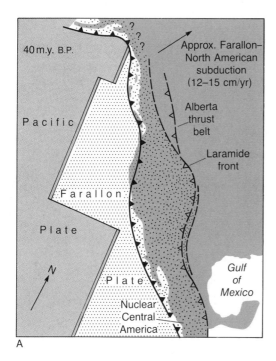

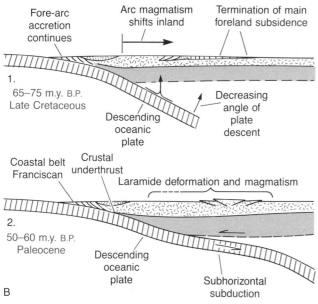

Figure 13-11. A. Inferred plate configuration and plate motions, western North America and east Pacific Basin, Paleogene, approximately 40 m.y. B.P. B. Plate interactions along Cordilleran margin of North America. 1. Late Cretaceous. Note the decreasing angle of oceanic plate descent and attendant eastward shift of arc magmatism; 2. Paleocene. Note subhorizontal subduction and Laramide deformation and magmatism.
(From W. R. Dickinson, 1979, Cenozoic Plate Tectonic Setting of the Cordilleran Region in the United States, Figs. 1B, 2B, 2C, *in* Cenozoic Paleogeography of the Western United States: Pacific Section SEPM Paleogeography Symposium, vol. 3. Reproduced by permission of Pacific Section, Society of Economic Paleontologists and Mineralogists)

On the basis of interpretations of the distribution of arc magmatism and the potassium gradient in arc-volcanic rocks (Fig. 13-10C), it is inferred that the increase in plate convergence rate was accompanied by a pro-

gressive decrease in the angle of dip of the subducted slab (Fig. 13-11B). As the angle of subduction decreased to perhaps only 10° in the Early Paleogene, arc magmatism crept inland and became greatly diminished in magnitude, a behavior that probably can be attributed to a change in the locus of melting beneath the arc. The place at which the descending slab penetrated the asthenosphere shifted progressively farther eastward with respect to the westward movement of the overriding continent (Fig. 13-12). As the thin zone of arc magmatism migrated inland, back-arc tectonism also shifted eastward and changed in structural style from "thin-skinned" thrusting in the miogeosynclinal belt to the characteristic cratonic basement-cored uplifts in the central Rockies. *In this context, Laramide deformation is viewed as a response to the effects of rapid plate convergence and plate descent at an abnormally shallow angle.*

The arrival and collision of exotic terranes, as a mechanism that contributed to the Laramide deformation, should not be ruled out. As suggested by plate tectonocists Peter J. Coney, David L. Jones, Allan Cox, and Myrl Beck,

> Laramide folding and faulting might be an expression of the final phases of exotic terrane collision with North America representing the "tightening up" of a "soft," fragmental continental crust composed of terranes accreted during the Mesozoic. Interaction of the terranes with the adjoining ancient crust would presumably cause rotations, uplifts, and overthrusts in a broad band encompassing both the North American craton and the terranes themselves. The driving force of such a process would have been the continuing subduction of oceanic plates under North America.*

As mentioned in the previous chapter, the Mesozoic and early Cenozoic Cordillera of western North America has often been referred to as an "Andean-type" margin. Eastward subduction, much of which was of the "flat-slab" style, during the late Mesozoic and early Cenozoic and during the Neogene—beneath the continental margins of western North America and South America, respectively—has produced remarkably similar structures and suites of tectonic provinces. Thus the Cordilleras of North and South America represent, respectively, deeply eroded and active examples of the same kind of orogenic system.

Rocky Mountain Basins

Uplift of the Rocky Mountains during very late Cretaceous and early Cenozoic brought to a close the final vestiges of marine conditions in the western interior of North America. Downwarping between the Rocky Mountain uplifts resulted in *intermontane basins* that became

*1982, The Growth of Western North America, p. 84: *Sci. American,* vol. 247, no. 5.

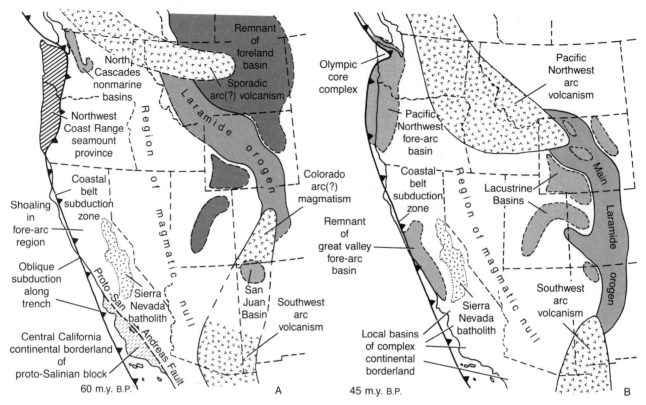

Fig. 13-12. Main tectonic features associated with subduction along Andean-type western continental margin of North America. A. Early Paleogene. B. Eocene.
(From W. R. Dickinson, 1979, Cenozoic Plate Tectonic Setting of the Cordilleran Region in the United States, Figs. 8, 10, pp. 10, 11, *in* Cenozoic Paleogeography of the Western United States: Pacific Section SEPM Paleogeography Symposium, vol. 3. Reproduced by permission of Pacific Section, Society of Economic Paleontologists and Mineralogists)

the dumping grounds for thick accumulations of Paleogene nonmarine sediments in alluvial fan, stream, and lake environments (Fig. 13-13). Intermontane basin sediment fills have produced important economic mineral deposits including petroleum, coal, and uranium.

Of particular interest regarding hydrocarbons are the Green River, Piceance Creek, and Uinta Basins in adjacent parts of Wyoming, Colorado, and Utah (Fig. 13-13). These basins provide instructive examples of nonmarine depositional models: During the Eocene they were occupied by a complex of shallow lakes in which accumulated the delicately laminated shale, chemically complex carbonate mudstones, and dolostones of the Green River Formation. The Green River Formation (Fig. 13–14) is famous for its fish fossils and the oil shales that have figured prominently in recent discussions regarding America's petroleum reserves. The oil shales actually consist of a hydrocarbon substance called **kerogen,** which is an early-stage oil. The kerogen is a product of bacterial transformation of organic matter derived from plant and animal life in the lakes. Abundant organic matter accumulated because of anaerobic conditions that existed at and below the sediment–water interface in the lake-bottom environments (Fig. 13-15).

These Green River kerogen-rich rocks are petroleum source beds. Geological history has not favored maturation of the oil and has not permitted expulsion of hydrocarbons into more porous and permeable carrier and reservoir layers. Usable petroleum is obtained only after an elaborate and expensive heating process to reduce the viscosity of the kerogen. At present, these rocks, even though estimated to contain hundreds of billions of barrels of oil, do not constitute a viable source of petroleum to meet the nation's fossil fuel requirements.

By Oligocene time, the intermontane basins (Fig. 13-16) were filled with sediments, and the last pulses of Rocky Mountain Laramide deformation had ceased. In places, flat-lying upper Eocene deposits unconformably overlie tilted and folded older Eocene and Paleocene strata. At a few localities, Paleozoic and Mesozoic rocks have been thrust over lower Eocene conglomerates. By the end of the Oligocene the Rocky Mountains had been nearly buried in their own debris and their summits were worn down to peneplain surfaces. Voluminous amounts of erosional debris had accumulated along the front of the Rocky mountains in a great eastward-sloping apron (Fig. 13-17). Oligocene sediments of the White

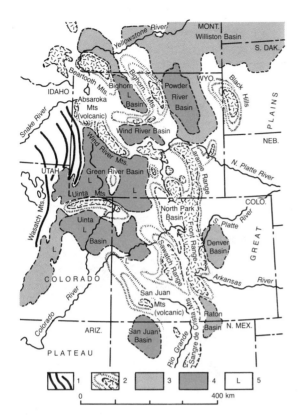

Figure 13-13. Central and southern Rocky Mountains showing uplifts and basins of Paleogene. 1, Folds and fault blocks in geosynclinal sediments; 2, Laramide uplifts east of geosynclinal area with outcrops of pre-Paleozoic basement rocks in their higher parts; 3, basins that received Paleocene sediments; 4, basins in which Eocene sediments were deposited over Paleocene sediments; 5, areas of lake deposits, mainly of Eocene age.
(From Philip B. King, 1977, *The Evolution of North America,* Fig. 74, p. 116. Copyright © 1959, rev. ed. © 1977 by Princeton University Press. Reproduced by permission of Princeton University Press)

River Group, exposed in the spectacular badlands of South Dakota and in erosional escarpments of northwestern Nebraska, have yielded abundant fossils of a diverse assemblage of mammals, including, among others, the giant titanotheres and sheeplike oreodonts (Fig. 13-18) (see Chapter 15). During the late nineteenth century, persistent and seasoned teams of bone hunters combed these beds in search of mammal bone treasures. Stratigraphic sections in the intermontane basins as well yielded tons of mammal fossils, making the Rocky Mountain region one of the most productive areas in the world for specimens and data on Paleogene mammals.

Igneous Activity in the Cordillera

All through the Rocky Mountain region, from Mexico to Alaska, intrusive masses were emplaced during Late Cretaceous and Early Cenozoic. Much of this intrusion of granite and related rocks was accompanied by copper, lead, zinc, iron, gold, and silver mineralization that later led to extensive mining activity and spawned many of the colorful mining camps and towns of the American West. The vast copper deposits of Bingham, Utah; Morenci, Arizona; Santa Rita, New Mexico; the silver of the Comstock lode and Tonopah, Nevada; the gold of Cripple Creek, Colorado; the Black Hills; Goldfield, Nevada; and other famous localities are the results of disseminated and widespread arc magmatism that also produced the Henry Mountains laccolith in Utah and the picturesque Devils Tower in Wyoming (Fig. 13-19).

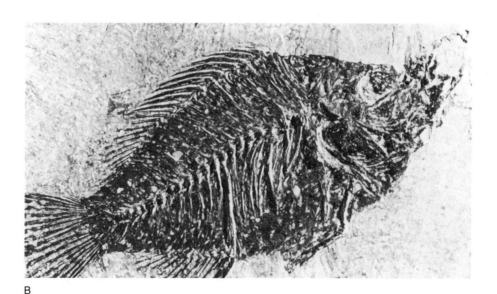

A   B

Figure 13-14. Eocene Green River Formation. A. Outcrop near Rifle, Colorado. Dark layers are kerogen-rich mudstone. B. Fossil freshwater fish from outcrop in southwestern Wyoming.
(A, photo by J. Cooper; B, photo by Wards Natural Science Establishment)

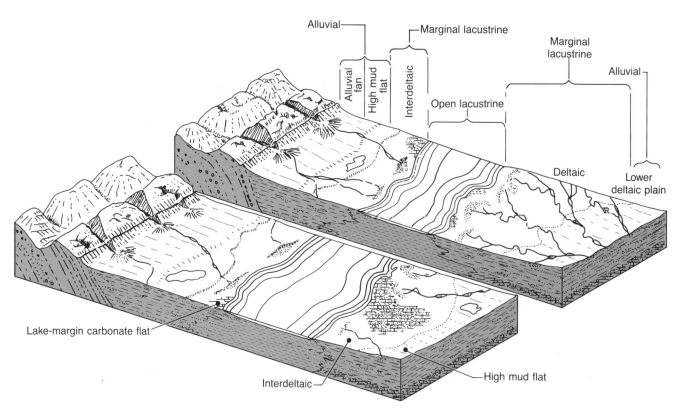

Figure 13-15. Distribution and interpretation of depositional environments of open-lacustrine, marginal-lacustrine, and alluvial facies of western Lake Uinta, Uinta Basin, Utah. Vertical lines: carbonate muds; diagonal lines: grain-supported carbonates; heavy dots: sandstone. Width of Lake Uinta in diagram is approximately 40 km; vertical exaggeration 15–20×.
(After R. T. Ryder, T. D. Fouch, and J. H. Elison, 1976, Early Tertiary Sedimentation in the Western Uinta Basin, Utah, Fig. 6, p. 502: *Geol. Soc. America Bull.* vol. 87, no. 4. Used by permission of author)

In addition to the intrusive activity, widespread eruptive centers formed in the Absaroka Range of Montana and Wyoming during Eocene and Oligocene time and covered a vast area that includes the present Yellowstone Plateau (Fig. 13-20A) with lavas and volcanic breccias. Amethyst Cliff in the eastern part of Yellowstone Park exposes a stratigraphic section of sedimentary beds interlayered with volcanic ash deposits and flows in which are preserved some twenty-seven levels of fossilized tree stumps, trunks, and leaf impressions (Chapter 4). Volcanic activity in the Yellowstone area continued into the Miocene. The geysers (Fig. 13-20B) boiling mud pots, and hot springs are graphic testimony that volcanic heat, the so-called *Yellowstone hotspot,* still lingers below the surface.

Canyon Cutting

Early Neogene history of the Rocky Mountain region involved a broad, gentle regional arching called **epeirogeny.** This upwarping inaugurated a new erosion cycle by elevating the previous erosion surface and causing the streams to be rejuvenated. Previously buried mountain ranges were exhumed and carved anew by streams seeking base level. Large volumes of erosional debris were carried beyond the ranges to form the Miocene and Pliocene beds of the high plains. Remnants of several Paleogene intermontane basin fills—for example, North and South Park Basins, Colorado—were elevated to high levels on the flanks and within several of the ranges. Spectacular canyons, such as Royal Gorge of the Arkansas River, Black Canyon of the Gunnison River (Fig. 13-21A), and Wind River Canyon (Fig. 13-21B) and Bighorn Canyon of the Bighorn River were cut during the epeirogeny by streams whose courses were superposed across the structural grain of Rocky Mountain ranges. This superposition of major streams is particularly well illustrated in the Wyoming Rockies (Fig. 13-22). The Rocky Mountains that we see today were formed structurally during the Late Cretaceous–Eocene Laramide orogeny, but the topography reflects a history of upwarping followed by erosion, with major streams

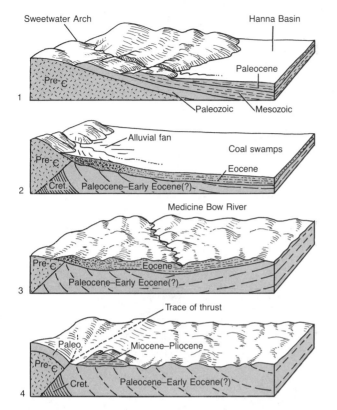

Figure 13-16. Stages in the development of the north flank of the Hanna Basin, Wyoming. 1. Early Paleocene. 2. Eocene. 3,4. Recent. (From S. H. Knight, 1951, The Late Cretaceous–Tertiary History of the Northern Portion of the Hanna Basin, Carbon County, Wyoming, Pl. III, Figs. 1-4, p. 51: Wyoming Geol. Soc. Guidebook, 6th Ann. Field Conf., Reproduced by permission of Wyoming Geological Society)

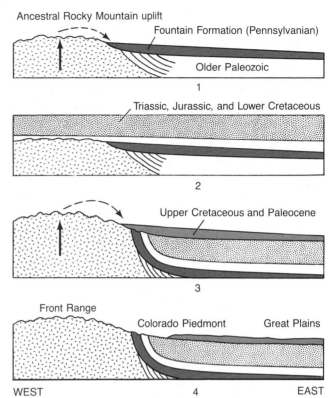

Figure 13-17. Structural evolution of the Front Range of Colorado. 1. Late Paleozoic; ancestral Rocky Mountain uplift (see Chapter 11). 2. Late Mesozoic, when range was tectonically quiet and buried. 3. Early Paleogene, immediately following Laramide Orogeny; formation of clastic apron of Great Plains. 4. Present relations following Cenozoic regional uplift and dissection. (From Philip B. King, *The Evolution of North America,* Fig. 66, p. 108. Copyright © 1959, rev. ed. © 1977 by Princeton University Press. Reprinted by permission of Princeton University Press)

incising deep canyons. The finishing touches of the final scene have been magnificently etched by glacial ice during the Pleistocene (Chapter 16).

The Colorado Plateau Province includes adjacent parts of Utah, Arizona, New Mexico, and Colorado (Fig. 13-23). Flat-lying to moderately tilted Paleozoic and Mesozoic rocks indicate the region was not severely affected by the deformation that accompanied the Cordilleran orogeny. The region is characterized by plateaus and erosional escarpments that rise steplike to the north of its most famous feature, the Grand Canyon (Fig. 13-4). Mesas and deeply incised canyons expose rocks ranging in age from Proterozoic to Cenozoic (Fig. 13-24). Mesozoic formations, such as the sequence exposed in the Glen Canyon, Utah, area, once covered the region like a blanket but have been eroded during the Cenozoic. The erosion was accomplished by streams rejuvenated during late Cenozoic epeirogeny that affected a major part of the Cordilleran region. Exploration of the plateaus and canyons by venturesome scientific surveys after the Civil War (see beginning of the chapter) con-

tributed new concepts of erosion in arid regions and inaugurated a fascinating chapter in the history of geologic investigations in western North America.

THE CONTINENT'S TRAILING EDGE

The Atlantic Margin

After the end of the compressional deformation of the Appalachian orogeny and following extensional block faulting during Late Triassic (Chapter 12), the Appalachian Mountains were eroded by Cretaceous time to an almost flat surface. The modern topography of the Piedmont, Blue Ridge, Valley and Ridge, and Appalachian Plateau Provinces (Fig. 13-18) has evolved through several cycles of late Mesozoic and Cenozoic erosion. Early Cenozoic epeirogenic uplift of the Appalachian region caused major streams to become incised; resultant

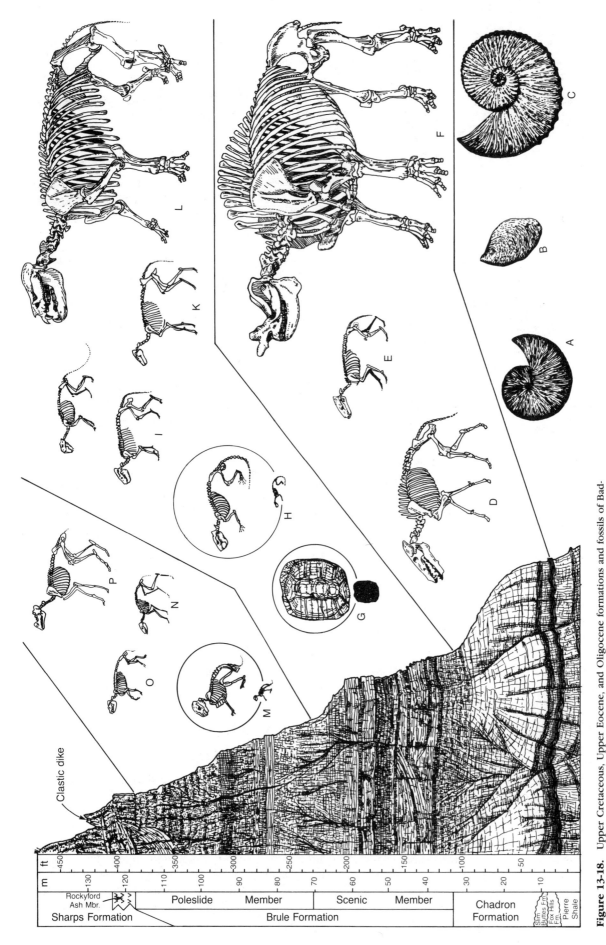

Figure 13-18. Upper Cretaceous, Upper Eocene, and Oligocene formations and fossils of Badlands National Park, South Dakota. A, C. Upper Cretaceous ammonites. B. Bivalve mollusc. D–F. Early Oligocene mammals (D, entelodon; E, oreodont; F, titanothere). G–L. Early Late Oligocene tortoise and mammals (G, land tortoise; H, squirrellike rodent; I, saber-toothed cat; K, *Mesohippus*, three-toed horse; L, amphibious rhinoceros). M–P. Mid-late Oligocene mammals (M, archaic rabbit; N, chevrotainlike ruminant; O, oreodont; P, deerlike ruminant. (From G. J. Retallack, 1983, A Paleopedalogical Approach to the Interpretation of Terrestrial Sedimentary Rocks: The Mid-Tertiary Fossil Soils of Badlands National Park, South Dakota, Fig. 4, p. 826: *Geol. Soc. America Bull.*, vol. 94. Reproduced by permission of author)

A

B

C

Figure 13-19. Features related to Late Cretaceous and Tertiary igneous activity in the Cordillera. A. Bingham copper pit, Utah. Copper mineralization associated with Late Cretaceous intrusions. B. Looking south at Henry Mountains, southern Utah, a laccolith. C. Devils Tower, Wyoming, a volcanic plug. (Photos by J. Cooper)

A

B

Figure 13-20. Yellowstone area. A. Grand Canyon of the Yellowstone; carved in Paleogene–early Neogene volcanic rocks of Yellowstone Plateau. B. Old Faithful Geyser, Yellowstone National Park. (Photos by J. Cooper)

A

B

Figure 13-21. Deep canyons in the Rocky Mountain region. A. Black Canyon of the Gunnison River, Colorado. B. Wind River Canyon, Wyoming. (Photos by J. Cooper)

367

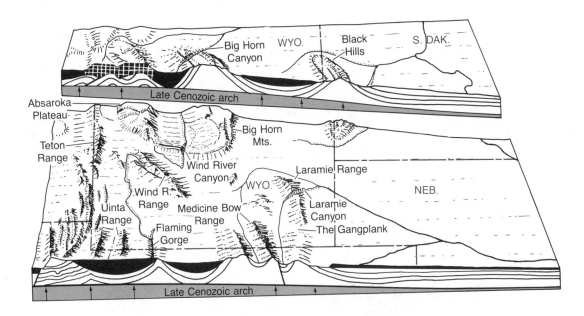

Figure 13-22. Rocky Mountain structures cut by superposed streams.
(From C. A. Dunbar, and K. Waage, *Historical Geology,* Fig. 17-12, p. 415. Copyright © 1969 by John Wiley
& Sons, Inc., New York. Reprinted by permission of John Wiley & Sons, Inc.)

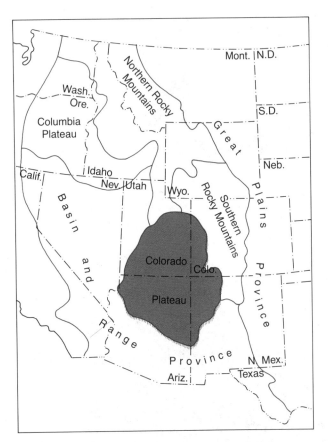

Figure 13-23. Western United States showing location of Colorado Plateau geologic province.

downcutting and erosion etched out the weaker rocks and created local relief on the more resistant rocks. This differential erosion is particularly well illustrated in the Valley and Ridge Province, where streams seeking new base level carved extensive lowlands in the weak formations of the folded belt. Elevated remnants of old erosion surfaces suggest that several such cycles of upwarping, erosion, and peneplanation occurred in the Appalachians during the Cenozoic. Today we see only the roots of Appalachian folds (Fig. 13-25), structures inherited from the late Paleozoic deformation of the Appalachian miogeosynclinal sedimentary succession. The topographic mountains are products of a complex cyclic Cenozoic erosional history. What caused the epeirogenic upwarping in the Appalachian region? The complete answer is not known, but, as in the Rocky Mountain area, upwarping probably was related to isostatic adjustments of continental lithosphere.

The Appalachian belt has been tectonically stable since the beginning of the Jurassic, although occasional earthquakes, such as the one in Charleston, South Carolina, in 1872, testify to renewed activity on some faults. A seaward-thickening wedge of terrigenous sediments comprises the Cenozoic section of the Atlantic coastal plain and continental shelf (Fig. 13-26A). The great bulk of this material was derived from erosion of the Appalachians.

Remember that beginning in the Jurassic and continuing during the Cretaceous, the North American continent moved away from Europe and Africa. We already have reviewed Mesozoic sedimentation along this drifting margin. Cenozoic deposits along the eastern seaboard and in the Gulf Coast region (Fig. 13-26B) represent the addition, during the last 65 million years, of sediments on the tectonically passive trailing margins of the continent.

Figure 13-24. Canyons of the Colorado Plateau. A. Grand Canyon, Arizona. B. Goosenecks of the San Juan River, southeastern Utah. C. Bryce Canyon, Utah. D. Colorado River, Utah. (Photos by J. Cooper)

The Gulf Coast

In the Gulf Coast, outcrop sections of fluvial and nearshore marine sedimentary sequences change facies seaward in the subsurface to more offshore marine-shelf deposits (Fig. 13-27). Important oil and gas accumulations occur in sandstone facies that lie updip from deeper marine organic-rich mudstones. The sandstones represent barrier-bar, delta-fringe, beach, and stream-channel environments. The porosity and permeability of these sandstone facies provide a hydrocarbon reservoir potential that is in sharp contrast to that of the downdip source-bed mudstone facies of relatively low permeability.

A most interesting petroleum entrapment situation is related to peculiar structures called **salt domes,** which are more abundant in the Texas and Louisiana Gulf Coast than in any other place in the world. The domes are vertical, fingerlike projections of intrusions called **diapirs** that have penetrated the continental-shelf sedimentary wedge (Fig. 13-28). Some have moved close to the surface (Fig. 13-29), such as the famous Spindletop Dome in Texas. The salt in the structures has come from the Louann Salt, a thick Jurassic evaporite

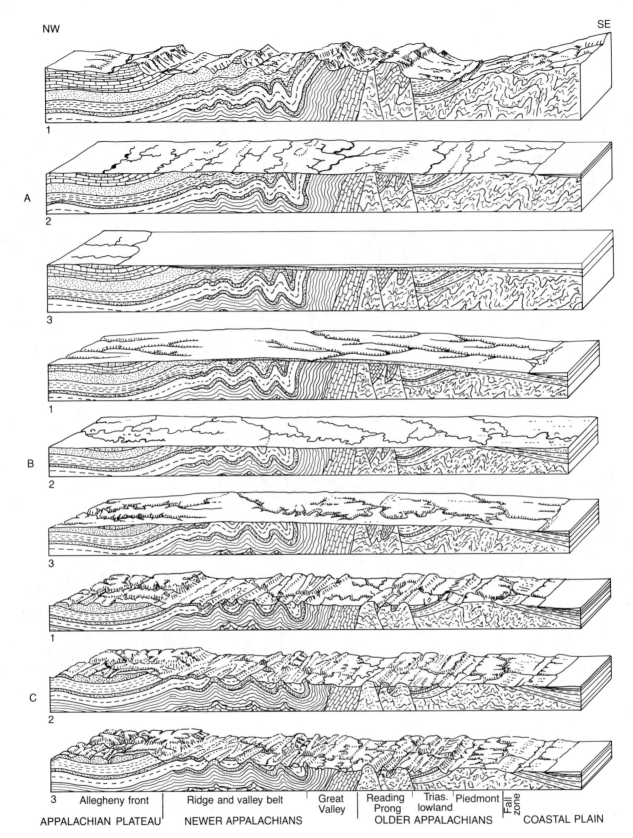

NW

SE

1

A

2

3

1

B

2

3

1

C

2

3

| Allegheny front | Ridge and valley belt | Great Valley | Reading Prong | Trias. lowland | Piedmont | Fall zone |
| APPALACHIAN PLATEAU | NEWER APPALACHIANS | | OLDER APPALACHIANS | | | COASTAL PLAIN |

Figure 13-25. Landscape evolution of Appalachian region in Pennsylvania. A. Mesozoic: 1, rejuvenated Appalachians after deposition of Newark Supergroup, Early Jurassic; 2, Mesozoic erosion; 3, encroachment of Cretaceous sea and deposition of coastal-plain strata. B. Cenozoic, Paleogene: 1, epeirogenic upwarping and rejuvenation of drainage with superposition of south-flowing streams, Late Cretaceous–Paleogene; 2, erosion and peneplanation; 3, regional epeirogenic upwarping in Late Paleogene. C. Cenozoic, Neogene: 4, Dissection of earlier erosion surfaces; 2, 3, Continued sculpturing of landscape to form present-day Appalachian Plateau, Ridge and Valley (valleys eroded in shales; ridges developed on more resistant limestones and quartzites), crystalline Appalachians and Piedmont, and Coast Plain physiographic provinces. Note superposition of streams, entrenched meander patterns, and relationship between topography and rock type.

(From Douglas Johnson, 1931, *Stream Sculpture on the Atlantic Slope,* Figs. 1-9, pp. 15, 17, 19: Columbia University Press, New York. Reproduced by permission of Columbia University Press)

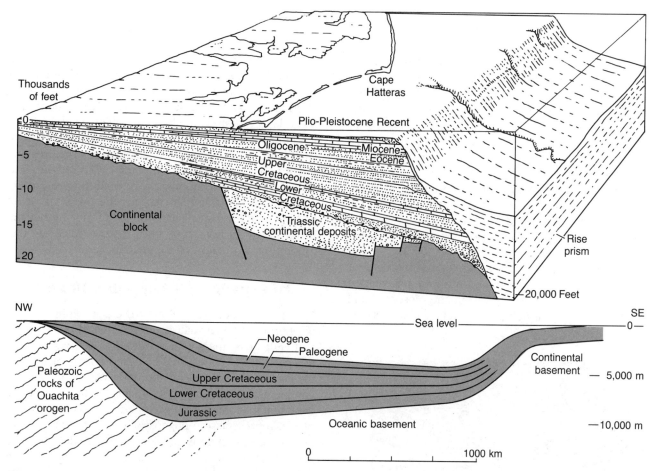

Figure 13-26. Continental shelf wedge deposits. A. Modern continental-shelf sedimentary-terrace prism (miogeosyncline) along eastern United States coast in vicinity of Cape Hatteras, North Carolina. Cretaceous and younger strata dip seaward and thicken at continental slope. The sedimentary prism was deposited on a Lower Cretaceous peneplain, and basement rocks include pre-Paleozoic and Early Paleozoic igneous and metamorphic complexes of the crystalline Appalachians. Late Triassic–Early Jurassic (Newark) rift basins developed during initial rifting of North America from North Africa about 180 m.y. B.P. B. Cross section across Gulf of Mexico from Texas to Yucatan showing sedimentary wedge of Gulf of Mexico geosyncline, variation in thickness of Mesozoic and Cenozoic sediments, and relation to basement complex of Ouachita orogenic belt.
(A, from R. S. Dietz, and J. C. Holden, 1974, Collapsing Continental Rises: Actualistic Concept of Geosynclines—A Review, Fig. 5, p. 20: *SEPM Spec. Pub.* 19. Reproduced by permission of Society Economic Paleontologists and Mineralogists; B, data from several sources)

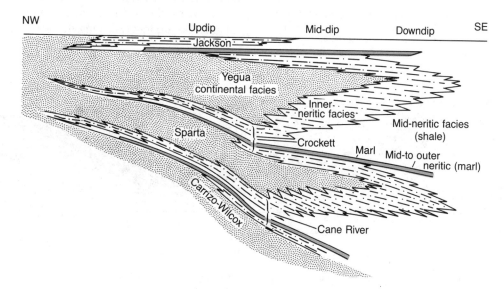

Figure 13-27. Cenozoic facies relationships and cyclic sedimentation, Gulf Coast region.
(From S. W. Lowman, 1949, Sedimentary Facies of the Gulf Coast, Fig. 23, p. 1972: Am. Assoc. Petroleum Geologists Bull., vol. 33. Reproduced by permission of American Association of Petroleum Geologists)

unit that was deposited during the early stages of formation of the Gulf of Mexico Basin (Chapter 12). Hydrostatic loading caused by subsequent deep burial of the Louann Salt beneath the thick Cretaceous and Cenozoic section has resulted in plastic behavior, upward flow, and diapiric intrusion of the less dense salt. The original loci for flowage were probably irregularities on the upper surface of the salt; such irregularities would have been accentuated by compaction during sediment loading. In addition to having influenced the entrapment of recoverable hydrocarbons (Fig. 13-30), the domes themselves are sources of commercial salt and sulfur.

With regard to hydrocarbons, it is noteworthy that the Gulf Coast, both onshore and offshore, has yielded large amounts of oil and gas from the Mesozoic and Cenozoic continental-shelf and coastal-plain succession, whereas the Atlantic coastal-wedge sequence has contributed little. Much of this discrepancy may be a result of the comparatively small concentration of exploration effort along the Atlantic seaboard. However, a significant oil and gas venture could be in the offing for the Atlantic margin: Recent geophysical profiles across the Atlantic shelf have identified attractive subsurface structures that might be favorable hydrocarbon traps. Our nation's critical energy needs may push us beyond the present environmental constraints to test for the presence of oil and gas in the Atlantic offshore. Gas-producing but immature hydrocarbons have been discovered in the Baltimore Canyon region of the Atlantic margin.

An Actualistic Geosyncline Model

American geologists Robert Dietz and John Holden have proposed an actualistic model wherein the Atlantic and Gulf Coast margins are viewed as geosynclines. In this model, the continental-shelf and coastal-plain sedimentary wedge represents the miogeosyncline, and the continental-rise prism of sediments, at the toe of the continental slope and developed on oceanic lithosphere, represents the eugeosyncline. The Atlantic margin-type geosyncline is considered the model for those ancient geosynclines that developed marginal to a rift ocean on the trailing edge of a drifting continent. The Atlantic miogeosynclinal continental-terrace prism was deposited on peneplaned continental crust that was formed by continental accretion—during the Acadian orogeny—of an earlier eugeosynclinal rock assemblage. The actualistic model envisions the possibility of the present eugeosynclinal sedimentary prism collapsing in the future and becoming welded to the continent by plate tectonics in the same fashion. In this light, Dietz and Holden envision the simplified scenario shown in Fig. 13-31 for the opening, closing, and reopening of the Atlantic Ocean, incorporating the notion of Atlantic-mar-

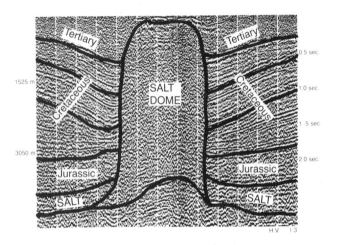

Figure 13-28. Reflection seismic profile from the Gulf Coast region showing salt dome and penetrated Mesozoic and Lower Tertiary stratigraphic section.
(From Morris S. Petersen, and J. Keith Rigby, INTERPRETING EARTH HISTORY, 3rd ed., Fig. 4-5, p. 29. © Wm. C. Brown Publishers, Dubuque, Iowa. All Rights Reserved. Reprinted by permission)

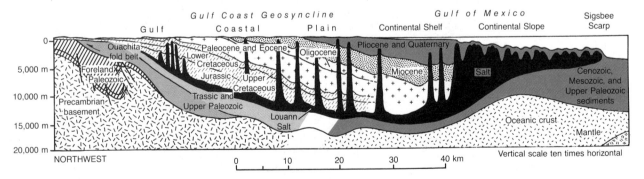

Fig. 13-29. Cross section across Gulf of Mexico Basin showing Jurassic Louann Salt and piercement salt domes in overlying sedimentary section.
(From P. Lehner, 1969, Salt Tectonics and Pleistocene Stratigraphy on Continental Slope of Northern Gulf of Mexico, Fig. 43, p. 1473: *Am. Assoc. Petroleum Geologists Bull.,* vol. 53. Reproduced by permission of American Association Petroleum Geologists)

gin geosynclines and the formation of marginal orogenic belts (orogens), an illustration of the *Wilson cycle*. This cycle also explains how the North American continent has grown by lateral accretion of orogens, a phenomenon that James Dwight Dana had described during the 1870s as part of his geosynclinal cycle.

THE CONTINENT'S LEADING EDGE

Plate Collisions

In marked contrast to the broad continental shelf along the trailing-edge Atlantic and Gulf Coast margins, the shelf along the western margin of the continent is much narrower. This is because it is the more tectonically dynamic, leading edge of the westward-drifting continent and has been engaged in collision with the eastern, leading edge of Pacific Ocean basin lithosphere since Mesozoic time. Today, deep-marine environments are present a short distance offshore, a setting that has been prevalent throughout the Cenozoic. This is evidenced by thick sections of deep-marine sedimentary rocks in the coastal ranges and basins of Washington, Oregon, and California. These deep-water sequences, commonly developed as turbidite-fan complexes, represent a continuation of Cretaceous depositional patterns. Many of these sections change facies rather rapidly in a landward direction to shallow-marine and even continental deposits. As discussed in Chapter 12, from early Mesozoic to late Paleogene, the western edge of North America was developed as an Andean-type margin with a margin-edge subduction zone and associated magmatic arc. Subduction activity along this margin was responsible for the deformation associated with the Cordilleran orogeny.

Perhaps no segment of the entire circum-Pacific rim has undergone a more complex tectonic evolution during the Cenozoic than the Cordilleran region of the western United States, and perhaps no other region has been the subject of more intense scrutiny and tectonic analysis than the Cordilleran belt. In 1970, Tanya Atwater, then a graduate student in geophysics at Scripps Institute of Oceanography, proposed a model to explain the implications of plate tectonics for the Cenozoic evolution of western North America. Her detailed analysis of sea-floor magnetic anomaly patterns revealed that at least three oceanic plates, including the now almost completely subducted *Kula Plate,* the *Farallon Plate,* and the *Pacific Plate,* interacted with the western margin of North America during the last 80 million years (Fig. 13-11, 32). Extrapolating back in time by unspreading and unsubducting various segments of plates, Atwater and others, during the decade of the seventies, recon-

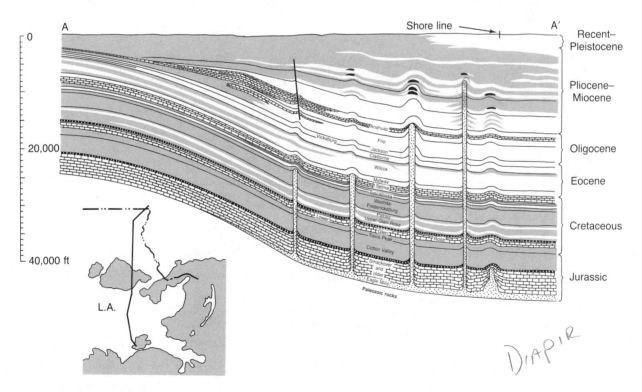

Figure 13-30. Cross section through Louisiana showing Gulf Coast sedimentary section and salt domes. Note petroleum accumulations (dark bands) in traps created by doming of sediments. (From J. B. Carsey, 1950, Geology of Gulf Coast Area and Continental Shelf, Fig. 2, p. 362: *Am. Assoc. Petroleum Geologists Bull.,* vol. 34. Reproduced by permission of American Association of Petroleum Geologists)

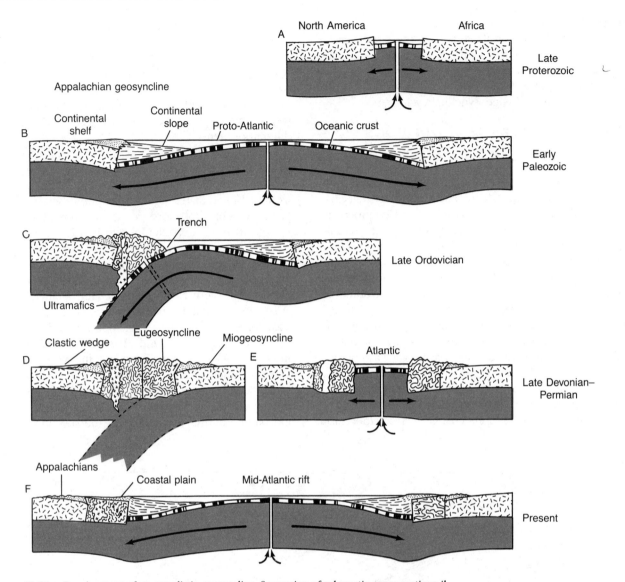

Figure 13-31. Development of an actualistic geosyncline. Succession of schematic cross sections illustrating inferred opening and closing of Iapetus, and reopening of modern Atlantic Ocean basin—the Wilson cycle. A. In late Proterozoic, ancient North America (Laurentia) and North Africa (proto-Gondwana) were rifted apart to form the proto-Atlantic (Iapetus) ocean. B. By the process of sea-floor spreading, Iapetus continues to open and geosynclinal sedimentary wedges develop along passive trailing continental margins during early Paleozoic. C. Iapetus begins to close in Late Ordovician as rift spreading ceases, and a subduction zone is produced along the margin of Laurentia. Continued subduction collapses the eugeosyncline, resulting in Taconic orogeny. D. Late Devonian–Permian; Iapetus finally closes, with continent–continent collisions during Acadian and Allegheny Orogenies. Suture belt is the locus of squeezed-up pods of ultramafic rock (ophiolites). E. About 180 m.y. B.P.; the present Atlantic Ocean basin reopens. F. Today, the central North Atlantic is opening at a rate of nearly 3 cm/ yr, and new margin-geosynclinal sedimentary prisms are forming along the passive, trailing continental margins.

(After R. S. Dietz, and J. C. Holden, 1974, Collapsing Continental Rises: Actualistic Concept of Geosynclines—A Review, Fig. 7, p. 24, *in* Modern and Ancient Geosynclinal Sedimentation, R. H. Dott, Jr., and R. H. Shaver, eds.: *SEPM Spec. Pub.* 19, Reproduced by permission of Society of Economic Paleontologists and Mineralogists)

structed the changing, almost kaleidoscopic pageant of plate configurations that have most influenced western North America during the late Cretaceous and Cenozoic.

From late Cretaceous to late Paleogene, the Farallon Plate was subducted beneath the entire Cordilleran margin (Fig. 13-11); its remnants continue to be subducted beneath the western edge of North America off northern California, Oregon, and Washington and off southern Mexico. During the late Oligocene, the Pacific Plate interacted with the North American Plate as part of the Farallon Plate and a segment of the **East Pacific Rise** spreading ridge were overridden by the edge of the

Figure 13-53. Volcanic mountains within magmatic arc setting, Cascade Range, northwestern United States and Central America. A. Mount Baker, Washington. B. Mt. Hood, Oregon. C. Looking north from top of Mt. Lassen to Mt. Shasta, California. D. Volcan Santiago, near Lake Atitlan, Guatamala. (Photos by J. Cooper)

the right-lateral San Andreas fault system. Remember that once the triple junctions between Farallon Plate segments and the Pacific and North American Plates had been established, their migration produced right-lateral transform motion, which replaced subduction. However, north of Cape Mendocino, which marks the terminus of the San Andreas Fault, subduction of the Juan de Fuca segment of the Farallon Plate has continued through the Neogene (Fig. 13-52); the Cascade volcanic arc is part of an arc-trench system that lies north of the San Andreas Fault. The volcanic chain of southern Mexico and Central America is part of an arc system lying south of the

San Andreas system.

Although the Cascade Province of andesite volcanism can be explained by subduction, no suitable explanation is yet available for the eruption of immense Neogene basalt fields in the back-arc region in the Pacific Northwest; these Columbia Plateau lavas remain as puzzling as they were before the advent of plate-tectonic models.

Most of the Cascade volcanic activity occurred during the Pliocene and Pleistocene, although some, such as the 1914–1917 eruptions of Mt. Lassen and the 1980 eruptions of Mt. St. Helens, has continued to the present. Such recent activity suggests that major volcanic peaks are sleeping giants and that Pacific Northwest volcanos are only dormant, not extinct.

GLOBAL TECTONICS

A major tectonic revolution, involving large-scale shearing motions and block faulting, has dominated the tectonic style of western North America and the northern Pacific Ocean Basin during the last 30 million years. The impetus for tectonic behavior has been interaction between the North American and Pacific Plates.

But this tectonic disturbance of the eastern Pacific Basin and western North America is only a part of late Cenozoic changes that were of global importance. Throughout the entire Pacific rim, plate interactions have fostered heightened tectonic activity; earthquakes and volcanic activity in Japan, Alaska, California, and Central and South America are testimony to the continuation of this unrest. Outside the Pacific region, one of the most active mobile belts during the last half of Cenozoic has been the Tethyan region between old Gondwana and Laurasia, where continued plate interactions resulted in formation of the Alps and Himalayas in late Cenozoic.

The great Alpine–Himalayan system of southern Europe and Asia displays plastically deformed Mesozoic and Cenozoic rocks in structures that we do not see in older mountain belts such as the Appalachians because of dramatic differences in erosional history. The Himalayan Mountain system was formed during the collision between India and southern Asia. The subcontinent of India is an excellent example of an accreted terrane of continental origin. The block that is now India was rafted along on spreading and subducting oceanic lithosphere into collision with the Eurasian continent. The converging lithosphere blocks telescoped some 800 km or more along thrust faults, forming the Himalayas and a lithosphere twice the thickness of normal continental crust. In the 40 million years since the initial collision, the Indian subcontinent has continued to move northward, tightening up the suture zone and shoving Asian crustal rocks to the north and east. The continuing convergence has caused major disruptions far into China and has resulted in many devastating earthquakes. In the words of John McPhee, in his book *Basin and Range,* if the essence of plate tectonics could be captured in one sentence, it would be "There is marine limestone on top of Mount Everest." The Alps resulted from collisions between northern Africa and southern Europe, and even today, Mediterranean islands such as Cyprus, which is experiencing uplift, attest to the squeezing between two major lithospheric plates. Indeed, structural disturbances of great magnitude occurred with amazing frequency during late Cenozoic (Fig. 13-54, 55).

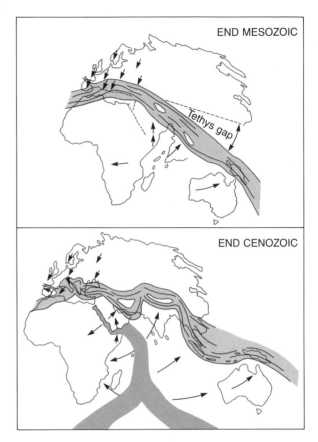

Figure 13-54. Plate interactions and deformation in Tethyan region. Light-gray bands represent orogenic belts; dark gray represents sea-floor spreading centers. Short arrows show continental displacements; long arrows represent magnetization directions.
(From R. H. Dott, Jr., and R. L. Batten, *Evolution of the Earth,* Fig. 17.31, p. 479. Copyright © 1981 by McGraw-Hill Book Co., New York. Reproduced by permission of McGraw-Hill Book Co.)

Figure 13-55. Closing up of Tethys Ocean and collision of India with southern Asia to form Himalayan belt.

LIFE OF THE CENOZOIC
The Age of Mammals

Mammals inherited the earth from the reptiles and diversified rapidly in an initial adaptive radiation that filled many niches left vacant by extinction of Mesozoic reptiles. As will be elaborated more fully in Chapter 15, two mammal groups—marsupials and placentals—that had emerged during the Cretaceous had a major influence on early Cenozoic mammal evolution. The real mammal success story inolved the placentals, which, probably because of their better organized brain, generally won out in competition with other kinds of mammals. An early placental group, the shrewlike insectivores, was the stem stock from which the other placental orders evolved during the Paleocene and Eocene. By the Eocene, early ancestors of most modern mammal orders were present, including such familiar groups as rodents, carnivores, odd- and even-toed hoofed mammals, elephants, whales, and even primates, the order to which apes and man (see Chapter 17) belong. In a pattern reminiscent of many other groups, placental mammals underwent an initial phase of "trial-and-error" and "experimental" evolution during which time a number of archaic groups became extinct. Real stabilization of mammalian evolution took place during the Neogene. The tremendous diversification of flowering plants—the angiosperms—had a profound effect on the evolution of animal groups such as insects and birds and particularly mammals. This is quite evident in the Miocene when prairie grasses evolved and influenced a change in many groups of mammalian herbivores from browsers to grazers. This change is well illustrated by the pattern of horse evolution in which changes in dental patterns and limb and toe structure (Chapter 15) can be viewed.

During the dawn of mammal evolution, the supercontinent Pangaea had begun to split apart. Through the Cenozoic, the pieces moved and slipped farther apart, drifting passively on their moving plates. Terrestrial life forms had been comparatively uniform across the supercontinent, but during the Cenozoic each drifting land mass developed its own environmental conditions. Such was the situation in North America: The birth of the Rockies and the formation of the Great Plains opened up new niches for enterprising mammals. In Australia, the critical timing of its isolation as an island spared the indigenous marsupial fauna from early Cenozoic competition with placentals. Plate motions and resulting continental separations were responsible for more endemism and comparatively greater diversification among the total world mammal fauna than we see in Mesozoic land reptiles, which have a decidedly more cosmopolitan complexion.

The Marine Realm

Cenozoic fragmentation and isolation of continents created the best of all possible worlds for marine invertebrates. After the end-of-Mesozoic extinctions that wiped out the ammonites and several other major groups, the composition of invertebrate faunas gradually took on a modern look as familiar groups of bivalve and gastropod molluscs, bryozoans, echinoderms, worms, arthropods, and corals evolved successfully (Fig. 13-56). Marine invertebrates probably have a greater total diversity today than during any time in geologic history. This modern diversity is the result of isolation and north–south configuration of land masses and continental shelves. This situation, which has evolved steadily through the Cenozoic, has influenced climatic patterns and important changes in oceanic circulation. More than thirty modern marine-shelf benthic-invertebrate provinces have been defined. These provinces are separated longitudinally by continental masses and ocean basins, which provide major barriers to migration and mixing. The provinces are partitioned latitudinally principally by temperature barriers and their boundaries coincide with those of major climate belts. These temperature boundaries also have created migratory barriers, thus facilitating geographic speciation and formation of endemic faunas. In contrast, during the Jurassic, for example, there were fewer than a half dozen separate provinces; faunas were much more cosmopolitan, resulting in lowered total diversity (Chapter 6).

Foraminifera (Fig. 56I–M), diatoms, radiolarians, dinoflagellates, coccolithophorids, and other single-celled groups have flourished through the Cenozoic. Foraminifera are widespread in Cenozoic sediments and have been studied extensively because of their use in biostratigraphic correlation in subsurface sections and in reconstructing paleoenvironments and *paleobathymetric* patterns. Certain planktonic forms have allowed precise correlation of deep-basin with shallow-marine sections. Spores and pollen also have served as useful microfossils in stratigraphic studies. Because of their subsurface stratigraphic value, all of these microfossil groups have proved useful in the search for petroleum.

"TWENTY SECONDS 'TIL MIDNIGHT": THE PLEISTOCENE AND HOLOCENE

If all of geologic history were scaled down to a 24-hour day, the latest 2 million-year interval of the Cenozoic—the Pleistocene and Holocene Epochs—would consume only the last 20 seconds! The Pleistocene is characterized by great continental glaciers that covered much of the northern hemisphere, moving southward from the Arctic region on at least four separate occasions. The diverse effects of Pleistocene glaciation and its possible causes are treated in Chapter 16; we mention it here only to put this fascinating epoch in its historical context. The onset of frigid climates during the Cenozoic is

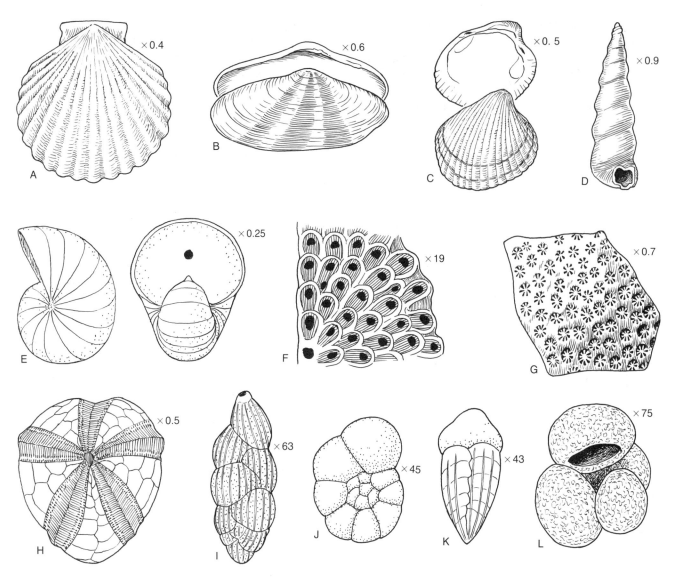

Figure 13-56. Representative Cenozoic marine invertebrate fossils. A–C. Bivalve molluscs. D. Gastropod mollusc. E. Nautiloid cephalopod. F. Bryozoan. G. Scleractinian coral. H. Echinoid echinoderm. I–L. Benthic Foraminifera.
(A–C, E–H from *Treatise on Invertebrate Paleontology,* University of Kansas and Geological Society of America; D from M. S. Petersen and J. K. Rigby, 1982, *Interpreting Earth History,* Pl. 8, p. 115: Wm. C. Brown Publishing Co.; I–K from R. M. Kleinpell, 1938, Miocene Stratigraphy of California, Pls. XX, XIV, Figs. 17, 12a, 3a: *Am. Assoc. Petroleum Geologists;* L from R. Z. Poore et al., 1981, Microfossil Biostratigraphy and Biochronology of the Relizian and Luisian Stages of California, Pl. 1, *in* The Monterey Formation and Related Siliceous Rocks of California: Pacific Section SEPM.)

indicated by cooler marine temperatures and lowered sea levels during the Oligocene. Pleistocene glaciation may well have been strongly influenced by late Cenozoic plate motions that resulted in the North Pole being surrounded by continents.

During the Pleistocene, alpine glaciation occurred in the higher altitude regions of the Rockies, Cascades, and Sierra Nevada. The powerful erosive force of glacial ice is graphically exhibited in such places as Yosemite Valley, California; Glacier National Parks, Montana and Canada; Rocky Mountain National Park, Colorado; and the Grand Tetons of Wyoming (Fig. 13-57), among others—testimony to ice having sculptured the final scene. Today some alpine glaciers remain, but they are small in comparison to their Pleistocene counterparts, and their numbers have been greatly reduced.

Figure 13-57. Glacially influenced landscapes. A. Grand Tetons, Wyoming. B. Grand Tetons and terraces along Snake River, Wyoming. C. The incomparable valley—Yosemite, California. D. Peyto Lake and lacustrine delta of outwash plain, Columbia ice fields area, Canadian Rockies, Alberta. (Photos by J. Cooper)

Pleistocene glacial and interglacial phases had tremendous impact on marine sedimentation. Great volumes of sediment were carried by major streams to the shelves of the continent's trailing edge and to the turbidite fans of its leading edge. Much of the construction of the gigantic Mississippi River delta occurred during the Pleistocene, and submarine fans filled the trench between the Juan de Fuca Rise and the Cascade magmatic arc (Fig. 13-35). Also, rainsoaked areas of the continent south of the ice front were the sites of major landslides and debris flows.

The Pleistocene was also a time of major tectonic uplift and deformation in California, a consequence of continued interactions between the Pacific and North American Plates along the San Andreas fault system. Vertical uplift, overthrusting, and overturning of Pleistocene sedimentary beds were commonplace. Many of the anticlines and faults that trap hydrocarbons in California

Neogene coastal basins developed during the Pleistocene wave of deformation; elevated marine terraces veneered with Pleistocene sediments bear evidence of major coastal uplift. Volcanic activity in the magmatic-arc belts north and south of the San Andreas Fault is evidence of continued subduction along much of the continental margin during the Pleistocene. Crater Lake, Oregon, occupies the collapsed caldera of an andesite volcano that blew its top during the Pleistocene.

The Holocene Epoch is generally considered to be the last 10,000 years: the time since retreat of the last great continental ice sheet and beginning of postglacial sea-level rise. Could the Holocene be just another interglacial phase, with more ice advances in the offing? Or will all the polar ice melt, thus raising sea level worldwide another 100 m, with severe consequences for coastal cities? This provocative subject will be treated in Chapter 16.

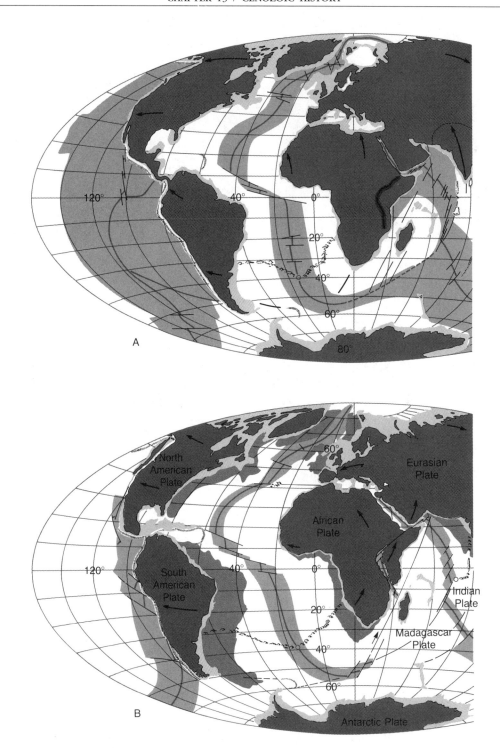

Figure 13-58. Present and future plate configurations. A. Modern world. B. World 50 million years from now; present-day continental positions shown in light gray. Note the change in position of the sliver of North American continent west of the San Andreas Fault.
(From R. S. Dietz, and J. C. Holden, 1970, The Breakup of Pangaea, pp. 38, 39: *Sci. American,* October. Reproduced by permission of W. H. Freeman and Co., New York)

During late Pliocene, human beings evolved in Africa (Chapter 15) and by late Pleistocene had entered the scene on the North American continent, probably by way of the Bering land bridge between Asia and North America. Humans quickly made their presence felt and

through hunting activities may have been instrumental in causing the extinction of a number of large mammals about 30,000 to 10,000 years ago. At first, human activities were geographically confined and compatible with the physical environment, as exemplified by the Mesa

Verde cliff dwellings that blend into the deltaic sandstone ledges. But later, since the era of European discovery and colonization, the industrial revolution, and the automated age, human beings have spread across the land from shore to shore, at times obtrusively.

Despite the presence and influence of *Homo sapiens,* the natural geologic forces of the earth will continue unabated, uncaring for those who choose to live near the San Andreas Fault or in the shadow of a Guatamalan volcano. Time and process will march on, and 50 million years from now plate tectonics will have molded and shaped an earth that may look something like that portrayed in Figure 13-58. In the context of exotic terranes, the narrow slice of continent west of the San Andreas Fault—coastal California and Baja California—will have accreted along the continental margin of Alaska. How would a scientist recognize the discontinuity between the "native" Alaskan rocks and the exotic California rocks? Comparison of rock sequences across major faults, of dissimilarities in the fossils of plants and animals, and of different magnetic characteristics would reveal the discontinuity and allow reconstruction of our present geography.

SUMMARY

The traditional subdivision of the Cenozoic Erathem into Tertiary and Quaternary Systems has been abandoned by many geologists who favor the more equal time-stratigraphic subdivisions of Paleogene and Neogene. The epochs of the Cenozoic periods do have universal use and retain the basis for definition prescribed by Charles Lyell in the 1830s. Cenozoic chronology is particularly well calibrated because of biostratigraphic ties between terrestrial mammals, marine invertebrates, and marine microfaunas, together with radiometric dates and recognition of magnetopolarity units in ocean basins and continents.

The Laramide phase of the Cordilleran orogeny continued from Cretaceous into early Paleogene and affected the eastern part of the Cordilleran belt. Uplift of pre-Paleozoic basement in the southern and central Rockies and folding and thrusting of Proterozoic, Paleozoic, and Mesozoic rocks from Alaska to Mexico were related to continued subduction along the continental margin. Laramide deformation began about 80 million years ago with onset of oblique and unusually rapid subduction of the Farallon oceanic plate accompanied by shallowing of the subduction zone. As the angle of subduction gradually drecreased, arc magmatism and back-arc tectonism migrated eastward.

Laramide uplift of the Rocky Mountains during late Cretaceous and early Paleogene caused the last vestiges of interior seas to recede from the North American continent. The Cretaceous interior seaway was replaced by Rocky Mountain ranges and intermontane basins. Paleogene deposits of the Rocky Mountain basins and the Great Plains are the sites of some of the richest fossil mammalian faunas in the world.

Late Cretaceous to early Paleogene was also a time of intense volcanic activity and intrusion of small plutons—a manifestation of broadly disseminated magmatic-arc activity associated with the Laramide phase of the Cordilleran orogeny. Much of the intrusive activity was accompanied by formation of economic metal deposits. Epeirogenic upwarping in late Paleogene and early Neogene inaugurated a new erosion cycle. Previously peneplaned and buried mountain ranges were exhumed and some were cut across by superposed streams, which carved spectacular deep canyons. The Colorado Plateau, largely spared by Cordilleran orogeny deformation, was intricately dissected into a landscape of plateaus, benches, mesas, buttes, arches, and deep canyons by streams rejuvenated during Neogene epeirogeny.

In the Appalachian region, several cycles of Cenozoic epeirogenic uplift likewise rejuvenated streams that carved valleys in weak strata and sculpted ridges in more resistant rocks. The present-day Appalachian Mountains expose the deeply eroded roots of an earlier, more majestic range. Mesozoic and Cenozoic erosion cycles in the Appalachians produced great volumes of sediment that was carried to the Atlantic coastal plain and continental shelf—the passive, trailing margin of the continent.

The Atlantic and Gulf Coast margins of North America are actualistic geosynclines. The coastal plains and continental shelves represent the miogeosyncline, and the continental slope and rise represent the eugeosyncline. The Atlantic–Gulf Coast model provides insight on those ancient geosynclines that developed marginal to a rift ocean on the trailing edge of a drifting continent.

In contrast to the passive trailing Atlantic and Gulf Coast margins, the Pacific margin of North America has involved interaction of continental and oceanic plates on a grand scale. One of the most dramatic manifestations of this plate interaction is the San Andreas fault system, which is one boundary between the North American and Pacific Plates. According to the Atwater model, about 30 million years ago the San Andreas Fault was initiated by collision and overrunning of a segment of the East Pacific Rise spreading ridge by the edge of the North

American continent. The San Andreas lengthened during the Neogene by northward and southward migration of the Mendocino and Rivera triple junctions, respectively.

The San Andreas fault system, a giant megashear and transform plate boundary, has played a key role in the tectonic evolution of California. The right-lateral shearing motion between the North American and Pacific Plates has been responsible for uplifting the coast ranges and fragmenting the continental crust into a great number of differentially moving blocks. The splintered margin of the North American Plate in California provided a setting for development of dynamic, fault-bounded basins that received thick accumulations of turbidites. Several, such as the Los Angeles Basin, became sites for major hydrocarbon accumulations. Other manifestations of Neogene plate tectonics activity in the western Cordillera include opening of the Gulf of California rift and pronounced block faulting to form the Basin and Range Province. Extensive Neogene magmatic-arc volcanism in the Cascades of the northwest and Mexico and Central America to the south is related to underflow of oceanic lithosphere north and south of the San Andreas transform.

Throughout the entire Pacific rim, plate interactions have generated earthquakes and volcanic activity in Japan, Alaska, California, Mexico, and Central and South America. Formation of the Alpine–Himalayan Mountain belt during the Cenozoic is evidence that dramatic plate motions were occurring over much of the globe during Cenozoic history.

Cenozoic life is highlighted by the rise to dominance of the most advanced class of organisms—the mammals. Continental separations, promoting genetic isolation and provincialism, were responsible for increased diversification of the total world mammalian fauna. Rapid diversification of angiosperm plants paralleled the radiation of insects and birds. Cenozoic fragmentation and isolation of continents, together with ecologic partitioning of north–south oriented continental shelves, promoted provincialism and diversity levels of marine invertebrates to a degree previously unattained in the Phanerozoic.

Pleistocene history was characterized by northern hemisphere glaciation and higher altitude alpine glaciation, the culmination of a global refrigeration trend that began in the Oligocene. Pleistocene glacial and interglacial stages had a profound influence on marine sedimentation owing to abundant precipitation, increased runoff, and delivery of detritus to the continental shelves and ocean basins. Neotectonic activity in the Pacific coast region produced dramatic uplift and deformation over a short period of time. Late Pleistocene witnessed the arrival of humans in North America—the main migratory route from Asia being the Bering land bridge. Their hunting activities may have contributed to extinction of many large land mammals between 10,000 and 30,000 years ago.

The Holocene Epoch, comprising the last 10,000 years since retreat of the last glacial ice sheets, may be just another interglacial stage. Return of glacial conditions to grip the northern hemisphere or melting of present polar ice caps to raise sea level will have dire environmental consequences. Fifty million years from now, southern California may be an accreted terrane of southern Alaska.

SUGGESTIONS FOR FURTHER READING

Burke, K. C., and Wilson, J. Tuzo, 1976, Hot Spots on the Earth's Surface: *Scientific American* Offprint No. 920, W. H. Freeman, San Francisco.

Dalrymple, G. B., Silver, E. A., and Jackson, E. D., 1973, Origin of the Hawaiian Islands: *Am. Scientist,* vol. 61, pp. 294–308; also *in* B. J. Skinner, ed., *Earth's History, Structure, and Materials:* Readings from *American Scientist,* 1980, William Kaufmann, Los Altos, CA.

Dietz, R.S., 1972, Geosynclines, Mountains, and Continent-Building: *Scientific American* Offprint No. 899, W. H. Freeman, San Francisco.

Heezen, B. C., and MacGregor, I. D., 1973, The Evolution of the Pacific: *Scientific American* Offprint No. 911, W. H. Freeman, San Francisco.

Hsu, K. J., 1972, When the Mediterranean Dried Up: *Scientific American* Offprint No. 904, W. H. Freeman, San Francisco.

James, D. E., 1973, The Evolution of the Andes: *Scientific American* Offprint No. 910, W. H. Freeman, San Francisco.

Jones, D. L.; Cox, Allan; Coney, Peter; and Beck, Myrl, 1982, The Growth of Western North America: *Sci. American,* vol. 247, no. 5, pp. 70–128.

Kurten, Bjorn, 1969, Continental Drift and Evolution: *Scientific American* Offprint No. 877, W. H. Freeman, San Francisco.

McPhee, John, 1980, *Basin and Range:* Farrar, Straus, & Giroux, New York.

Molnar, Peter, and Tapponier, Paul, 1977, The Collision between India and Eurasia: *Scientific American* Offprint No. 923, W. H. Freeman, San Francisco.

Powell, J. W., 1961, *The Exploration of the Colorado River and Its Canyons:* Dover Publications, Inc., New York. (Unabridged and unaltered republication of the work first published in 1895 by Flood and Vincent under the title, *Canyons of the Colorado.*)

Rosenfeld, C. L., 1980, Observations on the Mount St. Helens Eruption: *Am. Scientist,* vol. 68, pp. 494–509; also in B. J. Skinner, ed., *Earth's History, Structure, and Materials:* Readings from *American Scientist,* 1980, William Kaufmann, Los Altos, CA.

Valentine, J. W., and Moores, E. M., 1974, Plate Tectonics and the History of Life in the Oceans: *Scientific American* Offprint No. 912, W. H. Freeman, San Francisco.

14

MATCHSTICKS TO MAGNOLIAS:
The Evolutionary History of Plants

KEY TERMS

Vascular plants
Tracheophytes
Xylem
Phloem
Chloroplasts
Stoma
Pith
Epidermal layers
Cambium
Cortex
Thallophyta
Primary producers
Embryophyta
Spores
Terminal Sporangia
Sporophyte
Gametophyte
Gymnosperms
Growth Rings
Angiosperms
Dicotyledon
Monocotyledon
Spores
Pollen
Palynology
Dendrochronology

THE REMAINS OF AN ANCIENT PETRIFIED FOREST

Historical evidence indicates that fossilized plants have been known for over 3000 years and that many remains served as curios in the cabinets of collectors. However, for the most part these interesting objects, along with the remains of invertebrates and vertebrates, were considered sports of nature as late as the eighteenth century (recall the story of Johann Beringer described in Chapter 4). Scientific study of plant fossils began in the early 1800s, and in 1818 the Reverend Henry Steinhauer published the first descriptions of fossil plants in which binomial nomenclature was used to name the specimens. If you recall our discussion of taxonomy, binomial nomenclature was used for animals by Linnaeus in the 1750s—almost 60 years earlier. Evidently the taxonomic study of plant fossils lagged behind study of other branches of paleontology.

Beginning in the 1820s the study of paleobotany advanced rapidly. Outstanding for his early contributions was Adolphe Brongniart, who combined understanding of botanical classification with thorough knowledge of the distribution of living plant taxa. Although recognizing the great differences between living and fossil floras of Europe, he was a catastrophist and applied Cuvier's hypothesis of sudden extinctions to explain the abrupt changes in flora. To his credit, Brongniart also recognized that knowledge of the environmental conditions affecting living plants could be used to interpret paleoenvironmental conditions that determined the distribution of fossil taxa and that plants could be reliable indicators of ancient climatic conditions. As paleobotanist Erling Dorf of Princeton University has pointed out, Brongniart can be rightfully considered the "Father of Paleobotany" for these early insights into the study of plant fossils.

Many early reports of plant fossils involved specimens recovered from strata associated with coal deposits of Carboniferous age. This is not surprising, for coal, which supplied fuel for the industrial revolution, has been a major source of energy for over 200 years. Interest in coal deposits and their formation fostered many studies of the associated plant fossils. In North America the most important deposits containing coal-bearing strata are Pennsylvanian rocks extending in a band parallel to the Appalachian Mountains and Cretaceous rocks in the Rocky Mountain and Colorado Plateau areas.

Less important from an economic standpoint but of major scientific interest and aesthetic beauty are plant fossils found in the Chinle Formation in the Petrified Forest National Park. The Chinle, of Late Triassic age, is exposed widely in the Colorado Plateau. These distinctive rocks consist of variable and often beautifully colored shales, sandstones, and siltstones. Those of you visiting Petrified Forest or Painted Desert

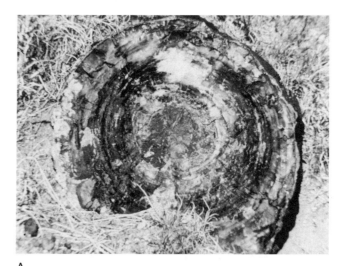

A

B

Figure 14-1. Petrified wood. A. Completely silicified log from the Chinle Formation at Petrified Forest National Park, Arizona. B. Thin section showing well-preserved internal cellular structure of silicified wood.
(A, photo by R. Miller; B, photo courtesy of Critter Creations, Inc.)

in Arizona and Capital Reef National Park in Utah (Fig. 12-34) as well as many other locations can observe this well-known formation.

The Chinle has been the object of study by many geologists since the 1850s, but was first formally named by H. E. Gregory in 1917. Over the years, many changes in formal status of these rocks were proposed by geologists working in the region, providing an excellent illustration of stratigraphic methodology and principles. A discussion of these changes and of other studies involving the Chinle is provided by *Investigations in the Triassic Chinle Formation,* published by the Museum of Northern Arizona.

Scientific interest in the Chinle resulted in part from its excellent and widespread exposures. More important was evidence indicating that these rocks were deposited in a variety of paleoenvironments— ponds, lowland swamps and floodplains— and that ancient paleoclimates were warm and humid and gradually changed through time, eventually becoming more arid. These conditions combined to produce the abundant red beds of the formation and helped preserve more than 20 species of molluscs and arthropods, 16 genera of fish, amphibians, and reptiles, and, most impressively, over 50 species of plants. As discussed in Chapter 12, the Chinle can be correlated with other Triassic rocks in the United States, allowing accurate reconstruction of the paleogeography of that time.

Most plant fossils preserved in the Chinle occur as impressions of leaves, stems, and reproductive parts. The most spectacular remains are the petrified stems, some of which are quite large. Indians had discovered the remains of these fossils, as evidenced from arrowheads and axes fashioned from the silicified logs, but the first written report of these spectacular remains was that of Army Lieutenant James H. Simpson in 1850. Numerous subsequent studies indicated that the most commonly preserved logs are conifers and that they consist of agate, jasper, and chalcedony, which are microcrystalline varieties of quartz. Replacement of the buried logs occurred very slowly by action of groundwater containing siliceous solutions and minute amounts of impurities that became incorporated into the mineral crystal structure and imparted beautiful colors. Replacement was so precise that thin sections of some of the logs show very well-preserved internal cellular structure of the original wood (Fig. 14-1).

By the 1890s many exposures of petrified wood were known from areas in northern Arizona and vast amounts were collected by scientists and the public for study and as curios. A mill for crushing the silicified wood for use as an abrasive was built but apparently never went into operation. Concern that the fossils would soon disappear prompted the Arizona Territorial Legislature (Arizona was not yet a state) to petition Congress to preserve the area. Lester F. Ward of the U.S. Geological Survey examined the region and wrote a report in 1900 proposing that the area be preserved as a national park. In December 1906 President Theodore Roosevelt declared the region a national monument; in 1962 more area was added and President John Kennedy declared the entire location a national park.

THE WORLD OF PLANTS

Characteristics of Plants

Our goals in this chapter are fourfold: (1) to describe the basic classification of divisions within the plant kingdom, (2) to discuss the origin and subsequent evolutionary history of plants, focusing on highlights of various groups, (3) to provide an introduction to the study of microscopic plants, and (4) to consider the geologically useful applications of fossil plants.

A definition provides the appropriate entry into our subject. In Chapter 5, plants are listed as multicellular organisms that manufacture food from inorganic materials. A more generalized definition of plant in the *Glossary of Geology* states: "An organism generally capable of manufacturing food from inorganic substances by photosynthesis." This definition would allow us to include as plants organisms currently classified in the kingdom Monera as well as in the kingdom Plantae.

However, monerans lack a cell nucleus and are single-celled organisms considerably distinct from true plants. For our purpose we will modify the glossary definition and restrict the term plant to single-celled or multicellular organisms containing a cell nucleus and capable of photosynthesis. Thus we are primarily interested in the origin and history of the kingdom Plantae and especially of those plants known as **vascular plants.**

Land surfaces of the Earth are covered by a myriad of vascular plants, and at least some species are found growing in all but the most extreme environments. Examples of harsh conditions that inhibit or prevent plant growth are the salt pan in Death Valley, the shifting sand dunes in parts of the Sahara and other deserts, and the ice cap of the Antarctic; in most other places there exists a profusion of green plants. In today's world, plants provide beauty but, more importantly, play a critical role in the biosphere, hydrosphere, and atmosphere.

Plants are the basic source of food for most terrestrial animals; they produce a significant amount of at-

A   B   C

Figure 14-2. Various types of vegetation which reflect prevailing climatic conditions. A. Various cactus species, characteristic of arid climates of high desert areas of the Southwest. B. Lush vegetation characteristic of tropical humid climates such as on Puerto Rico. C. Various evergreen conifers, characteristic of cool temperate climates.
(A, C, photos courtesy of R. Miller; B, photo courtesy of Fred Sundberg)

mospheric oxygen as a byproduct of photosynthesis; and their woody tissues provide a source of heat energy in many parts of the world. Remains of ancient plants that have been altered to various kinds of coal also provide an important energy source. In fact, as supplies of oil and gas become depleted, coal may again become a major source of energy, as it was in the 1800s and early 1900s.

Distribution of plants on the earth's land areas is controlled by conditions such as type of soil and overall climate. Compare, for example, the various types of plant species that exist in arid regions, such as deserts, with those that exist in tropical rain forests, in redwood forests, or on the Arctic tundra (Fig. 14–2). Differences in overall morphology—size, shape and size of leaves, type of bark, and other features—of plants living in different environments have been produced by natural selection operating on successive generations. Important controlling factors are the differences in climatic conditions that exist in different regions.

Information obtained from observations and studies of the characteristics and distribution of modern land plants provides important guides for interpretation of ancient plant fossils and also illustrates actualism. For example, the fossil record provides evidence that changes in plants may have influenced evolutionary adaptive radiations and extinction of animals (Chapter 7). Also, interpretation of distribution of ancient climatic zones, based on distributions of plant fossils, provides useful information to support the concept of plate tectonics (Chapter 1).

Morphology of Plants

Vascular plants are called **tracheophytes;** they occur in a wide variety of shapes and forms and can be recognized on the basis of a number of morphologic features they have in common. Except for some of the algal groups and the most primitive multicellular taxa, all higher plants contain a vascular system, some form

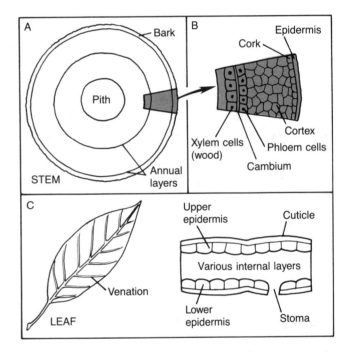

Figure 14-3. External and internal vascular plant structures. A. Stem cross section of two-year-old plant. B. Enlarged portion of stem showing various layers. C. Leaf and enlarged cross section; internal layers include vascular system and cells where photosynthesis occurs.

of root system, leaves, a cuticle epidermis, and woody tissues. Later in this chapter we will trace the evolution and development of these morphologic structures and describe how they relate to the successful colonization of land. Interestingly, representatives of some of the most primitive vascular plant groups survive, in some cases lacking one or more of these structures. These living fossils provide important clues that help paleobotanists unravel the details of plant evolution.

The vascular system of plants consists of internal tubelike structures made of cells called **xylem** and **phloem** that transport water and nutrients to various parts of the plant (Fig. 14–3). Root systems are basically

Table 14-1 Classification of subkingdom Thallophyta

Division	Common name	Geologic range
Chlorophycophyta	Green algae	Pre-Cambrian–Holocene
Rhodophycophyta	Red algae	Pre-Cambrian–Holocene
Phaeophycophyta	Brown algae (seaweeds)	Pre-Cambrian–Holocene
Chrysophycophyta	Diatoms	Jurassic(?)–Holocene
Euglenophycophyta	Englenids	Cretaceous–Holocene
Charophycophyta	Charophytes	Silurian–Holocene

a continuation of the stem. In most plants roots grow in soil, anchor the plant, and absorb water and nutrients from the soil. Leaves are the sites of food production; photosynthesis occurs there in **chloroplasts.** Small pores known as **stoma,** which allow for exchange of carbon dioxide, oxygen, and water vapor with the atmosphere, are found on the lower sides of leaves. The outer layers, especially epidermis and cork, provide protection from harmful parasites and excess ultraviolet radiation. In some trees, such as redwoods, a thick, well-developed epidermis also provides protection against fire and insect predators. Woody tissue forms part of the interior of the stem and supports the plant.

The internal structure of a plant stem is shown in Figure 14–3 A, B. The major parts are the central **pith,** woody tissue, and outer **epidermal layers** known as bark. Woody tissue consists primarily of xylem cells: dead cells that transport water from roots to other parts of the plant. The epidermis consists of a number of layers. The innermost of these consists of **cambium** cells and is the locus for plant growth. The next layer consists of living phloem cells, which transport nutrients. External to the phloem are the **cortex,** cork, and epidermal layers. Phloem, cortex, cork, and epidermis comprise what is commonly called the bark. Differences in thickness and structure of these layers are useful criteria for classification of various plant taxa.

Classification of Plants

With this brief introduction to plant morphology we can now consider classification of plants—not a simple task because of many conflicting concepts among botanists and paleobotanists. We will use a modified system combining some of the ideas of botanist Harold C. Bold and paleobotanists Henry N. Andrews and Thomas N. Taylor. In our classification, the plant kingdom is divided into two subkingdoms. The first, known as **Thallophyta** (see Appendix B), consists of unicellular or multicellular aquatic forms that lack true root systems, woody stems, or leaves. Thallophyte divisions are green, red, and brown algae, diatoms, euglenids, and charophytes (Table 14-1). These plants are classified on the basis of differences in the type of chlorophyll molecule

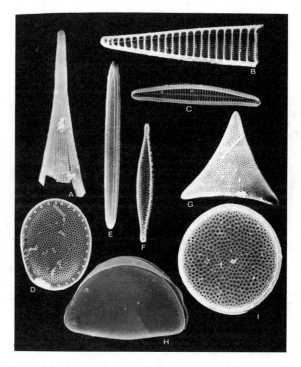

Figure 14-4. Scanning electron photomicrograph of marine diatoms.
(Reprinted by permission of the publisher from B. U. Haq and A. Boersma, *Introduction to Marine Micropaleontology*, Chap. 10, Fig. 3, p. 247. Copyright 1978 by Elsevier Science Publishing Co., Inc., New York)

and of other pigments found within cells and differences in other internal cellular structures.

Some thallophyte divisions are of major importance in the fossil record. For example, the diatoms (chrysophytes) (Fig. 14-4) have played an important role as **primary producers** in the oceans since Cretaceous time. During Cenozoic time they were important rock formers; diatomite is a rock consisting almost entirely of diatom remains. Diatomites are economically valuable as material for filters. The red algae (rhodophytes) and some green algae (chlorophytes) secrete calcareous walls and have been important contributors to wave-resistant organic structures such as reefs (Fig. 14-5) throughout Phanerozoic time. The mechanical breakdown of the calcareous wall structures of red and green algae provides considerable fine-grained calcareous ma-

Table 14-2   Classification of subkingdom Embryophyta

Division	Common name	Geologic range	Extant species
Seedless plants			
Rhyniophyta	—	Silvrian–Devonian	0
Zosterophyllophyta	—	Silvrian–Devonian	0
Microphyllophyta	Club mosses	Devonian–Holocene	1,100
Arthrophyta	Horsetails	Devonian–Holocene	20
Pteridophyta	Ferns	Devonian–Holocene	10,000
Seed-bearing plants			
Pteridospermophyta	Seed ferns	Devonian–Jurassic	0
Coniferophyta	Conifers	(?)Mississippian–Holocene	550
Cycadophyta	Sago palm	(?)Mississippian–Holocene	100
Ginkgophyta	Maidenhair tree	(?)Pennsylvanian–Holocene	1
Gnetophyta	—	(?)Permian–Holocene	70
Flowering plants			
Anthophyta	Flowering plants	(?)Jurassic–Holocene	300,000

Figure 14-5.   Calcareous reef-forming algae.
(Photo by R. Miller)

terial for the formation of limy sediments. These organically produced sediments may become lithified into fine-grained limestone, known as lime mudstone, which is found abundantly in all systems of the Phanerozoic. Today lime muds are accumulating in areas such as southwestern Florida and the Bahamas. Brown algae (phaeophytes), commonly known as kelp or seaweeds, attach to the bottom with structures called holdfasts. Some may grow to over 30 m in height and form forests in shallow shelf seas, which provide important habitats for many other marine-dwelling organisms. Euglenids and charophytes are found primarily in freshwater environments; the calcareous reproductive parts of charophytes are useful for correlation of Mesozoic and Cenozoic lake deposits.

The second subkingdom, **Embryophyta,** includes all terrestrial plants. Except for the less complex molds, liverworts, and mosses, all embryophytes have a vascular system. They are generally termed tracheophytes and

can be assigned to thirteen divisions. Table 14-2 provides a reference for our subsequent discussion of the evolutionary history of these land plants.

FOSSIL RECORD OF THE PLANT KINGDOM

Preservation

Concepts of evolutionary history and classification of land plants are dependent upon knowledge of living plants and interpretation of specimens recovered from the fossil record. Thus an understanding of methods and types of plant preservation is important. Before proceeding with our historical journey we will take a brief look at aspects of plant preservation.

Because they live mainly in terrestrial environments, many of which are located away from areas of sediment deposition, vascular plants are not preserved in the fossil record as commonly as are many invertebrate phyla. A further problem is the tendency for various parts of the plant to become separated at death and preserved in different places. Thus it is common that the seeds, leaves, roots, and stems of a single plant species are not found together. There are examples in paleobotanical taxonomy where separate specimens of fossilized roots, leaves, and stems—each assigned a different taxonomic name—were at a later date found together and determined to belong to the same species. The life of a taxonomist is sometimes difficult!

Plants are preserved as carbon films **(carbonization),** impressions, and mineralized replacements and as microfossils. A common type of plant preservation is that of carbonization. When volatile organic compounds of the plant—nitrogen, hydrogen, and oxygen—are driven off by heat and pressure after burial, carbon is left as a residue—a thin film of carbonized stems or leaves on bedding surfaces of rocks (Fig. 14-6). This

Figure 14-6. Carbonized remains of a leaf on shale of Carboniferous age.
(Photo by R. Miller)

Figure 14-7. Bituminous coal (left) and anthracite coal (right).
(Photo by R. Miller)

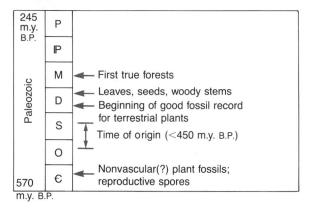

Figure 14-8. Major events in the early history of land plants. Appearance of flowering plants in the early Mesozoic is not shown.

slow process of carbonization can also transform plant remains into coal (Fig. 14-7).

Increasing heat and pressure →

Plant remains → Peat → Lignite → Bituminous coal (soft coal) → Anthracite coal (hard coal)

A second type of plant preservation is leaf or flower impressions, which usually form in fine-grained sedimentary rocks such as siltstone or shale or in volcanic ash beds deposited in lake or swamp environments. The leaves are compressed by the weight of overlying sediments and the structures are impressed in the sediments. When the sediments lithify the impression is preserved. These impressions may preserve very fine details of the leaf structures, but the fossil itself is two-dimensional.

Woody tissue of stems may become buried and subsequently completely mineralized by groundwater. The end result of this process is petrified wood, commonly consisting of microcrystalline quartz. If replacement occurs very slowly, molecule by molecule, the original cellular structure and other features of the wood may be preserved almost perfectly and can be studied in thin section (Fig. 14-1B). The beautiful colors of some petrified wood, such as the conifer trunks and logs in the Petrified Forest of Arizona, resulted from very small amounts of such elements as iron or copper in the groundwater. These elements were incorporated into the siliceous minerals that replaced the wood. Petrified wood is an example of a fossil that is beautiful and also very useful to paleobotanists because the plant structures provide taxonomic information and indicate the paleoclimatic conditions in which the organisms lived.

Microscopic-sized spores and pollen are also preserved in fine-grained sedimentary rocks (Fig. 14-24). These tiny particles commonly were wind-borne or transported by rivers out to sea, where they were buried in marine sediments. Some specimens were deposited hundreds of kilometers from land. Because the original organic outer layer was durable, these microfossils may be recovered as part of the undissolved residue remaining after acid treatment of the rocks. Spores and pollen also may be studied in thin section. Identification of species and inference of ancient climatic conditions can be made by studying the abundance and diversity of these microfossils and, as we will see, may provide clues to unravelling the early evolutionary history of tracheophytes.

Origins and Evolutionary Highlights

The fossil record of vascular plants records a number of important events (Fig. 14-8). The history of these plants extends back only about 450 million years, less than one-tenth of the total time of earth history and about 200 million years less than the first known metazoan (animal) fossils. The fossil record presently provides well-documented evidence of terrestrial plants that existed in latest Silurian time. This interpretation is based on evidence of body fossils containing preserved vascular systems of xylem and phloem cells and on reports of characteristic **spores.** The actual origin of vascular plants occurred earlier in time, as suggested by some questionable Cambrian vascular plant fossils and more plausible reports of Late Ordovician to Early Silurian spores. The Cambrian fossils do not appear to have vascular systems and have not been accepted by most paleobotanists as truly vascular in nature. Very recent reports describing the size and structural complexity of some Ordovician spores indicate they very probably were formed by vascular plants.

Although the fossil record indicates that vascular plants existed as early as Ordovician or possibly Cambrian time, only limited evidence points to their ancestors. So far this evidence indicates that the most likely available ancestors were the algal groups. For a number of reasons most paleobotanists believe that green algae were ancestral to vascular plants. First, the fossil record of this division extends back into the Ediacarian and possibly even earlier in time. The group had a long evolutionary history prior to the first appearance of a vascular plant and was therefore an available ancestral group. Second, green algae live in freshwater environments as well as in the oceans, permitting a much easier transition to terrestrial existence than would be possible for a similar group or groups that existed only in the marine realm. Third, the type of chlorophyll molecule and other biochemical characteristics of green algae are relatively close to those of vascular plants. Fourth, the life cycle and reproductive features of green algae are similar to those of the more primitive vascular plants.

On the basis of these similar characteristics and of available fossil evidence of early vascular plants, a simplified chart of plant evolution can be constructed (Fig. 14-9). Evidently one or more taxa of green algae made the transition to land by Late Ordovician and had evolved into primitive vascular plants. One of these early plant groups was the Rhyniophyta. This division is considered to be ancestral to most of the other vascular plant divisions. The other major group of early plants is represented today only by the club mosses.

Once they established a grip on land, primitive vascular plants began to fill available terrestrial niches at an ever-increasing pace. Between Late Silurian and Mississippian time, a span of approximately 75 million years, vascular plants developed many of their characteristic features: woody tissues, complex root structures, leaves, and seeds. Evolutionary history of vertebrate animals indicates a similar pattern. Colonization proceeded slowly as amphibians made the transition from water to land. Through Devonian and Carboniferous time many mor-

phologic and physiologic features evolved, allowing these organisms to cope with requirements for living and reproducing on land. The record of the first evolutionary radiation of land plants into previously unfilled terrestrial niches serves as a good example of adaptive radiation (Chapter 7).

Although they may have evolved earlier, root systems are first found in Lower Devonian rocks; leaves are known from Middle Devonian and seeds from Upper Devonian rocks. By the end of the Devonian Period plants had undergone an extensive evolutionary radiation onto land. Many diverse groups had appeared, and the first true forests developed. Evidence for the existence of forests is provided by well-preserved floras in terrestrial deposits of the Catskill clastic wedge near Gilboa, New York (Chapter 11). Similar floras found in what are now Belgium, Scotland, the Russian platform, and other areas, suggest that similar paleoclimatic conditions existed throughout these regions, perhaps because of their close proximity in Devonian time. Much of our knowledge of Devonian plants has come from the painstaking work of Professor Suzanne Leclercq of the University of Liege, Belgium. Her career as a paleobotanist has spanned 50 years.

The evolutionary radiation of post-Devonian plants was brought about by an advance in reproductive capabilities in the form of seeds. This event helped plants colonize highland environments as well as lowland swamps and fostered the development of widespread forests. The Carboniferous Period was the heyday of the development of these forests. Evidence of their exten-

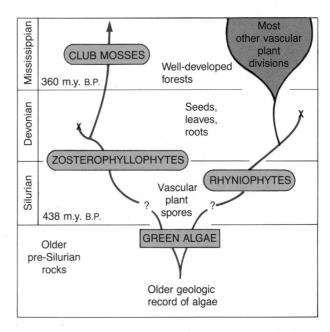

Figure 14-9. Origin and early evolutionary history of vascular land plants; includes some of the important morphologic features shown in Figure 14-8. Appearance of seeds was an important prerequisite to late Paleozoic adaptive radiation.

sive distribution is provided by major accumulations of coal in what are now the central and eastern United States and western Europe. During late Paleozoic time these areas were parts of ancient Laurasia. The coal-bearing rocks are parts of unusual sedimentary packages known as cyclothems (Chapter 11).

A major evolutionary event occurred in Mesozoic time, possibly during the Jurassic Period. This was the appearance of the first flowering plants, which subsequently became the dominant vascular plant group and have provided the base of the food chain for most terrestrial communities in late Mesozoic and Cenozoic time. Flowering plants underwent an adaptive radiation in the Cretaceous. Later, during the medial Cenozoic, many of these older taxa were replaced by new taxa such as grasses, shrubs, vegetables, and hardwood trees, many of which are common today. These changes in plants may have initiated the changes in mammals recorded in Cenozoic sedimentary rocks (Chapters 13 and 15).

EVOLUTIONARY HISTORY OF MAJOR PLANT TAXA

Seedless Vascular Plants

At this point we can trace a more detailed history of the plant divisions listed in Table 14-2. There are six divisions that produce spores as part of their reproductive cycle; they can be conveniently termed seedless vas-

cular plants. The oldest of these are the rhyniophytes and the zosterophyllophytes.

Rhyniophyte remains were first discovered about 60 years ago in the Middle Devonian Old Red Sandstone in Scotland (see Chapter 11). Because of excellent preservation of these fossils within silicified layers of the Rhynie Chert, internal structures such as xylem and phloem cells were recognized. The evolutionary significance of these plants was apparent. Fossilized specimens indicate that these plants lacked well-developed root structures and had no leaves. Reproductive organs were located at the tips of the stems and are called **terminal sporangia;** these sporangia were the sites of spore production. No seeds have been found associated with the plants. Modern reconstructions indicate that none of these plants were more than half a meter tall (Fig. 14-10).

The second group, zosterophyllophytes (Fig. 14-11), is represented by fossils found in Lower and Middle Devonian rocks. These plants also lacked true roots or leaves but had branching stems with the sporangia located along the axes of the stems rather than at the tips. This division is considered to be ancestral to the club mosses, which are known as Microphyllophyta or Lycopsida in some classifications. Club mosses have poorly developed roots and distinct stems and leaves. During the Carboniferous Period some genera attained heights

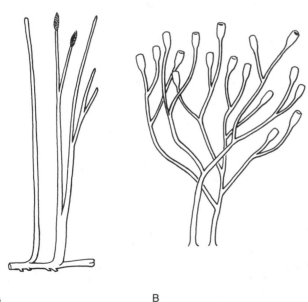

A   B

Figure 14-10.   Devonian vascular land plants. A. *Rhynia,* simple reconstruction. B. *Cooksonia,* simple reconstruction. Note lack of root structures, simple thin stems, and terminal sporangia. Width of stems 1.5 to 2.0 mm.

(A, from H. N. Andrews, Jr. 1966. Studies in Paleobotany. Copyright © 1966 by John Wiley & Sons, Inc. New York. Reprinted by permission of John Wiley & Sons, Inc.; B from D. Edwards, 1970, Fertile Rhyniophitina from Britain, Fig. 4b, p. 459: *Palaeontology,* vol. 13, pt. 3)

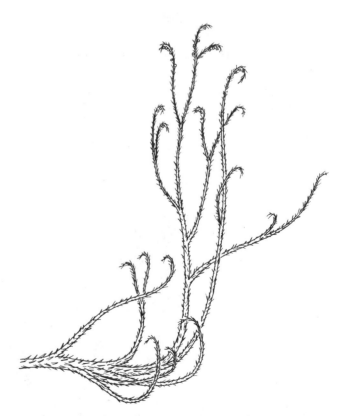

Figure 14-11.   Reconstruction of a zosterophyllophyte, *Sawdonia,* approximately 30 cm tall.

(After H. N. Andrews, 1974, Paleobotany 1947–1972, p. 186: *Annals Missouri Botanical Gardens,* vol. 61)

of up to 40 m with trunk diameters of 1 to 2 m (Fig. 14-12).

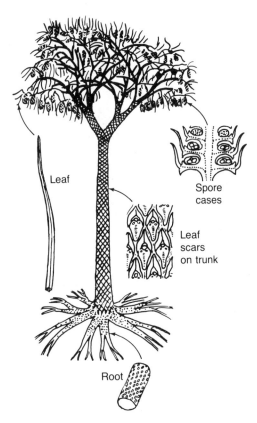

Figure 14-12. Reconstruction of the microphyllophyte (club moss) known as *Lepidodendron,* a common "scale tree" of Carboniferous age.
(From L. W. Mintz, *Historical Geology: The History of a Changing Earth,* Fig. 7.122, p. 183. © 1977 by Charles E. Merrill Publishing Co., Columbus OH)

It is interesting to speculate on the appearance of the earth's land surfaces in Late Silurian and Early Devonian time (Fig. 14-13). The only existing vascular plants were apparently rhyniophytes and zosterophyllophytes, and none of these were much over one-half meter in height. This early vegetation certainly provides a sharp contrast to today's vegetation. If we go back further in time, previous to the Late Ordovician there is no record of land plants at all, although nonvascular plants such as bryophytes and mosses probably existed. It is difficult to visualize our world without abundant plants, and certainly this lack of plant cover would have affected chemical weathering and rates of erosion on land surfaces. The lack of vascular plants as producer organisms at the base of terrestrial food chains (Chapter 7) would have placed severe constraints on the existence of other land-dwelling organisms. In fact, the first land-dwelling vertebrates did not evolve until the Devonian Period—after vascular land plants had become established.

Although most of the other plant divisions appeared during the Devonian Period, comparison of anatomical features among these groups suggests to most paleobotanists that the rhyniophytes were ancestral. The evolutionary relationships among these plant groups are diagrammed in Figure 14-14 and are discussed later.

The arthrophytes (Fig. 14-15) also first appeared in Devonian rocks. Fossil and living specimens have stems with horizontal joints and may have leaves in whorls at nodes. These plants lack well-developed root systems and produce spores which develop in sporangia located on upright stems. Some Paleozoic arthrophyte species were large trees and were common in Carboniferous forests. Today only a single genus—*Equisetum*—survives; it grows in moist environments.

Figure 14-13. Reconstruction of a Devonian landscape showing rhyniophytes dwelling in shallow-water swampy areas.
(From L. W. Mintz, *Historical Geology: The History of a Changing Earth,* Fig. 17.60, p. 428. © 1977 by Charles E. Merrill Publishing Co., Columbus OH)

The last division of seedless vascular plants, commonly known as ferns, is the most common of the six divisions today. Ferns have true root systems and well-developed distinctive leaves. In their reproductive cycle ferns go through a **sporophyte** and **gametophyte** generation (Fig. 14-16). Sporophyte adults produce haploid spores and are dominant in the life cycle in ferns and in other seedless vascular plants. To most of us, ferns are recognizable by their relatively distinct leaf shape and by the development of spores on the underside of the leaves. Ferns flourished in the widespread, warm, moist, swampy environments of the late Paleozoic but today they are more restricted. Some living species of tree ferns in Australia, New Zealand, and tropical areas may grow to a few tens of meters in height (Fig. 14-17).

Seed-Bearing Vascular Plants

Except for the division Anthophyta—the flowering plants—the remaining plant divisions can be described loosely as **gymnosperms** or, more commonly, as naked-seed plants. These divisions include seed ferns, confiers, cycads, ginkgos, and gnetophytes. Seed ferns were the dominant taxonomic group in the Carboniferous Period. It was originally assumed that these plants were the remains of true ferns, but discovery of seed-bearing fronds revealed their gymnosperm affinity. This division represents the most primitive seed-bearing vascular plants. These plants were abundant and diverse, but they became extinct in the Jurassic and were replaced by more advanced plants. Many of the well-preserved impressions in ironstone concretions of the famous Pennsylvanian Mazon Creek flora in Illinois represent leaves of these seed ferns (Fig. 14-18).

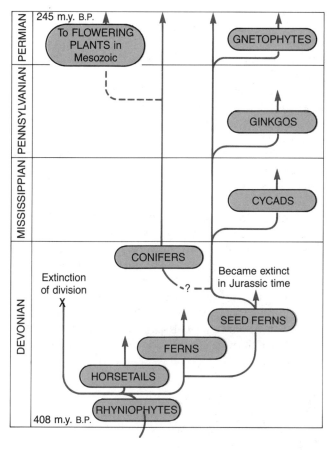

Figure 14-14. Evolutionary history of many vascular plant divisions; much of this information is speculative. Note that the flowering plants first appear in Triassic(?) or Jurassic rocks. Arrows indicate continuing geologic history.

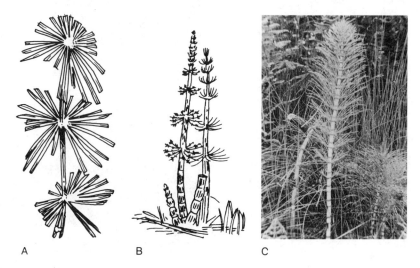

Figure 14-15. Arthrophytes. A. Leaf impressions of *Annularia;* Carboniferous. B. Reconstruction of *Calamites,* showing horizontal joints and round scars where smaller branches attached; Carboniferous. C. Modern *Equisetum* showing horizontal joints and sporangia.
(A, B from L. W. Mintz, *Historical Geology: The History of a Changing Earth,* Figs. 7.123, 7.125, p. 184. © 1977 by Charles E. Merrill Publishing Co., Columbus, OH: C, photo courtesy of Critter Creations, Inc.)

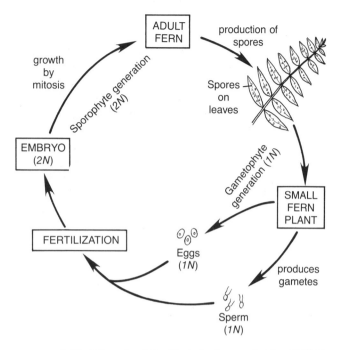

Distinctive seed ferns have been found in Permian and Triassic rocks of Australia, South Africa, South America, Antarctica, and India. These plants, dominated by the genus *Glossopteris* (Fig. 14-18C), had lanceolate leaves, were deciduous, and had **growth rings.** Such characteristics indicate the plants grew in temperate, or perhaps cooler, climates with distinct seasonal climatic changes. Other evidence, such as the presence of tillites, suggests these land areas—the southern megacontinent of Gondwanaland—were glaciated during the late Paleozoic. The *Glossopteris* flora seems to support this hypothesis.

Conifers are the most abundant gymnosperms today and are represented by about 550 species. They have well-developed root systems, woody stems, leaves, and in some cases, seed-bearing cones. Interestingly, modern conifers are most abundant in relatively harsh climates: very cold or dry regions or areas having poor soils. In the fossil record, however, they are found in deposits that apparently formed in milder climates. This suggests that they may have been displaced from many habitats by the flowering plants. Included in this group are evergreen trees such as pine, fir, juniper, and spruce (Fig. 14-19). Some unusual species such as the bristlecone pine, California coastal redwood, and giant sequoia are also conifers.

Figure 14-16 Life cycle of Pteridophyta. Ferns and other seedless vascular plants have an adult stage (sporophyte) that is diploid (2N) and produces spores. These grow and become a small gametophyte generation that undergoes meiosis and produces eggs and sperm (each haploid = 1N). These may unite (fertilization) to produce a zygote (diploid) which grows into a sporophyte adult stage.

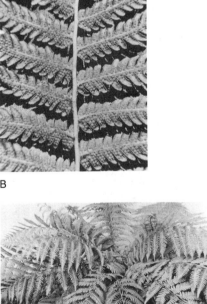

Figure 14-17. Ferns. A. Reconstruction of a tree fern *Psaronius,* about 8 m tall; Carboniferous. B. Fossilized fern leaves. C. Modern New Zealand tree fern, which may grow to heights of 10 to 15 m. (Morphology and Anatomy of American Species of *Psaronius,'* Fig. 1-108: from J. Morgan, 1959, Illinois Biol. Mon. 27, University of Illinois Press, Champaign IL. B, C, photos courtesy of Critter Creations, Inc.)

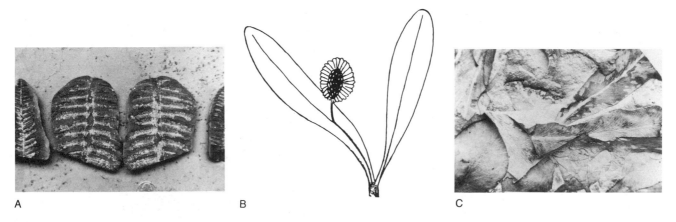

Figure 14-18. Seed ferns. A. Fossil leaves in ironstone concretions from Mazon Creek (Francis Creek Shale), Illinois, of Carboniferous age. B. Leaf with seed (arrow). C. *Glossopteris* leaves of Permian age from the Southern Hemisphere.
(A, C, photos courtesy of Critter Creations, Inc.; B from L. W. Mintz, *Historical Geology: The History of a Changing Earth,* Fig. 7.128, p. 185. © 1977 by Charles E. Merrill Publishing Co., Columbus OH)

Figure 14-19. Conifers. A, B. *Sequoia* (redwood tree) trunks in growth position and showing foliage and seed cone; Oligocene, Florissant Fossil Beds National Monument, Colorado. C. Living specimens of the coast redwood near Eureka, California. D, E. *Pinus,* living specimen and needle-shaped leaves. F. *Araucarioxylon* log; Triassic, Chinle Formation, Petrified Forest National Park, Arizona. G. *Araucaria* (star pine), living specimen about 2 m tall.
(A, D, E–G, photos courtesy of Critter Creations, Inc.; B, photo courtesy U.S. National Park Service; C, photo courtesy Fred Sundberg)

Cycads are short plants with a stem topped by a crown of leaves. They resemble pineapples with leaves when small and palm trees when larger (Fig. 14-20). In fact, their common name is sago palm, although true palms are flowering plants. Cycads were one of the dominant divisions during the Mesozoic and are thus

A   B

Figure 14-20. Cycads. A. *Cycadeoidea,* a reconstruction of Mesozoic cycadlike plant. B. Living cycad, *Cycas,* about 1 m tall.
(A from L. W. Mintz, *Historical Geology: The History of a Changing Earth,* Fig. 7.129B, p. 186. © 1977 by Charles E. Merrill Publishing Co., Columbus OH; B, photo courtesy of Critter Creations, Inc.)

A   B

Figure 14-21. *Ginkgo.* A. Living tree. B. Distinctive leaves.
(Photos courtesy of Critter Creations, Inc.)

associated with the giant ruling reptiles discussed in Chapter 12. Today about 100 species survive, mostly in tropical environments.

The ginkgos are represented by only one living species, *Ginkgo biloba,* commonly known as the maidenhair tree (Fig. 14-21). This species had nearly become extinct, but living specimens were discovered in China and have been spread around the world by humans. The species truly deserves the term living fossil and, along with the cycads, is one of the most primitive gymnosperms.

The Gnetophyta consist of three poorly known genera (Fig. 14-22). They tend to live in environments not conducive to preservation. For example, the genus *Ephedra* exists in arid to semiarid climates; thus the fossil record for the group is poorly known. However, pollen grains recovered from marine and lacustrine sediments and from sedimentary rocks indicate a geologic range extending back at least to the Permian Period. Most living and extinct species are relatively small.

Flowering Plants

The final division we will consider is the Anthrophyta, more commonly known as the **angiosperms,** or flowering plants. These organisms are the most advanced terrestrial plants and are characterized by flowers and seed-bearing fruits. Today they make up about 95 percent of living terrestrial plant species. Their brightly colored and often scented flowers attract birds and insects that transport pollen and fertilize the seeds. The seeds are covered or enclosed and some develop a fruit covering; various animals and birds eat the fruit covering and excrete the undigested seeds. Angiosperm pollen and seeds also can be transported by wind and water; the phenomenon of wind transport is well known to hay fever sufferers at certain times of the year! These reproductive methods have played an important role in allowing the rapid evolutionary radiation of the angiosperms.

The two main groups of angiosperms—**dicotyledons** (dicots) and **monocotyledons** (monocots)—are easily distinguished on the basis of differences in leaf venation (Fig. 14-23). Dicots are considered to be the oldest group. The exact time of origin of angiosperms is not yet known, but discoveries from the fossil record suggest they evolved from earlier seed plants in the Jurassic Period. In Cretaceous time they underwent an evolutionary radiation and by the end of the Cretaceous had become the most abundant and diverse land-dwelling plants. In the medial Cenozoic they underwent radiation and earlier taxa were replaced by groups more similar to those with which we are familiar.

Evolutionary changes in angiosperms during Cenozoic time were related to changing climatic conditions such as global cooling and increasing aridity. These changes, which began in the Oligocene Epoch, resulted in a gradual decrease in abundance of forests containing woody angiosperms (dicots) and an ecological replacement by grasses and shrubs, most of which are monocots. By Miocene and Pliocene time extensive grasslands had developed in areas that are now the prairies of the Great Plains. These climatic changes and resultant redistribution of plant taxa greatly affected the mammals. The fossil record indicates that major changes in this class of vertebrates occurred during Miocene time, with the extinction of many early Cenozoic browsing forms and the appearance of more agile grazing taxa (Chapter 15).

In this description of the various major taxonomic groups of plants we have noted the successive appearance of features that have allowed vascular plants to become progressively better adapted to a terrestrial mode of life. Of paramount importance was the first colonization of land by the seedless vascular plants in Silurian and Devonian time. Evolution of seeds allowed expansion into many new niches in late Paleozoic and Meso-

Figure 14-22. Living gnetophyte, *Ephedra,* from California hgh desert.
(Photo courtesy of Critter Creations, Inc.)

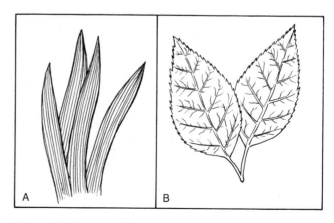

Figure 14-23. A. Parallel venation of monocot leaf. B. Radiating venation of dicot leaf.

zoic time. A significantly more successful reproductive feature was the evolution of protected seeds and flowers in the Mesozoic and subsequent radiation of flowering plants in the Cretaceous Period and then in the Cenozoic. These progressive reproductive and other morphologic features provide the basis for classification of plant taxa. Parallel types of changes can be seen in the development of land-dwelling vertebrates; it is evident that the evolutionary histories of terrestrial plants and vertebrates are closely related.

SPECIAL APPLICATIONS OF PALEOBOTANY: SPORES, POLLEN, AND TREE RINGS

Palynology

A somewhat different approach to the study of plants is the investigation of their reproductive parts, including **spores** and **pollen.** This subdiscipline of botany, called **palynology,** is a relatively new area of study. The term palynology was proposed as recently as 1944, although a few pioneers were publishing reports on plant microfossils in the late 1800s and early 1900s. Impetus for palynological studies was provided by the petroleum industry; plant microfossils are relatively easily obtained from surface outcrops or drill-hole samples and, like Foraminifera and other protist or animal microfossil groups, are used to correlate rocks. Spores and pollen also provide information on the distribution and diversity of plant species even where larger fossilized remains such as stems and leaves are not preserved. This information can, in turn, be used to interpret ancient climatic conditions.

Pollen and spore fossils are most commonly obtained as insoluble residues from rocks that have been dissolved in various acids. Most are recovered from fine-grained marine sedimentary rocks such as shale, siltstone, or lime mudstone. These remains of terrestrial plants are common in marine sedimentary rocks because they are durable and can be easily transported to the oceans by wind and water. Commonly, detailed interpretations of the taxonomic composition and diversity of ancient plant communities are made on the basis of studies of the distribution and characteristics of preserved pollen and spores.

Spores are reproductive structures of algae, mosses, club mosses, horsetails, psilophytes, and ferns; the extinct rhyniophytes and zosterophyllophytes also produced spores. Some plant taxa are heterosporous and produce two different sizes of spores. The smaller sized microspore is male and the larger megaspore is female. In other plants the spores are equal in size and are considered homosporous; these spores may be bisexual or may germinate into male or female plants and produce

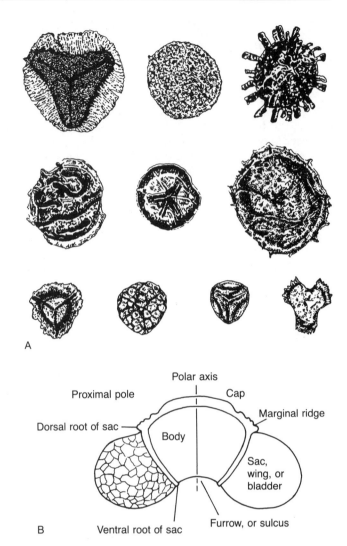

Figure 14-24. Palynological samples. A. Paleozoic spores. B. Pollen grain illustrating morphological features useful for identification. (A from H. N. Andrews, Jr., *Studies in Paleobotany,* Fig. 17-4, p. 451. Copyright © 1966 by John Wiley & Sons, Inc., New York. Reprinted by permission of John Wiley & Sons, Inc. B reprinted by permission of the publisher from B. U. Haq and A. Boersma, *Introduction to Marine Micropaleontology,* Fig. 3, p. 329. Copyright 1978 by Elsevier Science Publishing Co., Inc., New York)

gametes. The microspores in heterosporous plants evolved into the pollen grains of seed-bearing plants. Pollen is the male gamete in seed-bearing plants and in flowering plants (Fig. 14-24).

The oldest known complex spores are from Ordovician rocks. The spores predate any verified body fossils of vascular plants. Remember that the first commonly accepted tracheophyte stems are found in Upper Silurian strata. Thus these spores provide possible evidence of plant history earlier than Late Silurian. Some groups of algae and the fungi also produce complex spores, so possibly the Ordovician spores do not represent vascular plant remains. The final word on the antiquity of the first higher plants awaits further discoveries.

Dendrochronology

As noted earlier in this chapter, distribution of living plants provides valuable evidence useful for interpretation of events and conditions preserved in the fossil record. One aspect of this study is the relatively new technique—**dendrochronology**—that involves interpretation of tree rings.

The formation of new plant cells is controlled by climatic conditions: Larger thin-walled cells are added during optimum conditions in spring and early summer, and smaller, thicker walled cells are added near the end of the growing season in late summer and fall. A sharp boundary is apparent between the zone of small cells and the new larger cells added each subsequent year; these growth rings are most characteristic of plants growing in arid, temperate, or cold climates where there are distinct seasonal changes. Thus yearly growth and the climatic conditions are recorded by these annual rings.

Some giant sequoias have more than 4000 rings. Even more impressive are some specimens of bristlecone pines with over 4500 rings. These pines grow in the high peaks of the western United States. Not only do tree rings provide evidence of age, they also record climatic variations that occurred during the life of the tree. As indicated in Figure 14-25, excessively wet or dry years, or even fires, are recorded by tree rings. These events often occur on a regional scale, such as the droughts of the 1930s in the Midwest and the drought of 1976–1977 that affected many western states. Such conditions and events are recorded by tree rings, which then become a useful correlation tool. Tree rings have been used to trace prehistoric events back through the Holocene and latter part of the Pleistocene.

Dendrochronology can be coupled with radiocarbon dating to provide very detailed age determinations and correlation back several tens of thousands of years. This area of study promises to provide important evi-

Figure 14-25. Redwood stem illustrating growth rings. (Photo by R. Miller)

dence in correlation of Pleistocene and Holocene events and is useful in interpretation of climatic changes associated with glaciation and in providing age-datum lines for correlating fossil remains of humans.

SUMMARY

Nucleated single-celled or multicellular organisms using photosynthesis to obtain food are considered plants. Thallophyte divisions are commonly considered algae and are important primary producers. Such groups as diatoms, red algae, and calcareous green algae have left a relatively good fossil record and their remains may contribute to the rock record.

Embryophytes are known as vascular plants or tracheophytes; most have roots, stems, leaves, and seeds. However, early Paleozoic vascular plant fossils and a few living taxa lack one or more of these structures. Body fossils of seedless vascular plants are known from Upper Silurian rocks, and spores tentatively interpreted to be of tracheophyte origin are known from Upper Ordovician rocks. Major evolutionary advances occurred in the Devonian Period, with appearance of well-developed roots, leaves, and seeds. True forests containing seedless and seed-bearing taxa became widespread in the Carboniferous and may have provided a major impetus for the development and spread of amphibians and reptiles.

Appearance of flowering plants and their rapid diversification are highlights of plant evolution in the late

Mesozoic. Subsequent ecologic replacement during the Cenozoic with gradual reduction in hardwood forests and spread of grasslands affected the distribution of other taxa, in particular the mammals.

The distribution of land plants is controlled by climate and this has been applied to the fossil record to decipher paleoclimatic conditions. Characteristics of plant fossils—e.g., presence of absence of growth rings, shape and kind of leaves—have provided important clues for recognition of ancient tropical and cold climates, and the overall distribution of plant fossils during short intervals of geologic time is used to define ancient climatic zones. Interpretation of the displacement of paleoclimate zones on the continents has played an important role in our understanding of plate tectonics and the changes in positions of continents through time.

SUGGESTIONS FOR FURTHER READING

Andrews, H. N., 1961, *Studies in Paleobotany:* John Wiley & Sons, New York.

Banks, H. P., 1968, The Early History of Land Plants, *in* H. T. Drake, ed., *Evolution and Environment,* Yale University Press, New Haven, CT.

Bold, H. C., 1977, *The Plant Kingdom:* Prentice-Hall, Englewood Cliffs, NJ.

Breed, C. S., and Breed, W. J., eds., 1972, Investigation in the Triassic Chinle Formation: *Museum of Northern Arizona, Bull.* 47.

Dorf, E., 1964, The Petrified Forests of Yellowstone Park: *Sci. American,* vol. 210, no. 4, pp. 106-115.

Taylor, T. N., 1981, *Paleobotany:* McGraw-Hill, New York.

Tidwell, W. D., 1975, *Common Fossil Plants of Western North America:* Brigham Young University Press, Provo, UT.

15

THE MEEK INHERIT THE EARTH: Evolutionary History of the Mammals

CONTENTS

KEY TERMS

Occipital condyle
Endothermy
Monophyletic
Polyphyletic
Arboreal
Nocturnal
Molars
Omnivorous
Marsupials
Placentals
Asylum areas
Insectivores
Grazing
Browsers
Land bridges
Ungulates
Artiodactyls
Perissodactyls
Binocular vision
Omnivorous
Prehensile tail
Prosimii
Anthropoidea
Stereoscopic vision
Bipedal gait
Opposable thumb

MAMMALS AND EVOLUTIONARY THEORY

Study of the evolution of mammals is but one part of the disciplines of vertebrate paleontology. Throughout the history of this discipline spectacular and mundane discoveries of fossils have helped piece together interpretations of the origin and subsequent development of this class of chordates. As in other branches of paleontology, much time is spent in doing laborious field work that is occasionally spiced by discovery of new or exciting fossil evidence. Quite a number of highly independent and colorful individuals have been attracted to the field, as exemplified by E. D. Cope and O. C. Marsh (Chapter 12). Not surprisingly, the stories of some of these collectors and scientists provide a fascinating glimpse into the history of science.

Most of us know of Thomas Jefferson, but few know of his interests in paleontology. He had a collection of fossil bones and in 1797 published an article describing fossil mammals from Virginia. Because of his interest in natural science, he helped push through Congress in 1803 a bill supporting the now-famous Lewis and Clark expedition for exploration of the United States west of the Mississippi River. Published accounts of this expedition and others described the wonders of the West and helped prompt

further expeditions and settlement following the Civil War.

These subsequent exploratory expeditions are an adventure-filled chapter in the history of the Wild West and also hint at the beginnings of massive exploitation of western resources. Stories and specimens collected on these expeditions fostered increasing interest in the rocks and fossils of the various regions, and more serious collectors soon followed. In previous chapters we have recounted brief histories of some of the more colorful fossil collectors and scientists of the middle and late 1800s. Following in the footsteps of Joseph Leidy, Cope, and Marsh were many other paleontologists, many of whom began to concentrate their work in the western interior, particularly in badland areas on eastern slopes of the Rocky Mountains. The Cenozoic terrestrial deposits of these regions yielded some of the most abundant and well-preserved remains of early and medial Cenozoic mammals to be found anywhere in the world. A number of vertebrate paleontologists have spent a major part of their professional careers studying these remains.

Scientists such as Samuel Wendell Williston (1851–1918), Henry Fairfield Osborn (1857–1935), and William Berryman Scott (1859–1947) were some of the second generation of American vertebrate paleontologists. These men began their scientific

Figure 15-1. Expedition to the Gobi Desert in the 1920s; led by Roy Chapman Andrews of the American Museum of Natural History.
(Photo courtesy of the American Museum of Natural History. Smithsonian Institution Photo No. 251648)

416

studies amidst the turmoil created by the Cope–Marsh feud and they tended to support Cope. The careers of all three, and especially those of Osborn and Scott, extended well into the twentieth century. From 1900 to the mid-1930s they played a dominant role in vertebrate paleontology; Osborn in particular was extremely prolific and published over 400 papers during his career.

In 1891 Osborn became curator of a newly created Department of Mammals at the American Museum of Natural History. Later, as a trustee, he helped organize many of the great collecting expeditions sponsored by the museum. One famous expedition, to the Gobi Desert in Mongolia (Fig. 15-1), was led by Roy Chapman Andrews in the 1920s. On these expeditions many spectacular discoveries of reptiles and mammals were made, including the first dinosaur eggs and the first remains of Cretaceous mammals. These expeditions were quite adventurous: Automobiles were

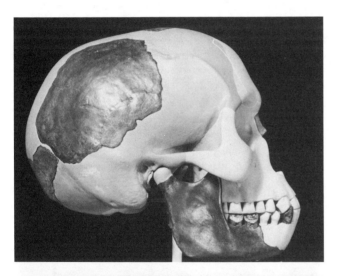

Figure 15-2. Skull and lower jaw with two crooked molar teeth of *"Eoanthropus dawsoni."* This skull is evidently that of a modern human *(Homo sapiens)* and the jaw is that of a modern orangutan. (Photos courtesy of British Museum of Natural History)

primitive and roads and maps were virtually nonexistent. Despite these arduous conditions vast collections were brought back to the museum and numerous specimens were pieced together. Many skeletons of large extinct mammals and dinosaurs found in museums throughout the world are replicas of specimens that were painstakingly reconstructed by these scientists.

A further aspect of Osborn's career was his interest in assembling a collection of recent and fossil horse remains and compiling all the literature related to the evolution of the horse family (Equidae). This turned into a monumental task involving many collectors and staff at the museum. Many publications resulted from this work, but Osborn never completed his major monograph on horses. This task was accomplished in 1951, sixteen years after Osborn's death, when the important book *Horses* was published by paleontologist George Gaylord Simpson. In tribute to the orignal work Simpson acknowledged that his publication was "on a scale incomparably more modest than was anticipated by Osborn and his staff."

Simpson is perhaps one of the most outstanding and well-known vertebrate paleontologists of this century. He has published many scientific and semipopular books on evolution and is especially well known for his studies on Cenozoic mammals, particularly his intensive study of the evolutionary history of horses. Simpson used the extensive detailed evidence from horse lineages to support his ideas on evolution. His concepts represent a combination of Darwin's ideas and more modern ideas of paleontology and genetics. The distribution of fossil horses in the rock record illustrates a complex series of morphological changes that represent genetic response to environmental change through nearly 55 million years and serves as a fine introduction to the study of mammals.

Along with these major discoveries and successful scientific careers are occasional episodes that illustrate another facet of scientific study: occasional fabrication of data or outright hoaxes. Most scientific hoaxes are rapidly recognized and brought out in the open, but some are not uncovered for years. Such an example is the Piltdown hoax.

Toward the end of 1912, a discovery of human fossils was reported in the prestigious British journal *Nature*. A skull and lower jaw with two teeth were found by a fossil collector named Charles Dawson in river gravels at Piltdown Common, near Sussex, England. So began the saga of what became known as Piltdown Man. This famous fossil was given the name *Eoanthropus dawsoni*. It had a skull very human in appearance, but the lower jaw was apelike and contained two relatively small, evenly worn molars. For about 40 years after discovery of the bones, anthropologists struggled to place Piltdown Man into a framework of human evolution, generally with little success. Then, in 1953, comprehensive restudy, including carbon-14 dating of the fossils, revealed many peculiarities. It was recognized by anthropolgist Kenneth Oakley that the remains were a hoax. In fact, the bones had been artificially colored and the teeth had been filed smooth and set into the jaw crooked (Fig. 15-2 on page 417). The skull was from a recent human being and the jaw was that of a modern orangutan!

The author of this elaborate and successful hoax has never been determined. A number of eminent British geologists—including A. W. Sollas, Professor of Geology at Oxford University in the early 1900s—have been suggested as perpetrators, but none ever confessed and all are now dead. A new twist has recently been added to this mystery with the addition of still another suspect in the hoax. The authors of an article in the September issue of *Science 83* present much circumstantial evidence that links Sir Arthur Conan Doyle, creator of Sherlock Holmes, to the hoax. They submit that he had the knowledge, abilities, and perhaps the motivation to plant the specimens. However, as with other suspects proposed through the years, he never confessed, and so the plot thickens. Indeed, this was a rather elaborate hoax and certainly one that had a long run; it brings to mind the hoax perpetrated on Johann Beringer in the late 1700s (Chapter 4).

INAUSPICIOUS BEGINNINGS

Introduction

During the last 65 million years of the earth's history—the Cenozoic Era—mammals have evolved into a wide variety of species that have filled many terrestrial and marine niches. Mammals have been the most abundant and diverse class of terrestrial vertebrates during the Cenozoic. Not surprisingly, this era is known as the age of mammals. However, the evolutionary history of this class of vertebrates is much longer, extending back to Late Triassic time, about 190 million years ago. The origins and subsequent evolutionary history of mammals through the Mesozoic and Cenozoic Eras provides a fascinating chapter in the history of life.

Study of reptile history (Chapter 12) indicates they were dominant consumers in terrestrial and marine communities and were successful for most of the 160 million years of the Mesozoic. Extinction of most reptile taxa in the Cretaceous provided opportunities for mammals, who underwent an evolutionary radiation from a few Mesozoic orders to nearly thirty orders by mid-Cenozoic time. Mammalian diversity can be illustrated by noting some of the orders that have appeared. The enormously abundant rodents (order Rodentia) exceed all other orders combined in diversity of species and are very widespread on land areas. Marine mammals such as the whales (order Cetacea) returned to the sea, as did the marine reptiles. One group, the bats (order Chiroptera), have developed flight but have not become abundant, probably because of competition with birds. Lemurs, tarsiers, monkeys, apes, and humans (order Primates), although not very diverse, represent an unusual and, to us, very important order.

The history of mammals provides good examples of many evolutionary patterns. In particular we recognize adaptive radiation in the Paleogene, replacement of archaic taxa by more modern forms in Miocene time, and the overall mechanisms of evolution illustrated by the 55-million-year history of horses. The geographic distribution of mammals in the Cretaceous and during the Cenozoic also illustrates many effects of plate tectonics on evolutionary history of major taxa. In this chapter we will discuss (1) the origin and early evolutionary history of mammals, (2) the adaptive radiations of new orders in the Cenozoic, and (3) the significance of the fossil record of horses.

Mammalian Characteristics and Classification

Mammals can readily be distinguished from reptiles by a variety of physiological and anatomical features. Some of these features are not easily recognized in the fossil record when comparing primitive mammals with reptiles. This is an indication of the close evolutionary history of the two groups in early Mesozoic time. We can recognize mammals on the basis of a bony palate between the nose and mouth; a single lower jaw bone; the presence of complex dentition, including a variety of teeth such as incisors, canines, premolars, and molars; three very small inner ear bones known as the hammer, anvil, and stirrup; and the double **occipital condyle.** These features, and in particular the single bone of the more durable lower jaw and teeth, are most useful to paleontologists because they are more easily preserved than physiological characteristics. Physiologically, mammals have a high metabolic rate that produces internal body heat, a condition known as **endothermy.** They also have a four-chambered heart, mammary glands, and external insulation such as hair or fur. As discussed in Chapter 11, the presence of some of these features may not be unique to mammals. Some evidence from fossil reptiles suggests that archosaurs may also have been endothermic; evidence of preserved archosaur epidermis having external insulation in the form of hair and feathers has been reported.

Many morphologic features used to recognize mammals also can be used to classify them. Some orders, as listed by vertebrate paleontologist Alfred S. Romer, are most common in Mesozoic rocks (Table 15-1). Some orders known from Mesozoic rocks are more common in Cenozoic rocks and are also included in Table 15-2. Fossils representing the Mesozoic orders are rare and their remains consist mainly of teeth and a few jaw and skull fragments. Most were relatively small and none of the species so far known was larger than a modern house cat. A great many more Cenozoic orders are known, and many, such as whales and ungulates, attained very large body sizes. As a result, a much better fossil record exists for Cenozoic mammals. Many mammalian evolutionary lineages, including the classic example of horse evolution, have been worked out. All provide support for evolutionary theory.

Table 15-1. Mammal orders of the Mesozoic

Taxon	Geologic range
Subclass Uncertain	
Docodonta	Triassic–Jurassic
Triconodonta	Jurassic
Subclass Allotheria	
Multituberculata	Jurassic–Eocene
Subclass Theria	
Symmetrodonta	Jurassic–Cretaceous
Pantotheria	Jurassic–Cretaceous
Marsupialia	Cretaceous–Holocene
*Condylartha	Cretaceous–Eocene
*Creodonta	Cretaceous–Pliocene
*Insectivora	Cretaceous–Holocene

*Placental mammals. Note that the marsupials and the placentals first appear in rocks of Late Cretaceous age.

Table 15-2. Mammal orders of the Cenozoic

Taxon	Common examples	Geologic range
Subclass Allotheria		
Multitub erculata		Jurassic–Eocene
Subclass Theria		
Marsupialia		Cretaceous–Holocene
Condylartha		Cretaceous–Eocene
*Creodonta		Cretaceous–Pliocene
*Insectivora	Shrews	Cretaceous–Holocene
*Taeniodontia		Paleocene–Eocene
*Tillodontia		Paleocene–Eocene
*Astrapotheria		Paleocene–Miocene
*Amblypoda		Paleocene–Oligocene
*Edentata	Sloths, armadillos	Paleocene–Holocene
*Lagomorpha	Rabbits	Paleocene?–Holocene
*Rodentia	Rodents	Paleocene–Holocene
*Primates	Primates	Paleocene–Holocene
*Carnivora	Dogs, cats	Paleocene–Holocene
*Litopterna		Paleocene–Pleistocene
*Notoungulata		Paleocene–Pleistocene
*Chiroptera	Bats	Eocene–Holocene
*Cetacea	Whales	Eocene–Holocene
*Artiodactyla	Deer, cattle	Eocene–Holocene
*Perissodactyla	Horses	Eocene–Holocene
*Proboscidea	Elephants	Eocene–Holocene
*Sirenia	Sea cows	Eocene–Holocene
*Tubulidentata		Eocene–Holocene
*Hyracoidea	Hyrax	Oligocene–Holocene
*Pholidota		Oligocene–Holocene
*Embrithopoda		Oligocene
*Demostyla		Miocene–Pliocene
Subclass Prototheria		
Monotremata		Pleistocene–Holocene

*Placental mammals. Note that most first appear in rocks of Paleocene-Eocene age.

Origin of Mammals

By use of teeth, skulls, and bones, the parts most likely to be preserved in the fossil record, it is possible to provide a close estimate of when the mammals first evolved. By careful study of the record, paleontologists have reconstructed the events leading up to the first mammals and also have traced their subsequent history through the Mesozoic and Cenozoic Eras. This early history is intricately interwoven with the history of reptiles, with major changes in land plants, and with the relative movements of the lithospheric plates through the 200 million years since the Triassic.

Evolutionary relationships between mammals and reptiles are illustrated in Figure 15-3. Reptiles first appeared in the Early Pennsylvanian and were descended from labyrinthodont amphibians. One subclass, the Synapsida, became the most diversified and abundant reptile group of the late Paleozoic. Therapsids, an order of synapsids, are very abundant in Permian rocks. Fossils reveal a number of mammalian anatomical characteristics such as a secondary bony palate and a single lower jawbone. Also, preservation of epidermis of some therapsids suggests that in life the animals may have had fur and whiskers, which are insulators and indicate the possibility that this group was endothermic. These reptiles dominated terrestrial environments until the close of the Permian Period, when approximately 85 percent of them became extinct. Remains of these creatures have been recovered from rocks on the southern-hemisphere continents of ancient Gondwanaland, but some are also known from red beds in Texas and New Mexico. The red beds were deposited in lacustrine and floodplain environments under arid climatic conditions perhaps similar to those existing in the Southwest today.

During the Triassic, remnants of mammallike therapsids that survived the Permian extinctions underwent significant evolutionary changes and a variety of new lineages appear in the rock record (Fig. 15-4). Toward the end of the Triassic at least one of these new lineages and possibly more had crossed a threshold and had made the transition from mammallike reptile to true mammal. Other lineages did not make the transition, and by the end of Triassic time all the therapsids were extinct. As Figure 15-5 suggests, only one therapsid lineage evolved into true mammals, but herein lies controversy. If all mammals descended from one lineage of therapsid reptiles they would be considered to have a **monophyletic** origin. If the mammals evolved from two or more lineages of therapsids they would have a **polyphyletic** origin. To distinguish between monophyletic or polyphyletic origin for mammals, considerable fossil evidence must be available; abundant jaws, skeletal material, and especially teeth are the main prerequisites. Such material is abundant for mammallike reptiles but is lacking for the early mammals, and at present there is

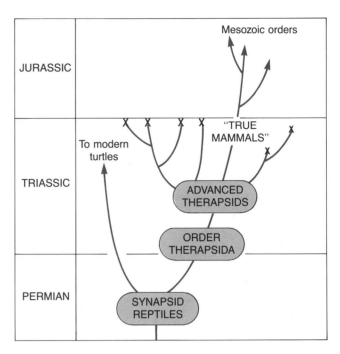

Figure 15-3. Late Paleozoic to Early Mesozoic mammallike reptiles leading to the first true mammals, which appear in rocks of Late Triassic age.

no conclusive evidence for monophyly or polyphyly. Compounding the complexity of the reptile–mammal evolutionary transition is evidence that specific mammalian characters appeared at different times in different evolving lineages. However, the morphological features of teeth and other skeletal evidence suggest to some vertebrate paleontologists that mammals evolved from a single lineage of small-sized, carnivorous therapsids. The genus *Cynognathus,* although belonging to the cynodont lineage of therapsids, is commonly cited as having features transitional between reptile and mammal.

As discussed in Chapter 11, it is difficult to distinguish between amphibians and the earliest reptiles in fossils of Early Pennsylvanian age. This same difficulty plagues interpretation of transitional fossils that led from therapsid to mammal in the Late Triassic. Furthermore, study of such transitional fossils indicates that the first mammals were probably distinct from therapsids primarily in physiological features which are not usually preserved; not until somewhat later in mammalian history do distinctive skeletal differences appear. These difficulties, while presenting challenging problems for paleontologists, are excellent examples of the process of evolution because they document the transitional nature of evolutionary changes.

Mammals survived the wave of extinction at the end of Triassic time, as a result of a number of characteristics. Endothermy allowed them to retain a small body size and still produce enough body heat to remain ac-

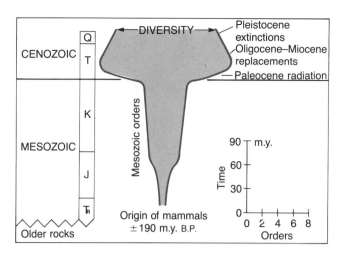

Figure 15-4. Evolutionary history of the class Mammalia. The chart illustrates diversity at the order level. A major adaptive radiation occurred early in the Cenozoic. Evolutionary replacement of Cretaceous and many early Cenozoic orders occurred during latest Eocene to Miocene time. Climatic changes and emergence of humans may have contributed to Pleistocene extinctions of many mammal lineages.

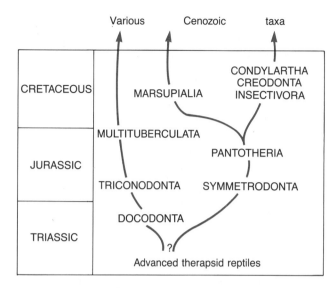

Figure 15-5. Interpretation of evolutionary history of Mesozoic mammals based primarily on remains of teeth and lower jaws (see Fig. 15-6).

tive, even in cold climates. The distinctive differentiation of mammalian teeth permitted a diet that could include a wide variety of foods. Further evidence suggests that many early mammals were **arboreal** and probably were **nocturnal.**

The large-scale evolutionary history of mammals for the last 190 million years can be divided into three main phases (Fig. 15-4). An early phase included the time of origin and early history during Mesozoic time. This interval represents about two-thirds of the entire history of mammals; at that time they were uncommon, small, and subordinate to the dinosaurs, and evolutionary opportunities favoring major diversification did not exist. A change in fortunes occurred with the extinction of dinosaurs. This paved the way for a second phase, represented by a rapid evolutionary radiation in the early Cenozoic, followed by an interval of extinction in mid-Cenozoic time. The third phase was another adaptive radiation in the medial Cenozoic, during which time most of the modern taxa appeared. We will consider these phases next.

Mesozoic Mammals

Despite extensive search of the rock record for over 100 years, Mesozoic rocks have yielded fossils representing only nine orders of mammals. These orders are distinguished from one another on the basis of shape of the **molars,** or cheek teeth, primarily because these teeth are by far the most commonly preserved hard parts and because they show recognizable morphologic differences. Early mammals were small, on average

about the size of modern rats, with a few perhaps reaching the size of small house cats. Their bones were small, fragile, and easily broken—and easy to overlook in the rock record, especially if they occur with huge reptile bones. Until the recent discovery of skull and bone fragments of Cretaceous pantotheres at Bug Creek, Montana, few skeletal parts of Mesozoic age had been discovered. Construction of evolutionary history (Fig. 15-5) is based upon teeth and jaws, most of which have been found in North America, Europe, and eastern and central Asia. An interesting picture can be visualized by comparing tools and techniques used to discover and collect huge dinosaur bones many meters long and tiny (3 mm to 10 mm in diameter) mammal teeth from the same collecting sites. Certainly the collecting of each type of fossil presents its problems!

The various orders of Mesozoic mammals are differentiated on the basis of crown position, crown complexity, and outline of molar teeth (Fig. 15-6). The shape of these teeth supports suggestions that primitive mammals had an **omnivorous** diet, probably of the insectivorous type; their teeth are adapted for grinding various types of food such as insects, seeds, and eggs.

Of greatest interest to us are the pantotheres because this order is considered to be ancestral to the more advanced **marsupials** and **placentals.** The other orders of Mesozoic mammals represent evolutionary dead ends, although one, the multituberculates, was very successful in terms of longevity. Multituberculates survived well into the Cenozoic but did not lead to more advanced groups. In contrast to other early mammalian orders, multituberculates had molars with longitudinal rows of cusps and were probably herbivorous.

Before considering the history of Cretaceous mammals we will review some events that occurred during the period. As discussed in Chapter 12, the breakup of Pangaea began in the Triassic and had progressed considerably by Late Cretaceous time. This fragmentation affected the number of terrestrial habitats and altered climatic patterns. These changes are recognized also in vascular plants—seed-bearing gymnosperms were replaced by the flowering plants. Finally, and perhaps most importantly, the end of the Cretaceous was the time of dinosaur extinctions. These events all influenced mammalian evolution in the Cretaceous and helped set the stage for subsequent explosive radiation of mammals in the Cenozoic.

MAMMALS TAKE CENTER STAGE

Refugees from the Mesozoic

Mammal fossils from earliest Cenozoic—the Paleocene Epoch—show a dramatic increase in diversity. A quick reference to Table 15-2 indicates that only five orders survived the Cretaceous extinctions, but, more significantly, ten new orders of placental mammals appeared. Of the holdovers, multituberculates represent what Romer described as a highly specialized but very successful order. They hold the longevity record for mammalian orders, having survived from Cretaceous to Eocene time, a span of nearly 100 million years! As noted previously, these mammals were herbivorous and may have filled niches later occupied by the rodents. In the early Cenozoic some species attained large size, a characteristic common to the evolutionary history. of many herbivorous vertebrate lineages. Extinction of the multituberculates in the Eocene may have been due to competition with placentals.

Marsupials, another Cretaceous order, have a pouch for nursing newborn, a distinctive pelvic structure, an inturned lower jawbone, and three premolar and four molar teeth (Fig. 15-7). Geologic history of this order

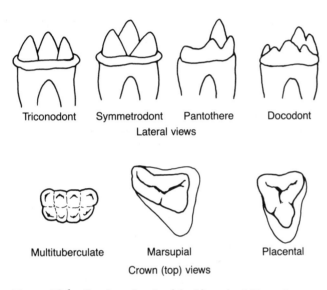

Figure 15-6. Sketches of molar (cheek) teeth of Mesozoic mammals. Note the distinct cusp pattern used to distinguish among the various orders.
(Modified from A. S. Romer, *Vertebrate Paleontology*, 3rd ed., Fig. 307, p. 199: University of Chicago Press. © 1966 by the University of Chicago. All rights reserved)

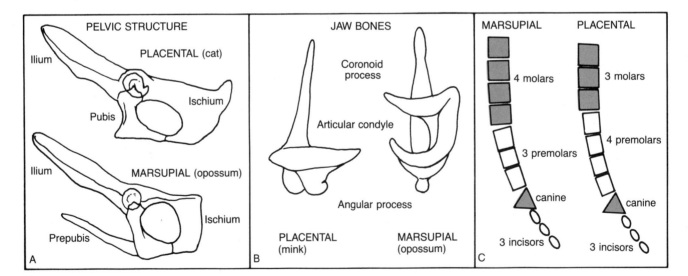

Figure 15-7. Marsupial and placental anatomy. A. Pelvic bones illustrating forward projection in marsupials. B. Lower jaw from rear showing inturned process characteristic of marsupials. C. Lower right tooth row indicating the differences in number of molar and premolar teeth in marsupials and placentals.
(A modified from A. S. Romer, 1962, *The Vertebrate Body*, Fig. 142, p. 190. W. B. Saunders Company, Philadelphia PA. B modified from A. S. Romer, *Vertebrate Paleontology*, 3rd ed., Fig. 312, p. 202: University of Chicago Press. © 1966 by the University of Chicago. All rights reserved.)

illustrates unusual evolutionary patterns. During the Cretaceous the marsupials had a nearly worldwide distribution, but they were apparently outcompeted in the early Cenozoic by rapidly diversifying placental orders and today are not common in most parts of the world. In Australia and South America, however, marsupials were not affected by the placentals and have remained diversified; during the Cenozoic these land areas have become **asylum areas** for the order.

This pattern of marsupial survival is closely related to plate tectonics. By Cretaceous time the breakup of Pangaea had progressed to the point that Australia and South America had become isolated from other land masses before placental mammals could become established. Thus marsupials, free from competition with placentals, evolved into a wide variety of species. Many of these, such as the marsupial dog, wolf, and various rodentlike animals, are similar to placental counterparts and fill equivalent niches in terrestrial communities. Today most South American marsupials are extinct because of competition with placentals that migrated southward across the Panamanian land bridge that formed in the Late Neogene, about 3 million years ago (Chapter 13). One exception to these extinctions is the opossum, a marsupial that has managed to survive in South America and even migrate to and spread in North America. Survival of this creature is undoubtedly related to its adaptibility; opossums can eat almost anything, are a prolific producer of young, and are able to survive in a wide variety of climatic conditions. Other organisms with these characteristics have also survived over long intervals of geologic time; examples of other adaptive generalists are turtles, cockroaches, and some genera of inarticulate brachiopods.

Early Cenozoic Placental Radiations

The second phase of mammal evolutionary history began in early Cenozoic time with the appearance of placental mammals. Placentals are by far the most abundant and diversified group of Cenozoic mammals. They are characterized by having a placenta for prenatal development, distinctive pelvic bones, a relatively enlarged brain, and four premolar and three molar teeth on each side of the jaw (Fig. 15-7). Three primitive orders first appear in uppermost Cretaceous rocks, and one, the **Insectivores** (shrews, moles, hedgehogs), is probably the ancestral stock that gave rise to many other orders. This rapid evolutionary radiation of placentals occurred during the Paleocene and Eocene Epochs, when eighteen new orders appeared (Fig. 15-8).

A brief scan of the placental orders shown in Figure 15-8 indicates that these mammals rapidly diversified into habitats left vacant by the demise of the major reptile groups. For instance, placentals of the order Carnivora include terrestrial types such as dogs, cats, bears,

raccoons, and hyenas, and marine forms such as seals (Fig. 15-9). This order underwent adaptive radiation in the Oligocene following the decline of a more primitive order, the creodonts, which had occupied carnivore niches in Paleocene and Eocene communities. The rodents, generally known as gnawing mammals, are the most abundant order and include diverse groups such as rats, mice, squirrels, porcupines, and beavers. Ro-

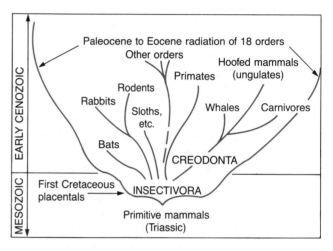

Figure 15-8. An early order of placentals, the insectivores, appear to have given rise to many Paleocene and Eocene orders.

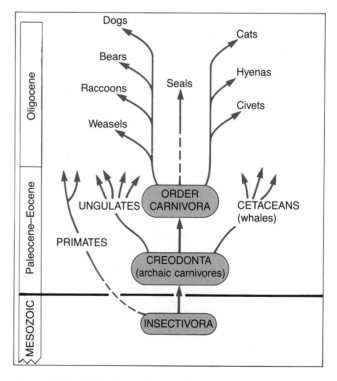

Figure 15-9. Evolutionary history of some placental orders. Carnivores replaced the older creodonts in Oligocene time. The pattern illustrated by these placental carnivores followed a pattern similar to that exhibited by carnivorous reptiles in the Jurassic and Cretaceous Periods. Seals represent aquatic forms (suborder Pinnipedia); the other groups are terrestrial (suborder Fissipedia).

dents underwent an expansion in the early Cenozoic and have continued to diversify to the present time. Hooved mammals, known as ungulates, are the most important large herbivorous placentals. They are characterized by hooved feet and by well-developed molar teeth which are very efficient for grinding vegetation. Many of these herbivorous mammals attained large sizes during the Cenozoic; many such as the titanotheres and other extinct primitive ungulates were at least 3 to 4 m in length and weighed several metric tons.

Other orders of placentals first appeared during the early Cenozoic, and some of these are of general interest. The cetaceans include most of the marine mammals such as whales and porpoises. First appearing in the Eocene as relatively small organisms, they rapidly increased in size. Today some, such as the blue whale, may exceed 30 m in length and are the largest animals known. Although whales have diverse feeding patterns today, early cenozoic species were carnivores and probably derived their dentition pattern from their creodont ancestors.

Also of interest are the primates. Although not abundant in the fossil record, this order includes humans among its ranks. The order evolved from a primitive insectivore ancestor during the Paleocene and subsequently diversified into a number of lineages. Tracing of this evolutionary history is difficult because of a notoriously poor fossil record. As we will see, a number of different opinions are currently under consideration.

Medial and Late Cenozoic Placental Evolution

The end of the Eocene Epoch is marked by extinction of many orders and families of mammals that had appeared in the Cretaceous and early Paleocene. Fossils found in central Oregon, the Rocky Mountain basins, and the great Plains indicate that during Oligocene and especially Miocene time many new mammal taxa appeared in the fossil record. These new taxa were much more similar to living mammals than were the earlier forms. One notable aspect of this medial and late Cenozoic radiation was the development of increasingly distinctive mammal faunas on different continents. We already mentioned the marsupials in Australia and South America. By Miocene time considerable differences can also be seen in placentals of North America and Eurasia. Some examples of these differences are: (1) By late Miocene, primates had disappeared from North America but were evolving new lineages in Europe and Africa. (2) Ungulates such as the horse family were evolving rapidly in North America but were absent from Europe. (3) Divergence of ungulate lineages occurred on these separated land masses.

As you may suspect, this increasing diversity and divergence was stimulated by opening of the Atlantic Ocean. Increased continental fragmentation produced geographic isolation and thus fostered increasing genetic divergence. The evolutionary changes in mammals may be linked to changes in climate, vegetation, and plate tectonics. During Miocene time widespread angiosperm grasslands developed, and new taxa of **grazing** mammals evolved, especially in North America and parts of Africa, to replace the more primitive early Cenozoic **browsers.** By Pleistocene time a combination of events produced widespread climatic cooling, which culminated in glaciation in the northern hemisphere and affected many mammal taxa. Also, fluctuations in sea level associated with waxing and waning of large ice sheets exposed and submerged **land bridges** across the Bering Straits in the Arctic and the Isthmus of Panama between North and South America. When emergent, these land bridges allowed increased migration of placentals betwen Asia and North America and between North and South America; when submerged, these areas acted as barriers to migration of terrestrial mammals. The myriad of mammalian changes that began in the latest Pliocene and accelerated in the Pleistocene Epoch has continued into the Holocene. Significant aspects of these changes have been the appearance of humans and the rapid extinction of larger mammals such as the giant ground sloths, cave bears, mammoths, mastodons, and sabertooth cats, to name a few. That this trend is continuing can be observed by noting the rapidly increasing number of species included on lists of endangered species.

A MODEL OF MAMMALIAN DIVERSITY

Even-Toed and Odd-Toed Ungulates

Fossils of a variety of herbivorous placentals, loosely grouped as **ungulates,** provide an example of mammalian evolutionary patterns during the Cenozoic. Ungulates are a diverse group of mammals that possess in common a wide variety of morphological features. In general they are represented by medium- to large-sized, herbivorous, hoofed mammals with well-developed, complex grinding teeth. They probably evolved from creodonts (Fig. 15-10), a group of Late Cretaceous to early Cenozoic carnivorous mammals of relatively small size. The first ungulates, known as condylarths, show modification of their molar teeth into grinding types but still retain skeletal features of their carnivorous creodont ancestors. Other ungulates appear in the record by medial Cenozoic time and represent most major lineages of modern ungulates. Two orders that we will consider in greater detail are even-toed ungulates,

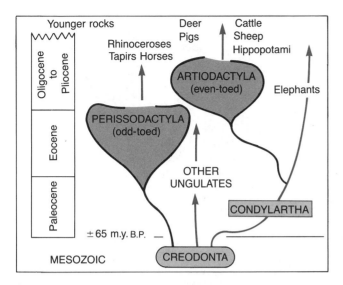

Figure 15-10. Ungulate evolutionary history through the Cenozoic. Maximum diversity of the odd-toed and even-toed orders occurred at different times during the Cenozoic.

called **artiodactyls,** and odd-toed ungulates, or **perissodactyls.** Other ungulates are elephants and the South American notungulates, most of which are extinct.

Artiodactyls are the most diversified order of ungulates existing today. They include most familiar groups of herbivorous mammals: deer, cattle, pigs, and sheep, among many others. The fossil record indicates that primitive artiodactyls had evolved from condylarths by Eocene time and subsequently underwent rapid diversification, reaching maximum diversity in late Cenozoic. During Oligocene and Miocene time the most abundant artiodactyls belonged to a group known as oreodonts (Fig. 15-11). Numerous skull and other skeletal remains of these sheeplike animals have been recovered from terrestrial deposits of the White River Group in the badlands of South Dakota. Diagnostic anatomical features include the existence of either two or four toes on each foot, distinctive articulation of the ankle bones, and teeth somewhat like those of their early Cenozoic carnivorous ancestors. Except for the hippopotamus and large hogs, most are well adapted for running and live in semiarid grassland environments such as plains and prairies. The evolutionary history of the group indicates a progressive increase in overall size and reduction in toes on each foot from four to two (Fig. 15-11C).

Perissodactyls also first appear in the lower Eocene fossil record. This group subsequently underwent rapid diversification into five main branches, including three that exist today: tapirs, rhinoceroses, and horses (Fig. 15-12). Odd-toed ungulates underwent evolutionary trends paralleling those of the even-toed ungulates: (1) a gradual increase in size, (2) adaptation for running, with a consequent decrease in toes on each hoof from five to

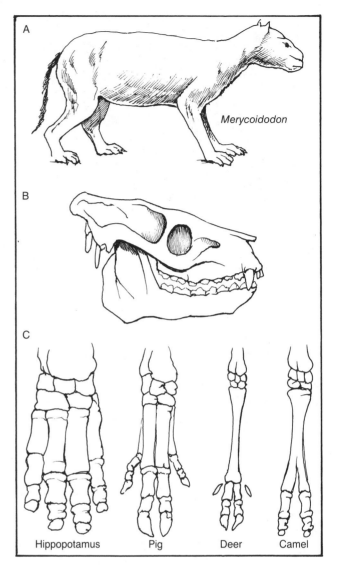

Figure 15-11. Even-toed ungulates. A. Reconstruction of Oligocene sheep-sized oreodont. B. Skull of Oligocene genus *Oreodon.* Note well-developed high-crowned teeth and placental mammal dentition (skull 12.5 cm long). C. Comparison of forelimb foot bones in modern artiodactyls.
(A redrawn from E. H. Colbert, *Evolution of the Vertebrates,* 3rd ed., Fig. 147, p. 408. Copyright © 1980 by John Wiley & Sons, Inc., New York. Reproduced by permission of John Wiley & Sons, Inc., and Lois M. Darling. B modified from A. S. Romer, *Vertebrate Paleontology,* 3rd ed., Fig. 408, p. 280: University of Chicago Press. © 1966 by The University of Chicago. All rights reserved. C modified from A. B. Howell, *Speed in Animals,* Fig. 20, p. 159: University of Chicago Press. © 1944 by The University of Chicago. All rights reserved)

three and in some groups to only one, and (3) development of complex molar teeth for grinding vegetation. In contrast to artiodactyls, perissodactyls attained their maximum diversity in the early Cenozoic. Of interest, because they grew to sizes approaching that of the dinosaurs, were titanotheres and chalicotheres (Fig. 15-13). Both of these groups were abundant in the early Cenozoic but have become extinct.

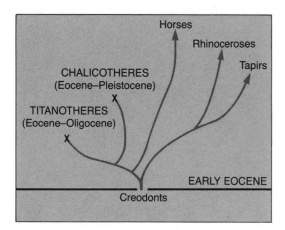

Figure 15-12. Cenozoic lineages of odd-toed ungulates. Early Cenozoic titanotheres and chalicotheres reached large sizes but are extinct. The other lineages are represented by a few living taxa.

Evolutionary History of Horses

Another group of perissodactyls, the horses, have received intensive study by such vertebrate paleontologists as H. F. Osborn and G. G. Simpson. These men studied the fossil record of horses for many years and gradually recognized many major evolutionary trends and characteristics within this lineage (see beginning of the chapter). Their published papers are classic studies and illustrate many patterns and processes of evolution we have discussed in previous chapters.

The horse family first appears in the fossil record in rocks of early Eocene age which are about 50 to 55 million years old. Except for the modern horse of the genus *Equus,* all other taxonomic groups within the family are extinct. As indicated in Figure 15-14, horse evolution has proceeded in a steplike manner with many side branches and lineages. Only a few of these branches are indicated on the diagram. A major pattern is the repeated trend for evolutionary radiation into many lineages, indicating the complexity of horse phylogeny. It is fortunate that the fossil record of this group is good and that Osborn, Simpson, and other collectors have been so meticulous. By piecing together the fossil record, most of the stages of various lineages can be reconstructed. It is evident that modern *Equus* represents just one branch within a lineage rather than the result of any directed evolutionary trend. It will be interesting to see whether the history of horses remains an example of phyletic gradualism as interpreted by Simpson or is found to support the idea of punctuated equilibrium of Niles Eldridge and others (Chapter 7).

Many morphological traits within the horse family underwent changes that can be traced through the fossil record. The most useful of these features are increase in length of the legs, progressive reduction in number of toes from four to one on each foot; increase in tooth size and complexity; increase in skull size and cranial

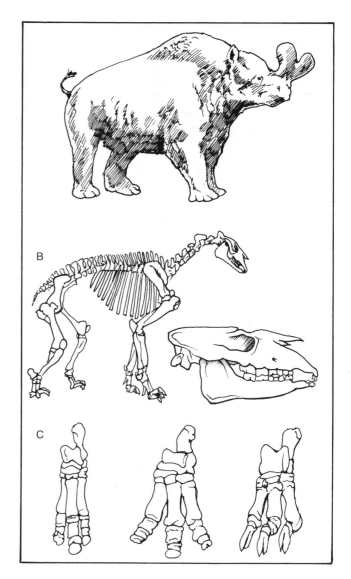

Figure 15-13. Odd-toed ungulates. A. Reconstruction of an Oligocene titanothere of the genus *Brontotherium,* which was 3 to 4 m long. B. Skeleton (3 m long) and skull of a Miocene chalicothere of the genus *Moropus.* C. Rear-limb foot bones of various odd-toed ungulates.
(A, C redrawn from E. H. Colbert, *Evolution of the Vertebrates,* 3rd ed., Figs. 133, 136, pp. 377, 385. Copyright © 1980 by John Wiley & Sons, Inc., New York. Reproduced by permission of John Wiley & Sons, Inc., and Lois M. Darling. B modified from A. S. Romer, *Vertebrate Paleontology,* 3rd ed., Figs. 387, 395, pp. 267, 270: University of Chicago Press. © 1966 by The University of Chicago. All rights reserved)

capacity, indicating an increase in brain size; and overall increase in body size (Fig. 15-15). Such evolutionary changes can be explained as representing genetic changes that occurred rapidly in populations as a response to changing environmental conditions; this would support the punctuated-equilibrium model. This would mean that small local populations underwent selective pressures at various times through the Cenozoic. For example, reduction in number of toes, increases in overall body size, and development of complex molar teeth did not occur simultaneously. Some of these

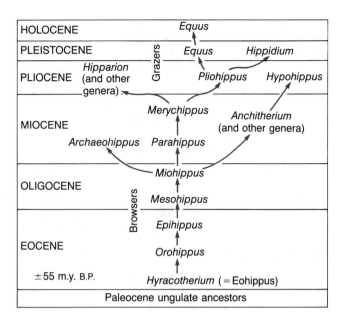

HOLOCENE			Equus		
PLEISTOCENE		Grazers	Equus	Hippidium	
PLIOCENE	Hipparion (and other genera)		Pliohippus	Hypohippus	
MIOCENE	Archaeohippus	Merychippus / Parahippus		Anchitherium (and other genera)	
OLIGOCENE		Miohippus / Mesohippus			
EOCENE		Browsers	Epihippus / Orohippus		
± 55 m.y. B.P.			Hyracotherium (= Eohippus)		
Paleocene ungulate ancestors					

Figure 15-14. Cenozoic horse evolution, based mainly on fossils from North America. Evolutionary radiations occurred in Oligocene, Miocene, and Pliocene times. Only the genus *Equus* is extant.

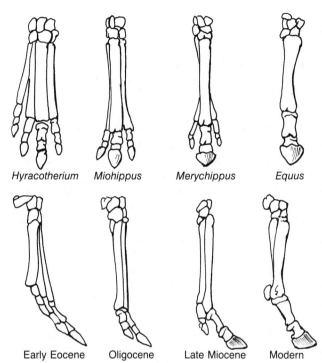

Hyracotherium   Miohippus   Merychippus   Equus

Early Eocene   Oligocene   Late Miocene   Modern

Figure 15-15. Reduction in number of toes in horses illustrates one of a number of evolutionary patterns evident from the Cenozoic fossil record of the group.
(Redrawn from A. S. Romer, *Vertebrate Paleontology*, 3rd ed., Fig. 383, p. 263: University of Chicago Press. © 1966 by The University of Chicago. All rights reserved)

changes provided opportunities for expansion and diversification, especially evident in the later Cenozoic.

The earliest members of the horse lineage, found in lower Eocene rocks, were relatively unspecialized odd-toed ungulates now known as *Hyracotherium (Eohippus* in earlier texts). These were small-dog-sized animals with three functional hooved toes on each foot and unspecialized teeth designed for browsing on shrubs and bushes. During Eocene and Oligocene time there was a slight increase in body size and development of more complex molar teeth. At least four different genera have been named. Beginning in Miocene time horses got larger, lost some of their toes, developed larger teeth, and evidently were becoming adapted to more open grasslands. At least eleven genera of horses representing four separate lineages are known from Miocene and Pliocene rocks. By this time horses had become more modern in appearance and were larger, with well-developed grinding molar teeth and large skulls with an increase in brain size. Each foot had only one toe and the animals were clearly adapted for life in open grassland environments such as the Great Plains of the United States. Fossils representing the lineage leading to the modern genus *Equus* first appear in Pliocene rocks.

Changes in tooth size and shape are characteristic of horse evolution, and because teeth are durable they provide an important source of paleontologic evidence. Horses have developed larger molars, increased the complexity of the crown pattern on the grinding surface of the molar teeth, and formed a thicker enamel coating on the tooth. These changes are related to the change from the browsing habit of Eocene, Oligocene, and early Miocene species to the grazing habit of later Miocene, Pliocene, Pleistocene and modern species and probably represent adaptations for better grinding of tough angiosperm grasses.

Another major adaptive feature was the increase in size and complexity of the brain. This is particularly evident in Pliocene and modern forms and parallels the increase of body size. Both of these changes are related to the gradual adaptation for living in open grasslands, where speed and agility coupled with increased body size and mental alertness would provide protection from predators.

Eocene and Oligocene horse fossils are known in Europe and North America. Horse evolution from Miocene to Pleistocene occurred almost entirely in North America. Well-known collecting areas are located in the Great Plains, Oregon, and Florida. Famous Pleistocene localities such as the Rancho LaBrea Tar Pits in California and other localities in Texas, Idaho, and Kansas have yielded many excellent specimens. Thus evolutionary history of horses is well documented by fossil evidence from extensive Cenozoic continental deposits (Chapter 13). During the Pleistocene, horses migrated to Eurasia over the Bering land bridge but may have become extinct in North America. They were reintroduced in the 1600s during colonization by Spain, France, Portugal, and England.

EVOLUTIONARY HISTORY OF PRIMATES

Although the fossil record of primates extends back to the Paleocene Epoch, it is important to realize that the record of this order is poor when compared with that of marine invertebrates or of many other mammalian orders. Consideration of the environments in which primates dwell provides an explanation for this scanty evidence. Highlands, forests, and plains areas are not good sites for preservation of fossils, but these are regions in which most primates have lived throughout the Cenozoic. Through considerable effort, however, enough evidence has been collected to provide a reasonably clear picture of primate evolution (Fig. 15-16). Much of the evidence consists of fragmentary bones and skulls and well-preserved teeth. These remains are found in lake deposits, volcanic ash beds, fluvial sands and gravels, and cave deposits.

Primates commonly have five-digit hands and feet, **binocular vision,** and distinctive teeth characteristic of an **omnivorous** diet. Other anatomical features found in some taxa include a **prehensile,** or grasping, tail; a pelvic structure and bones to support semierect-to-erect posture, allowing bipedal locomotion; and a relatively large skull and brain capacity compared to total body weight. Organisms containing some or all of these features include fossil and living lemurs, tarsiers, monkeys, apes, and humans. Lemurs and tarsiers are considered primitive primates and are included in the suborder **Prosimii;** more advanced primates such as monkeys, apes, and humans are classified in the suborder **Anthropoidea.**

Extant prosimians retain early characteristics of the primates in that they are arboreal, are generally omnivorous, have five-digit hands and feet, and have partially binocular vision. Today prosimians are found only in tropical regions of Asia, Africa, Madagascar, and India. However, the fossil record indicates these organisms and their ancestors had a much wider distribution earlier in the Cenozoic and were abundant and diversified on most continents during the Paleocene and Eocene Epochs. Decrease of once-diverse prosimians probably corresponds with gradual continental fragmentation, northward drift of some land masses out of mild tropical climatic zones, and appearance and diversification of higher primate groups. In this example we again see the effects of plate tectonics on distribution of organisms; similar patterns occurring in invertebrates, plants, reptiles, and other groups of mammals have been described in previous chapters.

Advanced primates include Ceboidea and Cercopithecoidea, or new world and old world monkeys, and Hominoidea, which includes apes and humans. These more advanced groups have relatively larger brains and

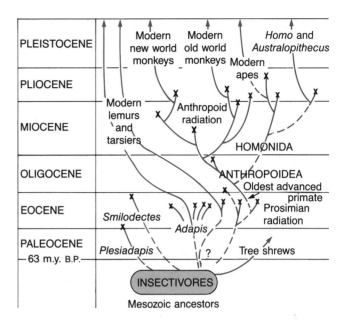

Figure 15-16. Evolutionary history of primates. Taxa noted in text are named. Dashed lines represent uncertain evolutionary relationships or lack of fossil evidence. "X" indicates extinct lineage.

stereoscopic vision because both eyes are positioned at the front of the skull; some are ground-dwelling rather than arboreal species. Distribution of modern monkeys and apes—known from South and Central America, Africa, India, and Asia—is much more widespread than that of prosimians. The fossil record indicates that monkeys existed during Oligocene time. By that time, however, they possessed distinct anatomical features, suggesting that they may have originated as early as Eocene time.

Apes are most abundant today in Africa and southeast Asia and include four groups commonly termed gibbons, chimpanzees, orangutans, and gorillas. Anatomically and physiologically, apes rather than monkeys more closely resemble humans; consequently some apes, such as chimps, are commonly used in experiments in lieu of human subjects.

Humans are represented today by only one species, *Homo sapiens,* and are adapted to a ground-dwelling habitat despite an occasional aching back and fallen arches. Humans are characterized by erect, or upright, posture, a **bipedal gait,** an enlarged brain, distinctive canine teeth, well-developed **opposable thumbs,** and, of course, our distinctive social structures. Our species has a worldwide distribution and a population of about 4 billion; thus *Homo sapiens* is the most abundant and widespread primate species.

Homo sapiens, named by Linnaeus in 1858, is considered to have appeared about 300,000 years ago, and fossils are known from all parts of the world. The oldest subspecies, here termed *H. sapiens neanderthalensis*

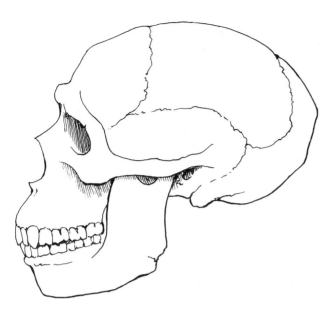

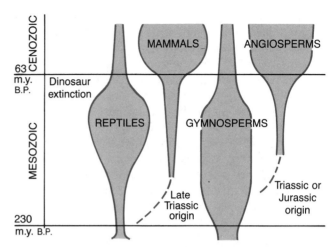

Figure 15-18. Relative diversity of major terrestrial vascular plants and vertebrates in Mesozoic and Cenozoic time.

Figure 15-17. *Homo sapiens.* Cast of neanderthal-type specimen found by three priests from La Chapelle-aux-Saints in southwestern France in 1908. A nearly complete skeleton was also found with tools in a shallow grave. Age of specimen 40,000 to 75,000 years; cranial capacity 1620 cm³.
(Photo courtesy of Harry Nelson)

(Fig. 15-17), is very well known and is named for the Neander Valley location where the fossils were first discovered in 1856. Neanderthals were originally considered a separate species, *H. neanderthalensis.* The change in classification indicates a change in our concepts of these fossils and provides an interesting story. The first reconstruction of neanderthals was based on specimens from France and pictured a short, squat, and quite brutelike individual. Restudy of the fossils, however, indicated that this original specimen was, in fact, afflicted with severe and perhaps crippling bone disease! Further discoveries from various parts of the world have altered our concepts of neanderthals. This subspecies had an erect posture, had a brain size averaging approximately 1500 cubic centimeters, formed and used well-shaped stone tools, apparently had developed social structures, and developed speech. In Europe the group is considered to have become extinct approximately 35,000 years ago.

The first modern humans, *Homo sapiens,* appeared between 35,000 and 50,000 years ago during the last ice age (Chapter 16). Anatomically, they were somewhat more robust than living humans. Within the last 50,000 years humans have evolved complex social structures, have developed writing, and have invented many complex tools and machines.

The most recent evolutionary history of *H. sapiens* seems to have been cultural rather than physical. In the last few hundred years we have been influenced by de-

velopment of many sources of energy, by the industrial revolution and automation, and by an unprecedented population explosion. Our civilizations have developed, or perhaps evolved, at an exceedingly rapid rate; in fact it has been suggested that we have advanced too far too fast and are on a collision course with extinction because of the global political and environmental problems we have created. The future evolution of our species no doubt will continue to be largely cultural. As the noted geneticist G. L. Stebbins has suggested, humans are in an evolutionary position *culturally* that is analogous to that of invertebrates during the earliest Phanerozoic, when a wide range of habitats were available. Perhaps we are on the threshold of a major cultural adaptive radiation or explosive evolution that will carry our species beyond present national and global crises to unimagined limits. What scenario would you envision for our future?

TERRESTRIAL PLANT AND VERTEBRATE COMMUNITIES

The history of reptiles and mammals is closely tied to the history of terrestrial vascular plants and we can speculate on the evolutionary relationships of these three groups (Fig. 15-18). During much of Mesozoic time in North America, gymnosperms such as cycads and conifers were abundant and diverse and existed in terrestrial environments of warm, moist climate (Chapter 12). During Cretaceous time gymnosperm diversity decreased and angiosperms greatly increased in abundance and diversity. By the Late Cretaceous angiosperms had become the dominant land plant division and formed the base of most terrestrial food chains.

Mesozoic reptile-dominated communities flourished under these conditions. Complex food chains culminating in large land-dwelling reptiles were well established by the Late Triassic and Jurassic. Jurassic rocks such as the Morrison Formation contain diverse and abundant remains of gigantic dinosaurs. During the Cretaceous, dinosaurs remained abundant, but there were many changes in the herbivorous forms. For example, hadrosaurs (duck-bills) appeared. They had a complex array of grinding teeth, perhaps adaptations for feeding on angiosperm plants. The dinosaurs became extinct relatively abruptly near the end of Cretaceous time.

The dominance of angiosperms continued into Cenozoic time. Many taxa continued from the Cretaceous, and a wide variety of shrubs and woody trees appeared. By mid-Cenozoic many new types of angiosperms such as grasses had evolved and forests began to decline in abundance. Changes in mammals in the Oligocene and Miocene seem to be closely related to these changes in vegetation; appearance of many new ungulates adapted for grazing rather than browsing provides evidence of these changes. In the late Cenozoic, extensive grasslands appeared in areas such as the Great Plains, large-scale deserts developed with specialized plants, and climatic changes related to glaciation affected the distribution of plants. Significant changes occurred in mammals. Many large species have become extinct, while rodents, including the ubiquitous rat, have become very abundant. Recently a variety of species of the primate *Homo* appeared. One of these species, *Homo sapiens*, has become very abundant and has greatly influenced conditions on the earth. Thus, in the last 10 million years of the earth's history climatic changes have altered community organization and food chains. The record of climatic change, culminating in late Cenozoic glaciation, is the subject of the next chapter.

SUMMARY

Mammals play a major role in terrestrial communities today and have done so throughout the 65-million-year history of the Cenozoic Era. However, the fossil record indicates mammals originated in Late Triassic time, perhaps 190 million years ago. Their ancestors were therapsids, or mammallike reptiles, whose geologic history began in late Paleozoic time and ended with extinction at the end of Triassic time.

Mesozoic mammals are known mainly from teeth, jaws, and a few skulls and bones. From this evidence nine orders have been recognized, none of whose members were larger than a modern house cat. Characteristically they had a single lower jawbone with a variety of teeth, a bony palate, a double occipital condyle, and, presumably, physiological characteristics such as endothermy.

Pantotheres, with a geologic range of Jurassic to Cretaceous, appear to have been ancestral to the various Cenozoic orders. In the early Cenozoic, placental mammals underwent rapid diversification and by the end of Eocene time had become dominant consumers in terrestrial communities. Mammalian fossils in Oligocene and Miocene rocks indicate that earlier forms had become extinct and that new families, more similar to those living today, had appeared. One obvious change was the appearance of grazing herbivores representing orders such as even-toed and odd-toed ungulates. Even-toed ungulates of deer, cattle, and sheep lineages became very widespread.

One lineage of odd-toed ungulates, the horses, has received intensive study aided by large collections of fossil remains that represent the 55-million-year history of the group. Studies of horse phylogenetic trends have provided evidence of major evolutionary patterns. Changes in various anatomical features—increasing body size, reduction in toes, and increase in size and complexity of molar teeth—illustrate various episodes of adaptive radiations and change in feeding habits from browser to grazer. These patterns indicate horses became progressively more adapted to grassland environments through the Cenozoic.

Another Cenozoic mammal order, the primates, includes our own species, *Homo sapiens*. Not surprisingly, the evolutionary history of this order is of considerable interest. Primitive primates are known from Paleocene rocks; as for most primate taxa, fossil evidence is scarce and often fragmentary. Fossils of more advanced primates are known from upper Eocene rocks but are more numerous in Oligocene strata. Approximately two million years ago the first representatives of our genus, *Homo,* appeared in Africa and approximately 300,000 years ago *H. sapiens* appeared. Within the past 30,000 to 50,000 years most changes in our species have been cultural rather than physical.

SUGGESTIONS FOR FURTHER READING

Colbert, E. H., 1980, *Evolution of the Vertebrates,* 3rd ed., John Wiley, New York.

Hopson, J.A., 1967, Mammal-like Reptiles and the Origin of Mammals: *Discovery,* vol. 2, no. 2, pp. 25–33.

Isaac, G., and R. E. F. Leakey, eds., 1979, *Human Ancestors:* W. H. Freeman, San Francisco.

Leakey, M. D., 1979, Footprints in the Ashes of Time: *Nat. Geographic,* vol. 155, no. 4, pp. 446–458.

Leakey, R. E., 1976, Hominids in Africa: *Am. Scientist,* vol. 64, pp. 174–178.

Millar, R., 1972, *The Piltdown Men:* St. Martins Press, New York.

Romar, A. S., 1966, *Vertebrate Paleontology,* 3rd ed.: University of Chicago Press.

Simpson, G. G., 1951, *Horses:* Oxford University Press, New York.

Tattersall, I., and N. Eldridge., 1977, Fact, Theory and Fantasy in Human Paleontology: *Am. Scientist,* vol. 65, pp. 204–211.

16

THE BIG FREEZE:
Glaciers and Glaciation

CONTENTS

KEY TERMS

Erratic boulders
Moraines
Till
Interglacial
Polarity event
Boundary stratotype
Lake varves
Isostatic rebound
Pluvial lakes
Submarine canyons
Land bridge
Heat budget
Greenhouse effect

BIRTH OF A "PREPOSTEROUS HYPOTHESIS"

Among the writings of most of the world's religions are descriptions of a widespread catastrophic event involving floodwaters. In western religion this event is called the Noachian deluge, or great flood; it is described in the Old Testament. Apparent evidence for such a flood was provided by displaced boulders now called **erratic boulders,** some of which are very large. Erratics were known to have been transported many miles from their areas of origin and deposited in a haphazard fashion. Well into the 1800s distribution of these boulders in northern and central Europe was attributed to action of very strong currents of water and mud or to boulder-laden icebergs drifting on the floodwaters. As could be expected, these ideas fit very well with the concept of neptunism advocated by Abraham Werner and his students in the late 1700s (Chapter 3). But there are major problems with this explanation.

Recalling our previous discussions of the concept of actualism, and considering our current knowledge of the science of meteorology, it seems highly unlikely that a flood of worldwide scale would occur. Floods are produced by runoff from excessive rainfall. Because rainfall is part of the hydrologic cycle and occurs when the atmosphere is saturated with water vapor, there would be no mechanism to produce water vapor if rain were falling worldwide. So there does not seem to be a scientific mechanism to produce enough rainfall for such a global flood. Furthermore, where would the waters recede to after the flood?

At the beginning of the 1800s, many major geological concepts were undergoing birth and early development, and controversy was the rule. In Europe, much controversy raged over an explanation for the origin of rocks and to many geologists the idea of a worldwide flood still held considerable attraction. To those scentists who observed huge erratics, polished and often grooved surfaces of exposed bedrock in some areas, and large mound-shaped hills and ridges of poorly sorted sediments, the idea of a great flood seemed to provide a reasonable explanation.

An alternative scientific explanation for the distribution of erratics and also for the occurrence of flooding had been proposed as early as 1787. However, it took 50 years before scientists seriously considered this new concept and perhaps another 25 years for it to achieve general acceptance. This concept has become known as the glacial theory.

John Playfair, most noted for his explanations of the Huttonian theory, suggested in 1802 that glaciers rather than floodwaters could have been the cause of these geomorphologic features: "For the moving of large masses of rock, the most powerful engines without doubt which nature employs are the glaciers. . . . In this manner, before the valleys were cut out in the form they now are, and when the mountains were still more elevated, huge fragments of rock may have been carried to a great distance. . ."* However, his idea was virtually ignored for 30 years. This example brings to mind other ideas that were proposed and either rejected or forgotten. Two such examples are those of Gregor Mendel and his experiments in the mid-1800s leading to the concept of genetics (Chapter 7), and of Alfred Wegener in the early 1900s and his concept of continental drift that led to plate tectonics (Chapter 1).

As the science of geology progressed into the middle 1800s, explanations for natural events relying on worldwide flooding or other catastrophic events gradually lost credibility and were replaced by other explanations. Beginning in the 1820s Jean de Charpentier and Ignace Venetz revived interest in the concept of glaciation. Venetz, a Swiss engineer, proposed that the peculiar features on the Swiss plains were produced by an ancient extension of alpine glaciers. This hypothesis seemed outrageous to most scientists. Louis Agassiz, a Swiss biologist, went to investigate the evidence. By 1837, he had become convinced by the evidence described by Venetz and by his own observations in the field and decided to present a paper to the annual congress of Swiss scientists. In this paper he suggested a number of rather startling ideas: First, he said that movement of glacial ice, not floodwaters, was responsible for all purported diluvial phenomena. Second, he proposed that glaciation was not a local phenomenon but was part of a major change in climate that had affected all of Europe. Third, and the most startling, he stated his belief that in the recent past vast ice sheets had covered much of Europe and that only isolated mountain peaks rose above the ice in northern Europe.

Agassiz, already a well-respected scientist in Europe by the late 1830s as a result of extensive studies of fossil fish, had a monumental task to convince the scientific community of the existence of former glaciers. The task Agassiz undertook was to

---

*From K. F. Mather and S. L. Mason, 1939, *A Source Book in Geology*, p. 137: Harvard University Press, Cambridge MA.

convince Europe's foremost geologists. William Buckland and Charles Lyell, for example, had by 1840 become staunch supporters. As John Imbrie and Katherine Palmer Imbrie noted in their book, *Ice Ages,* however, it took another twenty years before most British geologists accepted the glacial theory.

By the 1840s, Agassiz had not only been convinced by Venetz' evidence but became the leading proponent for the existence of widespread ancient glaciers. He traced the evidence of glaciation throughout Europe. Then in 1847 he accepted a teaching position in the United States at Harvard University, where he devoted many years to the study of similar glacial evidence in North America. He essentially started the branch of geology called glaciology. Agassiz travelled extensively, presented many lectures, and did much to convince skeptics in North America, as well as Europe, of the former existence of widespread glaciers on both continents. Along with these interests, he maintained his interest in fossil fish and founded the now-famous Museum of Comparative Zoology at Harvard.

From this discussion it is evident that Agassiz was a fine example of a naturalist. His formal training was in zoology, but he became renowned in other fields of natural science as well. In earlier chapters we have noted other well-rounded naturalists such as E. D. Cope, J. W. Powell, George Cuvier, and James Hutton.

In the 140 years since the pioneering work of Agassiz, geologists have amassed a vast amount of information about glaciers and glaciation. There is considerable evidence indicating that glaciers have advanced and retreated over much of the northern hemisphere a number of times during the Pleistocene Epoch. Thus, the Pleistocene has become popularly known as the ice age. We will consider the nature of historic and prehistoric evidence for glaciation, discuss its effects on the earth, and delve into hypotheses proposed to explain the phenomenon.

DISTRIBUTION OF GLACIERS

Active glaciers in North America occur in Alaska and Canada and to a lesser extent in the Sierra Nevada, Cascade Range, and Rocky Mountains of the United States. However, a variety of preserved physical evidence attests to the existence of much more widespread ancient glaciers. These features can be readily observed on visits to the Sierra Nevada and the Rocky Mountains. Somewhat less obvious but more widespread evidence can be observed in the midwestern and northeastern United States and Canada. Erosional and depositional features provide proof for the previous presence of vast ice sheets in these regions in the not too distant past. Similar features are known from northern Europe and from central and northern Asia. This evidence indicates that perhaps as much as two-thirds of the land area in the northern hemisphere was covered by thick sheets of ice (Fig. 16-1). In some areas, such as northern Canada, the thickness of these glaciers may have exceeded 3 km! The landscape in many areas of the northern hemisphere has been dramatically altered by glacial erosion or deposition. Today only remnants of these ice sheets exist: in Greenland and around the North Pole.

THE PLEISTOCENE ICE AGES

Development of a Glacial Chronology

By the early 1870s, about thirty years after Agassiz first proposed his hypotheses for glaciation, the concept of glacial theory had been accepted by most geologists.

Major research then focused on working out a history of the ice ages. Within ten years, mechanisms of ice formation and movement were recognized and geographic distribution of ancient glaciers was known. By mapping the distribution of **moraines,** erratic boulders, and other features, it was recognized that glaciers had in the past covered about 27 million sq km, or approximately three times the area covered by ice today; most of this area was in the northern hemisphere. A major discovery made during these initial studies was the cyclicity of ice advances and retreats. Subsequent careful mapping of the distribution of terminal moraines and correlation of **till** deposits indicated that at least four and perhaps more episodes of ice advance occurred during the Pleistocene Epoch. Each of these advances was followed by retreat of the ice: an **interglacial** episode (Fig. 16-2). Fortunately, each successive major advance of ice did not completely cover and remove evidence of the preceding episodes.

More recent studies have provided considerable refinement of the early evidence. In particular, a variety of radiometric dating techniques such as carbon-14, potassium-40:argon-40, uranium-234, thorium-230, and fission track (see Chapter 3) have provided a chronology for Pleistocene glacial and interglacial episodes and information for determining the age of the Pliocene–Pleistocene boundary. These dating techniques also indicate that each glacial advance lasted a different length of time and that advances and retreats occurred as irregular pulses over tens of thousands of years. The glacial features that are observed most readily today were produced by the most recent episode, the Wisconsinan glaciation. Maximum glacial advance occurred during the

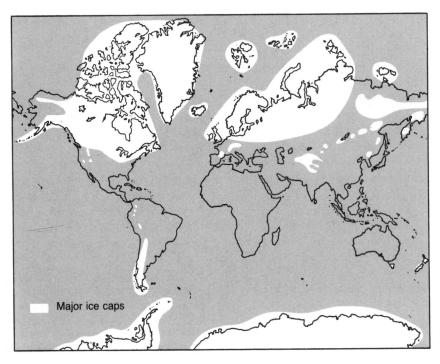

Figure 16-1. Extent of maximum glaciation during Pleistocene time.
(From L. W. Mintz, *Historical Geology: The Science of a Changing Earth,* Fig. 20.39, p. 535. © 1977 by
Charles E. Merrill Publishing Co., Columbus OH)

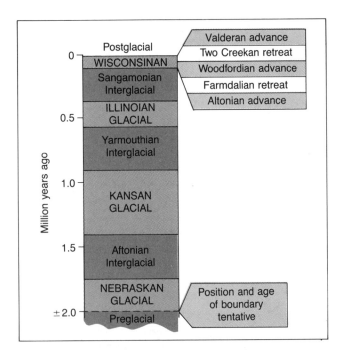

Figure 16-2. Glacial chronology for North America during the Pleistocene and Holocene Epochs.

Recognition of multiple glacial and interglacial events of the Pleistocene and development of radiometric chronology have provided valuable information for geologists. However, determination of the Pliocene–Pleistocene boundary has presented a number of difficulties. These problems illustrate a number of complications involved in trying to fit natural events into a preconceived set of ideas such as the geologic time scale.

How should the Plio-Pleistocene boundary be defined? Possible methods are: (1) Determine radiometric dates of volcanic rocks. (2) Establish the onset of the oldest glacial advance. (3) Note changes in land-mammal taxa. (4) Recognize changes in planktic foraminifera and isotopic content of deep-sea sediments. (5) Use changes in human lineages. (6) Choose a particular magnetic field reversal. (7) Combine two or more of these techniques. In actuality all of these have been used but difficulty arose when we recognized that few, if any, of these events occurred simultaneously (Fig. 16-4). A significant attempt to correlate the boundary by comparing all of these features was published by the American Association of Petroleum Geologists in 1978 *(The Geologic Time Scale).* There is by no means universal agreement, but many geologists use a figure of between 1.8 and 2.0 million years before present for the boundary. This number corresponds to the Olduvai magnetic normal **polarity event** within the Matuyama Polarity Epoch and to a widespread microfossil boundary. The Plio-Pleistocene **boundary stratotype** is in Italy.

Wisconsinan glacial age, and as Figure 16-3 indicates, ice extended as far south as 40° north latitude in parts of the Great Plains.

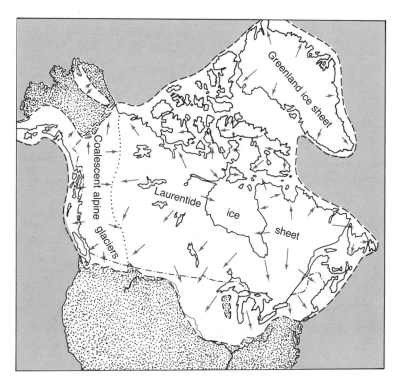

Figure 16-3. Maximum extent of ice sheets in North America during Wisconsinan glacial advance. Glaciers also occurred in mountainous regions of the western United States.
(From R. J. Foster, *General Geology,* Fig. 10.24, p. 204. © 1978, Charles E. Merrill Publishing Co., Columbus OH)

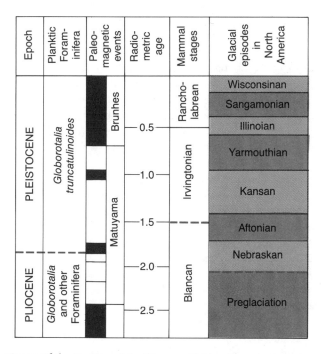

Figure 16-4. Position of the Pliocene–Pleistocene series boundary (heavy dashed line). Note the lack of synchroneity. Position of this boundary must be determined by agreement on what events are most significant and widespread.

A Record of Changing Climates

One significant aspect of Pleistocene ice age was climatic change on a global scale; glacial stages and interglacials were associated with worldwide decrease or increase in average temperature. These large-scale climatic changes affected temperature, humidity, rainfall, and wind patterns in regions far removed from direct influence of glacial ice. Expansion of cool temperatures to lower latitudes, formation of large lakes in currently dry desert areas of southwestern United States, and increased rates of erosion with development of large river systems all were results of climatic variations. We have no evidence to suggest that these variations have ceased; thus we may expect long-term changes to continue. Change is recognized by study of **lake varves,** tree-growth rings, coiling direction of Foraminifera, and changes in isotopic ratios in sediments.

Along with indicating climatic conditions, varved lake sediments provide a method for determining geologic time. Paired layers of lake sediments (Fig. 16-5) record seasonal fluctuations of freezing and thawing of the upper layers of the water. Each couplet represents sediments that accumulate during the year. The layer that forms during the winter when the lake surface is

Figure 16-5. Varved lake sediments deposited on glacial till and exposed in this core sample. Dark layers represent winter deposits and light layers represent summer deposits. Many years are represented by this sample from the Eocene of Colorado. Width of sample is 5 cm.
(From R. J. Foster, *General Geology,* Fig. 10.27, p. 205. © 1978 Charles E. Merrill Publishing Company, Columbus OH)

frozen is thinner, darker in color, has a higher organic content, and is finer grained than the layer that forms during the summer. The existence of varved lake sediments is, therefore, an indicator of rigorous climate with freezing winter temperatures; today lakes at high latitudes or at high elevations form yearly varves. The distribution of these sediments can therefore provide an indication of past climate conditions. The number of paired varves in a lake also records its duration. For example, many lakes in the northern United States contain varves representing about 8700 years of deposition and as such provide an indication of when the ice retreated from this region.

Seasonal or longer term fluctuations of climate are recorded by plants in the form of growth rings (Fig. 14-25). Most plants forming these yearly rings exist in temperate to arctic climates; they are rarely found in the tropics. Geographic distribution of fossil plants having such growth rings therefore can be used to map the extent and location of cool or cold climatic zones. In addition, a vertical succession of fossilized plant remains preserved in rocks at one location, where some layers contain plants with rings and some contain plants without rings, could indicate periodic fluctuation in climate. Spacing and thickness of the rings indicate dry or wet climatic cycles and these can be related to regional climatic changes (Chapter 14).

Change in climate is also recorded by changes in water temperatures of the world's oceans. During glacial advances, ocean-surface temperatures decrease; this change significantly affects many marine organisms. As a consequence some species of planktic foraminifera undergo changes that are quite spectacular, considering their microscopic size. In water temperatures above 10°C, one species, *Globorotalia truncatulinoides,* forms its shell in a right-hand coil; in water below 8–10°C this same species coils in a left-hand direction (Fig. 16-6). Since first noted by D. B. Ericson and other scientists in 1954, this geologic thermometer has been used to chronicle temperature changes in the oceans during the Pleistocene Epoch. Other species of foraminifers and other plankton may have disappeared because they could not tolerate the temperature changes. These changes in plankton in the water column are recorded by the sediments; the record of changes in taxa in the sedimentary record is used to recognize temperature changes that correspond with glacial and interglacial episodes.

Another possibility for a geologic thermometer was suggested by Harold C. Urey in the late 1940s. Beginning in the 1950s, work by Cesare Emiliani, Samuel Epstein, and others quantified Urey's idea. This bit of detective work involved the discovery that ratios of two isotopes of oxygen (oxygen-18 and oxygen-16) change

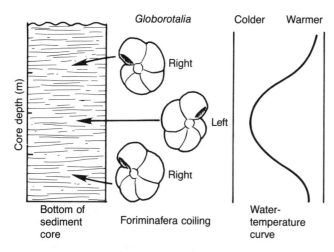

Figure 16-6. Coiling direction of the planktic foraminifer *Globorotalia truncatulinoides* is controlled by water temperature. Preservation of their tests in ocean sediments provides a clue to recognizing temperature changes.

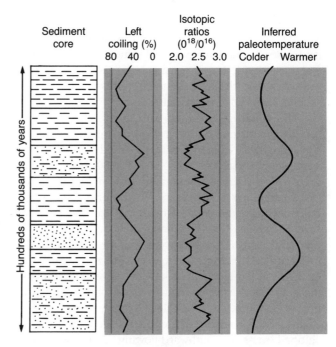

Figure 16-7. Simplified paleotemperature curve for ocean water based on oxygen-18:oxygen-16 ratios and coiling of planktic Foraminifera.

as ocean temperatures change. The amount of oxygen-16 decreases as the temperature of seawater decreases. Because Foraminifera as well as many other organisms construct their calcite tests from elements in the surrounding seawater, long-term fluctuations in ocean temperatures associated with glacial or interglacial episodes would affect the oxygen-isotope ratios in tests of successive populations of planktic Foraminifera. Accumulation of these Foraminifera in the sediments would provide a picture of changing climates (Fig. 16-7). The heavier isotope, oxygen-18, becomes more concentrated in seawater during times of high evaporation and high snowfall accumulation on the ice caps. The decreasing percentage of oxygen-16 and consequent gain of oxygen-18 in marine waters during times of glaciation is reflected in the geochemistry of shells in foraminifera that accumulate in sediments. Because of their high oxygen-18:oxygen-16 ratios, shells of deep-water benthic Foraminifera, living where water temperatures seldom rose more than a few degrees above freezing, also serve as sensitive indicators of glacial advances and retreats. Much of this evidence became available through the Deep Sea Drilling Project, which was begun in 1968 with funds from the National Science Foundation. As mentioned in Chapter 1, the DSDP has also furnished considerable evidence that has proven invaluable in constructing many hypotheses of sea-floor spreading rates and directions.

SOME UNUSUAL EFFECTS OF GLACIATION

As generally depicted in physical geology textbooks, movement and erosive power of large masses of ice produce many distinctive geomorphologic features. Many of these, represented by **V**-shaped valleys such as Yosemite Valley in the Sierra Nevada and peaks such as the Matterhorn in the Alps, are very beautiful. Others, such as fjords in northern Europe, glacial deposits in the Midwest, and the very large Great Lakes, represent excavation and deposition on a grand scale. Such features are widespread and often have important implications for interpreting geologic history. Of particular significance are formation of large lakes and deposition of lake sediments such as varves, formation of large volumes of clastic particles which may form major deltas, and development of submarine canyons and deep-sea fan deposits. Indirect effects of climatic change associated with glaciation are often no less significant and include distributional and evolutionary changes in plants and animals. Migration of hominids and other mammals in Pleistocene and Holocene time was and is strongly influenced by temperature and position of sea level. We will consider some of these conditions and events.

The Great Lakes and Pluvial Lakes

Figure 16-3 is a reconstruction of the ice sheets that covered North America during Pleistocene time; in some areas the ice attained a thickness of over 3 km. Formation of many valley and piedmont glaciers was a result of climatic change and uplift of mountain ranges that were produced by extensive tectonic activity that began in the Miocene. The Rocky Mountains and Sierra Nevada were extensively glaciated. Numerous large inland lakes formed from glacial runoff in the Midwest and these drained southward into the Gulf of Mexico by way of a vast drainage system, represented today by smaller rivers such as the Mississippi.

The Great Lakes, along with many other northern lakes, were formed by ice erosion during the latest Wisconsinan glacial episode and were filled with meltwater as the ice retreated some 8000 to 12,000 years ago. Figure 16-8 represents a time-lapse sequence of the Great Lakes beginning 11,000 years B.P.; the giant Lake Algonquin occupied the sites of modern Lake Michigan and Lake Huron and covered areas now occupied by Chicago and Detroit. By 9500 years before present the ice had retreated enough to expose eastern outlets to the Atlantic, and drainage reduced the size and levels of these ancestral lakes so that they took on a nearly modern appearance. As one might expect, this change in drainage had considerable effect on the previously developed massive southward-flowing drainage systems and resulted in a decrease of flow and sediment volume in those rivers flowing into the Gulf of Mexico. As the ice sheets retreated, an immense weight of up to 23 metric tons per square meter was rapidly removed from the underlying rocks of Proterozoic and Paleozoic ages. These rocks, which had been depressed by the weight of overlying ice, began to undergo **isostatic rebound,** which continues to the present day.

Although the climate of the southwestern United States is hot and dry and the region generally lacks major rivers and lakes, conditions were different during the Pleistocene glaciations. For example, during the Wisconsinan glacial stage, changing climate produced glaciers in many of the mountains. Meltwater as runoff and increased precipitation combined to produce a large number of inland lakes, some of which reached enormous sizes. These inland lakes, known as **pluvial lakes** (Fig. 16-9), partially or completely filled many of the intermountane basins of the Basin and Range Province. For those of you familiar with the extreme conditions of aridity found today in Death Valley, California, it is difficult to visualize the existence of a lake 150 km long and almost 200 m deep. This pluvial lake, named Lake Manley after one of the pioneers who crossed the valley, has completely vanished, but old shoreline terraces and lake deposits testify to its previous existence. Another pluvial lake was ancient Lake Bonneville; a remnant of this giant lake exists today in Utah as the Great Salt Lake. Unusually wet winters since 1980 have fed considerable water into Great Salt Lake, causing a rapid rise in water level and threatening to inundate large portions of the cities on its borders. Other notable examples of pluvial lakes include giant Lake Winnepeg in Canada and Lake Lahontan in California and Nevada. Both are represented today only by remnant lakes.

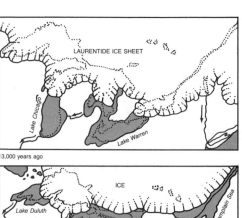

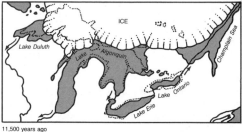

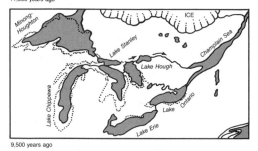

Figure 16-8. Sequence illustrating formation of the Great Lakes from erosion by glacial ice to subsequent filling by meltwater during melting of the Wisconsinan ice sheets.
(From V. K. Prest, 1970, Quaternary Geology of Canada, Fig. 7-6, pp. 90, 91, *in* Geology and Economic Minerals of Canada: *Dept. Energy, Mines and Resources Econ. Geology Rept.* 1, 5th ed.)

Deltas, Submarine Canyons, and Fans

During Pleistocene time an enormous amount of sediment was transported by large rivers that flowed from the eastern Rocky Mountains, the Great Lakes region, and the western Appalachians and emptied into the Gulf of Mexico. For example, at the mouth of the Mississippi River, the Pleistocene delta that prograded into the Gulf contains sediments exceeding 6000 m in thickness. These sediments reflect a very rapid rate of sedimentation that is probably not representative of av-

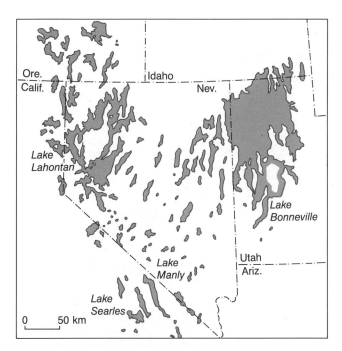

Figure 16-9. Pluvial lakes in the southwestern United States during the latest Pleistocene (Wisconsinan) ice age. Great Salt Lake is a remnant of Lake Bonneville and Death Valley contains evidence of ancient Lake Manley.
(From H. E. Wright, Jr., and D. G. Frey, eds., Quaternary Geology of the United States, Fig. 1, p. 266. Copyright 1965 by Princeton University Press, Princeton NJ. Reproduced by permission of Princeton University Press)

erage sedimentation rates through geologic time. During the Pleistocene, however, some of the most rapid rates of sedimentation in the world may have been recorded by this and other deltaic systems in various parts of the world.

During times of active glacial advance, vast amounts of water were locked up in ice, and sea levels were lowered worldwide by as much as 100 m. As the last ice age ended, approximately 8000 to 10,000 years ago, melting of these ice sheets raised sea level nearly 100 m. Many old shoreline features, including ancient beaches, wave-cut terraces, and stream valleys are now submerged. In addition, some prehistoric human settlements and campsites have been submerged. Other evidence of glacial activity during lowered sea-level conditions is provided by the existence of **submarine canyons** (Fig. 16-10). These canyons, which are especially distinctive along the west coast of the United States, may have originally begun as stream-cut channels eroded into parts of continental shelves that were emergent during intervals of lowered sea level. Some of these canyons were lengthened seaward, probably as a result of erosion by submarine turbidity flows. With rise of sea level in the last 8000 to 10,000 years, these canyons have been submerged but have been maintained and even enlarged by various processes of submarine erosion.

Another manifestation of accelerated erosion and deposition during the Pleistocene is abnormally thick submarine turbidite deposits that accumulated at the mouths of submarine canyons or at the toes of basin slopes. Coalescing fans along the toe of the Atlantic continental slope have contributed heavily to the sediments forming the continental rise.

Plant Extinctions and Animal Migrations

As we have described, climatic changes associated with glaciation had significant effects on physical characteristics globally and, on the basis of our discussions of evolution (Chapter 7), we would expect to see changes in plant and animal taxa in the Pleistocene record. In fact, cumulative changes in climate resulted in extinction of a variety of taxonomic groups, redistribution and migration of many taxa, and appearance of new species. This highly complex interaction of organisms and environment bears a closer look. You might also consider the following discussion in relation to the evolutionary history of humans (Chapter 15).

Evidence of the influence of changes in climate is afforded by the modern distribution of plants. From your own experiences you can recall the most obvious examples: abundance of cacti in arid climates; abundant grasslands of the prairies, which are now mostly converted to farmland; Arctic tundra and tropical rain forests. Each of these plant associations is the result of a set of climatic conditions and, to a lesser degree, of soil characteristics. Changes in climate from cooler glacial to warmer interglacial have influenced the distribution of many of these plant taxa. Cold-temperature floras spread during glacial episodes, whereas tropical and arid floras become more widespread during the interglacial episodes.

Redistribution of plants also affects the distribution of animals. Two examples that illustrate migration patterns of animals involve the history of the Bering Strait (Fig. 16-11) and the Isthmus of Panama. The Bering Strait in the north Pacific is today a region of shallow water between Asia and North America and is a barrier to migration by land. However, paleontologic evidence indicates that the two land areas were connected during early Cenozoic time, thereby affording terrestrial animals a migratory pathway. In the late Miocene the straits were submerged, thus curtailing terrestrial migrations. During low-water stages of Pleistocene glacial advances, however, the area was episodically exposed and again served as a migratory corridor. Humans migrated across from Asia to North America during an episode of lowered sea level.

The Isthmus of Panama is a thin strip of land that was uplifted tectonically in late Cenozoic time. Its width is greatly affected by sea level. The isthmus has provided

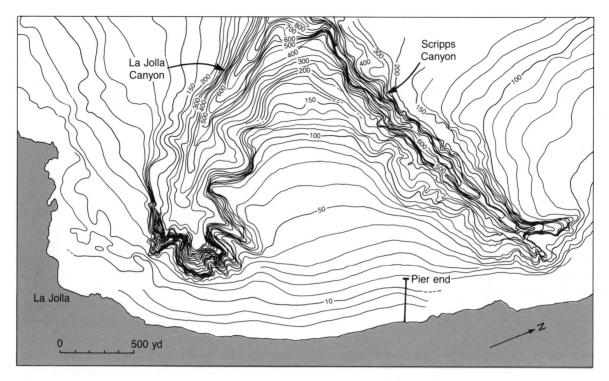

Figure 16-10.   Modern submarine canyons, such as Scripps Canyon extending offshore from La Jolla, California, probably began as river eroded channels during lowered sea level and currently serve as funnels for large amounts of sediment transported offshore into deeper marine environments. Such sediments may accumulate as deep-sea fan deposits.
(Adapted from F. P. Shepard, 1971, *The Earth Beneath the Sea,* Fig. 55, p. 117: Atheneum Press, reprinted by permission of Johns Hopkins University Press, Baltimore)

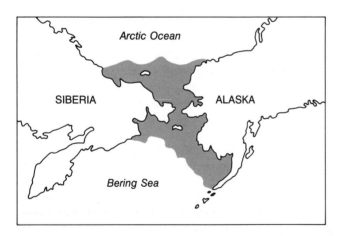

Figure 16-11.   The Bering Straits provided a wide migratory pathway for terrestrial vertebrates during lowered sea level but is completely covered during interglacial episodes such as are occurring today.
(Adapted from C. L. Matsch, 1976, *North America and the Great Ice Age,* Fig. 3, p. 101: McGraw-Hill, New York. Reproduced by permission of McGraw-Hill Book Company)

a convenient pathway for animal and plant migrations between North and South America and many mammal species migrated southward and northward between the two continents. The emergence of the isthmus, enhanced by lowering of sea level at times during the

Pleistocene, provided a **land bridge** that enabled many species to invade South America. Although some South American taxa were little affected by this invasion and some, such as ground sloths, armadillos, and opossums, extended their range northward, a major extinction occurred in South American mammals.

The fossil record of the Pliocene Epoch indicates relatively few extinctions in mammal taxa. By the beginning of the Pleistocene, however, major changes can be observed. Many taxa were severely reduced or became extinct and new mammals evolved and became adapted to the recurring severe ice-age climates. Examples that come to mind are the mammoths and mastodons, the Irish elk, and a variety of other giant species. Warmer climates also supported such well-known forms as the saber-tooth cats and giant ground sloths. Large numbers of these and many other taxa are preserved in the La Brea Tar Pits in southern California. The end of the Pleistocene and beginning of the Holocene is marked by extinction of many large mammals. Perhaps these extinctions were a product of climatic changes, but another influence of major significance may have been the spread of humans.

Medial and late Cenozoic climatic changes may have exerted significant selective pressure on the evolutionary history of primates. As we have previously noted, extensive grasslands had developed by Miocene time, probably in response to cooling and increasing aridity that had begun in the late Eocene. Coincidentally, the fossil record of the late Miocene provides examples of ground-dwelling higher primates, including ancestors of apes and humans. In Pliocene and Pleistocene time most hominid evolution occurred in Africa, a region of relatively warm climate, but during the Pleistocene, long-term fluctuations in rainfall would have corresponded to glacial and interglacial episodes. Appearance and early migration patterns of *Homo* were controlled by climatic conditions in much the same fashion as that of other terrrestrial animals. For example, it is probable that humans migrated northward from Africa into Europe and eastward into Asia during a medial Pleistocene interglacial stage. Evidence for existence of *Homo* in Europe and Asia is provided by preserved fossils and artifacts. Particularly well documented are the fossils of "Peking man" *(Homo erectus pekinensis)*. These fossils, along with evidence of fire and remains of plants and other animals provide evidence that caves near the modern city of Beijing (Peking), China, were inhabited from about 460,000 to 230,000 years ago during an interglacial stage. It is probable that descendants of this species in the form of *Homo sapiens* reached North America by crossing the Bering land bridge much later during lowered sea levels of the Wisconsinan glacial stage.

EVIDENCE OF PRE-PLEISTOCENE GLACIATION

Evidence obtained from a variety of disciplines during the last 100 years has allowed a broad understanding of the global effects associated with glacial and interglacial episodes. A variety of climatic indicators and distinctive geomorphologic and sedimentologic features are known to be associated with glaciation and can be used to recognize the extent of ancient ice advances. Geologists have applied this information to the rock record and have recognized several Phanerozoic and pre-Phanerozoic glacial episodes prior to the Pleistocene ice age.

The evidence from many widely scattered localities on several continents indicates that glacial episodes occurred in late Proterozoic to Ediacarian, early and late Paleozoic, and of course the Pleistocene (Fig. 16-12). These climatic interpretations are based on lithologic features such as glacial pavement and tillites, on preserved fossil plants that indicate cold climates, and on paleomagnetic data that indicate paleolatitudes of continents. A combination of this evidence plus that of other climatic indicators such as red beds, evaporites, and isotopic ratios has been used to construct paleotemperature curves for late Proterozoic and Phanerozoic time.

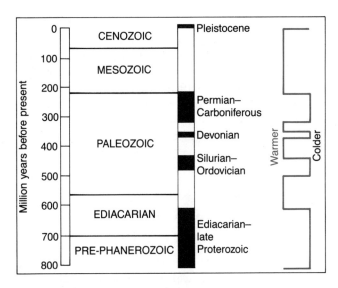

Figure 16-12. Paleotemperature curve for the past 800 million years indicates a number of major glacial episodes (dark color), including that of the Pleistocene Epoch.

As you might expect, accuracy of these curves diminishes with increasing antiquity of the record.

Recently reported evidence indicates that cooler climates and polar ice caps may have occurred more commonly in the past than previously thought. Such conditions, however, may not have been rigorous enough to produce major glacial episodes. For example, sedimentary rocks occurring between radiometrically dated lava flows on Iceland have yielded pollen of conifers, angiosperms, and other cold-temperature-adapted vascular plants that suggest a rapid drop in temperature between 9 and 10 million years ago. This palynological evidence indicates that cooling and possibly glaciation occurred in parts of the northern hemisphere as early as late Miocene, considerably predating the traditional Pleistocene ice ages.

Cooler climate conditions and the associated glacial episode of the late Paleozoic illustrate how evidence obtained from different disciplines can provide support for a hypothesis. In our discussion of late Paleozoic time, a reconstruction of Pangaea was described (Figure 11-68). Paleomagnetic evidence from volcanic rocks of this age indicates that much of the Pangaea supercontinent was located in high latitudes of the southern hemisphere, with the land masses generally clustered around the geographic South Pole. Sedimentary rocks of late Paleozoic and early Mesozoic age in what are now South America, Africa, Australia, and India contain glacial pavement and tillite. Associated plant fossils of the *Glossopteris* flora show by their leaf shapes and the presence of growth rings on stems that they were adapted to cool or cold climates.

Compared with graphs for older events, a relatively precise graph of temperature changes has been developed for the Pleistocene ice ages (Fig. 16-13) and provides a well-documented record of temperature fluctuations, especially for the last 500,000 years. It is evident that larger scale temperature changes are punctuated by small-scale pulses, thus providing a rather jagged curve. An interesting aspect of the graph is the cyclicity of climatic fluctuations. If this information is used to predict future climate, the present warm interglacial interval will cool as we begin to enter a new ice age.

GLACIATION— SEARCH FOR A CAUSE

The Earth's Orbit and Heat Budget

We have noted that climatic and geomorphic changes are associated with past episodes of glaciation and that these episodes are cyclic. Furthermore, although the earth has experienced major glacial events at least five times in the last 800 million years, much of this geologic time was characterized by mild climate. Knowledge of this climatic history and glaciation provokes some questions. How can we explain advances and retreats of the ice? What conditions are responsible for initiating glaciation? Are these factors the same for each glaciation, or could different sets of conditions occurring at different times have produced the same effect?

There are a few assumptions we can use to establish the groundwork for developing more complex ideas. One main assumption is that there has been only one source of external heat reaching the earth's surface for the last 4600 million years. This source is the sun. On the surface and within the interior of this star, nuclear reactions fuse hydrogen to helium to produce heat in the form of solar radiation. Incoming solar radiation provides a **heat budget** for the earth. We have evidence that indicates the sun's output has remained relatively constant through the Phanerozoic. Small variations in

solar radiation do occur and are associated with increased sunspot activity, but sunspots are of short duration and are not a significant factor in climatic controls over the tens or hundreds of thousands of years associated with major advances and retreats of the ice caps. However, it is possible that sunspot activity could act as a trigger mechanism in conjunction with other phenomena.

A second assumption is that the earth's orbit around the sun has remained a relatively constant elliptical pathway over geologic time. Thus, our distance from the sun fluctuates yearly from about 146 million km to 151 million km; these distances have remained nearly constant since early Archean time. Fluctuations in the rotational axis of the earth itself appear to be cyclic and may have an important bearing on global temperatures.

Search for an explanation of the ice ages has provided varied hypotheses. The search is still not complete, but we can distill a number of ideas and discuss some of them. Of the many proposed explanations, the following are the most common: (1) increase in tectonic activity and volcanic activity, (2) changes in atmospheric composition, (3) variations in ocean water circulation, and (4) astronomical fluctuations. Many of these possible explanations and others are discussed in detail in *Glacial and Quaternary Geology* by Richard F. Flint.

Changes in the Atmosphere

At intervals during the earth's geologic history, widespread tectonic activity has produced extensive mountain systems; this activity is most evident in but is not restricted to the late Paleozoic, the late Mesozoic, and the Cenozoic. Early in the history of glacial studies, an association of extensive mountain building with Pleistocene glaciation was recognized. Alteration of wind patterns resulting from the high elevation of many young mountain ranges, such as the Alps and Rockies, provides conditions favorable for formation of valley

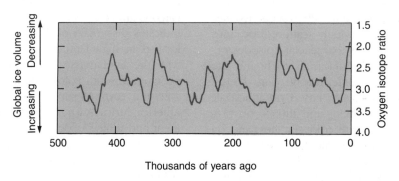

Figure 16-13. One example of a relatively detailed paleotemperature curve for the last 500,000 years; based on oxygen isotope ratios of Indian Ocean sediments.
(Adapted from ICE AGES: Solving the Mystery, Fig. 40, p. 169, by J. Imbrie and K. P. Imbrie, 1979, Enslow Publishers, Hillside NJ)

glaciers, and such glaciers may be extensively developed. These mountain systems, however, cannot provide the large-scale and cyclic climatic changes necessary to explain the waxing and waning of the Pleistocene ice sheets, so there does not appear to be a causal association between mountains and major glaciation.

The spectacular eruption of the volcanic island of Krakatoa in 1883 was heard 5000 km away in Australia, caused a large tsunami, and created a vast amount of dust that remained in the upper part of the atmosphere for many years. This atmospheric dust resulted in a measurable cooling of the earth's surface and led to the suggestion that episodes of glaciation could be initiated by large-scale eruption of one or more volcanoes with production of large quantities of dust. This idea held that such dust would decrease the amount of solar radiation reaching the earth's surface, thereby producing a cooler climate and initiating the onset of widespread glaciation. As the dust gradually settled, the climate would warm and the glaciers would recede.

A comparison of episodes of volcanic activity, including major caldera eruptions, fails to provide a close correlation between volcanism and widespread glaciation. However, a major volcanic eruption might trigger a glacial episode in conjunction with other causes. Just such an idea was discussed in a recent article in the journal *Science* by J. R. Bray,* who showed that major volcanic eruptions closely preceded the Pleistocene glacial stages and may have triggered them.

A related idea suggests that variations in carbon dioxide concentration in the atmosphere affect surface temperatures. Although there is some disagreement as to the actual climatic response, most scientists consider that an increase in CO_2 would cause temperatures to rise. This phenomenon, known as the **greenhouse effect,** has been discussed at length in many recent newspaper and magazine articles. Much concern has been expressed about the significant increase in atmospheric carbon dioxide produced by burning of vast quantities of wood, coal, and other petroleum materials in the last 250 years. There is as yet no evidence that episodic fluctuations of CO_2 have occurred in the past or that this could trigger glaciation. However, we do not know the overall changes to the earth's climate that could occur because of the induced greenhouse effect.

Variations in Ocean Water and Circulation

In 1956 Maurice Ewing and William Donn of Lamont Geological Observatory proposed an interesting hypothesis involving alternating episodes of freezing and thawing of water in the Arctic Ocean, to account for cyclic glaciation. In their model, an ice-free Arctic Ocean would provide a source area for evaporation of large amounts of water vapor that could then precipitate as snow and ice on polar land areas. The buildup of ice would cause a drop in sea level and shut off circulation between the Arctic and the North Atlantic Oceans. This shallowing and decrease in circulation of the Arctic Ocean would allow it to freeze, thus eliminating the source of moisture for glaciers. As the glaciers melted, sea level would rise and the Arctic Ocean would thaw. This process would theoretically be cyclic, but the hypothesis suffers from the fact that studies of Arctic bottom sediments provide no evidence that the Arctic Ocean was ice free at any time during the last few million years.

Astronomical Fluctuations

Our last hypothesis suggests that climatic fluctuations could be caused by variations in the earth's orbit. An early version of this idea was proposed in 1842 by the mathematician J. A. Adhemar, but his ideas, along with subsequent revisions by James Croll, were rejected by most geologists. Beginning in the early 1900s, the Serbian astronomer Milutin Milankovitch provided mathematical evidence of irregularities in the earth's orbit. From years of study, beginning in the early 1900's, he recognized a cyclicity to these irregularities and in the early 1940s proposed his hypothesis of orbital variations. Subsequent work by many scientists has substantiated and refined this evidence and provided a correlation between orbital variations and climatic conditions; these correlations are recognized in cores from the Deep Sea Drilling Program (DSDP).

Milankovitch proposed that recognized irregularities in the earth's orbit are cyclic and involve (1) ellipticity of the earth's orbit, with a 92,000- to 123,000-year bimodal cyclical variation; (2) inclination or obliquity of the earth's axis, with a 41,000-year cycle; and (3) precession of equinoxes, with a 21,000-year cycle (Fig. 16-14). Variations of these orbital conditions affect the amount of solar radiation received at the earth's surface and thus influence the global heat budget. Each factor by itself is not very significant, but at approximately 40,000-year intervals the three conditions reinforce each other and would be especially pronounced at higher latitudes.

The Milankovitch hypothesis has recently been strongly supported by John Imbrie and James D. Hays, geologists at Columbia University. Along with a team of scientists from around the world, they have provided paleontologic and isotopic evidence from many DSDP sediment cores that indicate close correlations of paleoclimatic changes with the astronomical cycles of Milankovitch. Further evidence indicates these cycles can be recognized in oceanic sediments nearly 10 million years old. Apparently, however, paleotemperature curves in-

---

*Pleistocene Volcanism and Glacial Initiation: *Science,* vol. 197, pp. 251–253.

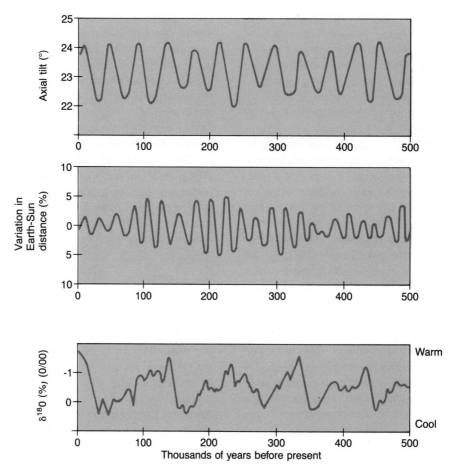

Figure 16-14. Curves comparing astronomical variations of precession of equinoxes (21,000-year cycle) and changes in inclination (41,000-year cycle) with variations of oxygen-18 content of deep-sea sediments.
(A, B, from W. Broecker and J. van Donk, p. 188, *Reviews of Geophysics,* vol. 8, 1970. Copyright by the American Geophysical Union: C from C. Emiliani, 1978, The Causes of the Ice Ages, p. 351, 353: Earth and Planetary Science Letters, Elsevier Science Publishers, Inc.)

dicate effects of the 100,000-year cycle were stronger over the last 2 million years than between 2 and 8 million years ago. It would appear that influence of Milankovitch cycles is enhanced at certain times by one or more other events; a recent idea is that size of the ice sheets themselves may influence the earth's heat budget.

Although each of the hypotheses discussed has provided a possible explanation for Pleistocene glaciation,

all have at least one drawback or fail to explain some aspect of glaciation. A strong possibility exists that a more complete and satisfactory answer lies in a combination of hypotheses. For example, combining plate motions and position of the continents with the patterns of the Milankovitch cycles might offer a solution. Possibly a new concept by itself or in conjunction with previously proposed ideas might provide the answers.

SUMMARY

Extensive work by Louis Agassiz beginning in the 1830s established a foundation for modern concepts of glaciation. Glacial pavement, erratic boulders, moraines, and other features were recognized in many areas of Europe, Asia, and North America and indicated existence of extensive ice sheets in these regions. Mapping of glacial features such as till deposits indicated the cyclic na-

ture of glacial and interglacial episodes and demonstrated that at least five major advances of ice sheets occurred during the last 2 million years. Maximum advance occurred during the Wisconsinan glacial age as ice covered much of North America as far south as 40° north latitude.

Along with waxing and waning of ice, other global changes occurred as a result of climatic variations. Formation of large pluvial lakes, major alteration of drainage systems, fluctuations in sea level, and other features have been documented by a variety of techniques. Of particular usefulness are studies of varved lake sediments, plant growth rings, variations in coiling directions of some species of planktic Foraminifera, and isotopic ratios in deep-sea sediments. All of these track changes in climate; they are especially significant if they can be used in combination.

Late Cenozoic climatic changes have affected evolutionary history and distribution of many plant and animal taxa. Adaptations to arid or cold conditions have been followed by large-scale extinctions. Sea-level changes have opened or closed migratory pathways and subsequently contributed to invasions and extinctions of taxa on various continents. It is likely that fluctuations during the Pleistocene influenced evolutionary history and distribution of hominids, including those of our own genus *Homo*.

Search for a causal explanation for cyclic glaciation has resulted in a number of diverse hypotheses. At present the most promising idea, proposed by Milankovitch, involves cyclic irregularities in the earth's orbit. These fluctuations combine at intervals and may influence the global heat budget enough to slightly cool or heat the atmosphere, thereby initiating either a glacial or interglacial stage. One appealing aspect of this hypothesis is its potential for explaining other episodes of glaciation noted in the geologic record of the last 800 million years. Further understanding of the Milankovitch cycles and collection of considerable data from modern weather studies may allow us to provide accurate short- and long-term climate predictions.

SUGGESTIONS FOR FURTHER READING

Agassiz, L., 1967, *Studies on Glaciers,* A. V. Carozzi, ed. and trans.: Hafner, New York.

Beaty, C. B. 1978, The Causes of Glaciation: *Am. Scientist,* vol. 66 no. 4, pp. 452–459.

Eddy, J. A. 1977, The Case of Missing Sunspots: *Sci. American,* vol. 236, pp. 80–92.

Flint, R. F., 1971, *Glacial and Quaternary Geology:* John Wilcy & Sons, New York.

Imbrie, J., and Imbrie, K. P., 1976, *Ice Ages:* Enslow Publishers, Short Hills, NJ.

Lurie, E., 1960, *Louis Agassiz: A Life in Science:* University of Chicago Press, Chicago.

Matsch, C. L., 1976, North America and the Great Ice Age: McGraw-Hill, New York.

Matthews, S. W., 1976, What's Happening to Our Climate: *Nat. Geographic,* vol. 150, no. 5, pp. 576–615.

Rukang, W., and Shenglong, L., 1983, Peking Man: *Sci. American,* vol. 248, no. 6, pp. 86–94.

GLOSSARY

Acadian orogeny Mountain-building event (*see* orogeny) affecting the Appalachian belt from South Carolina to Newfoundland during mid-Paleozoic time; associated with closing of the proto-Atlantic basin (Iapetus); European counterpart is the Caledonian orogeny.

Actualism Modified view of the concept of uniformitarianism; recognizes that earth processes of past and present are similar, but that rates, intensities, and relative importance of these processes have varied through time. *See* uniformitarianism.

Adaptive radiation Proliferation of new species (or higher taxa such as genera, families, etc.) from one or a few ancestral stocks; adaptive radiation is a major evolutionary pattern and has produced much of the present diversity of life forms.

Alleghenian orogeny Late Paleozoic mountain-building event (*see* orogeny) that completed the deformation of the Appalachian geosyncline; manifested mainly by folding and thrust faulting of rock of the Appalachian miogeosyncline; marks completion of the closing of the proto-Atlantic basin (Iapetus); European and North African counterpart is the Hercynian orogeny.

Allopatric speciation Model of formation of new species (speciation) where a reproductively isolated portion of a population changes rapidly through a succession of generations and becomes a genetically distinct new species; once formed these new species undergo little further genetic or morphologic change through time.

Anaerobic Capable of living in an environment characterized by a lack of free (uncombined) oxygen (anoxic environment).

Andesite The fine-grained igneous rock making up volcanic island arc chains and magmatic arcs.

Angiosperm Flowering plant, the most advanced type of terrestrial plant; characterized by flowers and seed-bearing fruits; seed is protected. *See* gymnosperm.

Angular unconformity Stratigraphic surface (erosional surface) between underlying deformed strata that dip at a different angle than the younger overlying strata. *See* unconformity.

Animal Multicellular organism (metazoan) that obtains nutrients by consumption of other organisms.

Anthropoidea Suborder of advanced primates, including monkeys, apes, and humans. *See* Prosimii.

Antler orogeny Mid-Paleozoic mountain-building event (*see* orogeny) in the Cordilleran geosyncline; best displayed in northern Nevada and Idaho (Roberts Mountains thrust).

Aphebian Era Earliest era of the Proterozoic Eon; 2500–1600 million years ago. *See* Helikian and Hadrynian Eras.

Arboreal Tree dwelling.

Archaeocyathid Extinct animal form of uncertain biological affinity, but generally classified as a separate phylum; probably intermediate between Porifera and Coelenterata; confined to the Cambrian Period.

Archean Earlier part of Cryptozoic time beginning with the oldest rocks of the earth's crust about 3800 million years ago and closing with the Kenoran orogeny 2500 million years ago. *See* Proterozoic.

Arthropod Organism belonging to most abundant and diverse group of animals; characterized by segmented body and jointed limbs; includes extant crabs, shrimps, spiders, insects, and extinct trilobites, among many others.

Artiodactyl Any of an order of even-toed (two or four toes on each foot) ungulates (hoofed mammals), including deer, cattle, pigs, and sheep. *See* perissodactyl.

Asthenosphere Weak "plastic" zone of the earth's interior structure characterized by relatively low seismic wave velocities; lies below the earth's lithosphere.

Astrobleme Crater on the surface of the earth that was formed by impact of meteoric or cometary debris.

Asylum area Geographic area in which an organism becomes isolated and is not affected by evolutionary trends that occur elsewhere; *example,* Australia and its distinctive marsupial mammals.

Atmosphere Gaseous layer surrounding the earth and held to the earth by gravity; composition 78 percent nitrogen, 21 percent oxygen, plus traces of argon, carbon dioxide, water vapor, and other gases.

Atomic number Number of protons in an atomic nucleus; this number characterizes each specific chemical element and determines the position of an element on the periodic chart.

Australopithecine One of a group of Pliocene–Pleistocene fossil hominids (humans) found in eastern and southern Africa.

Autotroph Producer organism. *See* producer.

Back reef Lagoonal area between reef and mainland; typically characterized by quiet water and deposition of limy muds.

Baltica One of six major continental blocks believed to have been in existence by Cambrian time; essentially equivalent to what is now Europe.

Banded iron formation (BIF) Iron precipitate deposits formed by the combination of ferrous iron in the sea with oxygen generated by early photosynthesizers; interbedded with chert; unique to strata of Cryptozoic age.

Base level Lowest elevation to which a portion of the earth's surface may be eroded; usually sea level.

Basin   Depressed area into which the surrounding slopes drain.

Benthos   Bottom dwellers; organisms that live on (epifauna) or within (infauna) a sea floor or lake bed.

Bentonite   Clay formed by the alteration in place of volcanic ash beds; may be significant time indicator because it is formed by individual eruptive events.

Binocular vision   Vision employing both eyes at once; ability to see with a depth-of-field vision; an important primate characteristic.

Biofacies   Sedimentary facies defined on the basis of fossil content rather than lithology.

Biogeographic province   Geographic area that supports similar flora and fauna; such an area is typically defined by environmental factors such as climate, temperature, topography, water depth, and salinity.

Biosphere   All living creatures of the earth; the array of life forms.

Biostratigraphic unit   Rock unit whose boundaries are defined on fossil content (one or more species or other taxonomic rank); fundamental biostratigraphic unit is the biozone.

Biozone   Succession of strata that is characterized by one or more fossil taxa; fundamental biostratigraphic unit.

Bipedal gait   Characteristic of walking with two feet.

Bivalve   Mollusc characterized by two shells typically of equal size and shape (some sedentary forms are asymmetrical); each valve nonbilaterally symmetrical; includes clams, scallops, and oysters.

Blueschist   Metamorphic rock characterized by the presence of the mineral glaucophane (blue amphibole); forms in a high-pressure, low-temperature metamorphic environment.

Boundary stratotype   Stratigraphic sequence that is a type section for a time-stratigraphic (chronostratigraphic) boundary; *example,* the Plio-Pleistocene boundary stratotype is in Italy. *See* stratotype, type section.

Brachiopod   Marine animal that secretes two shells (valves) typically unequal in size and/or shape; each shell bilaterally symmetrical; popularly called "lampshell."

Brain capacity   Amount of space (usually measured in cubic centimeters) within the skull that is occupied by the brain; has been used as a measure of development of primate evolution.

Browser   Land-dwelling animal that feeds on shoots, twigs, and leaves of trees and shrubs.

Bryozoa   Small colonial animals that secrete calcareous structures of various size and shape; popularly called "moss animal."

Cambium   Innermost portion of three epidermal layers (bark) of a plant stem; location for growth of the plant. *See* cortex.

Canadian Shield   Vast lowland rimming the Hudson Bay region and making up the eastern two-thirds of Canada, the United States margins of Lake Superior, and most of Greenland; the most extensive area of exposed Cryptozoic rock on the North American continent. *See* shield.

Carbonaceous film   Mode of fossil preservation by which a carbon residue forms the shape of the former living organism; most often found on bedding surfaces of dark shales.

Cast   Filling of a void left by the removal (dissolution) of a shell or other form of organic material. *See* mold.

Catastrophism   Concept that geologic changes were the result of sudden, violent, worldwide catastrophic events.

Cell   Mass of protoplasm enclosed within a membrane (often a cell wall) and usually including one or more nuclei (monerans lack a nucleus); basic unit of living organism.

Cephalopod   Mollusc characterized by a well-developed head with eyes, tentacles, and the ability to jet propel through the water; univalve shell takes various forms; includes extant squid, octopus, and chambered nautilus and extinct ammonoids.

Chloroplast   Site of the photosynthesis reactions of plants; located in the cells of most algae (not in cyanobacteria) and in the cells of the leaves of higher plants.

Chronozone   Stratigraphic unit defined by fossils (biozone); deposited during an interval of geologic time and thus has time-stratigraphic significance.

Class   Major subdivision in the classification of life forms; category just below phylum level.

Clastic wedge   Thick sequence of clastic sediments derived from and often deposited adjacent to a tectonic land mass.

Coacervate droplet   Very small organic globule that possibly existed as an intermediate form before appearance of living cells.

Collage   Assemblage of fragments of materials; collage *tectonics* refers to an accretion of fragments called microplates; much of western North America is viewed as a collage of accreted terranes.

Community   Association of organisms living in close proximity with one another; part of an ecosystem.

Conodont   Tiny tooth-shaped form of unknown zoological affinity; important Paleozoic fossil.

Consumer   Organism that is not able to directly produce its own food by photosynthesis, but consumes other organisms; animals and some protistans.

Convergent plate boundary   Boundary characterized by lithospheric plates moving *toward* one another; includes ocean–ocean convergence along island arcs, oceanic lithosphere underthrusting continental lithosphere, and continental collisions. *See* subduction zone.

Coral   Marine coelenterate that occurs as both a solitary individual and in colonies; secretes a calcareous external skeleton; reef builder.

Cordilleran orogeny   Mountain-building event (*see* orogeny) in western North America extending from Jurassic time into the early Cenozoic Era; began in the western part of the Cordilleran belt and progressed eastward through time. *See* Nevadan, Sevier, and Laramide orogenies.

Correlation   Matching of rock units on the basis of lateral continuity of lithology, typically on a local or regional basis; matching of rock units on the basis of similarity of age (biostratigraphic) on local, regional and worldwide levels; correlation of rock units on a physical basis often does *not* mean correlation in a time sense.

Cortex   Outermost portion of three epidermal layers (bark) of a plant stem. *See* cambium.

Cosmopolitan   Having a wide, unrestricted geographic distribution. Opposite, *see* endemic.

Craton   Relatively stable and immobile area of a continent; usually of large size and including both shield and adjacent platform; *example,* North American craton includes the Canadian Shield and the adjacent interior lowland.

Cratonic sequence   Large-scale, rock-stratigraphic sequence that represents major onlap-offlap cycles; bounded by unconformities of cratonwide extent.

Crinoid   Marine echinoderm characterized by a cup-shaped body and a series of branching arms; popularly called "sea lily."

Cross-cutting   Principle of relative dating; rock units, such as igneous intrusions, and geologic features, such as faults, that cut across preexisting rocks and structures are younger than the rocks/features they cut.

Cryptozoic   Largest span of geologic time; defined to include all of geologic history prior to the earliest evidence of visible animal life; in this text includes time prior to the Ediacarian Period; 4600–700 million years ago. *See* Precambrian.

Cyanobacteria   Blue-green bacteria.

Cyclothem   One of a series of rhythmically repetitious sedimentary sequences that include fluvial, brackish-water, and marine sediments; characteristically contains coal beds.

Dendrochronology   Interpretation of tree rings, especially in applications of tree-ring dating; coupled with radiocarbon dating provides very detailed age determination and correlation for the last several tens of thousands of years.

Desiccation crack Crack formed in fine-grained sediment, such as clay, due to shrinkage of the sediment as it dries.

Diamictite General term referring to any poorly sorted, noncalcareous, terrigenous sedimentary rock; term does not imply a specific depositional environment.

Diapir Intrusion of less dense rock material in a viscous solid state through overlying rocks (core of such an intrusion often is salt). *See* salt dome.

Diastem Depositional break with conformable strata; break represents a very short time (days, weeks, centuries) in a sequence of rocks representing essentially continuous sedimentation; not characterized by significant faunal or floral changes. *See* unconformity.

Diatom Single-celled plant that secretes a siliceous frustule (shell) of diverse and often ornate form; microscopic in size; important base of some food chains.

Diatomite Sedimentary rock composed mainly of the silica of diatoms.

Dicotyledon Plant belonging to one of two main groups of angiosperms (flowering plants); dicots are the more primitive group and evolved earlier (Triassic Period). *See* monocotyledon.

Disconformity Unconformable surface between essentially parallel strata; erosional surface is evident and represents a significant gap in the stratigraphic record; typically characterized by a change in fauna and/or flora. *See* unconformity, diastem.

Divergence of species Separation of species into two or more distinct new species.

Divergent plate boundary Boundary characterized by lithospheric plates moving *away* from one another; *example,* an ocean-floor spreading center such as the Mid-Atlantic Ridge.

DNA molecule (Deoxyribonucleic acid) Complex organic molecule (protein) that contains the genetic material of an organism; these molecules form genes located along strands (chromosomes) that control critical chemical reactions with cells that are significant in controlling physical characteristics.

East Pacific Rise Portion of the global oceanic ridge system that extends north–south on the eastern side of the Pacific Ocean floor.

Echinoderm One of a diversified group of marine animals that is typically characterized by five-rayed symmetry; starfish, sea urchins, crinoids.

Echinoid Mobile echinoderm characterized by globular body without arms or stem; sea urchins and sand dollars.

Ecosystem Complex interrelationships within a group of organisms and between the group and the environment; interaction of both the living and the nonliving components; *examples,* pond ecosystem, forest ecosystem.

Ectotherm Cold-blooded animal; organism lacks the ability to produce internal body heat and so body temperature fluctuates readily (reptiles).

Elsonian orogeny Mountain-building event (*see* orogeny) that occurred approximately 1300 million years ago in the Canadian Shield (during the Helikian Era).

Embryophyta Subkingdom of plants that includes the terrestrial plants; with the exceptions of mosses and liverworts, embryophytes are vascular plants (tracheophytes).

Endemic Restricted to a specific locality (biogeographic province); *example,* specialized finches of the Galapagos Islands. Opposite, *see* cosmopolitan.

Endotherm Warm-blooded animal; organism has the ability to produce internal body heat and thus maintain a near-constant body temperature (mammals).

Eon Largest abstract unit of geologic time (see Cryptozoic and Phanerozoic Eons); subdivided into eras. *See* eonothem.

Eonothem Highest ranking time-stratigraphic term; used for a sequence of rocks formed during a specific eon of geologic time; designation above erathem in rank; *example,* Phanerozoic Eonothem.

Epeiric sea Widespread shallow sea that inundates portions of the interior of a continent; brought about by sea-level changes affecting whole or major portions of continents.

Epeirogeny Broad, relatively gentle uplift or downwarp over a wide geographic region (as opposed to the more intense deformation of an orogeny).

Epidermal layer One of the three outer layers of the structure of a plant stem; bark. *See* cambium, cortex.

Epifauna Benthic (bottom-dwelling) organisms that live on a sea floor or lake bed.

Epoch Abstract unit of geologic time; subdivision of a period and subdivided into ages; *examples,* Late Cambrian Epoch, Early Ordovician Epoch; time equivalent of series. *See* series.

Era Abstract unit of geologic time; subdivision of an eon and subdivided into periods; *example,* Cenozoic Era, Mesozoic Era. *See* erathem.

Erathem Time-stratigraphic term for those rocks formed during a specific era of geologic time; designation above system in rank; *examples,* Cenozoic Erathem, Mesozoic Erathem.

Erect posture Standing/walking/running in an upright position.

Erratic boulder Boulder foreign to the area in which it is found; usually deposited by a glacier.

Eugeosyncline Outer belt (seaward side) of a geosyncline marginal to a craton; characterized by deep-water clastic sediments with associated volcanics. *See* geosyncline, miogeosyncline.

Eukaryote Life form with a cell(s) that contains an organized nucleus; includes protistans, fungi, plants, and animals. *See* prokaryote.

Eurytopic Able to survive and reproduce under a wide range of environmental conditions; wide tolerance of conditions. Opposite, *see* stenotopic.

Evolutionary theory Concept of organic evolution (development of life forms through time); supported by considerable biological and paleontological evidence. *See* organic evolution.

Exotic terrane Terrane in which the rocks are foreign to the general area; believed to have originated at some other location and moved into its present position by lithospheric plate movement. *See* suspect terrane.

Extinction "Dying out" of a taxon.

Facies tract Spectrum of laterally adjacent facies that are different but genetically related; *example,* adjacent subenvironments of stream channel, natural levee, and backswamp of a delta.

Family Subdivision in the classification of life forms; below the level of order and above the level of genus.

Farallon Plate Portion of Pacific Ocean lithosphere subducted beneath a margin of the North American Plate during the Cordilleran orogeny.

Fermentation Decomposition of complex organic molecules caused by the activity of yeast (fungi) and some bacteria in an oxygen-poor environment.

Flysch General term referring to marine rocks characterized by thinly bedded shales (term flysch derived from their fissile nature) alternating with coarse sandstones and graywackes; graded bedding common; often associated with rapid erosion of an adjacent rising land mass in the early stages of an orogenic event.

Food chain Relatively simple feeding sequence within a community of organisms; *example,* plankton eaten by small fish which are eaten by big fish. *See* food web.

Food web Complex interweaving of feeding patterns formed by the interlinking of individual food chains; *example,* connections between food chains in a pond and those in an adjacent meadow.

Foraminifera Abundant and diverse group of single-celled organisms (protistans) that secrete calcareous tests; many forms are microscopic in size.

Fore reef Seaward margin of a reef structure; typically composed of a slope of debris eroded from the face of the reef (reef talus).

Formation Mappable rock unit with distinctive physical properties by which it may be identified; the fundamental lithostratigraphic unit.

Fossil Remains of an ancient life form or traces of an organism preserved in rocks of the earth's crust (does not include remains less than about 10,000 years old).

Fossil fuels Energy-producing fuel derived from organic sedimentary deposits; coal, petroleum, natural gas.

Fossil succession Changes in fossil content of a chronological sequence of rocks due to evolutionary change through time.

Gametophyte Stage in the reproductive cycle of vascular plants during which the sex cells (gametes) are produced.

Gastropod Mollusc characterized by coiled univalved shells, each shell being a single chamber; snails.

Gene pool Collective genetic material of a population of one species of organism.

Genotype Genetic makeup of an individual organism.

Genus Subdivision in the classification of life forms; below the level of family and above the level of species. *Plural,* genera.

Geochronology "Science of earth time"; relates geologic events to time, especially by use of absolute time (radiometric time ascribing a specific number of years ago to an event).

Geochronometric unit Direct division of geologic time defined by chronometric age (radiometric dating); a time unit does not necessarily have a corresponding rock sequence or stratotype to which it is referred.

Geologic cycle Schematic portrayal of the interrelated way the earth works; includes the physical, chemical, and biological processes; hydrologic cycle, rock cycle, and tectonic cycle are all incorporated into the larger picture of the geologic cycle. *See* hydrologic, rock, and tectonic cycles.

Geosyncline Large-sized trough or basin marginal to a continental craton; gradually subsides through an appreciable span of time and receives a thick accumulation of sediments; many have subsequently become the sites of folded mountain belts. *See* eugeosyncline, miogeosyncline.

Gondwana One of six major continental blocks believed to have been in existence by Cambrian time; formed a southern hemisphere block that included South America, Africa, India, Australia, and Antarctica.

Graben Elongate depressed fault block; forms the relatively lower basins of basin-and-range topography. *See* horst.

Graded bedding Stratification characterized by an upward change in texture from coarse-grained sediment (base of stratum) to finer grained sediment (top of stratum); especially characteristic of deposition of a turbidity current. *See* turbidite.

Granitization Transformation of preexisting rocks into granite through recrystallization and/or replacement; essentially metamorphic formation of granite.

Graptolite One of a class of extinct, colonial, floating organisms of problematic zoological affinity, but usually classified as nonvertebrate chordates (hemichordates); important guide fossil in Paleozoic time (Ordovician through Devonian Periods).

Graze Feed on herbage: grasses on land and algae in oceans.

Greenhouse effect Concept that introduction of greater amounts of carbon dioxide into the atmosphere (such as from burning of fossil fuels) will cause a rise in atmospheric temperature due to absorption of solar radiation by carbon dioxide.

Greenstone belt Belt of ancient volcanic-sedimentary rock successions found in all Cryptozoic shield terranes of the world; name derived from the presence of green minerals in the lower part of the rock succession.

Grenville orogeny Mountain-building event (*see* orogeny) that occurred approximately 1000 million years ago in the Canadian Shield (Proterozoic Eon).

Group Lithostratigraphic unit next in rank above formation; includes two or more related formations.

Growth rings Tree rings indicating yearly growth; especially distinct in plants that grow in climates characterized by seasonal changes; also found in secreted calcareous skeletons of some invertebrates.

Gymnosperm "Naked seed" plant; the seed is not enclosed as in fruit-bearing plants; *examples,* cycad, conifer, ginkgo. *See* angiosperm.

Habitat Environment to which an organism is best adapted; the place where the requirements to support an organism are available.

Hadrynian Era Most recent era of time in the Proterozoic Eon; 900–700 million years ago. *See* Aphebian and Helikian Eras.

Heat budget Analysis of the distribution of solar radiation within the geosystem; based on amount of incoming solar radiation, amount of that radiation absorbed by clouds, and amount of that radiation reflected back into space by particulate matter in the atmosphere, among other factors.

Helikian Era Middle era of time in the Proterozoic Eon; 1600–900 million years ago. *See* Aphebian and Hadrynian Eras.

Heredity Genetic material that is distinctive for each species and is passed on to successive generations through the reproduction cycle. This genetic material plays a major role in the physical characteristics of individuals and species.

Heterotroph Consumer organism. *See* consumer.

Hiatus Time interval not represented by rocks of a stratigraphic sequence; span of time when nondeposition, erosion, or both occurred. *See* unconformity.

Homiothermy Maintainance of relatively constant body temperature; warm-bloodedness.

Horst Elongate uplifted fault block; forms the relatively higher ranges of basin-and-range topography. *See* graben.

Hudsonian orogeny Mountain-building (*see* orogeny) that occurred approximately 1800 million years ago in the Canadian Shield (Proterozoic Eon).

Hydrologic cycle Basic circulation of the hydrosphere (water) of the earth—ocean to atmosphere by evaporation, to land in the form of precipitation, back to ocean by runoff and other systems; includes many important subsystems such as river systems, groundwater systems, and glacial systems; component of the general geologic cycle. *See* geologic cycle.

Hydrosphere All forms of water associated with the earth: oceans, rivers, lakes, groundwater, glacial ice, and water vapor in the atmosphere.

Iapetus Early Paleozoic proto-Atlantic Ocean in existence by the beginning of the Cambrian Period; closed during Paleozoic time because of convergence of plates bearing Laurentia (northeastern North America and Greenland) and Baltica (Europe).

Induction Drawing of a conclusion on the basis of specific facts; reasoning from the specific (body of facts) to a general conclusion.

Infauna Benthic (bottom-dwelling) organisms that live within the sediments of the sea floor or lake bed; *example,* burrowing clams.

Insectivore Primitive mammal order exemplified by moles and hedgehogs and characterized by small size, nocturnal habits, and dependency on insects for food.

Interglacial Span of time between two glacial advances.

Invertebrate   Animal lacking a spinal column (backbone) or notochord (skeletal rod).

Isostasy   Concept of a theoretical balance between large portions of the earth's lithosphere, such as continental areas "floating" on more dense substratum.

Isostatic rebound   Uplift of a portion of the earth's crust following the removal of a great weight such as glacial ice. *See* isostacy.

Isotope   Form of a chemical element resulting from change of the number of neutrons in the atomic nucleus.

Jovian planet   Any of the outer planets of the solar system (located relatively farther from the sun): Jupiter, Saturn, Uranus, Neptune; gaseous planets composed predominantly of hydrogen. *See* terrestrial planet.

Kenoran orogeny   Mountain-building event (*see* orogeny) that occurred approximately 2500 million years ago in the Canadian Shield; closing event of Archean time.

Kerogen   Hydrocarbon substance (bitumen) dispersed in some sedimentary rocks, such as the so-called oil shales, that yield petroleum when heated.

Kingdom   Major subdivision in the classification of life forms; most inclusive ("largest") of the various subdivisions.

Lake varve   *See* varve.

Land bridge   Segment of land that connects two larger land areas; *example,* the Isthmus of Panama is a modern land bridge connecting the continents of North and South America.

Laramide orogeny   Youngest phase (late Cretaceous through early Cenozoic) of the more inclusive Cordilleran orogeny; most evident in the Rocky Mountain provinces. *See* Cordilleran orogeny.

Lateral continuity   Principle that sedimentary strata, when originally deposited, are three-dimensional and extend laterally in all directions until they thin to a zero-thickness edge or terminate abruptly against the margin of the depositional basin.

Laurentia   One of six major continental blocks recognizable by Cambrian time; included present-day North America and Greenland.

Laurussia   Paleocontinent formed by middle Paleozoic suturing of Laurentia and Baltica.

Law of inclusions   Principle of relative dating that says fragments of rocks included within other rocks are older than the rocks that contain them.

Lineage   Succession of evolution-related species through geologic time; ancestry.

Lithofacies   Sedimentary facies defined on the basis of rock type (lithology).

Lithosphere   Earth's outer rigid rind that is fragmented and consists of a mosaic of plates that move over the underlying asthenosphere.

Lithostratigraphic unit   Rock unit defined on the basis of physical aspects (lithology) and not fossil content and, therefore, may not be the same age everywhere; fundamental lithostratigraphic unit is the formation.

Magmatic arc   Volcanic mountain chain along a continental margin; associated with a convergent plate boundary; *examples,* Andes Mountains of South America, Cascade Mountains of northwestern United States.

Magnetic anomaly   Any measurement of the intensity or the polarity of the earth's magnetic field that is a departure from the normal or expected measurement; *examples,* in terms of polarity, a positive magnetic anomaly measured in rocks indicates an ancient magnetic field parallel to the present field; a negative magnetic anomaly indicates the former field was oriented, in terms of north and south magnetic poles, opposite to the earth's present magnetic field.

Marsupial   Mammal that is characterized by the presence of an abdominal pouch to carry the young; *examples,* kangaroo, opossum.

Meiosis   Cell reproduction accomplished by replication of DNA strands (chromosomes) and two successive cell divisions; each resulting cell contains only half the genetic information of the parent. *See* mitosis.

Mélange   Deposit composed of a tectonic mixture of rock materials; fragments of various composition, texture, and age consolidated in a sheared matrix indicative of intense deformation.

Member   Lithostratigraphic unit that is a subdivision of a formation.

Mesosphere   Solid middle zone of the earth's interior structure; located between the plastic asthenosphere and the core of the earth.

Methanogen   Distinctive form of prokaryotic microorganism that grows by oxidizing hydrogen and reducing carbon dioxide to form methane; thrives in anoxic (oxygen-deficient) environments.

Microplate   Relatively small-sized lithospheric plate. *See* collage.

Milankovitch hypothesis   Hypothesis that climatic fluctuations causing glacial cyclicity are due to variations of the earth's axis and orbit.

Miogeosyncline   Inner belt (closer to the continent) of a marginal geosyncline; characterized by thick sequences of shallow-water sediments (similar in type to those deposited on the craton margin, but much thicker) and lack of major volcanic materials. *See* geosyncline, eugeosyncline.

Mitosis   Cell reproduction accomplished by replication of DNA and division into two smaller cells that are essentially identical to the parent cell. *See* meiosis.

Molar   One of the cheek teeth in mammals, located behind the incisors and the canine teeth; characterized by a wide-crowned surface suitable for grinding food.

Mold   Impression left in sediment after shells or other organic materials are dissolved or otherwise removed; both internal and external molds are formed, depending upon whether the form of the inside or outside of the shell is preserved.

Mollusc   One of an abundant and diverse group of animals popularly called "shellfish"; soft bodied and often with a hard shell, unsegmented, with a head and a muscular foot; mostly aquatic; *examples,* bivalves (such as clams), gastropods (snails), and cephalopods (such as squids).

Moneran   Single-celled organism lacking a discrete nucleus.

Monocotyledon   Plant belonging to one of two main groups of angiosperms (flowering plants); monocots are the more advanced group. *See* dicotyledon.

Monophyletic   Originating from a single ancestral lineage. *See* polyphyletic.

Moraine   Unsorted sedimentary debris transported by a glacier and, when the ice melts, deposited in hummocky mounds as terminal moraine, recessional moraine, lateral moraine, or medial moraine or deposited in an irregular sheet as ground moraine.

Mutation   Sudden changes in the chemical structure of genetic material of organisms that can be passed on to subsequent generations.

Natural selection   Evolutionary mechanism whereby environmental conditions affect survival rates of variations within a population of organisms; organisms best adapted to the environment will survive and produce viable offspring.

Nekton   Organisms that have the ability to swim and to control their movements in water.

Neptunism   Late eighteenth–early nineteenth century doctrine that *all* rocks were formed by precipitation from an original, primeval ocean; championed by Abraham Werner (1749–1817).

Nevadan orogeny Initial phase (late Jurassic through mid-Cretaceous) of the more inclusive Cordilleran orogeny (mountain-building event in western North America); characterized by intensive igneous activity including the formation of the Sierra Nevada batholith of California. *See* Cordilleran orogeny.

Niche Specific role or function that an organism has in a living community; *example,* in a shallow-marine environment worms fill the niche of deposit feeders and some bivalves (clams) fill the niche of filter feeders.

Nocturnal Active at night.

Nonconformity Unconformable surface that separates older crystalline rock (igneous or metamorphic) from younger, overlying sedimentary strata. *See* unconformity.

Nova/supernova Star that collapses inward (implodes) and produces a sudden increase in brilliancy; such an event produces heavy elements and blasts them into space where they become "seed" matter for the development of new stars and planets.

Nuclide Atom defined by the number of protons and neutrons in its nucleus.

Occipital condyle Bony knob (double structure in mammals) at the base of the skull; articulates with the first vertebra.

Offlap Vertical succession of sedimentary facies reflecting the regressive movement of the sea *off* the land; typically characterized by a fine- to coarse-grained sedimentary sequence (limestone–shale–sandstone). Opposite, *see* onlap.

Omnivorous Adapted to feeding on animals and plants.

Onlap Vertical succession of sedimentary facies reflecting the transgressive movement of the sea *upon* the land; typically characterized by a coarse- to fine-grained sedimentary sequence (sandstone–shale–limestone). Opposite, *see* offlap.

Ontogeny Process of growth (from young to old) in an *individual* organism. *See* phylogeny.

Ophiolite suite Association of ultramafic rocks (peridotite), mafic rocks (gabbro and basalt—often pillow basalts), radiolarian cherts, and rocks rich in serpentine, chlorite, and other metamorphic minerals that are all representative of oceanic lithosphere; an exposed ophiolite suite indicates uplifted deep-sea floor.

Opposable thumb Primate thumb that can be moved into a position opposite the fingers, thereby increasing dexterity.

Order Subdivision in the classification of life forms; category just below the level of class and above the level of family.

Organic evolution Process of unidirectional genetic change of life forms; controlled by genetics and natural selection.

Organic macromolecule Large molecule of organic compound formed during a series of chemical reactions that in early Archean time may have led to evolution of the living cell (may be similar to coacervate droplets).

Original Horizontality Principle that sediments are originally deposited in an essentially horizontal position (parallel to the earth's surface) because they accumulate on an essentially flat depositional surface (interface between sediments and water or air); tilted strata, therefore, imply postdepositional deformation.

Orogenic Belt Region of lithospheric mobility and unrest (relative to stable blocks of the crust) that has been deformed into mountain chains; typically long and linear regions (former geosynclines) characterized by folded and faulted rocks.

Orogenic front Boundary (contact) between orogen (deformation belt) and less disturbed rocks.

Orogeny Process of forming mountains; may involve folding, faulting, and igneous activity; often related to lithospheric plate convergence.

Outgassing Process of expelling gases from the earth's interior by volcanic activity.

Ozone O_3 chemical form of oxygen caused by bombardment of the normal atmospheric O_2 molecules with ultraviolet radiation in the upper atmosphere.

Paleoecology Branch of paleontology that is concerned with the relationships of ancient organisms to their environments and to one another.

Paleogeographic map Graphic representation of ancient environments, such as shorelines, basins, river systems, and mountain belts; on a larger scale may depict ancient continental-oceanic arrangements.

Paleontology Branch of geology that is the study of prehistoric life preserved as the fossil record.

Palynology Field of study within paleobotany that concentrates on the investigation of spores and pollen (reproductive parts of plants).

Pangaea "All lands"; single supercontinent resulting from the late Paleozoic global suturing of continental masses.

Panthalassa All-inclusive ocean that surrounded Pangaea; spanned the globe from pole to pole and encompassed nearly 300° of longitude.

Paraconformity Unconformable surface between essentially parallel strata; "hardly distinguishable from a simple bedding plane" and not characterized by a change in fauna or flora. *See* unconformity.

Period Abstract unit of geologic time; subdivision of an era and subdivided into epochs; *examples,* Cambrian Period, Ordovician Period; time equivalent of system. *See* system.

Perissodactyl Any of an order of odd-toed (one, three, or five toes on each foot) ungulates (hoofed mammals), including horses and tapirs. *See* artiodactyl.

Permineralization Preservation of fossils by deposition of mineral material in the pore spaces of organic hard parts, indurating the original organic material.

pH Measure of the acidity or basicity of a solution; pH 7 = a neutral solution, from 7 to 0 = increased acidity (increased concentration of H^+ ions), and from 7 to 14 = increased basicity (increased OH^- in solution).

Phanerozoic Most recent eon; defined to include all of geologic history characterized by conspicuous animal life (from the present to 700 million years ago); includes the Paleozoic, Mesozoic, and Cenozoic Eras; in this text begins with the Ediacarian Period.

Phenotype Physical characteristics of an organism as determined by its genotype (genetic makeup) and environmental conditions.

Phloem Type of cell in the vascular system of plants; transports liquids and nutrients to various parts of the plant. *See* vascular plant.

Photosynthesis Chemical synthesis of organic compounds from water and carbon dioxide using energy from sunlight and chlorophyll molecules; a function of monerans, protistans, and green plants.

Phyletic gradualism Model of formation of new species (speciation) whereby one species gradually evolves into another species through a span of time.

Phylogeny Process of change in taxa (species, genus, family, etc.) through geologic time; history of development of a collective *group. See* ontogeny.

Phylum Major subdivision in the classification of life forms; broad and inclusive category just below kingdom level and above class level. *Plural,* phyla.

Pith Central portion of the internal structure of a plant stem; woody tissue.

Placental Characterized by the presence of a placenta, the organ that unites a fetus to the maternal uterus.

Planetesimal Small, solid, and cold body in space; planets may have formed by accretion of a cloud of planetesimals (the dust-cloud or planetesimal hypothesis).

Plankton Organisms that float freely in water; *example,* diatoms.

Plant Unicelled or multicelled organism that possesses the ability to manufacture food from inorganic materials by photosynthesis.

Plate tectonics The process whereby the earth's outer rind consists of a mosaic of rigid pieces (plates) that move and interact. *See* tectonic cycle.

Pluvial lake Lake formed during a time of exceptionally heavy rainfall; used specifically for a lake formed during Pleistocene glacial advances.

Polarity event Change of polarity of the earth's magnetic field of relatively short duration; occurs during a polarity epoch; *example,* a small span of time when normal polarity occurs within an epoch of reverse polarity.

Pollen Microspores produced in seed-bearing vascular plants; usually appears as fine dust.

Polyphyletic Originating from two or more lineages of ancestors. *See* monophyletic.

Precambrian Often-used term referring to all of geologic time prior to the beginning of the Cambrian Period (conventionally placed at 570–600 million years ago); in this text supplanted by term Cryptozoic.

Prehensile tail Tail that has the ability to wrap around an object, such as in grasping a branch; characteristic of South American monkeys.

Producer Organism that is able to produce its own food and energy by the processes of photosynthesis; monerans, some protistans, and plants.

Prokaryote Single-celled, nonnucleated life form; *examples,* bacteria, cyanobacteria, and methanogens. *See* eukaryote.

Prosimii "Primitive" primates, including lemurs and tarsiers.

Proterozoic An eon, the younger part of Cryptozoic time; 2500–700 million years ago.

Protistan Single-celled organism having a nucleated cell.

Protoplanet "First planet"; orbiting mass of dust and gas clouds that gradually condensed into a solid planetary body; precursor of a present planet.

Punctuated equilibrium Model of formation of new species (speciation) whereby an evolving lineage has relatively rapid and substantial morphological change within "geologically short moments"; these "moments" of rapid change punctuate (separate) longer intervals of time when the characteristics of the lineage were relatively constant (had equilibrium).

Radioactive decay Spontaneous emission of atomic particles from the atom nucleus causing disintegration of the atom; *examples,* decay of uranium-238 to lead-206, decay of potassium-40 to argon-40.

Radioactive emission Emission of alpha (two protons and two neutrons), beta (electrons from the nucleus when a neutron is split into a proton and an electron), or other energetic particles to produce radioactive decay.

Radiogenic Produced by radioactivity; *example,* radiogenic lead.

Radiolarian Microscopic, single-celled organism (protistan) that secretes a complex, and often ornate, siliceous shell.

Radiometric Pertaining to the measurement of radiation; applicable to determination of age of a rock by measurement of radioactive decay of some elements (radiometric dating).

Recrystallization Fossil preservation by conversion of unstable forms of shell material (such as aragonite) to more stable forms (such as calcite); microstructure of the original shell is destroyed in the process; not restricted to fossil preservation, but also applicable to recrystallization of inorganic minerals.

Red bed Detrital sedimentary rock (most commonly sandstone and shale) whose particles are coated with iron oxides, usually the mineral hematite; most often formed in nonmarine environments due to thorough oxidation of the iron by free (uncombined) oxygen in the atmosphere.

Reef Wave-resistant structure built up from the sea floor by accumulation of skeletal material from marine organisms such as calcareous algae, corals, and bryozoans; important organic feature in the sedimentary rock record; complex ecologic community of reef-building organisms.

Regression Fall of sea level relative to the shore, with resulting movement of the sea *off* the land. Opposite, *see* transgression.

Replacement Fossil preservation by removal of original material in solution and concurrent deposition of new compounds in its place (such as in petrified wood); not restricted to fossils, but also applicable to replacement of inorganic minerals.

Reproductive isolation Individuals of one species are biologically unable to reproduce with individuals of any other species because of genetic differences, although these species may coexist in the same geographic area.

Respiration Chemical processes associated with breathing; removal of oxygen from the atmosphere and giving off of carbon dioxide.

Ripple mark Wave-like (undulating) form on the surface of sediment; caused by water or air currents flowing over the surface or by back-and-forth motion of waves.

Rock cycle Schematic portrayal of interrelationships among the three rock classes; illustrates the basic concept that new rocks are formed from older rocks; component of the general geologic cycle. *See* geologic cycle.

Salt dome Vertical, fingerlike projection of salt that has intruded into overlying rocks from a lower salt stratum. *See* diapir.

Seamount Isolated mountain rising from the ocean floor; typically a volcanic cone.

Sedimentary cycle Vertical succession of sedimentary rocks formed during a transgression–regression episode. *See* onlap, offlap.

Sedimentary facies Lateral variation in sedimentary rock units partly or wholly equivalent in age; deposits produced by different but laterally adjacent environments of deposition. Facies in a vertical sequence; *see* Walther's Law.

Septum Partition, such as a radiating plate of the skeletal structure of coral or the wall between the chambers of cephalopod shells. *Plural,* septa.

Series Time-stratigraphic term for those rocks formed during a specific epoch of geologic time; below system in rank; *examples,* Lower Cambrian Series, Pliocene Series. *See* epoch.

Sevier orogeny Middle phase (Cretaceous) of the more inclusive Cordilleran orogeny; characterized by back-arc thrusting. *See* Cordilleran orogeny.

Shield Cryptozoic nucleus of a continent typically containing the roots of ancient orogenic belts; relatively stable for a long span of time and topographically a low-lying region. *See* Canadian Shield.

Sonoma orogeny Permo-Triassic mountain-building event (*see* orogeny) expressed by structures in Nevada.

Speciation Evolutionary processes of species formation; *examples, see* allopatric speciation, phyletic gradualism, punctuated equilibrium.

Species Subdivision in the classification of life forms; least inclusive ("smallest") of the various subdivisions.

Sponge Invertebrate animal with two-layer body and a cellular level of organization with no organs; spicules are the predominant skeletal part of the sponge found in the fossil record.

Spore Asexual reproductive body produced by spore-bearing vascular plants; has the capability to develop into a new individual.

Sporophyte Stage during which spores are produced in the reproductive cycle of spore-bearing vascular plants.

Stage Time-stratigraphic term for those rocks formed during a specific age of geologic time; below series in rank; *examples,* Trempealeauan Stage, Franconian Stage (both divisions of the Late Cambrian Series).

Stenotopic Able to survive and reproduce only in a narrow range of environmental conditions; narrow tolerance of conditions. Opposite, *see* eurytopic.

Stereoscopic vision Perception of an object in three-dimensional form because of having two eyes with overlapping fields of view.

Stoma Pore in the epidermis of plant leaves; allows gaseous exchange (oxygen, carbon dioxide, water vapor) between the plant and the atmosphere. *Plural,* stomata.

Stratigraphy Science of layered (stratified) rocks, including spatial and time relationships of different strata, interpretation of dynamic depositional patterns, and organization of rock sequences.

Stratotype Type section for any formal stratigraphic unit; ideally a stratigraphic sequence of essentially continuous sedimentation (not interrupted by unconformities).

Stromatolite Sedimentary structure formed by the sediment trapping and binding of cyanobacteria; characterized by laminated, mound-shaped form.

Stromatoporoid Extinct reef-building animal of uncertain biological affinity, but generally classified as a form of Hydrozoa in phylum Coelenterata; fossils typically calcareous laminated masses.

Subduction zone Zone of descent of one plate margin (leading edge) beneath the edge of the adjacent plate at a convergent boundary; expressed topographically by an oceanic trench; *example,* subduction of the Nazca Plate under the western margin of the South American Plate along the Peru–Chile Trench.

Submarine canyon Sea-floor canyon that typically crosses the continental shelf and extends down the continental slope; acts as an avenue of transport of sedimentary debris to the deep-sea floor.

Supergroup Lithostratigraphic unit that is an assemblage of related groups or of formations and groups; associated groups often have significant similar lithologic features.

Supernova *See* nova.

Superposition Principle that in a sequence of essentially undisturbed layers of sedimentary rocks, the oldest layer is at the bottom and the youngest layer is at the top.

Suspect terrane Terrane in which the rocks are suspected to be foreign to the general region; suspected to have originated at some other location and moved into its present position as a result of lithospheric plate movement. *See* exotic terrane.

Suture Line or groove; line of junction of a septum in a cephalopod shell to the inside of the outer shell, a significant feature in classifying cephalopods; in a tectonic sense, the line of union at a convergent plate boundary; *example,* suturing of the continents together by convergence to form Pangaea.

System Time-stratigraphic term for those rocks formed during a specific period of geologic time; below erathem in rank; *examples,* Cambrian System, Triassic System. *See* period.

Taconic orogeny Late Ordovician mountain-building event (*see* orogeny) affecting the northern Appalachian belt (Taconic Mountains of eastern New York, Vermont, and central Massachusetts); associated with partial closing of the proto-Atlantic basin (Iapetus) due to plate convergence.

Taxon *Any* level of classification of organisms; *examples,* species, genus, family. *Plural,* taxa.

Taxonomy Classification of organisms.

Tectonic cycle Model that explains the major structures of the earth (such as ocean basins, folded mountain belts, volcanic mountain chains) as a result of the interaction of the earth's lithospheric plates as they move about on a weak asthenosphere ("plastic" layer); individual plates move away from one another, toward one another, or slide past each other; component of the general geologic cycle. *See* geologic cycle.

Terminal sporangium Reproductive spore-producing organ of seedless vascular plants; located at tip of stem or branch. *Plural,* sporangia.

Terrane Geographic area in which a specific rock type or assemblage of rocks is prevalent; *examples,* shield terrane, coastal-plain terrane.

Terrestrial planet Any of the inner planets of the solar system (located close to the sun); Mercury, Venus, Earth, Mars; rocky and earthlike in composition. *See* Jovian planet.

Tethys Ocean East–west seaway between Gondwana (continents of the southern hemisphere) and Eurasia; existed from Early Paleozoic to Early Cenozoic.

Thallophyta Subkingdom of plants including unicelled or multicelled aquatic forms that lack true root systems, woody stems, or leaves; *examples,* diatoms, brown algae (kelp).

Till Ice-deposited debris, typically nonsorted and nonstratified.

Time unit Abstract unit of geologic time; eon, era, period, epoch.

Time-stratigraphic unit Succession of strata formed during a specific unit of geologic time and defined by isochronous boundaries; eonothem, erathem, system, series, stage.

Tommotian Stage Time-stratigraphic unit especially significant because it contains the earliest known shell-bearing fossils—primitive molluscs, brachiopods, archaeocyathids, and others; lowest Cambrian, but predates first appearance of trilobites.

Trace fossil Preservation of an indication of an organism's presence and activity; structural parts of the organism's body are not present; *examples,* animal tracks, burrows.

Tracheophyte Plant having a vascular system of xylem and phloem.

Transcontinental arch Series of emergent areas trending in a northeast–southwest direction across the central part of the North American craton (Lake Superior to Arizona) during early Paleozoic time; these individual land areas are known collectively as the transcontinental arch.

Transform fault Strike-slip fault associated with lateral motion between lithospheric plates; *examples,* San Andreas fault zone in California, sea-floor fracture zones between offset segments of the oceanic ridge system.

Transgression Rise of sea level relative to the shore with resulting encroachment of the sea *onto* the land. Opposite, *see* regression.

Trilobite One of an extinct class of arthropods with a significant Paleozoic fossil record; characterized by trilobed body.

Triple junction Point that marks the complex intersection of three lithospheric plates; *example,* junction of Pacific Plate, North American Plate, and Cocos Plate.

Turbidite Sedimentary deposit produced by a turbidity current (generally developed as graded sandstone and mudstone); common in deep-sea fan environment. *See* turbidity current.

Turbidity current Dense, sediment-laden water current that moves rapidly downslope; common in submarine-canyon and basin-slope environments. *See* turbidite.

Type section Particular stratigraphic section that was used to *originally* define a stratigraphic unit and that acts as a world standard; *example,* formations of the Cambrian and Ordovician Systems have specific type sections in Wales. *See* stratotype.

Unconformity Buried surface of erosion or nonaccumulation of sediments; produces a gap in the sedimentary record. *See* angular unconformity, disconformity, nonconformity, paraconformity.

Ungulate Any of the hoofed mammals; *examples,* horse, elephant. *See* artiodactyl, perissodactyl.

Uniformitarianism Doctrine that ancient earth processes may be understood by studying present-day processes, that the same physical principles have existed throughout geologic time; concept ascribed to James Hutton (1726–1797); strict interpretation of the concept implies uniformity in rates, intensities, and importance of the various processes. *See* actualism.

Varve (lake varve) Paired layer of lake sediments that records an annual deposit; composed of a light-colored, typically thicker summer lamina and a dark-colored, typically thinner, finer grained winter lamina.

Vascular plant Plant that contains internal tubelike structures called xylem and phloem that form a vascular (liquid-conducting) system. *See* xylem, phloem, tracheophyte.

Vertebrate Animal that posssesses a spinal column (backbone).

Volcanic island arc Oceanic, volcanic mountain chain that parallels a sea-floor trench—both features are associated with plate convergence; results from igneous activity produced by subduction of a lithospheric slab into the asthenosphere; *examples,* Japanese islands parallel to the Japanese Trench; Aleutian Islands parallel to the Aleutian Trench.

Walther's Law Within a sedimentary cycle, the same succession of facies occurs laterally and vertically; the facies are products of environments that occur laterally adjacent to one another and succeed or precede one another in a sedimentary cycle.

Wilson cycle Sequence of plate tectonics events producing (a) the opening of an ocean basin by continental rifting, and (b) the closing of an ocean basin by continental collision; *example,* formation of the proto-Atlantic Ocean (Iapetus) and subsequent Paleozoic closing of the ocean to form the supercontinent Pangaea.

Xylem Type of cell in the vascular system of plants; transports water to various parts of the plant. *See* vascular plant, phloem.

Zone *See* biozone.

APPENDIX A:
The Periodic Table

Key:

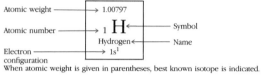

When atomic weight is given in parentheses, best known isotope is indicated.

1.00797 1 **H** Hydrogen 1s¹								
6.94 3 **Li** Lithium 1s²2s¹	9.012 4 **Be** Beryllium 1s²2s²							
22.99 11 **Na** Sodium [Ne]3s¹	24.31 12 **Mg** Magnesium [Ne]3s²							
39.10 19 **K** Potassium [Ar]4s¹	40.08 20 **Ca** Calcium [Ar]4s²	44.96 21 **Sc** Scandium [Ar]3d¹4s²	47.90 22 **Ti** Titanium [Ar]3d²4s²	50.94 23 **V** Vanadium [Ar]3d³4s²	52.00 24 **Cr** Chromium [Ar]3d⁵4s¹	54.94 25 **Mn** Manganese [Ar]3d⁵4s²	55.85 26 **Fe** Iron [Ar]3d⁶4s²	58.93 27 **Co** Cobalt [Ar]3d⁷4s²
85.47 37 **Rb** Rubidium [Kr]5s¹	87.62 38 **Sr** Strontium [Kr]5s²	88.91 39 **Y** Yttrium [Kr]4d¹5s²	91.22 40 **Zr** Zirconium [Kr]4d²5s²	92.91 41 **Nb** Niobium [Kr]4d⁴5s¹	95.94 42 **Mo** Molybdenum [Kr]4d⁵5s¹	(98)ᵇ 43 **Tc** Technetium [Kr]4d⁵5s²	101.1 44 **Ru** Ruthenium [Kr]4d⁷5s¹	102.9 45 **Rh** Rhodium [Kr]4d⁸5s¹
132.9 55 **Cs** Cesium [Xe]6s¹	137.3 56 **Ba** Barium [Xe]6s²	Lanthanides 57–71	178.5 72 **Hf** Hafnium [Xe]4f¹⁴5d²6s²	180.9 73 **Ta** Tantalum [Xe]4f¹⁴5d³6s²	183.8 74 **W** Tungsten [Xe]4f¹⁴5d⁴6s²	186.2 75 **Re** Rhenium [Xe]4f¹⁴5d⁵6s²	190.2 76 **Os** Osmium [Xe]4f¹⁴5d⁶6s²	192.2 77 **Ir** Iridium [Xe]4f¹⁴5d⁷6s²
87 **Fr** Francium [Rn]7s¹	88 **Ra** Radium [Rn]7s²	Actinides 89–103						

Lanthanides →

138.9 57 **La** Lanthanum [Xe]5d¹6s²	140.1 58 **Ce** Cerium [Xe]4f¹5d¹6s²	140.9 59 **Pr** Praseodymium [Xe]4f³6s²	144.2 60 **Nd** Neodymium [Xe]4f⁴6s²	(145) 61 **Pm** Promethium [Xe]4f⁵6s²	150.4 62 **Sm** Samarium [Xe]4f⁶6s²	151.9 63 **Eu** Europium [Xe]4f⁷6s²

Actinides →

227.0 89 **Ac** Actinium [Rn]6d¹7s²	232.0 90 **Th** Thorium [Rn]6d²7s²	231.0 91 **Pa** Protactinium [Rn]5f²6d¹7s²	238.0 92 **U** Uranium [Rn]5f³6d¹7s²	237.0 93 **Np** Neptunium [Rn]5f⁴6d¹7s²	(244) 94 **Pu** Plutonium [Rn]5f⁶7s²	(243) 95 **Am** Americium [Rn]5f⁷7s²

1.00797	4.0026
1 **H**	2 **He**
Hydrogen	Helium
$1s^1$	$1s^2$

10.81	12.01	14.00	15.999	18.998	20.179
5 **B**	6 **C**	7 **N**	8 **O**	9 **F**	10 **Ne**
Boron	Carbon	Nitrogen	Oxygen	Fluorine	Neon
$1s^22s^2p^1$	$1s^22s^2p^2$	$1s^22s^2p^3$	$1s^22s^2p^4$	$1s^22s^2p^5$	$1s^22s^2p^6$

26.98	28.08	30.97	32.06	35.45	39.95
13 **Al**	14 **Si**	15 **P**	16 **S**	17 **Cl**	18 **Ar**
Aluminum	Silicon	Phosphorus	Sulfur	Chlorine	Argon
$[Ne]3s^2p^1$	$[Ne]3s^2p^2$	$[Ne]3s^2p^3$	$[Ne]3s^2p^4$	$[Ne]3s^2p^5$	$[Ne]3s^2p^6$

58.70	63.54	65.38	69.72	72.59	74.92	78.96	79.91	83.80
28 **Ni**	29 **Cu**	30 **Zn**	31 **Ga**	32 **Ge**	33 **As**	34 **Se**	35 **Br**	36 **Kr**
Nickel	Copper	Zinc	Gallium	Germanium	Arsenic	Selenium	Bromine	Krypton
$[Ar]3d^84s^2$	$[Ar]3d^{10}4s^1$	$[Ar]3d^{10}4s^2$	$[Ar]3d^{10}4s^2p^1$	$[Ar]3d^{10}4s^2p^2$	$[Ar]3d^{10}4s^2p^3$	$[Ar]3d^{10}4s^2p^4$	$[Ar]3d^{10}4s^2p^5$	$[Ar]3d^{10}4s^2p^6$

106.4	107.9	112.4	114.8	118.7	121.7	127.6	126.9	131.3
46 **Pd**	47 **Ag**	48 **Cd**	49 **In**	50 **Sn**	51 **Sb**	52 **Te**	53 **I**	54 **Xe**
Palladium	Silver	Cadmium	Indium	Tin	Antimony	Tellurium	Iodine	Xenon
$[Kr]4d^{10}$	$[Kr]4d^{10}5s^1$	$[Kr]4d^{10}5s^2$	$[Kr]4d^{10}5s^2p^1$	$[Kr]4d^{10}5s^2p^2$	$[Kr]4d^{10}5s^2p^3$	$[Kr]4d^{10}5s^2p^4$	$[Kr]4d^{10}5s^2p^5$	$[Kr]4d^{10}5s^2p^6$

195.1	197.0	200.6	204.4	207.2	209.0	(209)	(210)	(222)
78 **Pt**	79 **Au**	80 **Hg**	81 **Tl**	82 **Pb**	83 **Bi**	84 **Po**	85 **At**	86 **Rn**
Platinum	Gold	Mercury	Thallium	Lead	Bismuth	Polonium	Astatine	Radon
$[Xe]4f^{14}5d^96s^1$	$[Xe]4f^{14}5d^{10}6s^1$	$[Xe]4f^{14}5d^{10}6s^2$	$[Xe]4f^{14}5d^{10}6s^2p^1$	$[Xe]4f^{14}5d^{10}6s^2p^2$	$[Xe]4f^{14}5d^{10}6s^2p^3$	$[Xe]4f^{14}5d^{10}6s^2p^4$	$[Xe]4f^{14}5d^{10}6s^2p^5$	$[Xe]4f^{14}5d^{10}6s^2p^6$

157.3	158.9	162.5	164.9	167.3	168.9	173.0	175.0
64 **Gd**	65 **Tb**	66 **Dy**	67 **Ho**	68 **Er**	69 **Tm**	70 **Yb**	71 **Lu**
Gadolinium	Terbium	Dysprosium	Holmium	Erbium	Thulium	Ytterbium	Lutetium
$[Xe]4f^75d^16s^2$	$[Xe]4f^96s^2$	$[Xe]4f^{10}6s^2$	$[Xe]4f^{11}6s^2$	$[Xe]4f^{12}6s^2$	$[Xe]4f^{13}6s^2$	$[Xe]4f^{14}6s^2$	$[Xe]4f^{14}5d^16s^2$

(247)	(247)	(251)	(252)	(257)	(258)	(259)	(260)
96 **Cm**	97 **Bk**	98 **Cf**	99 **Es**	100 **Fm**	101 **Md**	102 **No**	103 **Lw**
Curium	Berkelium	Californium	Einsteinium	Fermium	Mendelevium	Nobelium	Lawrencium
$[Rn]5f^76d^17s^2$	$[Rn]5f^97s^2$	$[Rn]5f^{10}7s^2$	$[Rn]5f^{11}7s^2$	$[Rn]5f^{12}7s^2$	$[Rn]5f^{13}7s^2$	$[Rn]5f^{14}7s^2$	$[Rn]5f^{14}6d^17s^2$

APPENDIX B
A Synoptic Classification of Organisms

All classifications are manmade schemes. Classifying anything—minerals, rocks, fossils, stars, or cars—into related groups is an endeavor to make order out of diversity—to put items into a series of cubbyholes that our minds can grasp. Classification is a tool of study. Being manmade, classifications do not meet with universal acceptance; thus synoptic classifications of organisms in different texts will vary. A key point in the classification of organisms is the thread of genetic relationships between groups; evolutionary concepts must be taken into account when assigning organisms to their groups. This is not just a descriptive classification, therefore, but phylogenetic as well; evolutionary relationships are honored as much as possible. As indicated in Chapter 4 (see discussion of conodonts), some fossil organisms are of unknown biologic affinity and are difficult to fit into the present classification scheme.

The synoptic classification that follows is not meant to include all known life forms, but rather includes the major organic groups and the important fossil forms discussed in this text.

Kingdom MONERAE (unicellular: nonnucleated cell structure)
  Division* Schizomycophyta—bacteria; Cryptozoic–Holocene
  Division Cyanophyta—cyanobacteria (blue-green bacteria), stromatolites; Cryptozoic–Holocene
Kingdom PROTISTAE (unicellular: nucleated cell structure)
  Phylum Protozoa
    Class Sarcodina
      Order Foraminiferoida—Foraminifera: most commonly with calcareous tests; Cambrian–Holocene; planktic forms; Cretaceous–Holocene
        Family Fusulinidae—fusulinids; Mississippian–Permian
      Order Radiolaroida—radiolarians; siliceous, perforate shell; Cambrian–Holocene

Kingdom PLANTAE (unicellular and multicellular [metaphyte]: nucleated cell structure; photosynthetic; see Chapter 14 for more detailed discussion)
  Subkingdom Thallophyta—primarily aquatic forms lacking true root systems; some unicellular algal groups
    Division Chlorophycophyta—green algae; Cryptozoic–Holocene
    Division Rhodophycophyta—red algae; Cryptozoic–Holocene
    Division Phaeophycophyta—brown algae (seaweeds); Cryptozoic–Holocene
    Division Chrysophycophyta—diatoms; Jurassic?–Holocene
    Division Euglenophycophyta—euglenids; Cretaceous–Holocene
    Division Charophycophyta—charophytes; Silurian–Holocene
  Subkingdom Embryophyta—terrestrial plants
    Division Rhyniophyta—Silurian–Devonian
    Division Zosterophyllophyta—Silurian–Devonian
    Division Psilophyta—Devonian–Holocene
    Division Microphyllophyta—club mosses; Devonian–Holocene
    Division Arthrophyta—horsetails and sphenopsids; Devonian–Holocene
    Division Pteridophyta—ferns; Devonian–Holocene
    Division Pteridospermophyta—seed ferns; Devonian–Jurassic
    Division Coniferophyta—conifers; Devonian?–Holocene
    Division Cycadophyta—cycads; Mississippian?–Holocene
    Division Ginkgophyta—maidenhair tree; Pennsylvanian?–Holocene
    Division Gnetophyta—Permian?–Holocene
    Division Anthophyta—flowering plants; Triassic?–Holocene
Kingdom ANIMALE (multicellular [metazoan]; nucleated cell structure; nonphotosynthetic)
  Phylum Porifera—sponges; Cambrian–Holocene
  Phylum Archaeocyatha—soft-tissue structure unknown; conical, double-walled skeleton; extinct; Early and Middle Cambrian
  Phylum Coelenterata; Ediacarian? Cambrain–Holocene
    Class Anthozoa—corals and sea anemones
      Order Tabulatoida—extinct forms without (or

---

*Division is a botanical classification term, generally equivalent to phylum of zoological classification.

with poorly developed) septa; horizontal tabulae supported the animal; important contributors to Paleozoic reefs; Ordovician–Permian

Order Rugosoida—extinct forms with septal development; includes horn corals; major Paleozoic reef builders; Ordovician–Triassic

Order Scleractinoida—coral reef builders of Mesozoic and Cenozoic times; predominantly colonial types; Triassic–Holocene

Class Hydrozoa

Order Stromatoporoida—extinct group of uncertain biological affinity; laminated masses of calcium carbonate; surficial star-shaped structures (suggesting to some taxonomists that stromatoporoids should be classified with Porifera); major mid-Paleozoic reef builders; Cambrian–Cretaceous

Phylum Bryozoa—colonial encrusting or branching forms; Ordovician–Holocene

Phylum Brachiopoda—marine forms with two-valved shell; Cambrian–Holocene

Class Inarticulata—two valves lacking tooth and socket hinge structure; chitino-phosphatic shell composition; Cambrian–Holocene

Class Articulata—two valves, with well-developed hinge structure; typically calcareous shell; fossils more abundant; Cambrian–Holocene

Phylum Mollusca—diverse group of both fossil and modern forms; Cambrian–Holocene

Class Bivalvia—two hinged, mirror-image shells; muscular, creeping foot; clams, oysters, scallops; Cambrian–Holocene

Class Gastropoda—most with single, coiled shell; muscular, creeping foot; distinct head; snails; Cambrian–Holocene

Class Cephalopoda—shelled or shell-less forms; shelled forms commonly with chambers and sutures; jet propulsion system of locomotion; extinct ammonoids comprise one of the most important guide-fossil groups; Late Cambrian–Holocene

Subclass Nautiloidea—simple sutures; Cambrian–Holocene

Subclass Ammonoidea—fluted sutures; Devonian–Cretaceous; suture types: goniatite, Devonian–Permian; ceratite, Mississippian–Triassic; ammonite, Permian–Cretaceous

Subclass Coleoidea; Mississippian–Holocene

Order Belemnoidea—belemnites; Mississippian–Eocene

Class Scaphopoda—curved, tusk-shaped shells open at both ends; common as Cenozoic fossils; Ordovician–Holocene

Class Polyplacophora—segmented shell of the chiton; Cambrian–Holocene

Class Monoplacophora—conical shell typically with several pairs of muscle scars on its inner surface; biologically an important form suggesting common ancestry of molluscs, annelids, and arthropods; Cambrian–Holocene

Phylum Annelida—segmented worms (earthworm an example); trace fossils as burrows and trails; Ediacarian? Cambrian–Holocene

Phylum Arthropoda—segmented bodies and jointed appendages; diverse forms; most abundant phylum—80% of all known animals; Ediacarian? Cambrian–Holocene

Class Trilobita—extinct trilobites; Cambrian–Permian

Class Crustacea—ostracodes, barnacles, lobsters, and crabs; Cambrian–Holocene

Class Insecta—insects; relatively rare as fossils, although the most abundant of living invertebrate classes; Silurian–Holocene

Class Arachnoidea

Order: Eurypterida—eurypterids; Ordovician–Permian

Phylum Echinodermata—internal skeleton; commonly five-rayed symmetry; Ediacarian? Cambrian–Holocene

Subphylum Crinozoa—mainly sessile (attached to the sea floor) echinoderms

Class Crinoidea—crinoids; Ordovician–Holocene

Subphylum Echinozoa—predominantly vagrant (not attached) echinoderms; Ordovician–Holocene

Class Echinoidea—sea urchins; Ordovician–Holocene

Class Asteroidea—starfish; Ordovician–Holocene

Phylum Hemichordata—some affinity to chordates; small dorsal stiffening rod

Class Graptolithea—extinct graptolites; Cambrian–Mississippian

Phylum Chordata—most often with a segmented vertebral column as in subphylum Vertebrata; diverse; Ordovician–Holocene

Subphylum Vertebrata

Class Agnatha—jawless fish; Ordovician–Holocene

Class Placodermii—primitive armored fish; Ordovician–Mississippian

Class Chondrichthys—cartilaginous fish (sharks); Silurian–Holocene

Class Acanthodii—may be ancestral to bony fish; spiny sharks; Silurian–Permian

Class Osteichthys—bony fish; Silurian–Holocene

Class Amphibia—amphibians; water-dependent stages in life cycle; Devonian–Holocene

Class Reptilia—reptiles; Pennsylvanian–Holocene (see Chapter 12 for more detailed discussion)

Subclass Anapsida—includes turtles; extinct cotylosaurs; Pennsylvanian-Holocene

Subclass Synapsida—includes extinct pelycosaurs and therapsids; Pennsylvanian–Triassic

Subclass Euryapsida—extinct marine forms including icthyosaurs and plesiosaurs; Triassic-Cretaceous)

Subclass Diapsida—most diverse subclass; includes lizards, snakes, crocodiles; extinct thecodonts, dinosaurs, pterosaurs; Triassic–Holocene

Class Mammalia—mammals; Triassic–Holocene (see Chapter 15 for more detailed discussion)

Subclass Prototheria—monotremes; Pleistocene?–Holocene

Subclass uncertain—includes docodonts and triconodonts; Triassic–Jurassic

Subclass Allotheria—includes multituberculates; Jurassic–Eocene

Subclass Theria—includes symmetrodonts, marsupials, insectivores, pantotheres, and all Cenozoic placental; Jurassic–Holocene

Class Aves—birds; Jurassic? Cretaceous–Holocene

Unknown affinity: "toothlike" conodonts—utilitarian order Conodontiferoida; phylum and class unknown; Cambrian–Triassic

INDEX